T0318665

This book is the first to give a comprehensive account of the fundamental properties of metal-oxide surfaces and their interaction with atoms, molecules and overlayers. It will act as a valuable reference for anyone interested in learning about the surface science of this fascinating class of materials, whose properties span the entire range from metals to semiconductors and insulators.

The surfaces of metal oxides are of great technological importance, and play a crucial role in an extremely wide range of phenomena. A detailed knowledge of their properties is central to the investigation of processes such as the environmental degradation of high-T_c superconductors, or in analyzing the nature of the bonding between grains of alumina in sintered ceramics. Metal-oxide surfaces are also of increasing importance in fields such as catalysis, the passivation of metal surfaces, and gas sensing for pollution monitoring and control.

As well as giving a general overview of the basic properties of metal oxides, an extensive and thorough compilation of the theoretical and experimental research which has been performed on well-characterized single-crystal oxide surfaces is provided, thus making the book suitable for those graduate students and established researchers in materials science, chemistry, physics or geology who have an interest in metal-oxide surfaces.

THE SURFACE SCIENCE OF
METAL OXIDES

THE SURFACE SCIENCE OF METAL OXIDES

VICTOR E. HENRICH

Eugene Higgins Professor of Applied Science
Yale University

and

P.A. COX

Fellow of New College and
Lecturer in Inorganic Chemistry
Oxford University

CAMBRIDGE
UNIVERSITY PRESS

CAMBRIDGE UNIVERSITY PRESS
Cambridge, New York, Melbourne, Madrid, Cape Town, Singapore,
São Paulo, Delhi, Dubai, Tokyo, Mexico City

Cambridge University Press
The Edinburgh Building, Cambridge CB2 8RU, UK

Published in the United States of America by
Cambridge University Press, New York

www.cambridge.org
Information on this title: www.cambridge.org/9780521566872

First published 1994
First paperback edition (with corrections) 1996

A catalogue record for this publication is available from the British Library

Library of Congress Cataloguing in Publication Data

Henrich, Victor E.
The surface science of metal oxides
p.   cm. Includes bibliographical references.
ISBN 0-521-44389-X
1. Metallic oxides – Surfaces.   2. Surface chemistry.   I. Cox, P. A.
QD181.01H46   1994
546′.7212-dc20   93–18566 CIP

ISBN 978-0-521-44389-0 Hardback
ISBN 978-0-521-56687-2 Paperback

Contents

Preface

If the number of publications and the existence of specialized journals are good guides to the success of a field, surface science has developed remarkably over the last two decades or so. The routine use of ultrahigh-vacuum equipment and the development of many new techniques for studying the outermost atomic layers of well-characterized crystals have contributed to an enormous increase in our understanding of surface physics and chemistry. A very high proportion of this work has been performed on metals, although semiconductors such as silicon and III–V materials have received increased attention as the field has progressed. Metal oxides, in spite of their great scientific and technological interest, have been relatively much less studied; we discuss some of the reasons for this in Chapter 1. In spite of a late start, the number of groups now active and the volume of publications in the field of oxide surfaces are considerable. However, oxides are still barely represented in the otherwise excellent textbooks available in surface science, and this is the reason for the present volume. It seems to us that the field is sufficiently well-established that it is worthwhile to take stock of what is now known. At the same time, by pointing out the many things that are not understood, we hope to provide directions for future research, in a field that we are confident will continue to grow.

We hope this book will be useful to those graduate students and established researchers who already have some background in surface science ideas and techniques, but who are not familiar with oxides. Each of the main chapters contains introductory sections that cover essential aspects in outline before a detailed discussion of results is presented. We have not described in detail the various experimental techniques of surface science, as there are other books that cover these; but we have tried to explain what is *different* about oxides, both in terms of their basic electronic and chemical properties, and the ways in which these necessitate changes in experimental technique or interpretation. In discussing results, we have referred to nearly a thousand papers and reviews, and, although we do not claim to be complete, we have tried to cover significant developments up to the beginning of 1992. We must point out, however, that we have primarily limited ourselves to work on single crystals and have not attempted to discuss the enormous literature (particularly in the catalysis field) on oxide ceramics and high-surface-area powders.

It would be impossible to write a book such as this without a good deal of help. Much of this has come from colleagues and students who have made our own research so enjoyable over a period of many years: they are too numerous to mention here, but their names appear frequently in the list of references. Our respective institutions, Yale and Oxford Universities, have contributed both by supporting our research and by the generous provision of sabbatical leaves, which has given us the time to travel and to work on this project. We are especially grateful to Victor Bermudez, Russ Egdell and Wendy Flavell, each of whom read a complete draft of the book and made many penetrating and constructive comments. We have tried to accommodate their suggestions, but obviously they cannot be held responsible for any remaining flaws. Special thanks are also due to Richard Kurtz for giving us his 'MacSurface®' program to draw the surface models and for patiently assisting us as we learned how to use the program. Finally we thank our families, without whose support and encouragement this book could not have been written.

V.E. Henrich   *Yale and Oxford*
P.A. Cox

1

Introduction

The metal oxides constitute a diverse and fascinating class of materials whose
properties cover the entire range from metals to semiconductors and insulators.
Their surfaces play crucial roles in an extremely wide range of phenomena. The
environmental degradation of high-T_c superconductors; bonding between grains of
alumina in sintered ceramics; the passivation of metal surfaces against corrosion;
catalysts for the partial oxidation of hydrocarbons; solid-state gas sensors for pollution monitoring and control; the failure of dielectric materials because of surface
conductivity; the stability of electrode/electrolyte interfaces in fuel cells: all of
these are dependent upon the properties of metal-oxide surfaces or the interfaces
between metal oxides and other materials.

For all of their technological and scientific importance, our understanding of the
basic physics and chemistry of metal-oxide surfaces lags a decade or more behind
that of metals and semiconductors; the reasons for this are discussed in § 1.2
below. However, over roughly the past fifteen years, an ever increasing number of
research groups have begun to study the properties of well characterized metal-
oxide surfaces, both experimentally and theoretically. This is shown graphically in
Fig. 1.1, which plots the number of papers published per year on fundamental surface-science studies of metal oxides (the data are based upon papers referenced in
this book). The field is now sufficiently well-developed that it is appropriate to
step back and view it in overall perspective, looking for trends in surface properties as well as details of specific materials and systems. That is the purpose of this
book.

1.1 Technological importance of metal-oxide surfaces

Before examining in detail what has been learned about the properties of metal-
oxide surfaces from experimental measurements and theoretical calculations on
well-characterized single-crystal samples, which is the approach taken in this

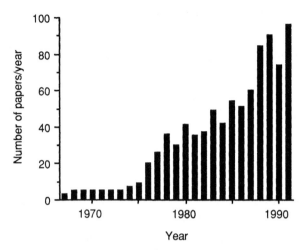

Fig. 1.1 The number of publications/year, based upon papers referenced in this book, in the
area of the properties of well-characterized metal-oxide surfaces.

book, it is useful to consider some of the reasons that make oxides of technological
interest, since they determine in large measure what properties are studied. The
importance of metal oxides in catalysis is profound. Most commercial catalysts
consist of microscopic particles of a metal supported on a high-surface-area oxide.
The two most commonly used oxide supports are SiO_2 and Al_2O_3 (the former is
not a metal oxide and will not be considered here), although other metal oxides are
sometimes used. Metal catalysts supported on such non-reducible oxides generally
exhibit little effect of the presence of the support, but in some cases either the
interaction between the catalyst and the support, or the existence of complementary reactions taking place on the metal and the oxide, play an important role in
catalysis. When catalyst metals are supported on reducible transition-metal
oxides, strong interactions can occur that alter the catalytic behavior of the metal;
this is discussed in § 7.6.

   Metal oxides are also themselves catalysts for a variety of commercially important reactions. Sometimes a metal oxide is used in its pure form (e.g., bismuth
molybdate for the oxidation of propene to acrolein or acrylonitrile), while in other
cases the oxide is supported on another oxide, and the catalytic activity results
from the interaction between the two oxides (e.g., V_2O_5 supported on TiO_2 for the
selective oxidation of hydrocarbons and the selective catalytic reduction of nitro-
gen oxides). All of these catalytic reactions proceed by complex mechanisms that
involve a range of different chemical interactions; these include acid/base reactions as well as oxidation and reduction steps where lattice oxygens are lost and
gained by the substrate. Chapter 6 is concerned with the chemisorption of mole-

cules on oxides, and we shall see there that fundamental understanding of chemisorption and catalysis is still in quite a primitive state. In supported systems, the interactions that take place at oxide/oxide interfaces are not at all well understood, and model systems will have to be studied in order to unravel the important aspects of the interactions and their effect on chemical reactions; research to date in this area is discussed in § 7.7.

A major landmark in the study of transition-metal-oxide surfaces occurred in 1972 with the report by Fujishima and Honda[1] that TiO_2 could be used as a catalytic electrode in a photoelectrolysis cell to decompose water into H_2 and O_2, without the application of an external voltage. The single-crystal transition-metal-oxide surface research that was stimulated by that discovery is mirrored in the increase in the number of publications in Fig. 1.1 in the mid-1970s. While not yet commercially competitive, photoelectrolysis is still an active area of research in electrochemistry (see § 6.2.5).

Another important application of metal oxides is as gas sensors. The most studied oxides for this purpose are ZnO and SnO_2. Both of these have high bulk electrical resistivity when stoichiometric, although they normally behave as n-type semiconductors due to the presence of defects or deliberate doping. The adsorption of certain molecules bends the bands at the surface, producing changes in surface conductivity large enough to be readily measured; thus the surface conductivity can be used to monitor the presence of that molecule. In some cases the sensitivity and selectivity of the material can be improved by deposition of submonolayers of metal on the oxide surface.

In air, oxides are the lowest free energy state for all but a few of the metals in the periodic table. Therefore much of the corrosion of metals involves the formation of (generally unwanted) oxides. Studies of metallic corrosion are most appropriately performed by beginning with the metal and following what happens to it when it is exposed to an oxidizing environment. A vast literature exists on the oxidation of metals, none of which will be discussed here. Instead we approach the phenomenon from the other direction by considering nearly perfect surfaces of metal oxides, which are often the end products of corrosion. A knowledge of the properties of those end products is a necessary prerequisite to understanding the corrosion process.

The most recent impetus for the study of oxide surfaces has been the discovery of Cu-oxide-based high-T_c superconductors. Although superconductivity is a bulk phenomenon, issues of environmental degradation and electrical contacts, combined with the short coherence lengths in oxide superconductors, have raised important surface questions, and there have been some attempts to answer them by using vacuum-fractured or single-crystal samples. This is partially responsible for the larger number of papers since 1988 in Fig. 1.1.

1.2 Considerations in the study of oxide surfaces

There are several reasons why the study of metal-oxide surfaces lags so far behind that of metals and semiconductors. This state of affairs can be explained partly by the simple reluctance of research groups with well-established programs to enter a new field. But there are also some sound reasons for their reluctance, since the study of oxides by surface-science techniques does present a number of challenges that are more severe than those encountered with metals and semiconductors.

One major problem is that of sheer complexity. This is noticeable first of all with crystal structures. Even a relatively simple structure such as corundum (α-Al_2O_3, V_2O_3, etc.) has a ten-atom primitive unit cell, while V_2O_5 has fourteen atoms in its unit cell. One of the first goals of a surface-science study is to find out where the atoms are on a surface, but it is clear that the complexity of oxide structures can make this a formidable task. In fact, the number of properly determined oxide surface structures is very small.

Along with the complex crystallography comes a similarly complex combination of chemical and physical properties. Many elements, especially transition metals, display a range of possible oxidation states and hence a series of oxides with different compositions. The oxides of vanadium, for example, include not only VO, V_2O_3, VO_2 and V_2O_5, but also a wide range of intermediate phases, some of well-defined composition such as V_4O_9, others like VO_x being essentially non-stoichiometric, with $0.8 \leq x \leq 1.3$. This chemical complexity has many serious implications for surface science. It means that even bulk samples of many oxides may be difficult to obtain with reproducible composition and properties; and they will often have high defect concentrations, which may dominate the physical properties of even the purest available materials. Such difficulties will be greatly exacerbated at surfaces and undoubtedly contribute to the problems in surface preparation that are mentioned below. Another consequence of chemical complexity is the wide range of chemical interactions possible with chemisorbed molecules. Surfaces with apparently similar composition, but prepared slightly differently, can have totally different properties. Most surface scientists, of course, are familiar with this type of problem, but it is generally much more severe with oxides than with metals or semiconductors.

The electronic structure of metal oxides is also much more complex than that of most metals or semiconductors. For example, the bulk electronic structure of the late 3d-transition-metal oxides lies somewhere between itinerant and localized, and neither theoretical approach – the former of which has been developed for metals and semiconductors and the latter that is used to describe molecules – is entirely appropriate. The discovery of metal-oxide high-T_c superconductors has spurred a great deal of interest in the bulk electronic structure of such compounds, but as yet no comprehensive theory exists. With so little known about the bulk electronic structure of metal oxides, it is small wonder that studies of their surface electronic properties are so far behind those of many other materials. Even when good models of electronic structure do exist, the structural complexity makes it difficult to carry out good calculations. The number of atoms per unit cell in many structures, coupled with the reduced symmetry at the surface and the necessity to calculate many unit cells to separate surface and bulk effects, means that first-principles electronic structure calculations still require a great deal of computer time.

Another difficulty with oxides is a practical one related to the experimental techniques used by surface scientists. Many of the most interesting metal oxides are very good electrical insulators. MgO and Al_2O_3 are of tremendous importance in the ceramics industry, and both are wide-bandgap insulators that cannot be made conducting by doping or reduction. Many of the most powerful techniques of surface science involve the emission or absorption of charged particles, whether they be electrons or ions, at some point in the measurement process. Samples that have negligible bulk conductivity often cannot be studied by those techniques due to problems of surface charging. While there may be ways to circumvent that problem to some degree, the fact remains that insulators are far more difficult to study, and the range of experimental techniques that is available is smaller, than with metals or semiconductors.

The question of surface preparation is central to all surface-science investigations, and herein lies what is possibly the greatest difficulty of all. Compared with elemental solids, the preparation of nearly perfect surfaces of any compound is difficult. The establishment of geometric order on the surface is not sufficient, since the stoichiometry of the surface can differ from that of the bulk. For semiconductors such as GaAs or InSb, procedures have been developed to produce surfaces that are as nearly ideal terminations of the bulk structure as nature will allow. But the preparation of each separate crystal face of each different compound is a research program in itself. For metal oxides, this has only been done in a few cases.

Perhaps the best method by which to prepare nearly perfect surfaces is cleaving in ultrahigh vacuum (UHV). A fairly large amount of energy is dissipated in the cleaving or fracture of most materials, and that energy can cause disruption of the surface. Fractoemission studies show that photons, electrons, atoms and ions can all be liberated during fracture, so it would be naive to consider even a surface that cleaves easily to be an ideal termination of the bulk crystal structure. But cleaving or fracture is still more likely than other methods to yield well-ordered surfaces having nearly the composition of the bulk. It has recently become possible to actually look on an atomic scale at surfaces by using scanning tunneling or atomic force microscopy, but only a small amount of such work has yet been done on oxides. Hopefully in the near future it will be possible to prepare oxide surfaces

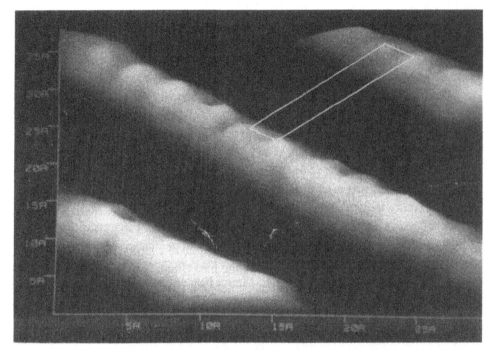

Fig. 1.2 Atomic resolution STM image of the TiO_2 (100)-(1 × 3) surface, with the surface unit cell (dimensions 3 Å × 15 Å) marked. The bright rows are ridges of atoms; this structure is discussed in § 2.3.2. [Ref. 2]

that we *know* correspond to surfaces modelled theoretically, but we are not yet at that stage.

The alternative to cleaving single-crystal surfaces is to cut samples along the desired crystal plane, polish them to remove as much roughness as possible, etch them to remove damaged layers left by the cutting and polishing processes and any surface contamination, and anneal them in UHV (or perhaps an O_2 ambient) to restore the geometric order of the surface. This is the method that has been used in most studies of single-crystal metal oxides; we will refer to surfaces prepared in that way as 'polished and annealed', or 'P&A' in tables. There is no reason to believe a priori that polished and annealed oxide surfaces will have the same structure as an ideal cleaved surface; in fact, numerous results show that not to be the case. It is thus necessary to characterize the stoichiometry and geometric structure of the surface as completely as possible before meaningful interpretations of data taken on it can be made. The techniques that are currently available to accomplish that goal are far from perfect, and in some cases they may even be misleading.

The development of scanning tunneling and atomic force microscopy (STM and AFM) is inaugurating a new era in surface characterization. Figure 1.2, for exam-

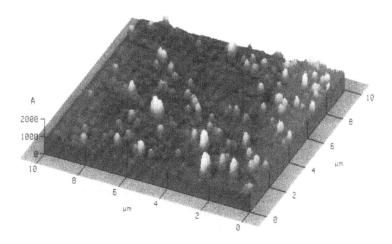

Fig. 1.3 Atomic force micrograph of a polished and annealed TiO_2 (110) surface prepared in ultrahigh vacuum. [Ref. 3]

ple, shows an STM image of a reconstructed TiO_2 (100) surface, in which the clearest features are rows of atoms having a spacing of 15 Å. Less obvious, but also detectable, are individual atoms within the rows, which have a 3 Å spacing.[2] Although results like this sometimes serve to confirm existing models, the ability to image surfaces on an atomic (or nearly atomic) scale has also shown the naiveté of many of our ideas concerning single-crystal surfaces. Figure 1.3 shows an AFM micrograph of a TiO_2 (110) surface that was prepared by standard surface-science procedures of ion bombardment and annealing.[3] It exhibited good low-energy-electron-diffraction (LEED) patterns, and photoelectron spectra indicated the absence of defect-derived states in the bulk bandgap. And yet the surface is hardly defect free! Much of the experimental work discussed in this book has been conducted on surfaces that may not be much more perfect than this one, although the data are interpreted in terms of nearly perfect surfaces, containing atomically flat planes and few steps. Until more atomic scale information is available concerning the actual topography and structure of metal-oxide surfaces from the use of such techniques, it will not be possible to compare experiments with more realistic models of surfaces than has been done to date.

1.3 Work functions of metal oxides

One of the very important properties of a surface is its work function, Φ, which is defined as the energy necessary to remove an electron from the Fermi level in a material and put it at rest in vacuum an infinite distance away from the material. (Practically, 'infinite distance' means distances beyond which the electron potential does not change with position away from the surface; this is usually tens or hundreds of ångströms.) The work function plays a dominant role in thermionic emission, Schottky barrier formation, photoelectron emission, etc. It can be measured in several ways, including thermionic emission, the low-energy cut-off in photoelectron emission, contact potential difference measured with a Kelvin probe, and retarding potential methods.[4]

The work function is an extremely sensitive measure of the state of a surface. In fact, it is *so* sensitive for metal oxides that its absolute value has little significance. Differences in step or point defect densities, traces of adsorbed molecules too small to be observed by any surface spectroscopy, slight changes in surface reconstruction, etc., can cause significant changes in Φ. Care must also be taken in comparing work functions determined by different experimental techniques, particularly if Φ is not uniform across the surface. For example, Kelvin probe measurements give an average value of the work function for the area of the surface covered by the electrode, while measurements of the low-energy cut-off of the secondary-electron spectrum in ultraviolet photoelectron spectroscopy (UPS) and other electron spectroscopies give the minimum value of Φ for the surface. Not only do different research groups measure different values on surfaces prepared in nominally the same manner, but a single group often obtains different values on different crystals, different cleaves, or different days. For that reason we will not stress the absolute values of work functions in this book. However, we would be remiss not to briefly summarize some of the experimental results that have been obtained on metal oxides; this is done in Table 1.1.

What do the results in Table 1.1 mean? Look, for example, at the first two entries under ZnO. Photoemission cut-off and field-emission measurements on ZnO (0001) surfaces that are both presumably atomically clean give very different values, although each measurement claims to have very small error limits. In theory, this should not be the fault of the different measurement techniques. Or look at the values for two photoemission measurements of the work function of polished and annealed $SrTiO_3$ (100), a surface that has been exhaustively studied by several groups. The very different values of Φ indicate that variables beyond the control, or knowledge, of the researchers play a large role. In Refs. 11 and 12, it was found that differences in annealing temperature and time could change the work function of ZnO surfaces by more than 2 eV, although in other ways the surfaces were essentially the same. Thus it is very difficult to know what absolute values of work functions on metal oxides really mean. The most dangerous situation occurs when only *one* determination of Φ has been made; people are then tempted to *believe* it! (There is a saying at the National Institute of Standards and

Table 1.1. *Measured work functions of clean metal-oxide surfaces*

Oxide	Face	Surface preparation	Work function (eV)	References
Non-transition-metal oxides				
ZnO	0001	UHV cleaved	4.25 ± 0.05	5, 6
		Field evaporation	3.15 ± 0.15	7
		P&A	4.3 ± 0.1	8
			3.9	9
			3.75 ± 0.05	10, 11
			3.2 to 3.7, depending upon surface preparation procedure	12
	$000\bar{1}$	UHV cleaved	4.95 ± 0.05	5, 6
			3.9 to 4.45, depending upon annealing temp	13
		Field evaporation	4.85 ± 0.15	7
		P&A	4.6 ± 0.2	8
			4.6	9
			4.35 ± 0.05	10, 11
			3.7 to 6.1, depending upon surface preparation procedure	12
	$10\bar{1}0$	UHV cleaved	4.64 ± 0.05	5
			4.5 to 4.7, depending upon annealing temp.	13
		P&A	4.37	14
			4.3	9
			4.05	10, 11
			4.4 to 5.0, depending upon surface preparation procedure	12
SnO$_2$	110	P&A	7.74	15
Transition-metal oxides				
TiO$_2$	110	P&A	5.5	16
			5.3 ± 0.1	17–19
			5.3	20
		Ion bombarded	5.1	20
			4.6	16
	100	P&A	5.9 ± 0.1	21, 22
		P&A, (1×3)	5.7	23
		P&A, (1×7), reduced	5.5	23
	441	P&A	4.6 ± 0.1	18, 19

Table 1.1. (*cont.*)

Oxide	Face	Surface preparation	Work function (eV)	References
SrTiO$_3$	100	P&A	4.2	24
			3.2	20
		Ion bombarded	2.8	20
	111	P&A	5.2	25
		Ion bombarded	3.4	25
Ti$_2$O$_3$	10$\bar{1}$2	UHV cleaved	3.9 ± 0.1	26–28
V$_2$O$_3$	10$\bar{1}$2	UHV cleaved	4.9 ± 0.1	29
V$_2$O$_3$ (1.5% Cr)	10$\bar{1}$2	UHV cleaved	4.2	30
		Ion bombarded	4.4	30
α-Fe$_2$O$_3$	0001	P&A	5.4 ± 0.2	31
		Ion bombarded	4.5 ± 0.2	31
NiO	100	UHV cleaved	4.4 ± 0.1	32
		P&A	4.65 ± 0.02	33
			4.40 ± 0.05	34
			4.35 ± 0.02	35

Technology that the only person who knows the exact temperature is the person who has only one thermometer.) Thus the data in Table 1.1 are intended more as a warning than as a source of useful information.

This is not to say that work functions of metal oxides are entirely useless. The *relative* changes in Φ that accompany the variation of some parameter of the surface can contain a tremendous amount of information, and such changes are often used in studies of adsorption, defect creation, etc. We will discuss work function changes wherever applicable, since they can be a very sensitive measure of the response of the surface. In some cases, such as chemisorption, quantitative information can be obtained from changes in Φ. But the absolute values will rarely be discussed, and many papers that study work-function changes do not even mention the absolute values for the reasons given above.

1.4 Scope and philosophy

This book is an attempt to bring together what is currently known about the fundamental properties of metal-oxide surfaces and their interaction with atoms, molecules and overlayers in order to provide a comprehensive background for anyone interested in learning about their chemistry, physics and materials science. While not a textbook, it is written to be accessible to students at the graduate level, as well as to researchers in various fields. This book is intended to give both a general overview of the properties of metal oxides and a comprehensive compilation of the research that has been done on well-characterized (primarily single-crystal) oxide surfaces. The former should be of use to scientists who have not worked extensively with oxides but who are interested in learning about their physics and chemistry; the latter serves as a reference source as well as an in-depth explanation of what is currently known about metal-oxide surfaces for use by researchers in the field. It is assumed that the reader has some knowledge of surface science. The experimental and theoretical techniques that have been applied to the study of oxides are discussed primarily from the standpoint of how their application to oxides differs from that for metals and semiconductors. Suitable references to books and articles describing the various surface-science techniques are included, but no extensive discussion is given here.

Although most technologically important applications of metal oxides involve their use in polycrystalline powder or ceramic form (metal oxides rarely occur in amorphous forms), the fundamental processes involved all occur at sites on (albeit small) single-crystal facets. Any understanding of real processes must thus necessarily be based upon an understanding of the basic physics and chemistry that occur on specific surface sites on single-crystal faces. It is the purpose of this book to consider the properties of metal-oxide surfaces at that very basic level. To as great an extent as possible, only experiments on well-characterized single-crystal surfaces will be considered, although it will occasionally be necessary, where single-crystal experiments are not available, to discuss results on polycrystalline samples. Theoretical calculations by their very nature are performed on perfect surfaces, unless defects are specifically introduced in order to explain the behavior of non-ideal surfaces. Thus only experiments performed on nearly perfect surfaces can be compared in any great detail with theoretical predictions, a process that is crucial to any in-depth understanding of materials.

Two books that are directly relevant to the material covered here deserve special mention. In *Transition Metal Oxides: Surface Chemistry and Catalysis* (Elsevier, Amsterdam, 1989), Harold Kung has treated the chemistry of metal-oxide surfaces in a somewhat different way than is done here. Less stress is placed on single-crystal surfaces, with most attention given to powder and polycrystalline (i.e., 'real') surfaces; more attention is also paid to surface reactions (i.e., catalysis) than to chemisorption. Since oxide surfaces are almost always hydroxylated in commercial environments, the interaction of adsorbed molecules with surface hydroxyl ions is a central element of his treatment. For the single-crystal surfaces studied in ultrahigh vacuum that are considered in this book, nominally clean surfaces rarely contain an appreciable concentration of OH^-. The other notable book is *Transition Metal Oxides: An Introduction to Their Electronic Structure and Properties* (Clarendon Press, Oxford, 1992) by P.A. Cox. It does not deal with surfaces at all, but it provides a more comprehensive treatment of the bulk electronic structure of transition-metal oxides than is possible in this book. It thus provides important background material for the study of oxide surfaces. We recommend that anyone who is seriously interested in learning about metal-oxide surfaces in detail consult that book as well.

The present book is organized in the following manner. Chapter 2, **Geometric structure of metal-oxide surfaces**, considers both the ideal atomic arrangements that would occur on the various faces of single-crystal metal oxides if they were simply a termination of the bulk crystal structure, and their actual geometric structure, as determined from comparisons of experimental measurements with theoretical models. Surface geometry is treated first since it is necessary to know what atoms are on the surface, where they are, and what their ligand coordination is in order to understand other surface properties. In Chapter 2 the lattice will be considered at rest, since lattice vibrations at temperatures well below the melting point of oxides are at most a slight perturbation of the static lattice, and they do not play a major role in determining either the electronic structure of surfaces or surface chemical properties. The vibration of atoms and the associated surface phonon spectra of oxides will be discussed in Chapter 3, **Surface lattice vibrations**. Surface vibrational properties are interesting in their own right, but in metal oxides the strength of surface phonon features in some spectroscopies also complicates the study of molecular adsorption. The electronic structure of clean metal-oxide surfaces will be discussed in Chapter 4, **Electronic structure of non-transition-metal-oxide surfaces**, and in Chapter 5, **Electronic structure of transition-metal-oxide surfaces**. In each chapter a brief introduction to the bulk electronic structure of that class of oxides will be given, since this is a necessary prerequisite to understanding surface electronic properties. The oxides are separated into two groups because many aspects of transition-metal oxides are much more complex than for non-transition-metal oxides: the differences include the availability of several oxidation states, the prevalence of defects, and many complexities of electronic structure that are generally less of a problem with the non-transition-metal oxides. By far the longest chapter in the book is Chapter 6, **Molecular adsorption on oxides**. This is because of the technological importance of gas/surface interactions on oxides and the corresponding amount of attention that has been devoted to them. Chapter 7, **Interfaces of metal oxides with metals and other oxides**, discusses the much smaller body of work that has been performed on those systems.

In order to manage the large amount of detailed information (and the correspondingly large number of references) that is contained in this book, extensive use has been made of tables. They serve the dual purposes of providing a way in which to readily see the trends in surface properties, as well as to list references to the relevant literature for each specific system. It is hoped that this format will prove useful to readers seeking either general or specific information. In most sections of the book, we have attempted to reference as much of the pertinent literature as possible so that readers interested in the details of specific systems can readily access the original articles, but we make no claim of completeness. In other sections we have included enough references to guide the reader to the relevant literature, but we have not attempted to be exhaustive. This compromise is necessary in order to keep the number of references manageable.

2

Geometric structure of metal-oxide surfaces

A knowledge of the geometric structure of metal-oxide surfaces is a necessary pre-requisite to understanding their other properties. For the most part, no one knows exactly where atoms are on the surfaces of metal oxides. But their bulk crystal structures are known very accurately, thanks to about a century of x-ray crystal-lography work on virtually every known crystalline material. We therefore begin with a brief review of the bulk crystal structures possessed by metal oxides. (We will only discuss those oxides whose surface properties have been studied either theoretically or experimentally to date.) Our approach to surface geometric struc-ture will then be to consider 'ideal' crystal surfaces, in which the bulk atomic arrangement is maintained up to and including the surface plane; these surfaces will be generated by imaginary cleavage along appropriate planes in the bulk crys-tal structure. Finally, the results of experimental measurements and theoretical calculations of how real surfaces differ from those ideal ones will be discussed.

In this chapter the lattice will be treated as though it were at absolute zero in that any motion of the atoms or ions in the lattice will be ignored. This is the appropri-ate approach since all of the experimental techniques that are used to determine atomic positions sample many atoms over a relatively long period of time; the only manifestation of lattice vibrations thus appears in Debye–Waller intensity factors. The vibrational properties of metal oxides will be considered in Chapter 3.

2.1 Bulk structures of metal oxides

The most useful starting point in understanding the structures of metal oxides is the **ionic model**.[36,37] The utility of this approach will be discussed again in later chapters concerned with electronic properties; for the moment, we shall take the view that the most important bonding forces are those that operate between the positively charged metal cations and the negative oxide (O^{2-}) anions. We antici-pate, therefore, that structures will be dominated by arrangements where metal

Table 2.1. *Oxide structures*

Formula	Coordination	Name	Symmetry	Examples
M_2O	2: linear	cuprite	cubic	Cu_2O
	4: tetrahedral	antifluorite	cubic	Na_2O
MO	6: octahedral	rocksalt	cubic	MgO, NiO
	4: tetrahedral	zincblende	cubic	(no oxides)
		wurtzite	hexagonal	ZnO
	4: planar	PdO	tetragonal	PdO
		tenorite	monoclinic	CuO
	4: pyramidal	PbO	tetragonal	SnO, PbO
M_3O_4	two M6 (oct); one M4 (tet)	spinel	cubic	Fe_3O_4
M_2O_3	6: octahedral	corundum	hexagonal	α-Al_2O_3, Ti_2O_3
MO_2	8: cubic	fluorite	cubic	ZrO_2, UO_2
	6: octahedral	rutile	tetragonal	TiO_2, SnO_2
M_2O_5	5 + 1; distorted	V_2O_5	orthorhombic	V_2O_5
MO_3	6: octahedral	ReO_3	cubic	ReO_3, WO_3 (distorted variants)
	6: distorted	MoO_3	monoclinic	MoO_3
AMO_3	M6, A12	perovskite	cubic (or distorted)	$SrTiO_3$, $BaTiO_3$ Na_xWO_3
	M and A6 (oct)	ilmenite	trigonal	$LiNbO_3$
A_2MO_4	M6, A12	layer perovskite (K_2NiF_4)	tetragonal or distorted	La_2CuO_4

ions are surrounded by oxygen and vice versa; this is indeed the major structural feature found in oxides. In some 'extreme' versions of the model, ions are regarded as strictly spherical entities of fixed radius, a viewpoint which has considerable limitations. As discussed later in § 4.1.1, the oxide ion O^{2-} does not exist in free space; it is stabilized by the Madelung potential of an ionic lattice. One might therefore expect its size to depend strongly upon the crystal surroundings. In spite of this problem, reasonably consistent sets of ionic radii can be derived for oxides.[37] In our discussion we shall not need to make use of quantitative ionic radii, although the idea that O^{2-} is a larger ion than most metal cations will be qualitatively useful; it is implied in the diagrams that appear later showing various surface structures.

2.1.1 Ideal crystal structures

The most important determinants of a crystal structure are the **stoichiometry**, or relative numbers of the different types of atoms present, and the **coordination** of

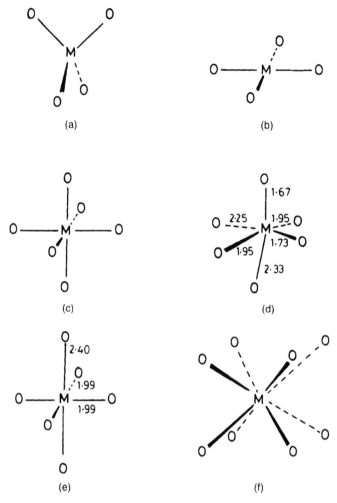

Fig. 2.1 Some of the coordination geometries of metal ions that are found in oxides. (a) Tetrahedral, (b) square planar, (c) octahedral, (d) distorted octahedral in MoO_3, (e) distorted octahedral in La_2CuO_4, and (f) cubic. Distances are in ångstrom units where indicated.

ions, i.e., the number of ions of one type surrounding another and their geometrical arrangement. A consideration of oxide structures suggests that metal-ion coordination numbers and geometries show reasonably systematic features that carry over between different structure types.[38] Common examples of metal coordinations are shown in Fig. 2.1. Most of the structures referred to later are listed in Table 2.1 along with some examples, and drawings of some important structures are shown in Fig. 2.2.

Six-fold octahedral coordination is the most common metal ion geometry in the

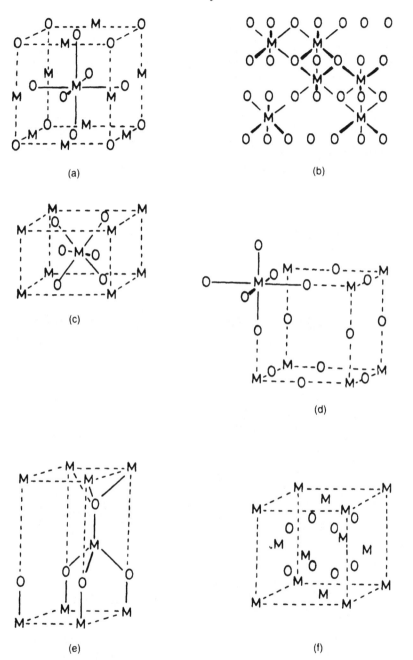

Fig. 2.2 Some important binary oxide structures. (a) Rocksalt, MO; (b) corundum, M_2O_3; (c) rutile, MO_2; (d) ReO_3; (e) wurtzite, MO; and (f) fluorite, MO_2. Crystallographic unit cells are shown except for (b).

oxides considered in this book; it occurs in the rocksalt structure found for many oxides of formula MO and in the other common binary structures such as corundum (M_2O_3) and rutile (MO_2). Higher coordinations are sometimes found with larger ions (for example, eight-fold in the fluorite structure of UO_2) and for pre-transition-metal ions in some ternary structures (e.g., twelve-fold for Sr^{2+} in $SrTiO_3$). The most common lower coordination is four-fold tetrahedral, as in ZnO (wurtzite structure), although Cu^+ in Cu_2O has an unusual linear coordination, and square planar arrangements are possible (e.g., Cu^{2+} in CuO), as are less regular geometries such as the four-fold pyramidal coordination of Pb^{2+} in PbO. Other irregular arrangements are also found; see, for instance, the two examples of distorted octahedral geometry found in MoO_3 and La_2CuO_4 illustrated in Fig. 2.1. These examples show some of the limitations of the 'hard-sphere' model, where ion size is expected to be the major factor controlling the coordination geometry. It appears rather that specific bonding factors related to the electron configurations of the metal ions are often involved. For example, the low coordination numbers found with Cu^+ and Zn^{2+} are characteristic of many post-transition-metal ions having the d^{10} configuration, irrespective of the relative ion sizes in a particular structure.[39] Pb^{2+} and Sn^{2+} are stable *lower* oxidation states of these elements (compared with Sn^{4+} in SnO_2), and the distorted non-centrosymmetric geometry is a common feature of ions with the $d^{10}s^2$ configuration; sometimes this is interpreted by postulating that the s^2 'lone-pair' of electrons effectively occupies a geometric site in the lattice. In transition-metal ions, other electronic factors operate. Tetragonally elongated octahedra such as those invariably found with Cu^{2+} are associated with a so-called **Jahn–Teller** distortion, driven by the arrangement of electrons in the d^4 and d^9 electron configurations; and the rather different type of distorted coordination in MoO_3 is also found with other d^0 ions, for example in V_2O_5. More details of these effects can be found in other books; some of them will be referred to again in Chapters 4 and 5 in connection with electronic structure.[40]

A different way of thinking about ionic structures is sometimes helpful. Many structures appear to be based on a close-packed or nearly close-packed array of oxide ions, with metal cations occupying 'interstitial' sites.[38] Both face-centered-cubic (fcc) and hexagonal-close-packed (hcp) arrays have octahedral and tetrahedral interstices available, in the proportion of one octahedral site and two tetrahedral sites per oxygen. Filling all octahedral sites in an fcc array generates the rocksalt structure. Doing the same in a hexagonal array gives the NiAs structure, which is electrostatically much less favorable than rocksalt because metal cations are arranged in chains with a considerably reduced cation–cation separation. Thus NiAs is a structure type adopted only by compounds such as sulfides, which are less ionic than oxides. On the other hand, filling a fraction of the octahedral sites can be done more effectively by starting with an hcp array than from fcc. In the corundum structure (as in the related ternary ilmenite structure), two-thirds of the sites are occupied. Although this results in a rather close approach between pairs of cations [see Fig. 2.2 (b)], a relatively easy distortion is possible which lowers the electrostatic repulsion. In a rather similar way, rutile can be derived from an hcp array of oxide ions by filling one-half of the octahedral sites; the way in which this is done does not preserve the hexagonal symmetry, however, and the distortion which minimizes electrostatic repulsion now leads to the tetragonal structure illustrated in Fig. 2.2 (c). Although the cations have six-fold coordination in the corundum and rutile lattices, the actual site symmetries are not fully octahedral: they are trigonal (C_{3v}) in corundum, and D_{2h} in rutile.

Filling of tetrahedral holes is similarly possible. Occupying all tetrahedral holes in an fcc oxide array gives the antifluorite structure of Na_2O; in the commoner fluorite structure, the arrangement of anions and cations is reversed. Filling one-half of the tetrahedral sites gives either the zincblende (spharelite) or wurtzite structures, depending upon whether one starts with an fcc or an hcp array, respectively. Zincblende is the common structure for many III–V and II–VI semiconductors such as GaAs and ZnS; however, this structure has a marginally lower electrostatic stability than wurtzite, and it is the latter which is found in the oxides BeO and ZnO. In some more complex structures, a mixture of sites can be occupied. Especially important is spinel, where one-half of the octahedral and one-eighth of the tetrahedral sites in an fcc oxygen array are occupied. In the mineral $MgAl_2O_4$, which gives its name to this structure, the divalent Mg^{2+} ions are tetrahedral and the trivalent Al^{3+} octahedral; this arrangement is known as the 'normal' spinel. Magnetite (Fe_3O_4), on the other hand, has the 'inverse' arrangement, in which Fe^{2+} is in octahedral sites and the Fe^{3+} ions are divided equally between tetrahedral and the remaining octahedral sites. Many spinels, in fact, are quite disordered, and the distribution of ions between the various sites may be strongly dependent upon the preparation and thermal history of the sample.

There is another important structure type that may be thought about in terms of filling of interstitial sites in another lattice. The ReO_3 structure, illustrated in Fig. 2.2 (d), is relatively open, having a large, vacant twelve-coordinate position at the center of the unit cell shown. In the ternary perovskite structure, this position is filled by another ion. Such occupancy may be only partial, as in the important series of sodium tungsten bronzes, Na_xWO_3, with $0 < x < 0.9$. WO_3 itself exhibits several distorted variants of the cubic ReO_3 structure, and distortions survive for $x < 0.4$ in the tungsten bronzes, although the compounds are cubic for higher Na contents. The structure may be regarded as a perovskite where the twelve-coordinate A site is only fractionally occupied.

2.1.2 Defects and nonstoichiometry

Thermodynamic arguments dictate that all crystals must contain a certain proportion of defects in equilibrium at non-zero temperatures; if the energy expended in creating a defect is ΔE and the associated increase in entropy of the system is ΔS, then the free energy $\Delta G = \Delta E - T\Delta S$ decreases for small defect densities since the entropy term outweighs the internal energy term.[41] Defects are especially important in our present account for two reasons. Firstly, defects may be much more common at surfaces than in bulk materials, and in oxides especially they may often be responsible for many of the catalytic and other chemical properties. Secondly, however, many transition-metal oxides have an unusually high concentration of defects even in the bulk solids; they are associated with the possibility of variable valence or oxidation state, which is a feature of transition-metal chemistry, and they exert an important influence on the physical properties of the solid compounds.[40]

Bulk defects may be classified either as **point defects** or as **extended** types such as line defects (dislocations) and planar defects. Elementary point defects are lattice vacancies and interstitials, as well as impurities or dopants that may occupy either substitutional or interstitial sites. In the elementary theory of defects in ionic solids, it is normally considered that such defects occur in combinations which preserve the overall electroneutrality of the solid. Thus we may have **Frenkel** (vacancy plus interstitial of the same type) and **Schottky** (balancing pairs of vacancies) types of defects. Impurities may be associated with vacancies or interstitials so as to balance the charge in a similar way: for example, in Li-doped MgO, Li^+ may substitute for Mg^{2+}, with the difference in charge being compensated by an appropriate number of O^{2-} vacancies.[42] Although these 'classical' defect types are certainly important in the compounds under consideration here, there is another equally important way in which charge can be balanced. This is by an alteration in the electronic structure at sites neighboring the defect, by the occupation of impurity levels, or by changing the charge on an ion. For example, the electrons remaining at an O^{2-} vacancy may be trapped at the vacancy site to give F or F^+ **centers**, as discussed in § 4.1.2. Electrons and holes may also act to change the oxidation state of a transition-metal ion. For example, oxygen deficiency in TiO_2 can be associated with electrons trapped at Ti^{4+} sites to give Ti^{3+}, and oxygen excess (normally, in fact, associated with metal vacancies) can increase the oxidation state of some atoms, as in $Fe_{0.9}O$, which may be considered to have a proportion of Fe^{3+} ions in addition to Fe^{2+}. Obviously such changes in electronic structure have implications for the properties of the oxides concerned; these are discussed in more detail in Chapters 4 and 5. At this stage, the essential point to be aware of is the existence of defects and the various types that can exist.

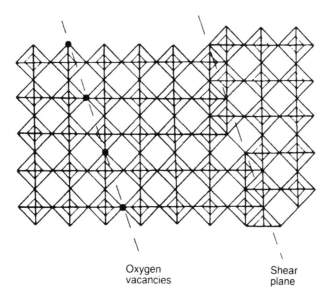

Oxygen
vacancies

Shear
plane

Fig. 2.3 Schematic diagram of the formation of a crystallographic shear (CS) plane in the
WO_3 lattice by the elimination of a row of oxygen vacancies.

Some structural aspects of defects on oxide surfaces are discussed later in this chapter.

The picture of isolated point defects may be appropriate for compounds with nearly perfect structures or stoichiometries. There is a great deal of evidence, however, that at higher concentrations strong interactions can occur between defects, causing them to cluster or order in various ways.[43,44] One such type of ordering leads to the formation of so-called crystallographic shear (CS) planes discussed below. Other types of aggregation have been found in non-stoichiometric oxides, and one might suspect that, because of the high concentrations of defects present at many surfaces, defect interactions may be especially important there. Some of the reconstructions found on oxide surfaces may be of this type, but there have as yet been essentially no systematic studies of these effects.

Dislocations are certainly present in oxides and, as with other types of solids, may be important in determining their mechanical properties. Their importance for surfaces is that they may often form sites where crystal growth takes place. However, this is not an aspect that we are considering in the present book, and from the point of view of the studies treated here, no work seems to have been done on dislocations.

Planar defects include the surface itself, and grain boundaries. Surfaces may obviously have many other defects associated with them, including point defects and steps (effectively a type of line defect), as discussed later. Another type of planar defect important in some transition-metal oxides is known as the **crystallographic shear (CS) plane**.[44] As mentioned above, point defects may cluster and aggregate in various ways: the formation of a CS plane can be regarded as the association of oxygen vacancies along a lattice plane, accompanied by a shear of the structure in such a way as to eliminate the vacancies altogether. This process (which is not meant to represent a genuine mechanism for formation of CS planes – that may be much more complicated) is illustrated in Fig. 2.3. CS planes form when some d^0 oxides such as TiO_2, V_2O_5, MoO_3 and WO_3 undergo oxygen loss. In well-equilibrated samples the planes are often regularly spaced in the crystal; the different possible spacings can give rise to series of phases of related stoichiometry, such as Ti_nO_{2n-1}, where n is proportional to the number of unaltered lattice planes between successive CS planes. Compounds of this type, known sometimes as **Magnéli phases**, will be mentioned again later.

2.1.3 Theoretical modelling

Theoretical modelling of structures is especially important in the context of defects and surfaces because of the difficulties involved in experimental structure determination. Indeed, for both types of situation there are many more theoretical calculations than good experimental results in the literature! Because the principles involved in these calculations are similar for both bulk solids and for surfaces, it is appropriate to discuss them briefly here.[41,42]

Theoretical models may be divided into two broad classes: those based on a quantum-mechanical calculation of the total energy, and those that use a parameterized interaction potential between ions. Such potentials may be empirically based, but the distinction between the two classes of theory is not absolute, because parameterized models sometimes use inter-ionic potentials derived from quantum mechanics. In all types of calculation, the goal is to predict the most stable structure by finding the atomic configuration which minimizes the total energy. This may be done in a perfect crystal by using periodic boundary conditions in three dimensions. For an ordered surface structure, one can use a slab several unit cells thick, invoking two-dimensional periodicity within the slab. Defects are more difficult to treat, especially when they are associated with charges that may exert a long-range force on the surrounding solid. Some type of 'embedding' technique is necessary to deal with local changes in geometry constrained by the rest of the solid. Although embedding methods do exist for quantum-mechanical calculations, to combine such a calculation with full energy minimization with respect to atomic positions is a formidable computational challenge, and in this area parameterized methods are probably essential.

Electronic structure calculations will be discussed in more detail in Chapter 4.

At this point it is worth noting that advances in computational power combined with efficient and accurate codes are now making serious *ab initio* calculations, using either local-density or Hartree–Fock formalisms, a real possibility even for systems with many structural parameters.[45] There is not yet much information from which to judge their accuracy directly in the field of oxide surfaces, but experience in other areas suggests that this could be fairly impressive. On the other hand, the older type of semi-empirical electronic structure calculation should probably be mistrusted in the area of structural prediction; although these methods may be very useful in the treatment of electronic levels, they are not generally well-adapted for calculating total energies as a function of geometry.

The qualitative success of the ionic model suggests an obvious parameterized scheme for calculation, using an inter-ionic potential that includes long-range Coulomb and short-range repulsive components. Experience in this area, and especially in the calculation of defects, has demonstrated, however, that it is essential to augment this simple idea by treating ions as *polarizable* entities.[41] The most fully developed model in this area is known as the **shell model**, since the ionic polarizability is incorporated by simulating each ion as a negatively charged 'shell', connected by a 'spring' to a positive core. The total (shell + core) charge is the normal ionic one, but both the distribution of charge between the shell and the core, and the force constant of the spring, are treated as parameters. Other parameters govern the short-range forces (overlap repulsion plus dispersion), and a full parameter set may be obtained by fitting various properties of the 'ideal' bulk solid: e.g., dielectric, vibrational and elastic properties, possibly supplemented (especially for short-range forces) by an appropriate quantum-mechanical calculation. Once the parameters are known, the energy of any configuration of ions may be calculated. Defects are treated by the Mott–Littleton method, in which the solid is divided into two regions. Immediately surrounding the defect, individual ions are treated explicitly using the appropriate interaction parameters. The rest of the crystal outside of this region is treated as a dielectric and elastic continuum, in which the inner region is embedded with an appropriate matching of ionic displacements, electric fields, etc., at the boundary. This method may obviously be applied to defects at surfaces, as well as within the bulk of the solid.

A number of calculations on perfect, stepped, and defective surfaces have been performed using the shell model, and some of the results will be discussed later. Apart from obvious questions such as the transferability of parameters derived from bulk solids to the different coordinations found at surfaces, there is an important limitation to these calculations. It must generally be assumed that ions do not change in electronic configuration, and that other electronic effects can be neglected. Unusual species such as O^- (which may be common on surfaces) are not easy to treat, and even more difficulty may arise with defects, such as F centers, where electrons are trapped at anion vacancies. For these reasons the types of defect configuration that have been treated are rather limited.

With what types of 'real' surfaces should theoretical models be compared? That is not an easy question to answer. Self-consistent calculations should give the *equilibrium* geometry and electronic structure of surfaces. For materials that cleave well [e.g., NiO (100)], the relevant surface is probably one cleaved in UHV, since other surface preparation procedures often result in non-stoichiometric surfaces. For materials that do not cleave well (e.g., TiO_2), a polished and annealed surface may have a smaller density of steps and in fact be closer to an ideal truncation of the bulk than a fractured surface. But even annealing will not necessarily result in an equilibrium surface in the sense of the theoretical models; the resultant surface depends upon annealing temperature and time, ambient atmosphere, cooling rate after annealing, etc. On balance, it is probably safer to assume that UHV-cleaved or fractured surfaces are closer to model surfaces than are annealed ones.

2.2 Experimental techniques for surface structure determination

While the bulk geometric structures of virtually all solids are known very accurately from x-ray diffraction measurements, the geometrical arrangement of atoms and ions on the surfaces of solids is only known accurately for a relatively small number of materials. The surface-sensitive technique that has been used most extensively to determine surface geometry is **low-energy electron diffraction (LEED)**.[4,46,47] As for x-ray scattering in bulk crystal structure determination, LEED is a delocalized probe that relies on the long-range order of the surface atoms to produce the constructive and destructive interference of the incident-electron wave that generates a diffraction pattern. Thus in order for LEED to give an accurate picture of the arrangement of surface atoms, the surface studied must be highly ordered over large areas (of the order of a few hundred Å). For an ordered surface, measurements of the current of electrons scattered into each LEED beam as a function of incident-electron energy – which are referred to as 'I-V' measurements – are theoretically capable of determining atomic positions both in the plane of, and normal to, the surface. The most accurate LEED structure determinations that have been performed are for metals and semiconductors, where the material either cleaves well (e.g., many semiconductors at room temperature and some metals at low temperatures) or where techniques have been perfected to produce nearly perfect surfaces by ion-bombardment cleaning and annealing.

Even if a sample were to have a perfectly ordered surface, there is another limitation to the accuracy of LEED surface structural determinations. In x-ray diffraction, the x-ray absorption cross-section is so small that one can assume that x-rays scatter only once during the diffraction process. It is then straightforward to interpret x-ray diffraction patterns in terms of a single-scattering theory (the 'kinematical' model), and even very complicated structures can be unraveled with a modest amount of computer time.[48] In LEED, however, the electron–atom scattering cross-section is extremely large over the range of electron energies used (about 20 to 300 eV), and an electron wave will scatter several times during its interaction with the crystal. (It is this strong electron–atom interaction, of course, that gives low-energy-electron spectroscopies their surface sensitivity; electrons traveling in a solid scatter both elastically and inelastically with a mean-free-path of only a few Å.) Kinematical theories therefore do not give a useful description of the scattering process, and a full multiple-scattering ('dynamical') theory must be used in order to determine the actual positions of atoms in the surface region from measurements of the direction and intensity of diffracted beams.[46] Dynamical theories involve large amounts of calculation, and only a few dozen specific cases have been treated in sufficient detail (both experimentally and theoretically) that the surface structure is known fairly accurately.[46] In contrast, several hundred surface structures have been determined qualitatively by LEED.[49]

The status of LEED determinations of the surface geometric structure of oxides lags far behind that for metals and semiconductors for several reasons. For one thing, only a few oxides cleave well along any crystal plane, with most oxides either exhibiting conchoidal fracture or having large step and defect densities on cleaved faces. Attempts to prepare oxide surfaces by ion bombardment and annealing of cut-and-polished faces generally lead to either reconstructed surfaces or surfaces whose stoichiometry is different than that of the bulk crystal. A second reason is that insulating oxides can be studied by LEED only for primary-electron energies where their secondary-electron yield (i.e., the number of electrons, both reflected primaries and true secondaries, that are emitted from the sample surface for each incident electron) is greater than unity;[50] this restricts the electron energy range over which LEED can be performed, thus limiting the amount of information that can be obtained. In addition, there are theoretical problems in treating many oxides in that their unit cells are relatively large, thus resulting in much greater computational complexities than for materials with simple unit cells. The net result is that only a few oxide surface structures have been determined by LEED, and none involving chemisorbed atoms or molecules! Most of the LEED work that has been performed on metal oxides consists of qualitative observations of the symmetry of diffraction patterns.

When considering LEED patterns it is important to bear in mind that they often contain less information than one may think. Most authors give the symmetry of the LEED patterns that they observe without much qualifying description. For example, a pattern that is described as '(1 × 1)' may refer to an essentially perfect surface having very low step and point defect density [e.g., cleaved NiO (100)], or it may refer to a surface that has a disordered monolayer on the surface and a large step density. The difference between those examples is in the 'quality' of the pattern (i.e., the size and brightness of the spots and the intensity of the background). Discussion of quality is often omitted in publications. Even when a description is given it is of limited use, since it is in any event subjective. What a '(1 × 1)' pattern *does* tell you is that there is no significant reconstruction of the surface, although there may still be major relaxation within each surface unit cell.

Higher-energy electrons can also be used for surface structure determination, but the longer electron mean-free-path decreases the amount of surface information obtained compared to that from the bulk. This can be partly overcome by using grazing incidence and emission geometries. This is the basis of **reflection high-energy electron diffraction (RHEED)**, where incident electrons having energies from a few keV to 100 keV are elastically diffracted at grazing incidence from surfaces.[4] The diffraction patterns are less straightforward to interpret than in LEED, but complementary information can be obtained. RHEED is much more sensitive than LEED to surface morphology, so it has found widespread use in monitoring epitaxial or thin-film growth on single-crystal surfaces. The technique has been applied in a few instances to determination of metal-oxide surface structures.

Ion scattering spectroscopy (ISS) is another technique that has proven very useful in surface structure determinations.[4] The experimental data are conceptually easier to interpret than in LEED since most ion–surface collisions involve only a single-scattering event (except for those at grazing incidence), so that kinematical theory is appropriate. This computational simplification also means that complicated surface unit cells can be treated without undue difficulty. Ion neutralization and surface charging effects can complicate the interpretation of data, however, and similar requirements of surface perfection and long-range order hold for ISS as for LEED. Only a few reports have been published of ion scattering determinations of oxide surface structures.

The **diffraction** of very low energy **He atoms** from surfaces has also been used in a few studies of metal-oxide surface structures. The surface potential seen by neutral He atoms is not the same as that seen by ions, so the information obtained by the two techniques is different.[4] But He scattering is extremely surface-sensitive, giving the corrugation of the surface potential that results from the geometric structure of surface atoms or ions. It can also be used to study highly insulating samples, since neither charge transfer nor neutralization is involved in the scattering process. To date the only metal oxide to which He atom scattering has been applied is MgO.

Two techniques that currently promise to yield a different type of information on the geometric arrangement of atoms on solid surfaces than that obtained by LEED or ISS, particularly in the presence of defects (which exist, of course, to some degree on any real surface), are **scanning tunneling microscopy (STM)** and **atomic force microscopy (AFM)**.[51–54] These are the only surface structural techniques that are truly local in that they image individual atoms and do not rely upon long-range order to produce a signal. To date, higher spatial resolution has been achieved with STM (about 1.5 Å lateral and 0.2 Å vertical), whereas AFM has only reached atomic resolution in a few special cases. Neither of these techniques has yet been used to a very large degree on well-characterized oxide surfaces. STM requires that a rather high current density flow between the tip and the sample, so it can be used only on materials that possess sufficient electrical conductivity (although it is not yet clear just what constitutes 'sufficient'). But the potential power of these techniques to determine even the geometric arrangement of atoms at point defect sites on surfaces (which *no* other techniques are able to address) make STM and AFM some of the most important tools for future surface structural studies. [STM is also theoretically capable of performing electron spectroscopy over areas comparable in size to individual atoms (or defects); this will be discussed in conjunction with surface electronic structure in succeeding chapters.]

Another class of techniques that has found limited application in oxide surface structural determinations is derived from **extended x-ray absorption fine structure (EXAFS)**.[4] EXAFS itself, which consists of measuring the x-ray absorption of a material as a function of x-ray energy, is a bulk technique owing to the large penetration depth of x-rays in most solids. However, if the x-ray absorption process is detected via some surface-sensitive mechanism such as Auger-electron emission, surface structural information involving interatomic distances and coordination numbers can be obtained. This technique, referred to as **SEXAFS**, is just now beginning to be applied to structural determinations of oxide surfaces.[55] EXAFS-like structure has also been observed in electron-energy-loss spectra for incident-electron energies above a loss threshold, arising from the backscattering of the excited electron by its surroundings.[56] The process, referred to as **extended energy-loss fine structure (EXELFS)**, has been used in studies of MgO.

2.3 Oxide surface structures

The goal of any surface structural analysis is to determine, as accurately as possible, the positions of all of the atoms in the surface and near-surface regions of the crystal. In order to interpret the results of experiments such as LEED or ISS, it is generally necessary to begin with a model of the surface structure and to refine it based upon the experimental data. In the case of metal oxides, only a few surfaces have actually been analyzed experimentally, so in many cases a model is all that we have. In this section we will present geometric models of some of the most important and most studied metal-oxide surfaces based upon hypothetical cleavage of bulk crystals. We will then review the experimental and theoretical work that has been performed on specific oxide surfaces in order to see how they compare with the models.

It is pedagogically useful to approach the geometric structure of crystal surfaces in two steps. First consider the surface as generated from the bulk crystal structure by means of cleaving along a given crystal plane, leaving the atoms in exactly the same positions that they had in the bulk. Then consider what relaxation or rearrangement of the surface and near-surface atoms takes place compared to their positions in the bulk crystal. We will consider only the former step in determining the 'ideal' crystal surfaces that could exist for various bulk crystal structures; the latter step will be discussed for specific materials based upon experimental data.

In theory one could create an infinite number of different surfaces from any bulk crystal by separating the crystal along any plane, no matter how large the Miller indices for that plane might be. What one would find on the atomic scale, however, is that most of the surfaces generated in this way were composed of terraces of low-Miller-index planes separated by steps of single atomic height in various directions. It is thus most important to consider the atomic geometry of low-index faces; the more complex stepped surfaces, which are called 'vicinal' surfaces when they are close in orientation to one of the low-index faces, can then be understood in terms of those faces combined with steps.

For any bulk crystal structure there are many low-Miller-index faces that could be formed. The surface energy for the various faces may be quite different, however, so that only certain ones will form preferentially or will even be stable. The surfaces that we will consider here are those that have been observed experimentally in studies of metal oxides or where attempts to prepare them on oxide crystals have been made.

It should be pointed out that the 'ideal' surface structures presented here have the stoichiometry of the bulk material by definition. The closest that one can expect to come to this type of surface in practice is to cleave a single crystal in UHV. An oxide surface prepared by cutting and polishing, followed by ion-bombardment cleaning and annealing in vacuum, could, and often will, have a different stoichiometry and structure. For that reason we will discuss measurements performed on UHV-cleaved single-crystal samples whenever possible; preferential sputtering will be discussed in § 2.6 below.

Many of the oxides that will be discussed here occur naturally as minerals, and a great deal can be learned about their properties from the vast mineralogical literature. One of the most important properties that is used in characterizing minerals is their cleavage or fracture, which occurs along planes having the lowest surface energies; those are the planes that are of primary interest to us here. (Cleavage should not be confused with 'parting', which refers to the breaking of minerals along planes of structural weakness. Parting occurs only in specimens that are twinned or have been subjected to certain types of pressure, while cleavage occurs repeatably in all specimens.[58]) However, some of the conclusions that might be drawn from mineralogy can be misleading. When considering the relative stability of various crystal faces in a material, it is tempting to look at which faces occur predominantly on natural specimens; this is referred to as the 'habit' or 'form' of minerals and is one of their most striking characteristics. The equilibrium shape of a crystal should be one that minimizes the total surface energy, and thus low-energy surfaces should predominate.[57] However, natural minerals need not be in an equilibrium configuration. Also, they did not form in a vacuum; they solidified from a matrix of other minerals, water, etc., and so their shapes were determined by *interfacial* energies rather than by true surface energies. That is the reason that the same mineral may have different habits when found in different geographical locations. An excellent example is periclase (MgO), which occurs most often naturally as octahedral crystals having predominantly (111) faces.[192] And yet the ideal MgO (111) surface is polar and unstable in vacuum (see § 2.3.1.2 below). It is nonetheless instructive to consider the properties of the mineral forms of metal oxides, and we will do so briefly where appropriate.

The experimental and theoretical results that have been obtained for the surface geometric structure of specific metal oxides are organized by crystal structure in Tables 2.2 through 2.7. References to the original literature are given in the tables and are thus generally omitted from the text. In most cases the bulk lattice parameters have been given,[59] from which the surface unit-cell dimensions can be obtained by simple geometrical arguments.

2.3.1 Rocksalt

The rocksalt crystal structure is the simplest oxide structure not only from the standpoint of the bulk arrangement of the atoms, but because the surface energy is far lower for the (100) surface than for any other.[103-105] For example, when Mg metal is burned in air or oxygen, the MgO smoke particles that are formed are almost perfect cubes having (100) faces; this is shown beautifully in Fig. 2.4, which is a low magnification transmission-electron microscopy (TEM) micrograph of such particles. Thus virtually all of the work on the surface properties of

Table 2.2. *Rocksalt oxides*

Oxide	Face	Surface preparation	Technique and results	Experiment	Theory
MgO $a_0 = 4.211$ Å	100	UHV cleaved	LEED I–V; nearly ideal termination	60–65	
			RHEED: agrees with LEED	66, 67	
			He atom diffraction: agrees with LEED	68–74	
		Annealed at 573 K	RHEED: significant rumpling	75–78	
			LEED I–V: as cleaved, little rumpling	64	
		Various	He scattering: UHV cleave superior to air cleave	79, 80	
		P&A in air	ISS: significant inward relaxation	81	
		Air cleaved	EXELFS: inward relaxation, O overlayer	56,81	
		–	Shell model calculations: small relaxation		82, 86
		–	HF calculation: little relaxation		87
		–	Electron–phonon coupling theory; predicts large relaxation		88
	110	Polished, not annealed	LEED: (1 × 1) pattern	89	
		P&A	LEED: (100) facets	89	
		–	HF calc: some relaxation; little rumpling		90
	111	Polished, not annealed	LEED: (1 × 1) pattern	89	
		P&A	LEED, SEM: (100) facets	89	
		Air-cleaved and annealed	LEED: (1 × 1) pattern	91	
CaO $a_0 = 4.811$ Å	100	UHV cleaved	LEED I–V: nearly ideal termination	62, 92	
		–	HF calculation: agrees with LEED		92
NiO $a_0 = 4.168$ Å	100	UHV cleaved	LEED I–V: nearly ideal termination	62, 93–95	
	111	P&A	LEED: (1 × 1) assumed stabilized by impurities	96, 97	
CoO $a_0 = 4.267$ Å	100	UHV cleaved	LEED I–V: nearly ideal termination	62, 98	
MnO $a_0 = 4.445$ Å	100	UHV cleaved	LEED: diffuse spots, charging problem	62, 99	
		P&A	LEED: (1 × 1), (2 × 2) and (6 × 6) patterns, surface O rumpling	100	
EuO $a_0 = 5.144$ Å	100	UHV cleaved	LEED: very disordered surface	62, 101, 102	

rocksalt oxides has been done on (100) surfaces. Attempts to prepare the rocksalt (110) and (111) surfaces have been partially successful, although they tend to facet to surfaces containing (100) planes in order to minimize the surface energy; in some cases they are stabilized by impurities.

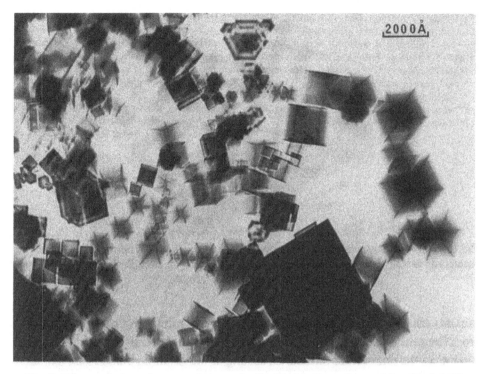

Fig. 2.4 Transmission-electron micrograph of cubic MgO smoke particles. [Ref. 248]

2.3.1.1 Rocksalt (100)

The ideal geometry of the (100) rocksalt surface is unambiguous, since the bulk structure in that direction consists of layers of atomically flat planes composed of equal numbers of cations and anions, and any imaginary cleave of the crystal would occur between two planes. Figure 2.5 shows a spherical ball model of a rocksalt surface oriented slightly off (100) so that both (100) terraces and a monoatomic step in the [010] direction can be seen. Missing cation and anion point defects are also shown on the uppermost terrace. (We will return later to a discussion of steps and point defects on this and other surfaces.) This type of model will be used throughout this book in order to visualize what various surfaces might look like on the atomic scale. In almost all cases the oxygen anions are larger than the metal cations, and they are represented by the large open circles, with the shading of the circles increasing for successive layers below the surface. The solid circles represent the cations. The ratios of the cation and anion radii are chosen to represent those characteristic of most of the oxides having that structure that will be discussed; in Fig. 2.5 the cation radius is taken to be one-half that of the anions. [As pointed out in § 2.1.1, the concept of ionic (or atomic) radii in

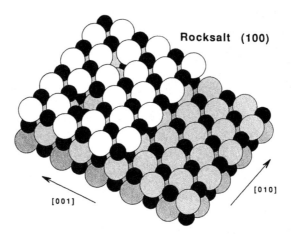

Fig. 2.5 Model of the rocksalt (100) surface. Large circles are O anions, small circles are metal cations. A [010] step to another (100) terrace is shown, as are both missing anion and missing cation point defects.

solids is an approximate one since it is not really possible to identify valence-electron charge as belonging to one ion or another. But the idea is very useful for visualizing crystal and surface structures, as long as the resulting pictures are not taken too literally.] The position of the ions is that of the bulk crystal structure, with no relaxation or reconstruction.

The ideal (100) surface is atomically flat (in the sense that all of the nuclei of the outermost ions lie in the same plane) and charge neutral; i.e., a slab of crystal having (100) faces on both sides will have no net dipole moment normal to the surface, regardless of the number of planes in the slab, and the slab will have zero net charge. Surfaces whose atomic planes are charge neutral are referred to as **non-polar** surfaces, whereas surfaces that do possess a net dipole moment are called **polar** surfaces. Polar surfaces are never truly stable, since the long-range electrostatic dipole fields that are present tend to drive surface reconstruction.[106,107] All rocksalt oxides cleave preferentially along the (100) plane, and some of them (e.g., NiO and MgO) cleave extremely well along that plane, yielding relatively flat and defect-free surfaces. The cations and anions on the rocksalt (100) surface each have five nearest-neighbor ligands, four in the surface plane and one directly below it in the second atomic plane. Only one nearest-neighbor ligand is missing compared to the bulk; it would lie directly above the surface ion. The two opposing crystal faces formed by an imaginary cleave are identical.

MgO (100)

More structural determinations have been carried out on the surfaces of rocksalt metal oxides than on any others. The most thoroughly studied oxide is MgO, which occurs naturally as the mineral periclase. Mineral specimens cleave extremely well along (100), although naturally occurring faces often include (111) and sometimes (110). Literally dozens of papers have attempted to determine the location of the ions on the surface of MgO either experimentally or theoretically. While there is general agreement among most of the studies as to the surface structure, some experimental results are in serious disagreement. Experiments have been performed both on surfaces cleaved in air and cleaved in UHV; essentially no differences in the arrangement of ions in the surface plane have been found between them, although He scattering measurements, which are extremely sensitive to the perfection of the surface, indicate that UHV-cleaved surfaces are of higher quality than those cleaved in air. One of the primary reasons for the difference is the effect of water on air-cleaved samples, which creates a high density of point defects on the surface.[79]

Several LEED studies have been performed on UHV-cleaved MgO (100) in which multiple-scattering calculations were used to analyze I-V data for both specular and non-specular beams. All studies are in agreement that the surface is very nearly an ideal truncation of the bulk structure. The best model for the surface corresponds to an inward relaxation of the surface plane (i.e., a decrease in the spacing between the surface plane and the second plane compared to the bulk value) of no more than 2.5 % and a rumpling of the surface (i.e., the motion of one type of ion relative to other types in a direction normal to the surface) of less than about 2 %, with the O ions moving away from the bulk relative to the Mg ions; there is a wider variation in values between the different experiments for the degree of rumpling of the surface than for the size of the relaxation. This model is consistent with the results of most elastic He atom diffraction measurements on MgO (100), although a direct comparison cannot be made since He atoms sense a different potential at the surface than do electrons.

Results in conflict with those above were obtained from measurements of Kikuchi patterns in RHEED from MgO (100).[75–78] Anomalous enhancement of Kikuchi patterns was observed when a vacuum-cleaved surface was heated to 573 K and then cooled back to room temperature; these changes were attributed to an outward motion of the surface O ions upon annealing. The magnitude of the resultant rumpling was estimated to be about 6 % of the atomic spacing, where it was assumed that the cleaved surface before annealing did not exhibit any rumpling. A theoretical model for these observations was proposed in terms of enhanced electron–phonon coupling and a reduction of the bandgap at the surface.[88] However, subsequent LEED work on surfaces prepared in the same manner found no significant differences in their surface structures,[64] and a comparison of RHEED measurements on UHV-cleaved surfaces with calculations of rocking curves for model surface geometries was also in agreement with the LEED results.[66,67]

Two other measurements using different techniques have concluded that the surface layer is significantly relaxed inward on MgO (100). Impact-collision ISS experiments deduced an inward relaxation of 15 % of the bulk spacing, although the MgO sample studied was prepared by polishing, etching and annealing at very high temperatures in air and thus may have a somewhat different structure than cleaved surfaces. EXELFS measurements on air-cleaved MgO (100) were interpreted in terms of an inward relaxation of 17 % compared to the bulk value. However, it was also necessary to include an extra layer of O atoms located at a distance of 2.2 Å above the Mg surface ions in order to obtain a good fit between theory and experiment; this is inconsistent with both LEED and He scattering results.

The geometrical structure of the MgO (100) surface has been addressed theoretically using various types of shell models and an *ab initio* Hartree–Fock crystalline-orbital linear-combination-of-atomic-orbitals (LCAO) approach. All calculations agree that any relaxation or rumpling of the surface plane should be of the order of a few percent or less, in agreement with the LEED results. Systematic features of the reconstruction of rocksalt surfaces (oxides and halides) as predicted by shell models include: 1) in the surface plane, the ion with the smaller polarizability relaxes inward relative to the ion with the larger polarizability; 2) the rumpling of the top layer is repeated in lower layers, but with decreasing amplitude and alternating sign; and 3) the average layer shift is inward for the top layer and alternates in sign in successive layers.[85] Changes in atomic position compared to the bulk are predicted to exist down to eight or ten atomic planes below the surface,[82] although no experimental techniques currently have the precision necessary to determine such changes.

CaO (100)

CaO occurs naturally in small amounts enveloped in blocks of lava at Vesuvius; it exhibits perfect cleavage along (100). Quantitative LEED I-V measurements on UHV-cleaved single-crystal CaO samples, coupled with dynamical diffraction analysis, have been used to determine the geometry of the (100) surface. The results are virtually identical to those for MgO (100) in that the surface is almost an exact termination of the bulk crystal structure. An upper limit of 2 % was placed on the rumpling of the surface, and the outermost atomic plane was found to relax inward by 1 % of the interlayer spacing. Shell-model calculations were also performed, which predicted a rumple of 5 % and a surface contraction of 3 %.

NiO (100) and CoO (100)

Of all the rocksalt metal oxides whose surfaces have been studied, NiO exhibits the highest quality cleavage. Cleavage occurs, of course, along (100) planes, and the energy required for cleaving is so small that care must be taken in handling samples. [In its naturally occurring mineral form bunsenite, however, NiO exhibits predominantly (111) faces in an octahedral habit.] The LEED patterns obtained on UHV-cleaved surfaces are of excellent quality, with sharp spots and extremely low diffuse background intensity.[32,191] Extensive LEED I-V measurements on UHV-cleaved samples have been made, and full dynamical calculations were used to interpret the spectra. The conclusions are similar to those for MgO (100) and CaO (100); no rumpling of the surface plane occurs on a scale detectable by LEED analysis, and the outermost plane of ions is relaxed inward by about 2 % of the interlayer spacing.

CoO also cleaves easily along (100), but the quality of the cleaves is not quite as good as that for NiO.[192] The same type of experimental and theoretical LEED analysis was performed on UHV-cleaved CoO (100) as on NiO (100), and the surface again looked like a termination of the bulk. Sample charging limited the incident-electron energy that could be used to greater than 170 eV, thereby restricting the data base and hence the accuracy of the determination. But once again the surface atomic plane was found to be at the ideal bulk location to within ± 3 % of the interlayer spacing.

MnO (100) and EuO (100)

Not all rocksalt metal oxides cleave as well as MgO or NiO. In fact, two of them are particularly bad actors. For MnO, which occurs naturally as the mineral manganosite, cleavage is only fair along (100), and its habit, like those of most other rocksalt oxide minerals, generally shows predominantly (111) octahedral faces. Artificially grown MnO crystals fracture moderately well along (100), but only diffuse LEED patterns are observed, even for UHV-cleaved samples. Surface charging at room temperature also limited the incident-electron energy in LEED measurements to greater than 250 eV; since this is too high to produce a useable data base, no quantitative measurements have been made. However, good LEED patterns could be obtained on polished and annealed samples at temperatures above approximately 480 K, and (1 × 1), (2 × 2) and (6 × 6) patterns were observed. The (1 × 1) patterns transformed into (2 × 2) and (6 × 6) at temperatures between 800 and 1000 K, but with no detectable change in surface stoichiometry. The high-temperature reconstructed surface was interpreted in terms of rumpling of surface O ions.

EuO is an even worse actor than MnO. One preliminary LEED study reported time-dependent patterns suggestive of an unstable surface. In another attempt, three different EuO single crystals were cleaved in UHV and, while charging was not a problem, only diffuse scattering was seen, with no trace of LEED patterns. Tests for surface contamination and electron-beam-induced oxygen desorption were negative, and it was concluded that the surfaces disorder immediately after cleaving.

2.3.1.2 Rocksalt (110) and (111): MgO

The rocksalt (110) surface is also non-polar, with equal numbers of cations and anions in each atomic plane parallel to the surface, although its surface energy is more than twice that for the (100) surface.[103,104] The (111) surface, on the other hand, is conceptually more complicated than either of the other two rocksalt surfaces in that the bulk crystal structure in that direction consists of alternating planes containing either all metal or all O ions. Thus individual (111) planes are not charge neutral, and an imaginary cleave along a single plane between layers of ions would yield one surface having a monolayer of cations as its outermost plane and the opposing surface having only O ions on the surface. Neither surface plane is charge neutral, so there would be a net dipole moment normal to each surface, which leads to a divergence in the surface energy.[105-108] One way to avoid the presence of a dipole moment and produce two equivalent cleavage faces would be to have one-half of the ions in a single (111) plane remain on each of the two surfaces,[109] but there is no evidence as to whether or not this occurs in practice.

Attempts to prepare the MgO (110) and (111) surfaces have been partially successful, but the surfaces tend to facet to structures containing (100) planes. (110) and (111) surfaces prepared by cutting and polishing in air did exhibit LEED patterns characteristic of the respective surface, but attempts to anneal or ion bombard the surfaces resulted in faceting to (100) faces; Figure 2.6 shows a scanning-electron-microscope (SEM) image of such a (111) surface after annealing at about 1400 K.[89] In another experiment an MgO (111) surface was prepared by fracturing in air; it also exhibited characteristic LEED patterns prior to any surface treatment.[91] There have been some reports in the literature of the preparation of stable (111) surfaces on NiO by polishing and annealing, but in all cases impurities were found to have segregated to the surface, presumably stabilizing the otherwise unstable, polar surface.[96,97]

Theoretical calculations of the structure of the MgO (110) surface have been performed using a Hartree–Fock approach within a crystalline-orbital LCAO formalism. The surface plane is predicted to relax inward by 6.7 % of the interplanar spacing, but to exhibit no more than 1.5 % rumpling. No experiments have been conducted to verify these predictions.

2.3.2 Perovskite

Another relatively simple cubic oxide crystal structure whose surface properties have been investigated is perovskite, ABO_3. While this structure is strictly cubic only for the high-temperature, non-ferroelectric phases of oxides, the distortions

Fig. 2.6 Scanning-electron micrograph of MgO (111) annealed at about 1400 K for 1 min.
[Ref. 89]

that occur at lower temperatures are slight and can be ignored to first order when considering surface geometric structure.[110] The two faces of the perovskite structure that have been studied experimentally are (100) and (111).

Two other important oxide crystal structures are closely related to perovskite and are therefore included in this section. As discussed in § 2.1.1, WO_3 and ReO_3 crystallize in structures that are only a slight distortion of the perovskite with all of the A cation sites vacant; in Na_xWO_3, Na ions occupy a fraction of the perovskite A cation sites.

2.3.2.1 Perovskite (100)

Figure 2.7 presents an ideal model of a perovskite surface oriented slightly off (100) in order to show the two types of (100) planes that can exist on this surface. All of the (100) planes in the crystal are atomically flat, but they alternate in composition between AO and BO_2. Both AO and BO_2 planes are charge neutral (nonpolar), since in perovskite the A cations are A^{2+} and the B cations are B^{4+}; the O anions are considered to be O^{2-}. Cleaving the bulk crystal along a plane separating

Table 2.3. *Perovskite oxides*

Oxide	Face	Surface preparation	Technique and results	Experiment
$SrTiO_3$ $a_o = 3.905$ Å	100	Fractured	LEED: (1×1) poor quality, improved by annealing	111–113
		P&A	LEED I–V: see text	114, 115
	111	P&A	LEED: (1×1); facets at higher temperatures	25
$BaTiO_3{}^a$	100	Various	LEED: see text	112, 116, 117
$WO_3{}^b$	100	P&A	LEED: (1×1) with splitting and triclinic distortion; twinned microdomains?	118
		Cut 7° off (100)	LEED: complicated, (111) facets and shear-plane structures	118
$Na_xWO_3{}^c$	100	P&A	LEED: (1×1), (2×1), (3×1) patterns associated with Na ordering	119–123
	110	P&A	LEED: (1×3) Na ordering	124

[a] Tetragonal at room temperature: $a_o = 3.98$ Å, $c_o = 4.01$ Å
[b] Distorted monoclinic structure at room temperature
[c] Cubic for $x > 0.4$, with $a_o \approx 3.8$ Å

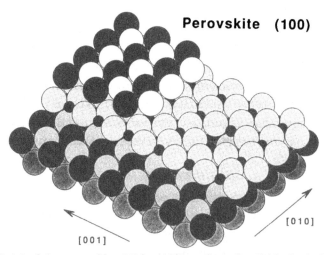

Perovskite (100)

[001]   [010]

Fig. 2.7 Model of the perovskite ABO_3 (100) surface. Small black circles are B cations, large dark circles are A cations. Two steps to other (100) terraces are shown, as is an O-vacancy defect in the BO_2 plane.

two of these atomic planes would result in one of the cleaved faces having an AO surface and the opposing one having a BO_2 surface. In practice, however, any real cleaved surface would consist of equal areas of terraces having each of the two compositions, separated by steps. The AO and BO_2 surfaces may, however, have different surface energies, so surfaces prepared by ion-bombardment cleaning and annealing might have different ratios of the two types of surfaces. On the AO surface, the large A cations have eight O ligands, compared with twelve in the bulk. On the BO_2 surface, the smaller B cations are five-fold coordinated with O ions. In discussions of the electronic structure of perovskite (100) surfaces below, interest will center on the BO_2 planes, since for most perovskite metal oxides the filled (empty) electronic levels of the A cation lie several eV below (above) the levels of the B and O ions in the vicinity of the Fermi level. An O-vacancy point defect is shown on the BO_2 plane and will be discussed later.

$SrTiO_3$ (100)

There have been several reported observations of qualitative LEED patterns for $SrTiO_3$ (100). Vacuum-fractured samples exhibit conchoidal fracture with characteristic (1×1) LEED patterns, Fig. 2.8 (a), although the spots are broader and the background intensity higher than for oxides such as MgO or NiO that cleave well. Annealing of fractured samples at about 1200 K significantly improves the quality of the LEED patterns, as shown in Fig. 2.8 (b). Polished and annealed $SrTiO_3$ (100) surfaces also exhibit very good (1×1) LEED patterns, although other spectroscopic measurements indicate that they do not have the stoichiometry of the bulk crystal.[111]

One quantitative LEED study has been performed on polished and annealed $SrTiO_3$ (100). As in the other studies, very good (1×1) patterns were obtained, although no other techniques were used to determine the surface stoichiometry. Full dynamical calculations were performed to analyze I-V data, and the resulting model of the surface contains equal areas of the SrO and TiO_2 planes, with the O ions relaxed away from the bulk by 0.16 Å on the SrO plane and by 0.08 Å on the TiO_2 plane. The spacing between the outermost cation plane and the plane below it was found to be 10 % smaller than the bulk value for the SrO termination, but 0 – 2 % larger for the TiO_2 termination. However, the lack of any independent check on the stoichiometry of the surfaces leaves this determination open to question, since it has been shown that, at least for ion-bombarded $SrTiO_3$ (100) surfaces, annealing to produce good LEED patterns results in a surface stoichiometry that is even farther from that of a UHV-cleaved surface than for the bombarded surface before annealing.[111]

SrTiO$_3$ (100);   E$_i$ = 95 eV

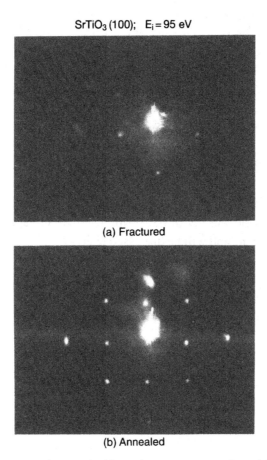

(a) Fractured

(b) Annealed

Fig. 2.8 LEED patterns taken at incident-electron energy, E_i = 95 eV for (a) UHV-fractured SrTiO$_3$ (100), and (b) the fractured surface after annealing at 1200 K.

BaTiO$_3$ (100)

The BaTiO$_3$ (100) surface has also been studied qualitatively by LEED. Measurements on annealed natural growth faces showed complex, streaked patterns for annealing temperatures below 1100 K.[116,117] Annealing at 1123 K produced (1 × 1) patterns characteristic of the bulk, but when the annealing temperature was increased to 1273 K, a ($\sqrt{5} \times \sqrt{5}$) structure appeared that was attributed to ordering of surface vacancies. Some samples that were treated to give the (1 × 1) pattern exhibited hysteresis of some of the beam intensities as a function of temperature between 300 and 923 K, and it was suggested that this might be explained by the existence of two surface layer geometries, one cubic and one tetragonal, as a function of temperature.

LEED observations have also been made on BaTiO$_3$ (100) samples that were bombarded with O$^+$ ions and subsequently annealed at 1273 K.[113] Such surfaces showed a low O-vacancy defect concentration and good quality (1×1) patterns. Samples that were bombarded with Ar^+ ions and then annealed at 1373 K exhibited fractional order spots indicative of two mutually perpendicular (2×1) LEED patterns.

No LEED I-V measurements or calculations have been performed for any $BaTiO_3$ surface.

WO_3 (100) and Na_xWO_3 (100)

The sodium tungsten bronzes, Na_xWO_3, have the perovskite crystal structure for Na compositions of $0.4 < x < 0.95$, with the Na ions occupying the A cation site. If all of the Na cations were to be removed without changing the WO_3 sublattice, the structure would be that of cubic ReO_3. In reality, a lattice distortion occurs for $x < 0.4$ that results in a monoclinic crystal structure. WO_3 may occur naturally as the mineral tungstite, although there is some question as to whether or not the mineral is a monohydrate. Tungstite cleaves well along (100) and less well along (110).

LEED patterns have been observed for annealed samples of WO_3 (100) and for surfaces cut 7° away from the (100) orientation. The patterns for the (100) surface were (1×1), but they exhibited a distortion from the expected monoclinic symmetry to triclinic. The diffracted beams were also split, which was interpreted in terms of twinned microdomains on the surface. Complicated LEED patterns were found for the surface that was cut 7° away from (100); they were interpreted in terms of faceting, with the formation of (111) planes. In addition, spots interpreted in terms of the formation of a shear plane structure on the surface were seen on the faceted surface.

More extensive LEED studies have been performed on Na_xWO_3 (100) surfaces. Samples having bulk Na concentrations in the range $0.5 < x < 0.9$ exhibited LEED patterns having the square symmetry characteristic of the bulk crystal, but with fractional order spots whose intensities varied with the bulk x value. The dominant pattern was (2×1), but (3×1) patterns were seen for small x values, and (1×1) patterns for large x values. The patterns were interpreted in terms of ordering of the Na ions into rows on the surface, with larger separation of the rows for lower surface Na coverages. LEED studies on a $Na_{0.8}WO_3$ (100) sample as a function of temperature showed that the surface undergoes a reversible sodium order–disorder transition at a temperature of about 580 K. The activation energy for Na diffusion on the surface was determined to be 17 kJ/mol, and the surface was found to have a lower Na concentration than the bulk.

LEED measurements were also made on an annealed $Na_{0.64}WO_3$ (110) surface. The surface stoichiometry was found to differ little from that of the bulk, but the presence of (1×3) patterns was interpreted in terms of ordering of the surface Na ions along rows parallel to the (001) direction, similar to the results for Na_xWO_3 (100).

Perovskite (111)

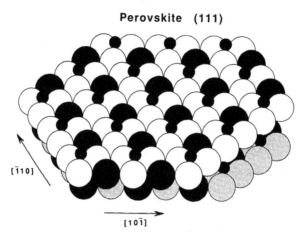

Fig. 2.9 Model of the unrelaxed perovskite (111) surface. See text for a discussion of the surface B-cation population.

2.3.2.2 Perovskite (111): SrTiO_3

The perovskite (111) surface is similar to rocksalt (111) in that the bulk crystal structure in that direction consists of alternating planes of only B ions and of AO_3 composition, so that individual (111) planes are not charge neutral. Hence an idealized cleave on a plane between two adjacent planes of atoms would yield one atomically flat surface having an AO_3 composition and the opposing surface having B ions lying above an AO_3 plane; both of these surfaces would be polar. Figure 2.9 shows a model of the latter type of perovskite (111) plane, where the surface B ions have only three of their six bulk O ligands. Two equivalent, non-polar cleavage faces could be produced by constraining one-half of the B ions to remain on each of the two surfaces; presumably the ions would be arranged randomly on a real cleaved surface. On perovskite (111) surfaces prepared by annealing, however, there could well be other, lower-energy reconstructions that would remove the net dipole moment and increase the ligand coordination of the surface cations, perhaps by rearrangement of more than just the top layer of ions.

Some experiments have been performed on polished and annealed $SrTiO_3$ (111) surfaces. Ar^+-ion bombardment at room temperature followed by annealing at 873 K yielded stable (1 × 1) patterns, but when the crystal was ion bombarded at 873 K, a complex LEED pattern characteristic of faceting was observed; the orientation of the facets was not determined.

2.3.3 Rutile

Two of the metal oxides whose surface properties have been most thoroughly studied, TiO_2 and SnO_2, have the tetragonal rutile lattice. Three low-index faces

Table 2.4. *Rutile oxides*

Oxide	Face	Surface preparation	Technique and results	References
TiO$_2$ $a_0 = 4.594$ Å $c_0 = 2.958$ Å	110	Various	LEED: (1×1) see text	16, 18, 125–129
			LEED: (2×1) see text	129–131
		P&A	MEED: consistent with bulk	132
	100	UHV fractured	LEED: (1×1)	125
		P&A	LEED: $(1 \times 3), (1 \times 5), (1 \times 7)$	16, 23, 133
			LEED and X-ray: (1×3), model involving (110) facets	133
			STM: (1×3), O vacancy and microfacet model	2
			AFM: facets	134, 135
	001	UHV fractured	LEED: (1×1)	125
			LEED I–V: models of (1×1) structure	136
		P&A	LEED: (011) and (114) facets	126, 127, 137
			AFM: (011) facets	134, 135
			STM and LEED: (011) and (114) facets and (111) planes	138
SnO$_2$ $a_0 = 4.737$ Å $c_0 = 3.186$ Å	110	Various	LEED, ISS: see text; nearly perfect surface on anneal in O$_2$ at 700 K	139–147
		Theory	Rumpling for either stoichiometric or reduced surface	148
	100	P&A	LEED: (1×1) sharp for anneals above 700 K	149
	101	P&A	LEED: (1×1) as for (100)	149
RuO$_2$ $a_0 = 4.51$ Å $c_0 = 3.11$ Å	110	Various	LEED: see text	150, 151
	100	Various	LEED: see text	151

have been studied: (110), (100) and (001). Models of these surfaces are shown in Figs. 2.10–2.12.

2.3.3.1 Rutile (110)

The (110) surface, shown in Fig. 2.10, is the most stable of the low-index faces of rutile.[106] The structure in the figure was formed by the imaginary cleave that

Rutile (110)

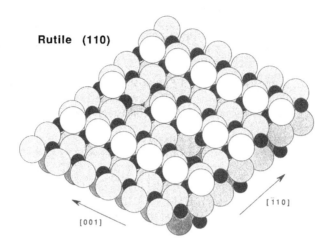

Fig. 2.10 Model of the rutile (110) surface. Two types of O-vacancy defect are shown.

reduces the nearest-neighbor ligand coordination of the surface ions by the smallest amount. (This would be equivalent to 'breaking the smallest number of interatomic bonds' in a covalently bonded crystal structure.) Both of the opposing
faces formed by this cleave are identical. The (110) surface is not atomically flat in
the sense of the rocksalt (100) or perovskite (100) surfaces; rows of 'bridging' O
ions lie above the main surface plane, which contains equal numbers of five- and
six-fold coordinated cations. Symmetric rows of O ions lie an equal distance
below the surface plane, resulting in a non-polar surface. Figure 2.10 also contains
two types of O-vacancy point defects, one in a row of bridging O ions and one in
the surface plane.

TiO_2 (110)

TiO_2 occurs naturally in relatively pure form as the mineral rutile, which exhibits
predominantly (110) natural growth faces, although (011) and (010) are also common. (Two more complicated polymorphs of TiO_2, anatase and brookite, also exist
in nature. The only surface-science study reported to date on either of them is a
measurement of surface phonon modes on anatase, which will be mentioned in
Chapter 3.) Cleavage in the mineral form is preferentially along (110), although
not of high quality. The (110) surface of single-crystal TiO_2 has been investigated
perhaps more thoroughly than any other metal oxide. It is thus surprising that no
quantitative LEED measurements have ever been performed on this surface,
although qualitative observations of LEED patterns have been made by many
groups. UHV-fractured (110) surfaces exhibit fair (1 × 1) LEED patterns, the quality of which can be improved a great deal by annealing in UHV or low partial pres-

sures of O_2 at temperatures above about 1000 K. [The LEED patterns are similar to those for $SrTiO_3$ (100) shown in Fig. 2.8.] Polished surfaces also exhibit very good (1 × 1) LEED patterns when ion bombarded and annealed in UHV. In fact, TiO_2 (110) is one of only a few metal-oxide surfaces that can be prepared by polishing, ion bombardment and annealing to give surfaces whose stoichiometry is the same as that of UHV-fractured samples and whose electronic properties are essentially the same as those of fractured surfaces. A (2 × 1) LEED pattern can also be observed from TiO_2 (110) depending upon surface treatment, although different groups report different recipes for producing it. The electronic properties of the surface appear to be essentially the same regardless of whether the pattern is (1 × 1) or (2 × 1).[130,131]

Preliminary medium-energy electron diffraction (MEED) measurements have been performed on polished and annealed TiO_2 (110) surfaces. While there was sufficient uncertainty in the data and analysis to preclude quantitative interpretation, the measurements were consistent with a bulk termination of the TiO_2 crystal structure.

Scanning tunneling microscopy is just beginning to be used to image metal-oxide surfaces. Most measurements have not given any real information on just where the atoms are on surfaces. However, STM images having almost atomic resolution have been reported on TiO_2 (110), (100) and (001) surfaces.[2,152–155]

SnO_2 (110)

Cassiterite, the mineral form of SnO_2, cleaves less well along (110) than it does along (100), although neither cleave is of very high quality. Natural growth faces are also mainly (110) and (100). Although SnO_2 has the same rutile structure as TiO_2, the differences in electronic structure result in a weaker bonding of O ions to the surface. This becomes manifest in the different structures that can exist on the SnO_2 (110) surface. On the TiO_2 (110) surface, the rows of bridging O ions are tightly bound, and the removal of a bridging O ion results in drastic changes in surface electronic structure (see § 5.2.2). On SnO_2 (110), on the other hand, the bridging O ions can be easily removed and replaced, depending upon surface treatment; this is discussed in § 4.6.2. Natural growth faces that were ion bombarded and annealed in UHV exhibited (4 × 1), c(2 × 2), (1 × 2) and (1 × 1) LEED patterns for various annealing temperatures. These LEED patterns have been interpreted in terms of ordered patterns of O vacancies.[141,144] Subsequently ISS was used to study the stoichiometry of the SnO_2 (110) surface, and it was found that annealing at 700 K in 1 Torr O_2 resulted in a nearly perfect (110) surface that contained the rows of bridging O ions.[140,144] (In more recent experiments, it was found that even better surfaces could be obtained by exposing the surface to an oxygen plasma at room temperature.[145]) When the nearly perfect surface is heated to temperatures below 800 K in UHV, the rows of bridging O ions can be removed. Heating above 800 K in UHV results in removal of all of the bridging O ions and some in-plane O ions (see the O-vacancy point defects in Fig. 2.10). The surface stoichiometry of SnO_2 (110) is thus much more sensitive to preparation procedure than is that of TiO_2 (110).

One theoretical study has addressed the geometric structure of both the stoichiometric SnO_2 (110) surface and the reduced surface from which all of the bridging O ions have been removed. Both surfaces are predicted to rumple slightly, with the surface cations remaining in approximately their bulk positions and the in-plane O ions moving about 0.4 Å outward.

RuO_2 (110)

Single crystals of RuO_2 have also been studied by LEED and other surface-science techniques. The surfaces of RuO_2 are found to be easily reduced in vacuum, exhibiting different ordered surface structures depending upon O_2 partial pressure, temperature and time of annealing. The RuO_2 (110) surface exhibits no LEED patterns after ion bombardment, but upon annealing at 723 K a reversible series of patterns is obtained. The first patterns observed upon annealing in O_2 are hexagonal and of poor quality. Continued annealing produces c(2 × 2) patterns, and after longer annealing times a (1 × 1) pattern finally appears. Annealing at 723 K in UHV reverses the order of the patterns as O desorbs from the surface. The (1 × 1) pattern has the same lattice parameters as the bulk crystal, and the other patterns are presumably due to ordering of O vacancies on the surface. The core-level binding energies of the Ru and O ions on the most heavily oxidized surfaces, as measured by x-ray photoelectron spectroscopy (XPS), have been interpreted in terms of RuO_3 formation, although this has been disputed.[156]

2.3.3.2 Rutile (100)

Figure 2.11 shows a model of the (100) rutile surface; opposing cleaved faces both have this structure. This is also a microscopically rough surface, with rows of O ions lying above the surface plane, but it is non-polar. All of the cations on this surface are five-fold coordinated with O ligands. One O-vacancy point defect has also been shown in Fig. 2.11.

TiO_2 (100)

The TiO_2 (100) surface has been prepared both by fracturing in UHV[125] and by polishing and annealing. The fractured surface exhibits fair (1 × 1) LEED patterns whose quality is comparable to those for fractured TiO_2 (110). Polished and annealed surfaces did not exhibit any (1 × 1) patterns; they were found to recon-

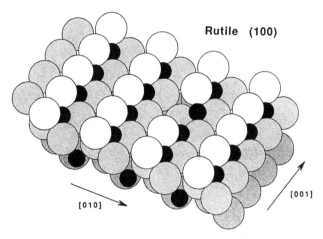

Fig. 2.11 Model of the rutile (100) surface, showing one surface O-vacancy point defect.

struct upon annealing and to exhibit (1×3), (1×5) and (1×7) LEED patterns, with the direction of reconstruction perpendicular to the rows of O ions.[16] A combination of glancing-angle x-ray diffraction and LEED has been used to determine the atomic positions for the (1×3) reconstruction,[133] and the data suggest a model in which the (1×3) symmetry is produced through the creation of (110) facets parallel to the [001] direction in Fig. 2.11. The model is reasonable since the (110) face is the most stable one in the rutile structure, and it allows for a natural explanation of the (1×5) and (1×7) reconstructions in terms of deepening of the troughs formed by the (110) planes. The STM image in Fig. 1.2 is of the (1×3) reconstructed surface and is consistent with a slight modification of this model, where Ti rather that O atoms are placed outermost on the ridges.[2] The bright rows in Fig. 1.2 are the ridges of atoms, identified as Ti because they appear to be associated with empty electronic levels more than with occupied ones. Faceting on a larger scale on TiO_2 (100) has been observed by atomic force microscopy.[134,135]

SnO_2 (100) and (101)

LEED studies were also performed on the (100) and (101) surfaces of SnO_2 prepared by polishing and annealing in UHV. The development of LEED patterns from an initially disordered, ion-bombarded surface was similar to that described above for SnO_2 (110), although only (1×1) LEED patterns were seen. As the annealing temperature was raised, the quality of the patterns improved until they became sharp for temperatures above 700 K. The development of the patterns was correlated with changes in the stoichiometry of the surface.

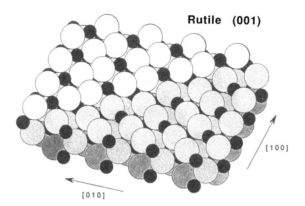

Fig. 2.12 Model of the unrelaxed rutile (001) surface, showing a step to another (001)
terrace.

RuO$_2$ (100)

The RuO$_2$ (100) surface has been studied in the same way as was the RuO$_2$ (110)
surface discussed above, and the results are very similar. Reversible changes in
LEED patterns were found for annealing at 723 K and 823 K in both O$_2$ and UHV.
(1 × 1), p(1 × 2) and c(1 × 2) patterns were all observed, with the p(1 × 2) stable
over the widest range of surface stoichiometry. XPS suggested that the surface Ru
ions were in an ionization state characteristic of RuO$_3$ on the most heavily oxidized surfaces.

2.3.3.3 Rutile (001): TiO$_2$

The (001) surface of rutile, in spite of the high symmetry of the lattice in that
direction, is the least stable of the low-index faces. Cleaving between adjacent
charge-neutral planes would produce two identical surfaces, whose ideal structure
is shown in Fig. 2.12; also shown is a step to another (001) terrace. The nearest-
neighbor ligand coordination of the surface cations has been reduced from six to
four. Although this surface is non-polar, the extremely low ligand coordination of
the surface cations favors reconstruction to increase that coordination. The surface
energy calculated for an unrelaxed TiO$_2$ (001) surface is twice as large as that for
TiO$_2$ (110).[106]

The TiO$_2$ (001) surface has been produced by fracturing a single crystal at room
temperature, but the (1 × 1) LEED patterns from that face are poor, and any
attempts to ion- or electron-bombard or anneal the surface result in disorder or
faceting. LEED I–V measurements have been performed for the UHV-fractured
surface at room temperature before any faceting had occurred [i.e., the surface
exhibited (1 × 1) LEED patterns], and two possible structural models for the sur-

face were proposed. One structure was very nearly bulk-like, with only small distortions of the outermost two atomic planes. The other exhibited large distortions in atomic positions in the second atomic layer, but little distortion in the first or third layers. The former model seems more likely on physical grounds, but it was not significantly different from the latter model based upon the quality of the fit to the data.

The ion-bombarded and annealed TiO_2 (001) surface has been carefully studied by qualitative LEED, and the facet planes are found to be (011) and (114). The (011) facets, on which all of the cations have five O ligands, form for annealing temperatures below about 1000 K. Annealing above 1300 K produced (114) facets, on which many of the cations still have only four-fold O coordination; however, the location of the LEED spots does not quite match that predicted for the ideal structure. In general, different facet geometries were observed on the surface simultaneously. This same type of faceting has also been observed by STM. In AFM studies of TiO_2 (001) surfaces annealed at 1573 K, the facet structure was replaced by the formation of a network of 50–300 Å-high ridges.[134]

2.3.4 Corundum

As the symmetry of the Bravais lattice is reduced, the corresponding bulk and surface crystal structures become more complicated. An important case in point is the trigonal corundum lattice, which is shared by α-Al_2O_3 and several transition-metal oxides. In spite of the reduced symmetry, however, the local ligand environment in the bulk of corundum is similar to that in the other structures discussed above in that the cations occupy distorted octahedral sites surrounded by six O ligands. The two single-crystal corundum faces that have been studied most extensively are $(10\bar{1}2)$ and (0001), although qualitative LEED patterns have also been reported for Al_2O_3 $(11\bar{2}3)$.[158]

Crystal structures that are closely related to corundum can be obtained by replacing one-half of the M cations by A cations and the other half by B cations, giving a compound of formula ABO_3. Different ways of ordering the A and B ions are possible, one being found in ilmenite, $FeTiO_3$, and another in both lithium niobate, $LiNbO_3$, and $LiTaO_3$. $LiNbO_3$ cleaves along the $(10\bar{1}2)$ plane, as do some corundum oxides.

2.3.4.1 Corundum ($10\bar{1}2$)

In the bulk of the corundum structure, one-third of the possible octahedrally coordinated cation sites are vacant. These vacancies are arranged on $(10\bar{1}2)$ planes, which is found to be the primary cleavage plane for Ti_2O_3, V_2O_3 and Cr_2O_3 (but not for corundum itself, Al_2O_3), presumably because it is the surface that reduces

2 Geometric structure of metal oxides

Table 2.5. *Corundum oxides*

Oxide	Face	Surface preparation	Technique and results	Experiment	Theory
Al_2O_3 $a_0 = 4.763$ Å $c_0 = 13.003$ Å	$10\bar{1}2$	P&A	LEED: 1/4 order reflections	157	
			LEED: only (2×1) patterns observed	158	
			RHEED and electron microscopy: (4×1) reconstruction in air	159, 160	
			LEED: (2×3) and ($\sqrt{2} \times \sqrt{2}$)R45° O-vacancy structures after high-temperature aneal	161	
		–	Defect-lattice calculation: slight relaxation		103
	0001	P&A	LEED: complex reconstructions	157, 158, 162–168	
		–	HF calculation: appreciable relaxation		45, 169
		–	Defect-lattice calculations: slight relaxation		103
	$10\bar{1}0$	P&A	HF and defect-lattice calculations: large relaxation (surface not seen experimentally)		45, 103
	$11\bar{2}3$	P&A	LEED: (4×5) reconstruction	158	
	$11\bar{2}0$	–	Defect-lattice calculations: slight relaxation		103
	$10\bar{1}1$	–	Defect-lattice calculations: slight relaxation		103
Ti_2O_3 $a_0 = 5.148$ Å $c_0 = 13.636$ Å	$10\bar{1}2$	UHV cleaved	LEED: excellent (1×1) patterns	26, 28	
V_2O_3 $a_0 = 5.105$ Å $c_0 = 14.449$ Å	$10\bar{1}2$	UHV cleaved	LEED: (1×1) rather poor quality	29	
α-Fe_2O_3 $a_0 = 5.035$ Å $c_0 = 13.72$ Å	$10\bar{1}2$	Various	LEED: depends on treatment, see text	170	
	0001	Air cleaved and annealed	LEED: (1×1) pattern	171	
		P&A	LEED: complex, see text	170, 172	

the nearest-neighbor ligand coordination of the surface ions by the smallest amount. Figure 2.13 shows a corundum surface oriented slightly off ($10\bar{1}2$), again showing two ($10\bar{1}2$) terraces, a monoatomic step separating them, and one O-vacancy point defect on the upper terrace. On defect-free terraces, each cation is surrounded by five nearest-neighbor ligands, four in the surface plane and one in the plane below. These incomplete O octahedra are alternately tilted with respect

Corundum (10$\bar{1}$2)

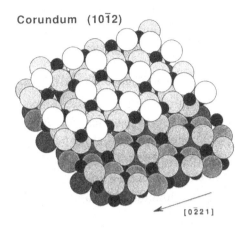

[0$\bar{2}$21]

Fig. 2.13 Model of the corundum (10$\bar{1}$2) surface, including a [0$\bar{2}$21] step to another (10$\bar{1}$2) terrace, and an O-vacancy defect.

to the macroscopic surface plane, resulting in a slightly corrugated surface. It is non-polar, and identical faces are formed on both sides of the cleavage plane. Although this surface lacks the high symmetry present on rocksalt (100), perovskite (100)-BO_2 and rutile (110) and (100) surfaces, the ligand coordination of the cations on corundum (10$\bar{1}$2) is similar to that on those surfaces.

[In some of the early work on Ti_2O_3, the (10$\bar{1}$2) plane was indexed as (047).[26,129,173] This was an inexact indexing that was based upon the binary-bisectrix-trigonal coordinate system used to describe rhombohedral elements such as Bi. It caused undue confusion and was rapidly abandoned.]

α-Al_2O_3 (10$\bar{1}$2)

One of the most important metal oxides in the ceramics industry is alumina, α-Al_2O_3, and yet relatively little effort has been devoted to trying to understand its surface structure. It occurs commonly as the mineral corundum, which does not cleave and exhibits a wide variety of natural growth faces. There have been three reported LEED observations on the α-Al_2O_3 (10$\bar{1}$2) surface. In one case surfaces prepared by polishing and annealing exhibited patterns having one-fourth order reflections, regardless of the thermal treatment used.[157] The quality of the LEED patterns, however, was quite good. In another report, only (2 × 1) LEED patterns were observed.[158] RHEED and reflection-electron microscopy have been used to study reconstruction of the Al_2O_3 (10$\bar{1}$2) surface in air, where a (4 × 1) restructuring was observed. In the third study, the surface was annealed at 2000 K; both (2 × 3) and ($\sqrt{2}$ × $\sqrt{2}$) R 45° patterns were observed as a function of annealing conditions, with the latter structure being the most stable.[161] These patterns were attributed to ordered arrays of O vacancies on the surface. Defect-lattice calculations for this surface predict only slight distortions from the ideal bulk termination, consisting mainly of small relaxation perpendicular to the surface.

Ti_2O_3 $(10\bar{1}2)$ and V_2O_3 $(10\bar{1}2)$

The corundum transition-metal oxide Ti_2O_3 cleaves extremely well along the $(10\bar{1}2)$ plane. Excellent (1×1) LEED patterns are seen at all electron energies, since Ti_2O_3 is either a semiconductor or a metal, depending upon temperature, and charging is not a problem. No LEED I-V measurements have been performed, however, so nothing is known about surface relaxation. V_2O_3 also cleaves preferentially along the $(10\bar{1}2)$ plane, although the quality of the resulting surface is not as high as that for Ti_2O_3. The macroscopic surface is often irregular, exhibiting conchoidal fracture, and the LEED spots are broader than those for Ti_2O_3. There are no published calculations of the geometric structure of any face of Ti_2O_3 or V_2O_3.

α-Fe_2O_3 $(10\bar{1}2)$

α-Fe_2O_3 occurs naturally as the mineral hematite and is a major source of iron ore. Hematite crystals do not cleave, and their predominant growth faces are (0001) and $(10\bar{1}1)$. The surface geometry of the $(10\bar{1}2)$ growth face of artificially grown α-Fe_2O_3 single crystals has been studied by LEED as a function of surface preparation procedure. After Ar^+-ion-bombardment cleaning of the surface, both the temperature and the O_2 partial pressure were varied as the sample was annealed. The LEED patterns observed from the $(10\bar{1}2)$ surface were found to depend more strongly on O_2 partial pressure than on temperature. In 10^{-6} Torr O_2, (1×1) patterns were observed for all temperatures between 923 and 1273 K, while annealing at the same temperatures in 10^{-10} Torr O_2 yielded (2×1) patterns. The diffracted beams were very sharp after annealing at or above 1173 K, but at lower temperatures streaking of the beams along the $(02\bar{2}1)$ direction was apparent in all of the patterns. The (2×1) patterns correspond to an increase in the surface unit cell dimension in the direction along the zig-zag rows of O ions (see Fig. 2.13).

Cr_2O_3 $(10\bar{1}2)$

Atomistic lattice simulation calculations of the surface energies of various faces of Cr_2O_3 conclude that the $(10\bar{1}2)$ is the most likely face to be observed experimentally.[174] This is borne out by the quality of cleaved $(10\bar{1}2)$ surfaces as determined by atomic force microscopy;[3] Figure 2.22 in § 2.4.1 below shows a micrograph of an air-cleaved surface whose terraces are large and atomically flat. No quantitative LEED measurements for any surface of Cr_2O_3 have yet been reported.

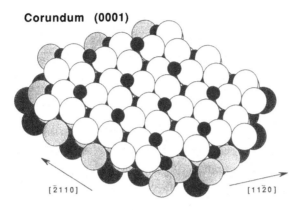

Corundum (0001)

[2̄110]   [112̄0]

Fig. 2.14 Model of the corundum (0001) surface.

2.3.4.2 Corundum (0001)

The other corundum surface that has been studied experimentally is the higher symmetry (0001). This is not a cleavage surface, however, and it has only been prepared by cutting, polishing and annealing. It poses conceptual problems similar to those of the perovskite (111) surface in that the atomic planes normal to the *c*-axis contain either all cations or all anions. It differs from the perovskite case, however, in that the reduced symmetry results in the planes of cations not being atomically flat, with some of the cations lying closer to the O ions in the plane above and some closer to the O ions in the plane below. One can imagine cleaving the crystal along a (0001) plane that lies between two planes of O ions. If the cations that occupy the space between the two O-ion planes are associated with the O-ion plane to which they were closer in the bulk, two identical surfaces, as shown in Fig. 2.14, are obtained. It is non-polar, although the surface cations have only three O ligands, one-half of their bulk ligand coordination. These surfaces would thus be expected to relax or reconstruct in some way that would increase the number of O ligands around the surface cations.

α-Al_2O_3 (0001)

Several qualitative LEED studies have been performed on the α-Al_2O_3 (0001) surface, but no I-V measurements have been made, and no work has been reported on single-crystal surfaces cleaved or fractured in UHV. Polished and annealed surfaces exhibit at least four different LEED patterns as a function of annealing temperature and time, two of which are quite stable. For annealing temperatures below about 1300 K, the most stable pattern is (1 × 1) with dimensions corresponding to within 5 % of the bulk lattice constant. Annealing at higher temperatures in UHV results in (2×2), $(3\sqrt{3} \times 3\sqrt{3})$ R $30°$ and $(\sqrt{31} \times \sqrt{31})$ R $\pm$ $\tan^{-1}(\sqrt{3}/11)$ for temperatures of about 1500, 1600 and 1700 K, respectively. The surface can be converted from one to the other by various treatments, including electron bombardment at LEED energies. The $(\sqrt{31} \times \sqrt{31})$) R $\pm \tan^{-1}(\sqrt{3}/11)$ surface is also stable and is believed to be O deficient, with a cubic Al-rich overlayer on the underlying trigonal substrate. No evidence for faceting was found in any of the reconstructions.

Ab initio Hartree–Fock calculations have been performed for the α-Al_2O_3 (0001) surface, and an appreciable relaxation is found, with the surface Al ions relaxing inward toward the underlying layer of O ions by as much as 0.4 Å. This is not unexpected since it brings the surface Al ions in closer contact with their three O ligands. More modest relaxation was predicted by defect-lattice calculations. Both types of calculation were also performed for an α-Al_2O_3 $(10\bar{1}0)$ surface, whose outermost plane consists entirely of O ions. Here both methods predicted large relaxation, with the O ions relaxing inward toward the next atomic plane. This surface has never been studied experimentally.

α-Fe_2O_3 (0001)

The predominant natural growth face of α-Fe_2O_3 is the (0001) basal plane.[171] Single-crystal (0001) faces were studied by LEED as a function of annealing temperature and O_2 partial pressure after being cleaned by Ar^+-ion bombardment. Before annealing no LEED patterns were visible, but when the sample was annealed between 1050 and 1273 K, several different types of LEED patterns, corresponding to different surface reconstructions, were observed. The reconstructions were modeled in terms of monolayers of various Fe oxides on the α-Fe_2O_3 substrate. (Simultaneous XPS measurements of the valence state of the Fe ions in the surface region were used together with the LEED results in formulating the models.) For low annealing temperatures, an Fe_3O_4 (111) layer appears to form on the surface. For slightly higher temperatures, a complex pattern appears that was attributed to multiple scattering across an $Fe_{1-x}O$ (111) / α-Fe_2O_3 (0001) interface. Only after prolonged annealing at 1173 K or above was a (1×1) pattern characteristic of the α-Fe_2O_3 (0001) surface observed. No theoretical calculations are available to check these models.

In Ref. 170 the observed surface stoichiometry and structure were compared with the bulk phases predicted by the Fe–O phase diagram. Over much of the temperature and pressure range studied the surface structure was different than the stable bulk phase. This is true in many cases for oxide surface structures, and, while it is not unexpected, it is important to remember that bulk phase diagrams can only be used as a guide and do *not* predict surface phases.

Although not included in Table 2.5, some LEED, STM and AFM measurements

Table 2.6. *Wurtzite*

Oxide	Face	Surface preparation	Technique and results	Experiment	Theory
ZnO $a_o = 3.250$ Å $c_o = 5.207$ Å	$10\bar{1}0$	P&A	LEED I–V: small inward relaxation of Zn	175–178	179, 180
			ARPES: agrees with LEED	181	
		–	Bonding calculations: agree with LEED		182–185
	$11\bar{2}0$	P&A	LEED: facetting and reconstruction	186, 187	
			LEED I–V: nearly ideal termination	175–177	
		–	Bonding calculations: significant inward relaxation of Zn		182–184
	0001	P&A	LEED: good (1 × 1) patterns	186, 187, 189	
			LEED I–V: very small inward relaxation	175, 188	179
	$000\bar{1}$	P&A	LEED: less stable than (0001), some faceting	186	
			LEED: good (1 × 1) patterns	187, 189	
			LEED I–V: no relaxation seen	175, 188	179

have been reported on (0001) surfaces of natural mineral specimens of hematite.[980–983] Using STM and AFM on air-cleaved surfaces, structures were observed that are consistent with the O–O interatomic spacing on the surface, and with the spacing between Fe atoms in the outermost atomic plane.

2.3.5 Wurtzite

In the wurtzite structure the cations are tetrahedrally coordinated with O ions in the bulk. This leads to different types of surface structures than those obtained for octahedrally coordinated oxides. Four low-index faces are important in the wurtzite structure: the non-polar $(10\bar{1}0)$ and $(11\bar{2}0)$ surfaces parallel to the *c*-axis (which are often called 'prism' faces), and the polar (0001) and $(000\bar{1})$ surfaces. Since the only metal oxide having the wurtzite structure that has been studied is ZnO, we will refer specifically to that oxide.

One particular theoretical study should be mentioned here. Tight-binding calculations of the surface atomic and electronic structure of two other less common polymorphs of ZnO – zincblende and rocksalt – have been performed and compared with wurtzite ZnO.[190] No experimental data on either of those polymorphs is available for comparison.

2.3.5.1 Wurtzite $(10\bar{1}0)$ and $(11\bar{2}0)$: ZnO

ZnO occurs naturally as the mineral zincite, although natural single crystals are rare. They cleave well along $(10\bar{1}0)$; the dominant natural growth face is $(000\bar{1})$.

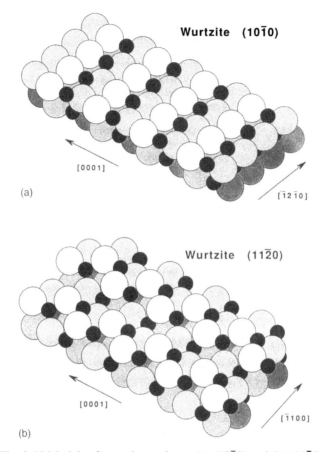

Fig. 2.15 Models of wurtzite surfaces: (a) (10$\bar{1}$0) and (b) (11$\bar{2}$0) .

Figures 2.15 (a) and (b) show ideal models of the wurtzite (10$\bar{1}$0) and (11$\bar{2}$0) surfaces, respectively. Each surface cation is missing one of its four O ligands, and
each surface O ion is missing one nearest-neighbor cation. The missing O ligand
would be in the fourth tetrahedral position; in discussing chemisorption it is sometimes useful to think of the surface cation as having a 'dangling bond' in that
direction. Both (10$\bar{1}$0) and (11$\bar{2}$0) surfaces are atomically flat and have equal numbers of cations and anions in the surface plane. The planes can be thought of in
terms of Zn–O pairs or dimers; on the (10$\bar{1}$0) face these dimers bond to ions in the
plane below rather than directly to each other, while on the (11$\bar{2}$0) face they bond
to each other as well as to the plane below and are arranged in zig-zag rows parallel to the c-axis.

The amount of theoretical and experimental effort devoted to determining the
positions of ions on ZnO surfaces is even greater than that for MgO, partly

because there are a large number of different stable crystal faces on ZnO. Early qualitative LEED experiments on annealed samples, including I-V measurements, determined that the $(10\bar{1}0)$ surface was the most stable one for ZnO, while the $(11\bar{2}0)$ was one of the least stable.[186] The $(11\bar{2}0)$ surface exhibited intense beams from $(01\bar{1}0)$, $(11\bar{2}2)$ and $(11\bar{2}\bar{2})$ facets after the surface cleaning procedure used, which included prolonged annealing at 773 K. Other measurements on ZnO surfaces prepared in different ways were in qualitative agreement with those results, in that fractional order spots arising from surface reconstruction were seen for the $(11\bar{2}0)$ face but not for $(10\bar{1}0)$.[187]

The quantitative work on ZnO surface structure includes extensive theoretical calculations and LEED I-V measurements on annealed surfaces of all four major low-index faces. The results indicate that the $(11\bar{2}0)$ surface is an ideal termination of the bulk structure within the accuracy of the measurements, while on the $(10\bar{1}0)$ surface the O ions remain in their bulk locations but the Zn ions relax inward by 0.4 Å. Similar conclusions were drawn for the $(10\bar{1}0)$ surface based upon angle-resolved photoemission measurements. More recent theoretical calculations of the geometric structure of ZnO $(10\bar{1}0)$ using an sp^3 model of the electronic structure predict an inward relaxation of the Zn ions relative to the O ions of 0.57 Å, within the error limits of the LEED determination; however, the same model predicts an inward relative displacement of the Zn ions of 0.54 Å for ZnO $(11\bar{2}0)$, which is in serious disagreement with experiment.

2.3.5.2 Wurtzite (0001) and (000$\bar{1}$): ZnO

The creation of polar surfaces on wurtzite crystals is subject to a constraint not encountered in the other crystal structures discussed above. The bulk crystal structure does not possess a center of inversion, and there is thus an inherent asymmetry along the *c*-axis. When the crystal is cleaved normal to the *c*-axis in such a way as to break the fewest number of interatomic bonds, two different polar surfaces are formed on the opposing crystal faces, each having only one type of ion in its outermost plane. The cation plane is outermost for the (0001) surface, which in ZnO is sometimes referred to as the (0001)-Zn face, or just the 'zinc face'. The O plane is outermost on the (000$\bar{1}$) surface, referred to as (000$\bar{1}$)-O, or the 'oxygen face'. In fact, in order that the entire crystal have no net charge, any slab of crystal that has a (0001) surface on one side must have a (000$\bar{1}$) on the other side; this is just what is observed for real crystals. These surfaces are shown in Figs. 2.16 (a) and (b). Since they are polar, they are expected to relax or reconstruct in some manner that will reduce the net dipole moment.

ZnO (0001)-Zn and (000$\bar{1}$)-O surfaces behave very differently due to the different atomic species that are outermost on the two faces. It is easy to distinguish the two faces on a crystal simply by their response to etching with HCl: the (0001)

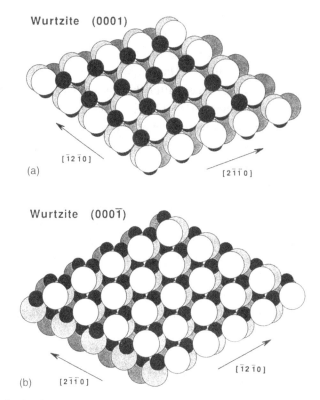

Fig. 2.16 Models of wurtzite surfaces: (a) (0001) and (b) (000$\bar{1}$) .

face is relatively unaffected, while the (000$\bar{1}$) face is rapidly etched and becomes visibly roughened.[193] Qualitative LEED data on annealed surfaces also indicated the (000$\bar{1}$) to be less stable because of the poor quality of the LEED spots and the appearance of additional beams resulting from faceting.[186] Other LEED measurements on annealed surfaces, however, reported similar quality and symmetry of the patterns on both of the polar surfaces.[187,189] LEED patterns were also taken on UHV-cleaved ZnO (0001) and (000$\bar{1}$) faces, and both exhibited excellent (1 × 1) patterns with low background intensity, also suggesting similar stability for the two faces in vacuum.[189]

Quantitative LEED I-V measurements, interpreted in terms of dynamical multiple-scattering calculations, have also been performed for ZnO (0001) and (000$\bar{1}$). The best fit to the data for the (0001) surface was with the outermost layer relaxed inward by 0.2 Å (presumably the small Zn ions increase their effective coordination by O ions in this way), while analysis of the data for the (000$\bar{1}$) face did not indicate any contraction of the outermost layer of O ions. The latter result is expected owing to the large size of the O^{2-} ions.

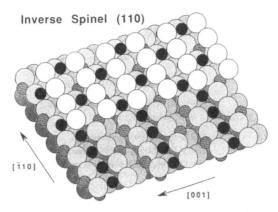

Inverse Spinel (110)

[$\bar{1}$10]

[001]

Fig. 2.17 Model of the spinel (110) surface. The cations in octahedral and tetrahedral sites are shown by small black and shaded circles, respectively. A step between the two types of (110) plane is shown.

2.3.6 Spinel

An important class of transition-metal oxides, of formula M_3O_4, has the cubic, although rather complicated, spinel structure. As mentioned in § 2.1.1, the metal cations reside in both octahedral and tetrahedral sites surrounded by O ions. The most common mineral form is magnetite, **Fe_3O_4**, which has the inverse spinel structure with Fe^{2+} in octahedral sites, and Fe^{3+} distributed equally between octahedral and tetrahedral sites. Fe_3O_4 does not cleave, but the predominant natural growth faces are (111) and (110). Few surface studies have yet been conducted on these oxides; although some experiments have been performed on the (110) surface of Fe_3O_4, these did not include any structural studies. The (110) surface exhibits the best fracture in UHV; its idealized geometry is shown in Fig. 2.17. Two types of (110) plane can exist on this surface, one having an MO_2 composition, with only octahedral cation sites in the surface plane, and the other having the composition M_2O_2, with one-half of the surface cations in octahedral sites and the other half in tetrahedral sites. The upper terrace in Fig. 2.17 is MO_2 and the lower is M_2O_2. If the M^{2+} and M^{3+} cations were distributed among the octahedral and tetrahedral sites on the surface in the same way that they are in the bulk, neither of the two planes would be charge neutral. Due to the symmetry of the crystal, (110) cleavage faces would be expected to have equal areas of the two types of planes shown in Fig. 2.17, separated by steps. Large dipole fields would be caused by the net charge on each type of surface plane, even though equal areas of the two types would result in the crystal having no net charge in the surface region. This is not a stable situation, and the surface should reconstruct or facet. However, to date no information is available on the actual surface structure of any spinel metal oxide.

Table 2.7. *Miscellaneous compounds*

Compound	Face	Surface preparation	Technique and results	Experiment	Theory
MoO_3	010	Various	LEED: (1×1) after heating in O_2 at 1000 K; sensitive to electron-beam damage	194, 195	
$Mo_{18}O_{52}$	100	Natural growth faces	LEED, RHEED: (010) terraces of MoO_3 with atomic-height steps	196, 197	
UO_2	111	P&A	LEED: nearly ideal termination	198–202	
			ISS: as LEED, but small U relaxation	202	
		–	Calculations: small inward relaxations predicted		107, 203
	110	P&A	LEED: reconstruction and facetting	199, 201	
		–	Calculations: some relaxation and rumpling		107, 203
	100	P&A	LEED: $c(2 \times 2)$	199, 204	
			ISS: ordered O overlayers	201, 204	
			Photoelectron diffraction: similar to bulk	205	
	Vicinal 111	P&A	LEED: regularly spaced steps	198, 200, 206, 207	
	Vicinal 100	P&A	LEED: less step ordering than vicinal (111); low-index facet formation	207	
Cu_2O	111	P&A	LEED: reconstruction; ordered O vacancies	208	
	100	P&A	LEED: reconstruction; dependent on treatment	208	
YBa_2Cu_3 O_{7-x}	001	UHV cleaved	LEED: (1×1) patterns, no reconstruction	209	
		Chemical etch	RBS: nearly bulk termination	210	
Bi_2Sr_2Ca Cu_2O_{8-x}	001	UHV cleaved	LEED: surface lattice identical to bulk superlattice	211–214	
Nd_2CuO_4	001	P&A	LEED: tripling of periodicity in [110] direction, sensitive to electron beam exposure	215	
$(NdCe)_2$ CuO_4	001	P&A	LEED: tripling of periodicity in [110] direction, sensitive to electron-beam exposure	215	
$(LaSr)_2$ CuO_4	001	P&A	LEED: $(\sqrt{2} \times \sqrt{2})R45°$ reconstruction	216	
La_2CuO_4	001	–	Defect-lattice calculation: close to ideal termination		217
	100	–	Defect-lattice calculation: extensive relaxation		217
V_2O_5	001	UHV cleaved	LEED: (1×1) patterns, but damages easily; no quantitative work	218, 219	
PbO	001	Air cleaved	LEED: (1×1) patterns with both tetragonal and orthorhombic forms	220	
γ-Al_2O_3	100	–	Calculations of relaxation		221
η-Al_2O_3	Various	–	Calculations		222

2.3.7 Other oxide surface structures

Table 2.7 shows a number of other metal oxides with less-common crystal structures, on which single-crystal surface studies have been conducted. **MoO₃** has an orthorhombic layered structure, with each layer comprised of two interleaved planes of MoO_6 octahedra; the layers are parallel to the (010) crystal plane. Only O ions are exposed on the surfaces of the layers, which results in weak van der Waals bonding between layers and in a relatively inert surface. Qualitative LEED measurements showed (1 × 1) patterns for MoO_3 (010) surfaces that had been either scraped in UHV and annealed, ion etched and re-oxidized at 1000 K, or air-cleaved followed by cleaning in O_2 at 1000 K in the UHV chamber; no other LEED patterns were observed. The MoO_3 (010) surface is very sensitive to electron-beam damage, however, and great care must be taken during LEED measurements or the surface will reduce to an oxide close to MoO_2. LEED and RHEED measurements have also been made on **$Mo_{18}O_{52}$**, which is an O-deficient structure whose (100) surface consists of MoO_3 (010) terraces combined with atomic height steps.

Another metal oxide whose surface geometry has been studied rather thoroughly both experimentally and theoretically is **UO_2**. Owing to the relatively large size of U ions, UO_2 crystallizes in the cubic fluorite structure, where each U ion has eight O nearest neighbors, and each O ion is surrounded by four U ions [see Fig. 2.2 (f)]. The mineral form of UO_2 is uraninite, which is the principle source for uranium. Natural single crystals are rare, exhibiting predominantly (111) faces with some (110) and (100); uraninite does not cleave. Three low-index UO_2 surfaces have been studied on polished and annealed single crystals: the non-polar (111) and (110) surfaces, and the polar (100) surface. Measurements have also been made on vicinal (111) and (100) surfaces.

Theoretical calculations of the geometry of the (110) and (111) faces conclude that both should be close to ideal terminations of the bulk, with only a small inward relaxation of the outermost atomic plane on (111) and some lateral rumpling of the (110). Very good (1 × 1) LEED patterns can be obtained by ion bombardment and annealing in UHV. ISS measurements on UO_2 (111) suggest that the surface is nearly an O-terminated bulk structure, but that the U ions in the outermost layer relax outward by 0.19 Å. UO_2 (110) displays a variety of reconstructions as a function of surface preparation, with (3 × 2), (3 × 3), c(2 × 2) and (1 × 1) LEED patterns observed. When heated to temperatures above 873 K, faceting of the surface occurs, with excess O appearing on the surface.

The polar UO_2 (100) surface exhibits c(2 × 2) LEED patterns, and ISS measurements gave evidence for ordered overlayers of O ions arranged in distorted bridge-bonded, zig-zag chains along [100] directions. This structure could not be con-

firmed theoretically because of the energy divergence in calculations of polar surfaces. The (100) surface has also been studied by photoelectron diffraction, and,
within the resolution of the technique, displacement of the surface O ions could
not be detected.

Cu_2O possesses the cuprite structure, with O ions at the corners and the center
of the cubic unit cell and Cu ions occupying four of the eight tetrahedral interstices. In this structure each Cu ion has only two O nearest neighbors, while each
O ion has four tetrahedrally oriented Cu ligands. Two of its surfaces have been
studied on single-crystal samples: the non-polar (111) and the polar (100). A (111)
surface that was determined to be stoichiometric could be prepared by ion bombardment and annealing to 1000 K in UHV; the LEED patterns for that surface had
(1 × 1) symmetry. Annealing an ion-bombarded (111) surface at lower tempera-
tures gave a surface that exhibited ($\sqrt{3} \times \sqrt{3}$) R 30° LEED patterns; the structure of
that surface was tentatively interpreted in terms of one-third of a monolayer of
ordered O vacancies. The same preparation procedure applied to the (100) surface
resulted in surface structures that depended upon the history of the sample. Both
($3\sqrt{2} \times \sqrt{2}$) R 45° and ($\sqrt{2} \times \sqrt{2}$) R 45° LEED patterns were observed, with the former attributed to a Cu-terminated surface and the latter, which was obtained after
higher-temperature annealing, to an ordered overlayer of O atoms. As expected for
a polar surface, the lack of a stable, repeatable structure mirrors the fundamental
instability of the surface.

There have been a number of surface structural studies of single-crystal high-T_c
superconducting Cu oxides that have addressed the stoichiometry and geometric
structure of the surface in order to determine whether electron-spectroscopic techniques such as photoemission really do see the bulk properties of the materials.
Rutherford backscattering spectroscopy employing ion channeling has been used
to study the geometric structure of chemically etched (001) surfaces of single-
crystal **$YBa_2Cu_3O_{7-x}$**. The data suggest that the surface is nearly an ideal termination of the bulk crystal structure, with the Y-containing plane outermost. LEED
measurements on UHV-cleaved $YBa_2Cu_3O_7$ (001) surfaces reached the same conclusion, with only (1 × 1) patterns observed. $YBa_2Cu_3O_7$ tends to lose oxygen in
UHV, and the LEED measurements that have been performed do not really
address that issue. Surfaces that exhibit (1 × 1) LEED patterns may still be O-deficient, since random removal of surface O ions, which do not scatter electrons as
strongly as do the higher-Z metal cations, would not destroy the symmetry of the
LEED patterns unless they caused the surface to reconstruct. The O vacancies
would contribute to the diffuse background, but it would be very difficult to determine the degree of surface reduction from the background intensity.

A large number of surface structural studies have been performed on
$Bi_2Sr_2CaCu_2O_{8-x}$, where the surface is less prone to O loss. Single-crystal sam-

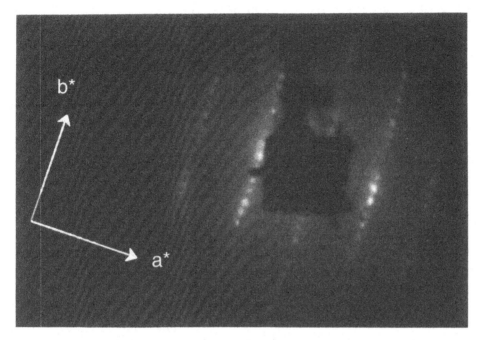

Fig. 2.18 LEED pattern obtained from the cleaved *a–b* surface of single-crystal $Bi_2Sr_2CaCu_2O_8$. Incident-electron energy, $E_i = 80$ eV. [Ref. 211]

ples cleaved along the (001) plane exhibit LEED patterns that reflect the bulk structure. In that oxide, a long-range ordering occurs along one of the crystal axes perpendicular to the *c*-axis. This long-range order is reflected in the symmetry of the LEED patterns, where the periodicity in that direction is found to be 27 Å. A LEED pattern for an electron energy of 80 eV is shown in Fig. 2.18; the long-range periodicity in the *b* direction in the crystal lattice is responsible for the small reciprocal lattice spacing in the *b** direction. Similar LEED results have been obtained for crystals cleaved in air and in UHV. One study concluded that the superstructure is localized in the Bi–O planes.[211]

LEED measurements have been performed on annealed (001) growth surfaces of Nd_2CuO_4, $(Nd,Ce)_2CuO_4$ and $(La,Sr)_2CuO_4$ single crystals. For the Nd compounds, the LEED patterns exhibited a symmetry characteristic of a tripling of the surface unit cell along the [110] direction. For the La compound, a ($\sqrt{2} \times \sqrt{2}$) R 45° LEED pattern is seen; no interpretation of the patterns was presented. The surfaces of both Nd materials were found to be sensitive to electron- or ion-beam exposure.

The surface structure of La_2CuO_4, one of the prototype compounds for the high-T_c superconducting Cu oxides, has been addressed theoretically by using defect-lattice calculations. Both the (100) and (001) surfaces were considered. The (100) surface was found to exhibit extensive atomic relaxation within the top few layers of the surface, while the (001) was close to an ideal termination of the bulk crystal structure. After relaxation of the (100) surface, its energy was approximately equal to that of th (001) surface.

Orthorhombic V_2O_5 has a two-dimensional layered structure and cleaves well along the basal (001) plane.[218,219] The (001) surface consists predominantly of O ions, since one-half of the V ions in a layer have an O ion directly above them, while the other half lie slightly below the main O-ion plane. Within a layer, each V ion has five O nearest neighbors. V_2O_5 damages very easily under electron bombardment,[219,223,224] so no quantitative LEED measurements have been performed on it.

Qualitative LEED measurements have also been performed on single crystals of both the orthorhombic (yellow) and tetragonal (red) forms of **PbO**. In both cases only the (001) face was studied, and the dominant LEED pattern was (1 × 1), exhibiting the same lattice constant as that of the bulk.

The structure of the catalytically most important alumina, **γ-Al$_2$O$_3$**, is not well determined but it is probably cubic, based on the spinel structure. The possible surface geometries of that structure and **η-Al$_2$O$_3$** have been considered, and the structural relaxation that should occur on the (100) surface of γ-Al$_2$O$_3$ has been calculated. However, there is no experimental data with which to compare, so the surfaces of these forms of Al$_2$O$_3$ are still a mystery.

The ultimate way in which one would like to determine surface crystal structure is by means of a local probe that could image individual atoms and measure their positions to within a small fraction of an inter-atomic distance. STM has achieved this goal in determining the positions of atoms in the direction normal to the surface plane for metals and semiconductors, but it has not yet achieved that resolution for oxides. (Some of the special considerations in using STM to study the surfaces of oxides and other ceramics are discussed in Ref. 225.) To date, close to atomic resolution has been achieved on the normally conducting oxides **$Rb_{0.3}MoO_3$**,[226,227] **$K_{0.3}MoO_3$**[228] **$Na_{0.9}Mo_6O_{17}$**,[227,229] and **$Rb_{0.05}WO_3$**[230] and on superconducting **$Bi_2Sr_2CaCu_2O_{8+\delta}$**,[231] **$Bi_2(Ca,Sr)_3Cu_2O_{8+\delta}$**,[232] and **$(Pb,Bi)_2Sr_2CaCu_2O_8$**.[233] The results are encouraging, but they have not yet been of real value in a surface structure determination on oxides.

2.4 Steps on oxide surfaces

Any real crystal surface – no matter how perfect the bulk single crystal, how well it cleaves, or how carefully the surface is prepared – will contain a large number of steps. Several of the surface structural models shown above include idealized

Stepped TiO$_2$(110)

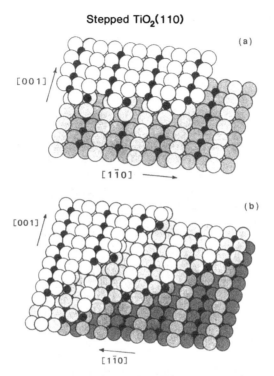

Fig. 2.19 Models of the rutile (110) surface with different types of steps. In (a) the steps are parallel to the [001] and [1$\bar{1}$0] directions. In (b) the edges of the upper and lower steps are contained in the (223) and the (443) planes, respectively. [Ref. 127]

steps, although only a few of the many possible step orientations are shown. This is because there has been very little experimental work on the properties of steps on oxide surfaces, and very little is known about them. The primary structural characteristic of steps is the reduced ligand coordination of step-edge ions. For example, terraces on the rocksalt (100) surface (see Fig. 2.5) have five-fold coordinated cations and anions, while that coordination is reduced to four at step edges. The step shown in Fig. 2.13 for corundum (10$\bar{1}$2) also has four-fold coordinated step-edge ions. However, the creation of steps such as those shown in Figs. 2.5 and 2.13 does not result in the creation of any net charge or dipole moment on the surface, so such stepped non-polar surfaces are still non-polar.

On metal oxides where the surface energies for various low-index faces are more nearly equal, step structures can become more complicated. Some of the possible geometries of stepped TiO$_2$ (110) surfaces have been considered in order to explain electron- and photon-stimulated desorption results.[127] Figure 2.19 shows several possible steps on (110) surfaces. Steps can contain four- and/or five-

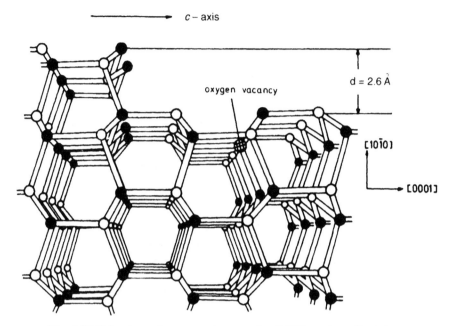

Fig. 2.20 Side view of a step on the wurtzite (10$\bar{1}$0) surface. [Ref. 234]

fold coordinated cations, and it is not clear from energetic arguments which types
would actually form. Chemisorption experiments have also been performed on
TiO$_2$ (441) surfaces, which have been modeled as (110) terraces three unit cells
wide, separated by (111) step edges.[18] In that model the terrace cations are five-
fold coordinated, but the step-edge cations have only four O ligands.

2.4.1 Experimental measurements on stepped oxide surfaces

Steps on non-polar surfaces of ZnO have been observed by LEED,[234–237] and their
orientation and height have been determined from the direction and magnitude of
the splitting of the LEED spots as a function of E_i, the incident-electron
energy.[238,239] ZnO samples that were cleaved parallel to the c-axis in UHV exhibited LEED patterns consistent with step arrays having a minimum step height of
2.8 Å, which is the height of one double layer of ions in that orientation (i.e., one-
half of the lattice repeat distance).[235] The ideal geometry of one such step is shown
in Fig. 2.20 for a stepped (10$\bar{1}$0) surface. However, samples cleaved perpendicular
to the c-axis always exhibited a minimum step height of 5.19 Å, which is the full
lattice repeat distance along the c-axis and corresponds to two double layers of
ions.[237] The creation of steps on ZnO (10$\bar{1}$0) samples that were annealed in UHV
has also been observed for annealing temperatures greater than 1100 K.[234] The

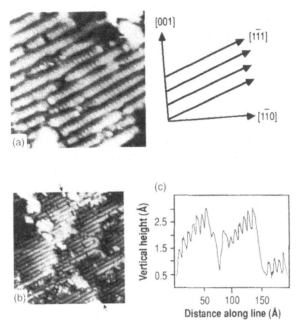

Fig. 2.21 (a) An STM image showing a 109 Å by 109 Å area of the (110) surface of heavily reduced TiO$_2$ (110). (b) A 246 Å by 246 Å area that includes several steps. (c) Profile taken along the line indicated by the arrows in (b), showing the height of the corrugations and the steps. [Ref. 153]

steps were found to run perpendicular to the *c*-axis, and the predominant step height found was 2.6 Å, close to that seen for UHV-cleaved surfaces. ZnO (40$\bar{4}$1) and (50$\bar{5}$1) surfaces were also prepared by cutting crystals a few degrees away from the (10$\bar{1}$0) plane and annealing at 800–900 K in UHV.[236] The terrace widths determined from the LEED spot splitting as a function of incident beam energy agreed with those expected for surfaces having that orientation.

Steps on vicinal UO$_2$ (111) surfaces have been studied by LEED and ISS.[198,200,206] Initially the ledge arrays were approximately equally spaced and parallel, but upon heating the surface decomposed irreversibly into (553) and (311) facets. The U ions at step-edge sites are apparently covered with excess O. Steps on vicinal UO$_2$ (100) surfaces did not order as well as those on the (111) surface, and heating again caused faceting, although the onset of faceting occurred at a lower temperature for the (100) surface than for (111).

Although STM and AFM have barely reached atomic resolution on oxide surfaces, they are able to image steps on oxide surfaces in greater detail than any other technique. The density, orientation and in-plane topography of steps on solid surfaces can be seen in STM and AFM without achieving lateral atomic resolution. Figure 2.21

1 μm

Fig. 2.22 AFM image of step structure on a cleaved Cr_2O_3 ($10\bar{1}2$) surface. The image area is $6 \times 6 \ \mu m^2$. [Ref. 3]

shows an STM image of, and a linescan across, an annealed TiO_2 (110) surface, where steps can be seen superimposed on a terrace structure which corresponds to bulk shear planes intersecting the surface; individual atoms are not being imaged on the terraces.[153] At much lower lateral resolution, Fig. 2.22 shows an AFM image of atomically flat terraces and steps on a ($10\bar{1}2$) cleavage face of Cr_2O_3.[3,135] The steps have a height of one unit cell in the direction normal to the ($10\bar{1}2$) plane; the primary step direction is [$\bar{2}201$], with occasional kinks along [$\bar{4}223$] directions also observed. (The [$\bar{2}201$] direction is equivalent to the [$0\bar{2}21$] direction shown in Fig. 2.13.) STM has also been used to image steps on TiO_2 (001)[240] and $Rb_{0.3}MoO_3$[226] surfaces in air. Polycrystalline ZnO has been imaged both in air and UHV, and steps are clearly visible.[310] Steps on single-crystal ZnO ($000\bar{1}$) surfaces have been imaged in UHV by STM,[241] and gross surface topographic features have been seen on ZnO (0001) in an aqueous electrolyte, although individual steps could not be imaged.[242] AFM has imaged steps and facets on NiO (100) and TiO_2 (100), (110) and (001) surfaces in air.[3,135]

STM cannot be used directly on the surfaces of highly insulating materials, but some attempts have been made to image insulator surfaces after coating them with a thin layer of Au.[243] Fractured surfaces of insulating MgO have been imaged after coating the surface with 20–60 Å of gold,[244] and, although atomic-height steps could not be resolved, larger steps were apparent.

High-resolution TEM has the resolution capability to image individual atoms in materials, and some experiments have been undertaken to study the atomic geometry of oxide surfaces. Care must be taken since the roughly 500 keV electrons incident on the surface can cause severe surface reduction, but it has still been possible to see atomic level details of steps and terraces, and cation redistribution has been followed in real time. The oxides that have been studied include: MgO;[245-248] $Bi_{0.1}WO_3$, Cr-doped TiO_2, VNb_9O_{25}, $(Mo,Ta)_5O_{14}$, Al_2O_3, NiO, UO_2 and U_4O_9;[249] Tb_7O_{12}, $Tb_{11}O_{20}$ and $Tb_{12}O_{22}$;[250] and La_2O_3 and La_2CuO_4.[251] Reflection-electron microscopy does not have the resolution of TEM, but it can be used to advantage to study steps on single-crystal surfaces. A good example is some recent work on Al_2O_3 surfaces.[252,253] Figure 2.23 shows a series of micrographs taken of the same area on a polished and annealed Al_2O_3 (0001) surface but for different azimuthal angles in order to determine the orientation of the facet faces on the steps.

2.4.2 Calculations of extended defects

While little experimental work has been done on the properties of steps on rocksalt oxide surfaces, several theoretical calculations of the relaxation of ions at and near step sites have been performed.[103,104,254-257] Various defect configurations have been considered, including clusters of vacancies on a terrace, straight steps, jogs and kinks in steps, and clusters of adatoms. Detailed ionic positions have been calculated, and trends are found in the motion of surface ions at these extended defects away from the positions that they would have for an unreconstructed termination of the bulk structure.[256] An appreciable relaxation of non-planar irregularities is found, in contrast to the virtually unrelaxed geometry of terraces. The most general trend is a smoothing of all irregularities, such as a contraction of the lattice at the upper edge of steps and a dilation at the bottom; this is shown graphically for the calculated positions of ions at steps on an MgO (100) surface in Fig. 2.24. Ions at outside-corner sites at kinks in steps, where their ligand coordination is even lower than that of ions on straight steps, suffer particularly large inward relaxation.

2.5 Point defects on oxide surfaces

The most important types of defect on oxide surfaces as far as electronic structure, chemisorption, catalysis, corrosion, etc., are concerned are point defects, such as cation or anion vacancies or adatoms. Measurements of the electronic structure of a variety of oxide surfaces have shown that the predominant type of point defect formed when samples are heated or exposed to electron, ion or photon beams consists of O vacancies; that is the reason that O-vacancy defects are included in most

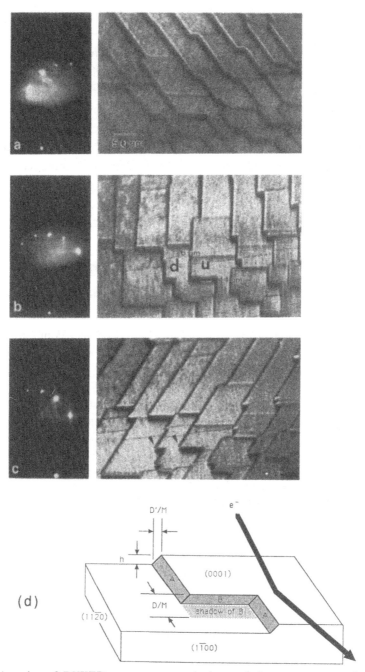

Fig. 2.23 A series of RHEED patterns, rotated views of the same area, and simulated images of a stepped Al_2O_3 (0001) surface. The foreshortening factor of all images is 24. [Ref. 253]

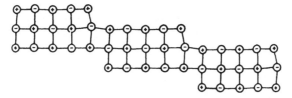

Fig. 2.24 Theoretically predicted atomic displacements around the step on a MgO (105) surface, drawn to scale. [Ref. 257]

of the surface models above. Cation vacancies certainly exist on oxide surfaces, but there is little direct evidence of their importance, and so we will concentrate on anion vacancies here. Point defects are primarily important when considering the electronic structure of surfaces, so most of the discussion of their properties will be in succeeding chapters. Here we will only discuss what is either known or speculated about their geometry. Adatoms on oxide surfaces will be discussed in Chapter 7.

The geometrical properties of O vacancies on octahedrally coordinated oxide surfaces are similar for the different bulk crystal structures and surface planes. Removal of a surface O ion reduces the coordination of adjacent cations by one. For rocksalt (100), four surface cations become four-fold coordinated, and one cation in the plane below becomes five-fold coordinated (see Fig. 2.5). For perovskite (100)-BO_2 and corundum ($10\bar{1}2$), two surface cations become four-fold coordinated with the removal of one O ion (Figs. 2.7 and 2.13). In the model of rutile (110) in Fig. 2.10, two O ions have been removed, one in the row of bridging O ions and one in the main surface plane. In the former case two six-fold coordinated cations become five-fold; in the latter case two five-fold surface cations become four-fold.

Point defects on the surfaces of wurtzite oxides are qualitatively similar to those on octahedral oxides, but the details of the sites are different. Removal of one anion (cation) from the ($10\bar{1}0$) face reduces the coordination of one adjacent cation (anion) from three to two and reduces the coordination of the two other adjacent cations (anions) from four to three. On the ($11\bar{2}0$) surface, removal of an anion (cation) reduces the coordination of two cations (anions) from three to two, and one cation (anion) from four to three. On the polar (0001) [or (000$\bar{1}$)] surface, removal of a cation (anion) reduces the ligand coordination of three anions (cations) in the plane below from four to three. Studies of the electronic structure of ZnO ($10\bar{1}0$) surfaces indicate that the predominant type of surface point defect consists of O vacancies,[258-260] as is the case for octahedral oxides.

It must be stressed that the models of O-vacancy point defects shown here represent only some of the conceptually simplest types of defects that one can imagine. Many other types of defect structures exist, but virtually nothing is known about their geometry. We will therefore consider only very simple defects in order to discuss their general properties. It is also important to note that no relaxation has been included in any of these models, and yet removal of a relatively large O^{2-} ion *must* result in significant relaxation of the surrounding ions. While there have been calculations of the energetics of point defect formation on oxide surfaces,[261-263] no results have been presented for the relaxation of the ions in the vicinity of the defects. There have also been several cluster calculations of the electronic structure of point defects on oxides that will be discussed in subsequent chapters, but all of those assumed an unrelaxed geometry for the defect. There is thus no theoretical guidance to assist in constructing more realistic models of surface point defects, so we will use the idealized models shown.

Although individual point defects on metal-oxide surfaces have not yet been imaged directly, impact-collision ISS has been used to determine the density of single vacancies and various kinds of vacancy clusters on MgO (100) by means of the intensity change of scattered ions with the angle of incidence of the primary ion beam.[264] Even though sharp (1 × 1) LEED patterns were observed for surfaces that had been polished and etched before UHV treatment, the ISS spectra indicated a high density of point defects (it was not stated whether the defects were missing cations or anions). The density of isolated vacancies and small clusters decreased as the temperature was increased above 1300 K, while larger clusters were stable to higher temperatures. Above 1650 K surface vacancies could not be detected, but above 1700 K thermal etch pits began to appear. Some additional experimental methods that have been used to study defects on oxide surfaces are discussed in Ref. 265.

2.6 Preferential sputtering

A corollary of the observation that O vacancies are the dominant type of defect formed under ion bombardment is that such bombardment (or 'sputtering') preferentially reduces oxide surfaces. Preferential sputtering is extremely material dependent and depends upon ion type, energy, angle of incidence, temperature, etc.[266] A wide range of preferential sputtering behavior is exhibited by metal oxides.[267,268] For MgO, Al_2O_3 and BeO, for example, there is little effect of inert-gas-ion bombardment on surface stoichiometry; cations and anions seem to be removed at essentially the same rate. The (110) surface of TiO_2, on the other hand, reduces under a few hundred eV Ar^+-ion bombardment to a stoichiometry close to that of Ti_2O_3, another stable oxide of Ti.[269-272] NiO (100) and CoO (100) reduce only a few percent under even 5 keV Ar^+-ion bombardment at room temperature, although when bombarded at temperatures above 500 K surface reduction can be severe.[32,273,274] Preferential sputtering is generally more pronounced for transition-metal oxides than for non-transition-metal oxides,[267] probably because many transition metals have a range of stable oxides with different compositions. However, the predominant effects have been found to be due to the difference in mass and surface binding energy between the metal and O ions.[268] [It is also possible to bombard oxide surfaces with oxygen ions (O^+ or O_2^+) to reduce the extent of preferential sputtering, but the surface is still damaged structurally as it would be for inert-gas-ion bombardment.] One of the major preferential sputtering headaches consists of the high-T_c Cu-oxide superconductors, where attempts to use UPS to study their electronic structure has been hampered by the difficulty of obtaining a surface that has the stoichiometry of the bulk.[275]

The occurrence of preferential sputtering severely limits the ways in which oxide surfaces can be prepared, and it has been the cause of misinterpretation of experimental data on more than one occasion. Stoichiometric, nearly perfect surfaces of elements are often easy to prepare in UHV by simply cleaning by inert-gas-ion bombardment, followed by annealing to restore the geometric order. (Although this process is conceptually easy and straightforward, it is often necessary to repeatedly alternate ion bombardment and annealing steps, and the total time necessary to obtain an atomically clean surface may be very long.) All compounds are susceptible to preferential sputtering, and care must be taken with metallic alloys and semiconductors in order to achieve a surface that is truly representative of the bulk crystal composition and structure. The preparation of stoichiometric, nearly perfect surfaces is even more difficult for oxides, and in only a handful of specific cases has it been shown that such surfaces can be prepared by sputtering and annealing. In general, sputtering removes O ions preferentially, and subsequent annealing, even at quite high temperatures, will not restore the bulk composition to the surface. This may be due to the high Debye temperatures of most oxides, so that the ions are relatively immobile even at high temperatures. For oxides containing two or more types of cation, sputtering often removes one type preferentially, and annealing will not usually restore the bulk stoichiometry [e.g., $SrTiO_3$ (100)[111]]. Therefore, in general the most reliable way to prepare a stoichiometric oxide surface is by cleaving or fracture in UHV.

In addition to preferential sputtering under ion bombardment, defects can be created on metal-oxide surfaces by electron or photon bombardment. The physics of the process is different than for sputtering, since neither electrons nor photons possess significant momentum; the process of breaking interatomic bonds and ejecting atoms from the surface thus involves primarily energetic considerations. Electron-beam damage to oxide surfaces can be severe, so much so that Auger spectroscopy and LEED cannot be performed on some oxides without partially reducing the surface. Since the processes of photon- and electron-stimulated desorption of lattice atoms are closely related to the desorption of adsorbed molecules, we will treat all aspects of such desorption together in Chapter 6.

A few oxides exhibit somewhat different behavior under electron or ion bombardment than do the majority of metal oxides. When Y-doped ZrO_2 (100) is bombarded with electrons, the surface becomes reduced, with some of the Zr^{4+} ions becoming Zr^0; there does not appear to be an associated reduction of the Y^{3+} ions.[276] However, ion bombardment, instead of also causing surface reduction, has been found to restore the surface stoichiometry; this unique behavior has only been observed in this material. Cu_2O (111) surfaces become slightly depleted in Cu under bombardment by ions with energies of a few keV;[277] electron bombardment, however, does not appear to alter the surface stoichiometry.

2.7 Surface segregation

In a binary oxide containing only one type of cation, deviations from bulk stoichiometry at the surface can be described simply in terms of surface reduction or oxidation. But when two or more types of cation are present, surface compositions can become more complicated. The field of surface segregation has been fully developed for the important case of metallic alloys; a variety of experimental techniques have been used to quantify surface composition, and a great deal of theoretical effort has also gone into explaining and predicting surface segregation. Similar phenomena are present in oxides, and a few experimental and theoretical studies have addressed the issue.

One case where surface segregation is important is when small amounts of an impurity (usually another cation) are present in the bulk. Heating of the sample may result in segregation of a high concentration of the impurity species to the surface. The most thoroughly studied system of this type is **Ca in MgO**, where ion-scattering and Auger spectroscopies have been used to monitor the segregation of 200 ppm Ca in the bulk to the MgO (100) surface.[278,279] The surface concentration of Ca was monitored as a function of temperature between 1173 and 1723 K, and the heat of segregation was determined to be 50.3 kJ/mol. Near 1200 K the equilibrium Ca surface concentration corresponded to about 20 % occupation of the surface cation sites. In a separate study, the scattering of 1 keV He atoms from the (100) surface of Ca-doped MgO was used to determine the position of the Ca ions in the surface layer.[280] The Ca ions were found to lie 0.4 Å above the Mg-O plane, while the Mg ions were found to lie in the Mg-O plane (i.e., the surface was not rumpled except at Ca sites) to within the accuracy of previous LEED determinations and theoretical predictions. The segregation of **Ca** atoms to the (100) surface of **$SrTiO_3$** has been shown to produce a p(2 × 2) reconstruction, as determined by LEED.[281] Experiments have also been performed on the segregation of **Ba** to the **MgO** (100) surface, and a saturation coverage of 0.3–0.4 monolayer was found for a wide range of bulk dopings.[282,283]

The segregation of Ca and a large number of other cations to the (100) surface of MgO has been treated theoretically in great detail.[103,254–256,262,283–295] Both (100) terrace and step sites were considered, and relaxation and rumpling of the lattice were included. The energies of cation substitution were found to depend on a combination of electrostatic, elastic and polarization forces, where any one of those might predominate in a given situation. The calculated heat of segregation of Ca on MgO (100) is in agreement with experiment. A theoretical survey of a number of impurity ions in MgO suggests that **Na**, **Ca**, **Sr**, **Ba** and **Ti** are likely to segregate to the (100) surface.[262] (Measurements of Sr segregation to the surface of MgO ceramic samples are in qualitative agreement with the theoretical results.[296,297]) The segregation of Ca ions to the (110) surface of MgO was also calculated, although no experiments have been performed for that face.[285]

The segregation of **Cr** to the (100) surfaces of both **NiO** and **CoO** has been studied on single-crystal samples by secondary-ion mass spectroscopy (SIMS) and XPS.[298] The samples, which were bulk doped with up to 0.7 at.% Cr_2O_3, were annealed at temperatures between 1173 and 1573 K for one week. Sputter profiles were then performed of the Cr concentration as a function of distance below the (100) surface. Cr segregated to the surface of both CoO and NiO, although the Cr-rich layer was much thicker on CoO than on NiO for the same annealing conditions and bulk concentrations; there is a question of the possible formation of a $CoCr_2O_4$ spinel in the surface region of Cr-doped CoO. For low Cr concentrations, the measured heats of segregation of Cr to the (100) surfaces of CoO and NiO were about 24 and 29 kJ/mol, respectively.

The surface cation ratios in solid solutions of **binary rocksalt oxides** have been addressed theoretically.[299] The divalent cations **Ni**, **Co**, **Mn**, **Fe** and **Mg** were considered, and it was suggested that the bulk strain energy was the most important factor in predicting segregation. Thus the larger cation is the one that will segregate to the surface.

Surface segregation has only been studied experimentally on single crystals of two non-rocksalt oxides: Al_2O_3 and V_6O_{13}. Auger spectroscopy and LEED were used to study the segregation of **Mg** to the (0001) and (10$\bar{1}$0) surfaces of Al_2O_3 into which Mg had been doped or ion implanted.[300–303] When the (0001) crystal surface was heated to between 1173 and 2073 K in UHV, no Mg could be detected on the surface, presumably because the rate of Mg evaporation from the surface was much greater than the diffusive flux of Mg to the surface. In order to observe the segregation of Mg to the surface, it was necessary to place two crystals face-to-face, heat them, and then separate them at lower temperature before analyzing their surfaces. This type of experiment simulates the segregation of impurities to grain boundaries, which is an extremely important effect in ceramics processing. In that situation as much as 5–10 % of a monolayer of Mg was found to segregate for temperatures above 1473 K. For the ($10\bar{1}0$) surface, however, no Mg segregation was found even when two crystals were placed face-to-face and heated in UHV. [Mg segregation to the Al_2O_3 ($10\bar{1}0$) surface was observed when the sample was annealed in air rather than UHV, but in that case the Mg could be reacting with adsorbates on the surface.]

Somewhat different results were obtained for **Ca** segregation to the **Al_2O_3** (0001) and ($10\bar{1}0$) surfaces.[300,302,304] Strong segregation of Ca to the ($10\bar{1}0$) surface was observed between 1573 and 1773 K, but no Ca segregation was found for the (0001) surface, even for face-to-face crystals at temperatures up to 1773 K. For the ($10\bar{1}0$) surface heated in air, it was found that, when both Mg and Ca were present in the bulk, the segregation of Mg prevented Ca from segregating. Segregation of Y, Ca and Fe was also reported on the Al_2O_3 (0001) surface.[305]

The segregation of **K** to the (001) surface of **V_6O_{13}** has been studied by ISS.[306] Segregation was observed to be strong at 800 K, where a bulk K concentration of < 100 ppm resulted in the segregation of about one K atom per V_6O_{13} (001) surface unit cell.

The segregation of **Mg, Ca, Fe, La** and **Y** to the (0001), ($10\bar{1}0$) and ($10\bar{1}2$) surfaces of **Al_2O_3** has been modeled theoretically, and it was found that the segregation characteristics were highly surface specific, reflecting the major differences in structure of the various faces.[283,293,294,307,308] The theoretical calculations do not agree with experimental results for Ca, since it is predicted that Ca will segregate to both the (0001) and ($10\bar{1}0$) faces. Atomistic simulations have also been performed for the segregation of **Y, La** and **Al** in **Cr_2O_3**, and **Sr** in **La_2CuO_4**,[217,293] for **Y, Mg** and **Fe** segregation to various faces of **Cr_2O_3**,[308] and for **Na** segregation to the surface of **Li_2O**.[294]

Surface segregation in metal oxides has been studied on polycrystalline samples of a number of oxides. Some of that work, including segregation in **NiO, CoO, Al_2O_3, SnO_2, MgO** and **ZrO_2** and their solid solutions, is reviewed in Ref. 309.

3

Surface lattice vibrations

The vibrational modes of surfaces are of interest for several reasons. On clean surfaces, they provide information about interatomic forces, which can be used in the interpretation of structural and dynamic properties. An even more important aspect is the study of adsorbates, where vibrational spectroscopy provides one of the major sources of information about the nature of adsorbed species.[4] Vibrational spectra of surfaces can, in principle, be obtained with a number of techniques: these include **infrared reflection/absorption spectroscopy (IRRAS or RAIRS)**,[311] **Raman spectroscopy**, as well as the surface-specific technique of **high-resolution electron-energy-loss spectroscopy (HREELS)**. In HREELS a low-energy incident electron beam (usually 1–20 eV) is scattered off a surface and the kinetic energy of the scattered electrons is measured in order to determine the energy of the inelastic loss (or gain) processes taking place.[312–314] One of the problems with ionic materials such as oxides is that their intrinsic phonon modes couple very strongly both to infrared radiation and to low-energy electron beams. As a result, surface-specific information can be very difficult to obtain, and even in HREELS the intrinsic spectra of the substrate can completely overwhelm adsorbate loss peaks. Most of this chapter is therefore concerned with intrinsic phonon spectra of metal-oxide surfaces, although some adsorbate studies are discussed in the final section.

3.1 Bulk phonons

The intrinsic surface vibrations of metal oxides are closely related to bulk modes. Therefore some aspects of these bulk phonons are discussed in this section, before the surface modes are described in § 3.2.

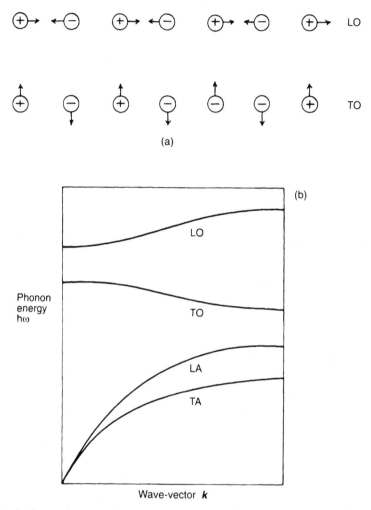

Fig. 3.1 (a) Ionic motion associated with LO and TO phonons in a binary crystal; (b) schematic dispersion curves for optical and acoustical phonons.

3.1.1 Phonon modes and dielectric properties

The normal modes of vibration of the atoms or ions in a crystal can be described either classically, in terms of the relative motion of each atom with respect to the other atoms, or quantum mechanically, in terms of fundamental excitations called **phonons**.[315] The energy of a single phonon ranges from essentially zero for long-wavelength acoustical modes, where all atoms or ions in each unit cell move in phase with one another, to several hundred meV for optical modes, in which different atoms in a single unit cell move in opposite directions. As with electronic states in band theory, phonons are characterized by their wave-vectors, k, determining the wavelength and direction of propagation within the crystal. For a crystal structure having N atoms per primitive unit cell, there are always three acoustic phonon branches and, if $N > 1$, there are $(3N - 3)$ optical branches, although for certain directions of propagation in crystals with sufficiently high symmetry, some of these may be degenerate. In all compounds, there are at least two atoms per primitive unit cell, so that some optical phonon modes are always present.

Unlike electronic states, phonons have polarization properties; they are called **transverse** when the atomic motion is perpendicular to the wave-vector, and **longitudinal** when it is parallel. For example, of the three optical phonon branches in a simple binary crystal such as MgO, two branches are transverse, and one is longitudinal: for high-symmetry directions of propagation such as along [001], the two transverse modes are degenerate. Figure 3.1 shows (a) the types of motion associated with transverse and longitudinal optical (TO and LO) phonons in such a solid, and (b) typical dispersion curves for these and the acoustic (TA and LA) phonons for wave-vectors [00ζ].

In ionic compounds such as metal oxides, strong internal electric fields accompany the relative motion of positive and negative ions. Acoustical modes do not generate electric fields since both positive and negative ions are moving in the same direction at the same time. For optical modes, however, positive and negative ions in each unit cell move out of phase with one another, resulting in local oscillating dipole electric fields within the crystal that have the period and frequency of the lattice vibration (about 10^{15} Hz). These optical modes couple very strongly to externally applied electric fields, and this coupling is important in many ways. It is the origin of the energy difference between LO and TO phonons, in the long-wavelength limit of zero phonon wave-vector. It also results in the strong interaction of optical phonons both with infrared (IR) electromagnetic radiation – hence the name 'optical' phonon – and with charged particles such as the electrons used in HREELS.

The properties of optical phonons are often interpreted in terms of a **dielectric function**, which shows how the dielectric 'constant' of the solid varies with the frequency of the applied electric field. A parameterized form commonly used in connection with optical phonons is

$$\varepsilon(\omega) \;=\; \varepsilon(\infty) + \sum_i \left(\frac{Z_i}{\omega_i^2 - \omega^2 - i\omega\gamma_i} \right). \tag{3.1}$$

Here $\varepsilon(\omega)$ is the dielectric function at frequency ω and $\varepsilon(\infty)$ is the high-frequency value, at frequencies well above all phonon modes (but below that associated with electronic excitations). The sum is over the different distinct groups of optical phonon modes (see § 3.1.2): Z_i is a parameter proportional to the 'dipole strength'

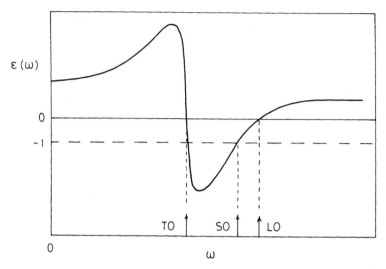

Fig. 3.2 Schematic variation of the real part of the dielectric function with frequency for a binary ionic crystal. The frequencies of the long-wavelength bulk (TO and LO) and surface (SO) modes are indicated.

of the mode i, ω_i the frequency of the transverse optical mode, and γ_i a 'damping factor' associated with anharmonic effects.

Figure 3.2 shows a sketch of the real part of $\varepsilon(\omega)$ for a crystal such as MgO with only two atoms per unit cell, and thus only one term in the sum in Eqn 3.1. The form of this curve is important in understanding the IR and HREELS discussion below. Bulk propagating modes occur when

$$\mathrm{Re}[\varepsilon(\omega)] \;=\; 0, \qquad\qquad (3.2)$$

and this happens, as shown, at the two frequencies labelled TO and LO, the former being equal to the ω_i parameter in Eqn 3.1. Between these two frequencies the real part of $\varepsilon(\omega)$ is *negative*, and, as discussed below, this corresponds to a region of high IR reflectivity by the solid. The other frequency, marked SO in Fig. 3.2, is the surface optical phonon frequency of the **Fuchs–Kliewer** mode, determined by the condition

$$\mathrm{Re}[\varepsilon(\omega)] \;=\; -1, \qquad\qquad (3.3)$$

and discussed below in § 3.2.2.

For crystals having more atoms per unit cell, and hence more optical phonon modes, the dielectric function will show a correspondingly larger number of oscil-

Table 3.1. *Dipole active phonon modes in some important crystal structures*

Structure-type	Symmetry	Polarization	Number of modes
Rocksalt	cubic	–	1
Fluorite	cubic	–	1
Perovskite	cubic	–	3
Wurtzite	hexagonal	$\parallel$	1
		$\perp$	1
Rutile	tetragonal	$\parallel$	1
		$\perp$	3
Corundum	hexagonal	$\parallel$	2
		$\perp$	4

lations of the type illustrated in the figure. Thus associated with each type of mode there will be a set of TO, SO and LO frequencies.

Equation 3.1 involves certain approximations and limitations. In the first place, it is valid in the long-wavelength limit, and thus is appropriate for IR and specular HREELS experiments where momentum transfer is very small. Secondly, it ignores anharmonicity effects except through the damping term γ_i and cannot be used to deal with overtone or combination bands appearing in IR (but see § 3.2.2 for HREELS multiphonon structure, which arises through a different mechanism). Finally, and most importantly, it assumes that the dielectric function is a *scalar* quantity, whereas in reality it should be represented as a symmetric second-rank tensor, with six components ε_{xx}, ε_{xy}, … , ε_{zz}. For cubic crystals, in fact, the dielectric properties are isotropic, and the scalar representation is adequate. For the important case of *uniaxial* crystal symmetries (tetragonal, trigonal and hexagonal), only two distinct values are required: ε_{zz} (normally denoted $\varepsilon_{\parallel}$) represents the response to electric fields polarized along the unique symmetry axis, and the equal components $\varepsilon_{xx} = \varepsilon_{yy}$ (denoted $\varepsilon_{\perp}$) that to fields perpendicular to this axis. The strengths and frequencies of phonon modes contributing to these polarizations will be different, so each dielectric component should be expressed as a different sum of the form given by Eqn 3.1. Clearly, lower-symmetry crystals are correspondingly more complex.

The number of distinct optical phonon modes contributing to different polarizations may be determined from the crystal structure and its symmetry by group theoretical arguments.[316] These calculations also show which modes are **dipole active**, i.e., contribute to the dielectric function. This latter point is important since in crystals of high symmetry some 'optical' phonon branches correspond to combinations of ionic motion that do not produce electric fields in the long-wavelength limit. Table 3.1 summarizes these results for some important cubic and uniaxial structures.

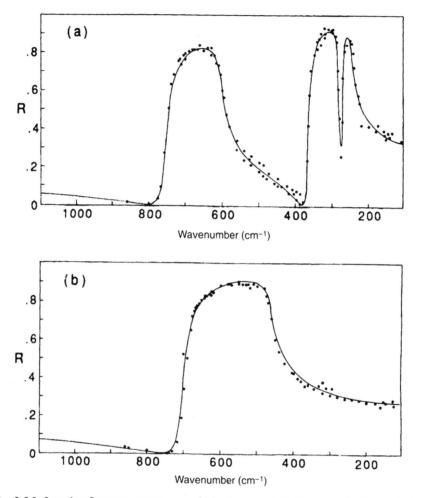

Fig. 3.3 Infrared reflectance spectrum of single-crystal SnO_2 for radiation polarized (a) perpendicular and (b) parallel to the *c*-axis. Points are experimental and curves are calculated. [Ref. 317]

3.1.2 Infrared reflection and absorption

Infrared spectroscopy is in most respects a *bulk* technique of limited applicability for studying the surface properties of oxide single crystals. Although there are exceptions to this general statement, it is important first of all to understand how IR reflection and absorption are related to the dielectric functions discussed in the previous section.[316]

In a solid where the magnetic susceptibility is negligible, the refractive index (which may be complex) is given by the square root of the dielectric constant

$$n^2 = \varepsilon. \qquad (3.4)$$

The reflection coefficient for radiation at normal incidence is given by

$$R = \left\{ \frac{|n-1|}{|n+1|} \right\}^2 . \qquad (3.5)$$

It follows that the reflectivity will be very high when n is nearly purely imaginary; this is precisely what happens when ε is negative, at frequencies between the transverse and longitudinal optical phonon frequencies, as discussed above (see Fig. 3.2). The strong **Reststrahl** bands associated with this effect are illustrated in Fig. 3.3, which shows the reflectivity of single-crystal SnO_2 measured in the two polarizations corresponding to $\varepsilon_{\parallel}$ and $\varepsilon_{\perp}$.[317] The analysis of reflectivity curves such as these is the normal method of determining the parameters of the dielectric function given in Eqn 3.1.

The interaction of IR radiation with crystals is in reality much more complicated than the above discussion suggests.[316] Vibrational modes mix with the radiation field, giving coupled excitations known as **polaritons**. These are important, for example, in understanding the IR reflectance from thin films. The shapes and sizes of particles may also be important for polycrystalline samples. Nevertheless, our simple treatment is adequate for understanding single-crystal samples of insulating oxides, and it illustrates the difficulty involved in using IR spectroscopy to study surfaces. The enormously strong reflection bands coming from the bulk of the sample interfere with weak bands that may be present due to surface modes, including those due to adsorbates. The use of IR spectroscopy to study adsorbates on metal surfaces depends on the substrate having a high and uniform reflectivity.[4,311] But this is not the case with oxides, except at frequencies well above or well below the optical phonon frequencies. Typically these are in the range 300–800 cm^{-1}, and so it is possible in principle to study adsorbates such as water and CO which have vibrations at much higher frequencies. This is commonplace on polycrystalline samples, but it has not been done yet with single crystals. As discussed later, surface-specific modes on clean crystals are not expected at frequencies higher than those of bulk modes, and such surface phonons therefore cannot be studied by IR.

So far we have assumed that the dielectric properties are determined entirely by vibrational (phonon) modes. The interaction with conduction electrons in metallic oxides, which opens up a more promising area for IR study of surfaces, is discussed below in § 3.2.3.

3.2 Surface phonons

Bulk phonon modes correspond to normal vibrational mode solutions for a crystal that is infinite in all three dimensions (or for which suitable periodic boundary

Table 3.2. *Surface phonon measurements on metal oxides*

Compound	Face	Technique	Type of phonon measured	References
MgO	100	He atom HREELS HREELS	Rayleigh dispersion FK Microscopic	69, 73 144, 320 321, 322
	Epitaxial 100 film	HREELS	H_2O, D_2O, CH_3OH adsorption	323
CaO	100	HREELS	FK	321
Al_2O_3	0001	HREELS	FK	324, 325
ZnO	0001	HREELS	FK	326
	$000\bar{1}$	HREELS HREELS	FK Coupling with plasmons in surface accumulation layer	326 327–329
	$10\bar{1}0$	HREELS	FK	312
SnO_2	110	HREELS	FK	330, 331
	Polycrystal.	HREELS	Effect of electronic carriers	332
TiO_2 (rutile)	110	HREELS HREELS	FK Effect of surface reduction	333 333–335
	100	HREELS	FK	336
TiO_2 (anatase)	(001) and (100)	HREELS	FK	337
$SrTiO_3$	100	HREELS HREELS	FK FK plus vibrations of adsorbed H_2O	338 339–341
V_2O_5	001	HREELS	FK	218, 342
NiO	100	HREELS	FK; deconvolution of overtones	341
	111	HREELS	FK; surface with Pb overlayer	97
WO_3 and Na_xWO_3	100	HREELS	FK; effect of electronic carriers, adsorbed H_2O	343–346
UO_2	111	HREELS	FK	347–349
	110	HREELS	FK	348, 349

conditions have been applied so that the surfaces effectively disappear). When the equations of motion are applied to a crystal that is infinite in two dimensions but finite in the third – for example, a semi-infinite crystal having one surface – new normal-mode solutions are found in addition to the bulk modes.[318] These surface modes have oscillatory, travelling wave solutions in directions parallel to the surface, but have an exponentially damped amplitude normal to the surface into the bulk of the crystal. They are localized in a layer roughly one phonon wavelength wide at the surface, so that the distance of penetration normal to the surface decreases with decreasing wavelength of the oscillatory solution (i.e., for increasing phonon wave-vector). These surface modes can be either acoustical or optical; the same criteria for the number of modes of each type apply as for bulk phonon modes.

Surface phonon modes may be observed experimentally in energy-loss experiments involving neutral atoms (particularly He) and electrons with low incident energy. Table 3.2 summarizes the literature in this area. The different types of modes observed are discussed in the following sections.

3.2.1 Surface acoustical modes: Rayleigh waves

Even in elements having only one atom per primitive unit cell, three acoustical surface phonon modes will exist. In order to be clearly identifiable as surface phonons, they must not overlap with the projection of the bulk phonon bands onto the surface plane in question;[318] those surface modes that do fall within the projected bulk bands are referred to as 'surface resonances'. In general there will only be small regions of reciprocal space in which true surface acoustical phonons will exist, and there is only one mode that will remain separated from the bulk bands down to zero wave-vector. That surface mode is referred to as a '**Rayleigh wave**' and has been well-known in the literature for over a century. It also appears as the solution to the elastic response of a semi-infinite continuum solid.[319]

Rayleigh waves are present in ionic compounds as well as elements, of course, and they have received some attention in metal oxides. Since both types of ion move in the same direction in acoustical modes, Rayleigh waves are not accompanied by electric fields and hence do not couple strongly to incident electrons. They are detected most easily in atom scattering from the surface, but they can also appear in off-specular HREELS, which is dominated by non-dipole impact scattering. The dispersion of surface acoustical phonon modes (i.e., the dependence of the phonon energy on wavelength) on MgO (100) has been measured by low-energy He atom scattering.[69,73,74] Using inelastic scattering of 20 meV He atoms from both air- and UHV-cleaved MgO (100) surfaces, the surface phonon dispersion could be measured out to a momentum transfer of 1 Å^{-1}.

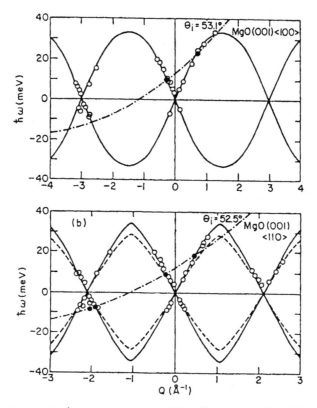

Fig. 3.4 He atom scattering measurements of the dispersion of Rayleigh (solid curve) and shear horizontal (dashed curve) phonon modes on MgO (100) for the (a) <100> and (b) <110> azimuths. [Ref. 74]

Figure 3.4 shows the measured and calculated dispersion of surface phonons in both the <100> and <110> azimuths for MgO (100). The experimental results are in good agreement with Rayleigh surface phonon theory and shell-model calculations using bulk parameters, thus supporting the idea that relaxation, rumpling and charge redistribution on MgO are negligible. Off-specular HREELS measurements on UHV-cleaved MgO (100) have also shown losses corresponding to Rayleigh modes.[322]

3.2.2 Surface optical (Fuchs–Kliewer) modes and HREELS spectra

Charged particles such as electrons interact strongly with optical rather than acoustic phonons, and the HREELS spectra of oxide surfaces are dominated by so-called **surface optical (SO) phonons** or 'Fuchs–Kliewer' modes, named after the authors who first described them in terms of lattice dynamics.[350]

The strongest HREELS losses in many solids are dominated by long-range electric fields associated with dipole-active excitations; the detailed theory of such losses is described elsewhere.[318] The most important results from the present point of view may be briefly described as follows. Consider a beam of electrons incident on a surface with normal velocity $v_\perp$. An energy loss of $\hbar\omega$ corresponds to excitation of a mode with frequency ω. In an angle-resolved experiment, the change in momentum parallel to the surface may also be determined, and by momentum conservation arguments can be related to k, the parallel wave-vector of the excitation. For an isotropic solid described by a dielectric function $\varepsilon(\omega)$, the theory shows that the probability of such a loss event during the scattering of an electron from the surface is given by

$$P(\omega, k) = \frac{4e^2 k}{\pi^2 h v_\perp^2 [k^2 + (\omega/v_\perp)^2]^2} \,\mathrm{Im}\, \frac{-1}{[\varepsilon(\omega) + 1]}. \qquad (3.6)$$

Under normal experimental conditions this function peaks rather sharply at low k values; that is, long-wavelength phonons are predominantly excited. The physical reason for this is that the electric fields outside the solid associated with such phonons die away less rapidly than those for short-wavelength phonons, so that the interaction with electrons is correspondingly stronger. The experimental consequence is that dielectric losses of this kind have rather small parallel momentum transfers, and so scattered electrons are confined to angles close to the specular direction. In spectrometers having a collection angle of a few degrees, it is a good approximation to assume that an effective angular integration of the loss spectrum occurs.[338] Performing the appropriate integration over k in Eqn 3.6 gives the total probability of loss at energy $\hbar\omega$ as

$$P(\omega) = \frac{2e^2}{h v_\perp \omega} \,\mathrm{Im}\, \frac{-1}{[\varepsilon(\omega) + 1]}. \qquad (3.7)$$

It can be seen that this function peaks at frequencies satisfying the condition given above (Eqn 3.3) for SO modes (Fuchs–Kliewer excitations). As shown in Fig. 3.2, these modes have frequencies slightly less than that of the bulk LO phonon.

The above discussion emphasizes that Fuchs–Kliewer excitations on insulators are determined by the bulk dielectric function; they therefore cannot be sensitive to detailed changes in bonding or lattice dynamics at the surface. The energy difference between bulk and surface optical phonons arises purely from electrostatic effects due to the different boundary conditions. Fuchs–Kliewer modes are surface localized, but, because of their long wavelength under normal excitation conditions, their exponential decay into the bulk of the solid is very long, typically over 100 Å.

Equation 3.7 is strictly valid only at absolute zero; at finite temperatures a correction for the Bose–Einstein thermal distribution of phonons is required. The most obvious consequence is that weak energy gain peaks appear in HREELS, corresponding to the de-excitation of a phonon already present in thermal equilibrium. The loss and gain peaks in HREELS are thus analogous to the 'Stokes' and 'anti-Stokes' bands seen in Raman spectroscopy.

Another important aspect of HREELS is that the probability of energy loss may be quite high, sometimes as much as one-half that of elastic scattering. This means that there is also a significant probability that successive loss events can occur during the time that a given electron is interacting with the surface. Thus strong multiple loss peaks can be observed.[326] The origin of these is quite different from multiphonon absorption in IR spectra, which result from anharmonic effects.[316] Multiphonon spectra in HREELS can be derived from the dielectric theory just discussed. When a solid has only one sharp Fuchs–Kliewer mode, multiple loss peaks should have intensities given by the Poisson distribution

$$\boldsymbol{P}_n \;\; = \;\; \frac{\mathrm{e}^{-x}\,x^n}{n!}, \tag{3.8}$$

where x is the predicted single-phonon loss probability, and $\boldsymbol{P}_n$ gives the probability of an n-phonon excitation in the full spectrum. In more complicated spectra, multiphonon structure may be calculated using the following straightforward algorithm:[341] take the Fourier transform $\mathbf{p}$ of the single-phonon loss spectrum calculated from the dielectric function by Eqn 3.7. Then the complete multiphonon spectrum is obtained as the Fourier transform of exp $(\mathbf{p})$. Inverting this procedure allows multiphonon structure to be removed from an experimental spectrum, as discussed later.

Figure 3.5 shows a HREELS spectrum for the single-crystal MgO (100) surface[320,324] and illustrates the principles just outlined. The elastically scattered electrons generate the peak at zero loss energy, and the loss peak corresponding to a single phonon created by the interaction of the incident electron with the surface lies at 651 cm^{-1}. Note that the single-phonon loss peak is one-fourth as large as the elastic peak, indicating the extremely strong cross-section for the electron–phonon scattering process. The other loss peaks are each at integer multiples of the single-phonon loss peak and correspond to multiple phonon excitations. A small surface phonon gain peak can also be seen at 651 cm^{-1} to the left of the elastic peak. The experimental spectrum is compared with a fit using the dielectric theory.

Only one fundamental surface phonon frequency is seen in the spectrum in Fig. 3.5 due to the high symmetry of the rocksalt (100) surface. In general, however, several different frequencies, and their associated multiple phonon losses, could be present in the spectrum. The dielectric theory of HREELS of anisotropic dielectrics is rather more complicated than that discussed above for isotropic com-

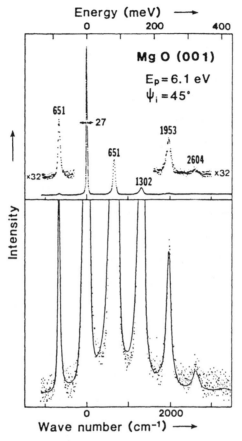

Fig. 3.5 HREELS spectrum of 6.1 eV electrons reflected from MgO (100). The lower panel shows the same spectrum on an expanded scale, with a fit to the data using dielectric theory superimposed. [Ref. 324]

pounds.[324,325] Consider, for example, a crystal with three principal components ε_{xx}, ε_{yy} and ε_{zz} of the dielectric tensor. Suppose that z is the surface normal, with electrons incident in the x–z plane. Then the dielectric function $\varepsilon(\omega)$ in Eqn 3.6 should be replaced by the effective value given by

$$\varepsilon' = \frac{\varepsilon_{zz}}{k}\left\{\frac{\varepsilon_{xx}\varepsilon_{yy}}{\varepsilon_{zz}^2}k_x^2 + \frac{\varepsilon_{yy}}{\varepsilon_{zz}}k_y^2\right\}^{1/2} \tag{3.9}$$

where k_x and k_y are the wave-vector components of the mode excited. Slightly simpler expressions can be given for uniaxial crystals, but in general one expects to see modes with all possible polarizations, the relative intensities of these depending on the crystal face and the direction of the electron beam.[324,325]

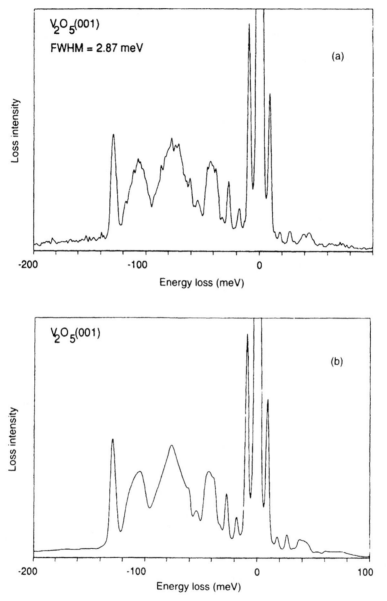

Fig. 3.6 (a) Experimental HREELS spectrum for V_2O_5 (001), and (b) theoretical simulation, optimized by adjusting optical parameters and adding three supplementary vibrational modes. [Ref. 342]

Figure 3.6 (a) shows an experimental HREELS spectrum for the low-symmetry V_2O_5 (001) surface, which is much more complicated than that for MgO (100).[218] In this complex, layered orthorhombic crystal structure, there are 39 bulk optical phonon modes! Nine different fundamental surface phonon frequencies can be seen in the experimental data; several phonon gain peaks are also visible on the opposite side of the elastic peak. Figure 3.6 (b) shows a calculated single-loss simulation of the spectra, which is seen to be in excellent agreement with experiment. Note that the intensities of the fundamental surface phonon losses in V_2O_5 are much smaller relative to the elastic peak than those in MgO; for this reason, no discrete multiple phonon losses are visible in the spectrum. However, multiple phonon losses contribute to the broad spectrum of losses above 150 meV.

Other HREELS measurements of Fuchs–Kliewer excitations are given in Table 3.2. As with the examples discussed above, good agreement is generally found with the predictions of the dielectric theory.

3.2.3 The influence of electronic carriers

Section 3.2.2 was concerned with insulating oxides containing no electronic carriers. As will be discussed in detail in Chapters 4 and 5, some oxides are metallic or can be doped to make them semiconducting. The screening effect of electronic carriers has an important influence on the HREELS spectrum, generally reducing the amplitude of optical phonon excitations compared to insulators.[344,345] An excellent example of the role played by conduction electrons in screening the electric field associated with surface phonons, and hence in decreasing the strength of the coupling of the phonons to low-energy electrons in HREELS, can be seen in experiments on Na_xWO_3.[343-346] Pure WO_3 is a good insulator, and the surface optical phonon losses exhibit an intensity comparable to that of the elastic peak. However, the addition of Na produces free carriers that can move in response to the internal electric fields and screen them. As shown in Fig. 3.7, the phonon loss peaks in $Na_{0.7}WO_3$ are one to two orders of magnitude smaller in intensity than they are in WO_3. The same phenomenon has been seen in Sb-doped SnO_2, where Sb doping produces free conduction electrons in normally insulating SnO_2.[332,345]

The simplest approach to the HREELS of metallic compounds is to add a term $\varepsilon_c(\omega)$ to the dielectric function given in Eqn 3.1 to represent the effect of electronic carriers. This is often approximated by the Drude form

$$\varepsilon_c(\omega) = \frac{-\omega_p^2}{\omega^2 - i\omega\gamma_p}, \qquad (3.10)$$

where ω_p is the (unscreened) bulk plasma frequency and γ_p is the conduction electron relaxation rate.[110] In the nearly-free-electron approximation, ω_p is related to the carrier concentration, n, and effective mass, m^*, by

$$\omega_p^2 = \frac{ne^2}{\varepsilon_0 m^*}. \qquad (3.11)$$

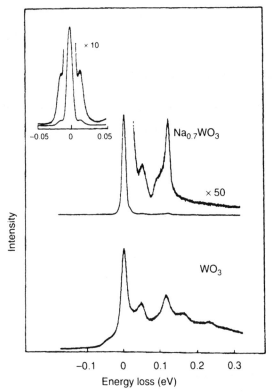

Fig. 3.7 HREELS spectra of WO_3 and $Na_{0.7}WO_3$. [Ref. 344]

An important consequence in energy-loss spectroscopy (ELS) is the appearance of surface plasmon excitations, as discussed in Chapter 4. However, the inclusion of such a term ε_c in the bulk dielectric function is not satisfactory in the phonon region, as the surface phonon losses, although weaker than in an insulator, are nevertheless considerably stronger than predicted. It appears therefore that these losses are excited predominantly in a surface layer, where the dielectric screening by conduction electrons is less efficient than in the bulk.[344] This could arise either because of the finite screening length, or because of a carrier-depleted surface layer.

The theoretical treatment of these surface screening effects has not reached a fully satisfactory level. One approach has been to use a simple two-layer model consisting of an unscreened, carrier-free surface dielectric layer on top of a metallic substrate.[344] That model accounts reasonably well for both the energies and intensities of the surface phonon losses observed in Na_xWO_3, but it runs into conceptual difficulties for flat-band materials.[351] A more sophisticated model combining the hydrodynamic dielectric function with a semi-classical infinite-surface-barrier model has also been used and is in quantitative agreement with HREELS phonon spectra, but it can only be used for flat-band materials and it does not describe electronic excitations of the material entirely correctly.[345]

A rather different effect can be found in ZnO. HREELS measurements have been made on the $(000\bar{1})$-O surface on which an n-type accumulation layer had been induced by exposure to atomic hydrogen, and then on the same surface after the accumulation layer had been removed by exposure to O_2.[327-329] The hydrogen-exposed surface exhibited two-dimensional conduction-band plasmons that coupled strongly to the surface phonons, while O_2 exposure eliminated the plasmon excitations. A detailed theory of plasmon–phonon coupling was developed that is in good agreement with experiment. Carriers can also be introduced into TiO_2 (110) surfaces by ion bombardment or high-temperature treatment. The creation of surface defects is associated with an increase in the conduction electron density at the surface. The HREELS of these surfaces has also been modelled by adding an electronic term to the dielectric function.[333,334] Unlike ZnO, however, the carriers are not free-electron-like, but occupy localized defect levels, probably associated with small-polaron formation, as discussed in Chapter 5.

The discussion above has concentrated on HREELS, but the screening of optical phonons also has important consequences in IR. Figure 3.8 (a) shows the reflectance spectrum of $Na_{0.6}WO_3$ measured in air.[352] The high metallic reflectivity, and the weak dips at around 750 and 950 cm^{-1} associated with TO and LO phonons, contrast greatly with the Reststrahl spectrum found for insulators (see Fig. 3.3). The experimental spectrum has been compared with calculations using both the two-layer and hydrodynamic models mentioned above [Fig. 3.8 (b)], and quite good agreement is found. However, the appropriate screening length (around 10 Å) is much larger than that required to fit the HREELS spectrum (around 1 Å). The difference may arise because the HREELS spectrum is measured in UHV and the IR in air: thus the 10 Å 'screening length' may represent the thickness of an actual depletion layer associated with O_2 adsorption. Similar results have been found for Sb-doped SnO_2.[353,354]

An important conclusion of this section is that both vibrational HREELS and IR reflectance measurements have considerable potential for investigating the electronic structure of oxides where carriers are present. A fuller exploitation of this area will depend partly on improvements in the theoretical modelling of the dielectric properties of surfaces with non-uniform carrier distributions.

3.2.4 Microscopic surface phonon modes

Rayleigh and Fuchs–Kliewer modes may be regarded as 'macroscopic' excitations in the sense that they can be predicted from the continuum (elastic or dielectric)

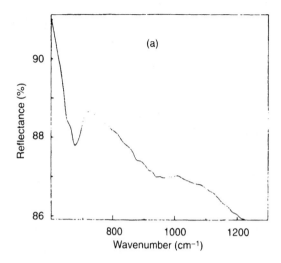

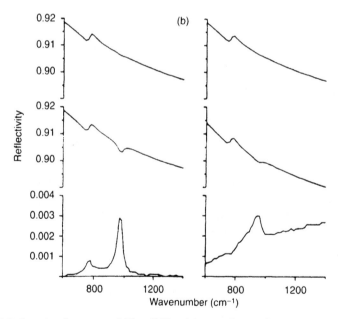

Fig. 3.8 Infrared reflectance of $Na_{0.6}WO_3$: (a) experimental spectrum measured in air; (b) calculated spectrum, using two different approximations (two-layer model left, hydrodynamic model right) and different screening lengths (1.2 Å above, 10 Å below). [Ref. 352]

properties of a solid, modified by the presence of a surface boundary condition. True **microscopic surface modes** are those which depend on details of the lattice dynamics of surface ions, possibly modified by changes in structure or bonding at the surface. Rather little work on such phonons has been reported for oxides.

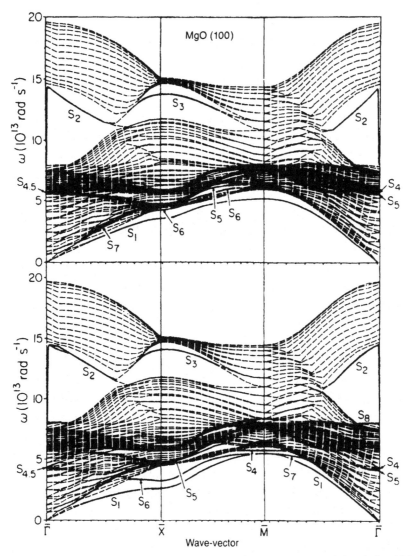

Fig. 3.9 Theoretical phonon dispersion curves for a 15-layer thick MgO (100) slab. Different shell models were used for the upper and lower panels. [Ref. 355]

Lattice dynamics calculations have been performed on finite-thickness slabs of MgO in order to determine the energy and dispersion of surface phonon modes.[287,288,295,355-357] Slabs containing about fifteen layers of atoms are usually sufficiently thick to give good values for the surface modes due to the short distance of penetration of the surface modes into the crystal. An example of the resulting dispersion relations is shown in Fig. 3.9, which presents the results of slab calculations using two different shell models for the inter-ionic potentials for

MgO (100).[355] The discrete surface phonon modes are marked, with mode S_1 being the Rayleigh mode referred to in § 3.2.1 above. Mode S_2, which has non-zero energy at the Γ point, is an optical surface phonon mode. The unlabelled dispersion curves are those of the bulk phonon modes.

Microscopic phonon excitations could be studied by neutral-atom scattering, or by off-specular HREELS when the strong dipole excitation of Fuchs–Kliewer modes is suppressed. Two off-specular HREELS measurements on MgO (100) have been reported, but they are in disagreement over both the energy and assignment of the loss peaks. One group saw a strong excitation at about 45 meV, which is assigned to the S_4 mode predicted by the slab calculation just discussed (see Fig. 3.9).[321] The other experiment, performed with higher resolution in both loss energy and wave-vector, showed a peak corresponding to the S_2 mode.[322] The discrepancy between the two spectra might be related to differences in surface preparation (air-cleave followed by UHV anneal, versus UHV-cleave) but deserves further exploration.

One cannot generally expect to study microscopic phonons on insulators by IR. Even if they are dipole active, their interaction with radiation will be very weak compared with the bulk effects discussed in § 3.1.2. Another problem can be understood by examining Fig. 3.9. Surface phonon modes generally appear in gaps in the projected bulk phonon spectrum or at energies below the bulk modes; they rarely occur at energies above the highest-lying bulk modes. [There is even a theorem, due to Rayleigh, prohibiting modes from having energies above the highest bulk modes, providing that the effective interatomic potentials (i.e., 'spring constants') are no larger at the surface than they are in the bulk.[318]] Since the infrared photons couple so strongly to bulk optical phonons, any possible contribution of the surface modes would be completely overwhelmed by absorption by bulk modes having the same frequency. The situation may be different in metallic oxides, where the excitations seen in both IR and HREELS are confined to a narrow surface layer, as discussed above.

One reason for interest in surface phonons is their relation to structural phase transitions and segregation effects. Some studies of surface segregation on metal-oxide surfaces have used calculations of the surface vibrational modes as a guide to the stability of a particular surface structure. Figure 3.10 shows the results of a slab calculation performed on an MgO (100) surface covered with a monolayer of segregated Ca ions, allowing a relaxation of the surface structure but retaining the (1 × 1) periodicity of the bulk.[287] The Rayleigh mode (the branch labelled '1') is seen to become soft, having an imaginary frequency over much of the Brillouin zone. This surface should thus restructure to one having a larger surface unit cell and lower energy. In this particular study it was suggested that a more appropriate surface structure would be a c($\sqrt{2} \times \sqrt{2}$) R 45° with one-half of the surface O ions

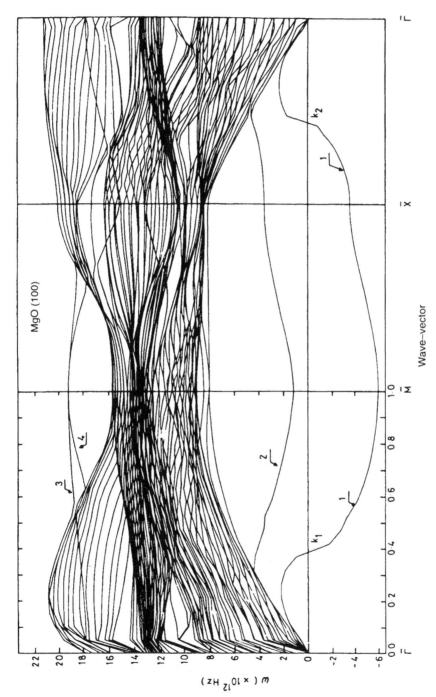

Fig. 3.10 Theoretical phonon dispersion curves for a monolayer of Ca on the surface of a 15-layer MgO (100) slab. [Ref. 287]

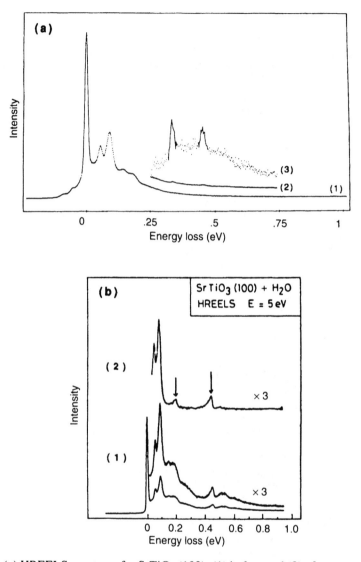

Fig. 3.11 (a) HREELS spectrum for SrTiO$_3$ (100): (1) before and (2) after exposure to 50 L of H$_2$O; (3) difference between (2) and (1). (b) HREELS spectrum for SrTiO$_3$ (100): (1) after exposure to H$_2$O but not deconvoluted, and (2) after deconvolution and smoothing to remove phonon overtones. [Refs. 339 and 341]

pushed high out of the surface. Even for that model, however, a small phonon anomaly remained in the lattice dynamics calculations, suggestive of a still longer-range rumpling of the surface.

3.2.5 The study of adsorbates

From the point of view of surface chemistry, the main reason for measuring the vibrational spectra of a surface is to study adsorbates, and to infer information about their structure and mode of bonding. IR spectroscopy has been widely used in this way on polycrystalline oxide catalysts,[358] but not apparently so far on single crystals. Recent work on other insulating substrates [for example, CO on NaCl (100)[359]] suggests that it should be possible to use IR reflection/absorption measurements to look at adsorbate modes, as long as they are significantly higher in energy than the optical phonons of the substrate.

HREELS is another important method for measuring the vibrational frequencies of adsorbed species. One of the problems with using this technique on metal oxides, as in the case of IR spectroscopy, is the presence of strong intrinsic losses. The most favorable cases are probably metallic oxides, where the intrinsic optical phonon losses are sufficiently weak that adsorbate vibrations can compete effectively. The molecular adsorption of water on $Na_{0.7}WO_3$ was studied in this way, and three adsorbate bands were seen: the symmetric stretching and bending modes of H_2O, and the vibration of water as a whole.[343] The (dipole-based) selection rules in this situation are the same as for elemental metal substrates: the vibrations appearing are those for which the dipole moment has a component normal to the surface.

On insulating oxides the loss peaks due to adsorbed moieties generally have an intensity only a few percent of that of the elastic peak, and their loss energies lie in the range of the multiple surface phonon losses. The presence of strong multiple-phonon loss peaks in the HREELS spectra makes it extremely difficult to see the adsorbate loss peaks, and attempts at separating them by subtracting the spectrum for a surface before adsorption from one after adsorption have been only partially successful. The result of that process is shown in Fig. 3.11 (a) for H_2O adsorption on $SrTiO_3$ (100).[339,340]

An alternative method has been developed that takes advantage of the theoretically predicted intensity variation of multiple-phonon loss peaks to remove them from the spectrum by means of Fourier transformation and deconvolution: this is essentially the reverse of the algorithm used to predict multiphonon spectra that was mentioned in § 3.2.2.[341] The single-phonon loss peaks cannot be removed, of course, but most of the important adsorbate loss features lie at energies in the vicinity of the two- to five-phonon losses. The effectiveness of this approach is shown in Fig. 3.12 for NiO (100). (Note that the elastic peak is not shown in the upper trace.) The improvement in signal-to-noise that can be obtained in identifying adsorbed species is shown in Fig. 3.11 (b) for the same adsorption system as in Fig. 3.11 (a). Another example of the effectiveness of phonon overtone deconvolution in ELS spectra of

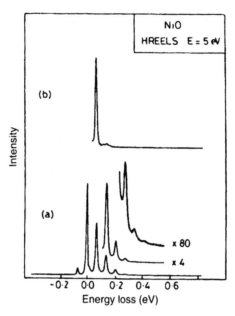

Fig. 3.12 HREELS spectra for NiO (100) (a) before and (b) after deconvolution and smoothing. [Ref. 341]

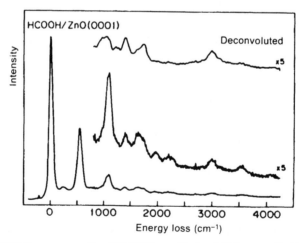

Fig. 3.13 HREELS spectrum for HCOOH on ZnO (0001) at 160 K before and after deconvolution. [Ref. 360]

adsorbates on oxides is shown in Fig. 3.13 for molecular HCOOH adsorption on ZnO (0001) at 160 K.[360] Here both phonon–phonon and phonon–adsorbate combination losses have been removed by means of the deconvolution process, permitting much more accurate identification of adsorbate loss energies.

The deconvolution method has only been used on a few systems to date, and it is not clear whether it will be as effective for all surfaces. But some such procedure is necessary in order to be able to accurately study adsorption on the surfaces of ionic materials by using HREELS. Alternative experimental methods for reducing the amplitude of phonon loss peaks (and especially of overtones) also show some promise in this area. One approach is to use higher-energy incident electrons (> 40 eV) and off-specular scattering;[361] another is to study thin oxide films on a metal. For example, the vibrational modes of H_2O, D_2O and CH_3OH adsorbed on an epitaxial MgO (100) film grown on metallic Mo have been successfully studied be HREELS.[323]

4

Electronic structure of non-transition-metal-oxide surfaces

Metal oxides exhibit a much wider range of electronic properties than they do surface geometries. Their electrical behavior ranges from the best insulators (e.g., Al_2O_3 and MgO) through wide-bandgap and narrow-bandgap semiconductors (TiO_2 and Ti_2O_3, respectively) to metals (V_2O_3, Na_xWO_3 and ReO_3) and superconductors (including reduced $SrTiO_3$, which is superconducting below 0.3 K, and high-T_c Cu oxides). Some metal oxides are ferroelectric ($BaTiO_3$) and others antiferroelectric (WO_3). Their magnetic properties include ferromagnetism (CrO_2), antiferromagnetism (FeO, NiO, etc.), paramagnetism (RuO_2) and complex mixed magnetic behavior ($Y_3Fe_5O_{12}$). Several metal oxides exhibit metal/non-metal transitions as a function of temperature (V_2O_3, VO_2, Ti_2O_3) or composition (Cr-doped V_2O_3). This range of behavior is clearly important in the context of surface properties. Surface electronic structure is interesting not only for its own sake, but also because it has a more-or-less direct influence on other behavior such as chemisorption.

The range of electronic structures of oxides is so wide that it is convenient to divide compounds into two categories. Non-transition-metal oxides, in which the cation valence orbitals are of s or p type, are more straightforward than transition-metal oxides, where the cation valence orbitals have d symmetry. The former are described in the present chapter; transition-metal oxides form the subject of Chapter 5.

4.1 Models of electronic structure

The non-transition-metal oxides discussed here can be divided into two groups:

(i) pre-transition-metal compounds, such as the alkaline-earth oxides and Al_2O_3, which are wide-bandgap insulators; and
(ii) post-transition-metal oxides ZnO and SnO_2, which have intrinsic bandgaps around 3–4 eV, but can become n-type semiconductors (or even metallic in the case of SnO_2) as a consequence of oxygen loss or doping with suitable donor impurities.

The following sub-sections review some aspects of bulk electronic structure and then discuss the ways in which the surface may be different from the bulk.

4.1.1 Bulk electronic bands

Conceptual approaches to the electronic structure of oxides may be divided into two categories. On the one hand, we may take a **localized** view of the electron configurations of atoms or chemical bonds.[362] The simplest of such pictures is the **ionic model**, widely used as a zero-order approximation. For MgO, this suggests that the filled orbitals of highest energy are the filled O 2p on O^{2-}, and that the lowest unoccupied ones are the empty Mg 3s orbitals on Mg^{2+}. A localized viewpoint is not necessarily a fully ionic one, however: it is possible to think of occupied bonding levels, formed from a mixture of metal and oxygen orbitals, and corresponding empty antibonding orbitals. This may be a preferable way of considering post-transition-metal oxides, where a great deal of chemical and physical evidence suggests that their ionic character is less than that of the alkaline-earth oxides. As far as its valence electrons are concerned, ZnO is essentially isoelectronic with other tetrahedral solids such as Si and GaAs. Localized orbitals may be constructed from tetrahedral sp^3 hybrids on both Zn and O; these overlap, forming occupied bonding and unoccupied antibonding combinations. Because of the very different electronegativities of Zn and O, the bonding orbitals are concentrated on the O atoms and the antibonding ones on Zn. The ionic model can therefore be regarded as the limit where bonding and antibonding orbitals become completely localized on oxygen and metal atoms, respectively.

A superficially very different picture is that of **band theory**. Here the occupied and unoccupied orbitals are regarded as fully extended Bloch states, constructed in accordance with the translational symmetry of the lattice.[110] An 'ionic' band structure would have filled (valence) bands composed of combinations of O 2p and (deeper in energy) 2s orbitals, while the empty (conduction) band would be made up of combinations of metal orbitals: for example, Mg 3s and 3p, or Zn 4s and 4p. As with the localized models, however, the band picture can incorporate variable degrees of mixing or **hybridization** between these bands, so that the valence band is in general composed of metal–oxygen bonding combinations of orbitals, and the conduction band of antibonding combinations.[362]

(The term 'hybridization' is used somewhat differently in the chemistry and physics literatures. To solid-state physicists, this word means any type of mixing of different atomic orbitals: for example, between the O- and Zn-based orbitals in the bonding in ZnO. Chemists, however, generally reserve the concept of hybridization to refer to mixing of atomic orbitals on the *same* atom: for example, the formation of sp^3 tetrahedral hybrids for use in the construction of bonding and antibonding orbitals in a tetrahedrally bonded solid such as diamond. The mixing of orbitals on different atoms would normally be called 'covalency' by a chemist. In this book we shall adopt the physicists' convention and use hybridization in a more general sense, although the term covalency will also be used, e.g., in the context of 'surface-enhanced covalency'.)

Although the localized and delocalized pictures appear to be fundamentally different, it is important to realize that they represent two alternative, and equally valid, views of the ***ground state*** of a closed-shell non-metallic solid. This is because the complete many-electron wavefunction, represented approximately in these pictures as a set of occupied single-electron states, is invariant under any non-singular linear transformation among these states. Occupied Bloch states in the band model can be constructed as linear combinations of localized orbitals; conversely, we can, if we wish, form localized combinations ('Wannier functions') from band states. The equivalence of the two viewpoints disappears, however, when an electron is removed or excited. This is important since any experimental measurement of the electronic levels in a solid must involve a perturbation of this kind. It now makes a difference whether we regard a hole as a missing electron in a Bloch state extending throughout the solid, or in one ion or localized bonding orbital. Simple 'independent-electron' interpretations of ionization and excitation experiments naturally tend to use the band picture, but there are many factors which may make such excitations more localized in character than band theory suggests: these include the Coulomb interaction between electrons themselves, and the interaction of electrons with lattice vibrations and defects.[40] We shall return to this problem at later stages. The important conclusion for the moment is that band and localized types of models may give equally valid representations of the ground states of the compounds we are considering, but that one needs to be very careful in using these different models to interpret experimental results.

The different models may be illustrated using MgO as an example. Figure 4.1 shows how the valence- and conduction-band levels could be derived using an ionic picture.[40,362] Column (a) shows the energy levels for electrons in the free ions. The O^{2-}/O^{-} level is at around 9 eV, *positive* with respect to the vacuum level because O^{2-} is highly unstable as an isolated entity. The other level, Mg^{+}/Mg^{2+}, is the negative of the second ionization energy of Mg. In order to stabilize the normal ionic configuration, it is necessary to put the ions into a lattice so that they experience the **Madelung potential** due to other ions. The magnitude of this potential is proportional to the Madelung constant used in lattice-energy calculations; the value is positive (giving a negative energy contribution for electrons) at O sites which have near-neighbor Mg^{2+} ions, and negative (destabilizing for electrons) at Mg sites.

The bandgap of MgO estimated with the Madelung potential alone is around 24 eV, very much larger than the experimental value of 7.8 eV. In order to get the

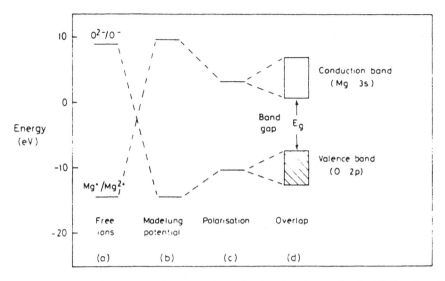

Fig. 4.1 Derivation of the valence- and conduction-band energies in MgO using the ionic model. Cation and anion energy levels are shown (a) for free ions; (b), (c), and (d) including Madelung potentials, relaxation, and bandwidths, respectively

correct magnitude, two further effects need to be considered [see columns (c) and (d) in Fig. 4.1]. The first is the electrostatic **polarization**, resulting from the deformation of electron distributions in an electric field. It is important in any solid when electrons are removed or added. Polarization lowers the energy of an added electron as shown; it also makes it easier to remove an electron, and so must be represented by raising the energy of an occupied level. Polarization terms may be estimated at various levels of sophistication, ranging from continuum electrostatic models to shell-model calculations of the kind referred to in Chapter 2. The other effect is the **bandwidth**, originating from the overlap of orbitals on neighboring ions. The magnitude of this cannot be estimated within the context of the localized ionic model alone; band-structure calculations or experiments are needed. Most estimates of the O 2p bandwidth are around 6 eV, and we may expect a somewhat larger value for the conduction bands based on the more diffuse Mg 3s orbitals.

The ionic model is at best semi-empirical, and certainly unsuitable for first-principles estimates of energy levels. This is particularly true when more-highly charged ions are involved, such as Al^{3+} or Sn^{4+}. The terms in the calculation now become individually much larger, and the final bandgap may be the difference between numbers as large as 50 eV or more. Nevertheless, all of the oxides discussed here have appreciable ionic character, and the different steps in the argument just given are important in considering the effect of defects and surfaces. It is crucial to note that O^{2-}, regarded as the 'normal' state of oxygen in an ionic lattice,

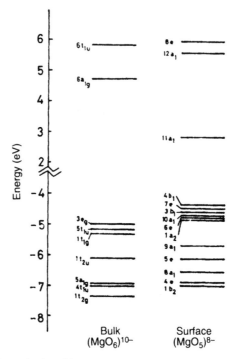

Fig. 4.2 Orbital energies calculated by the DV-Xα method for MgO clusters, representing the bulk (left) and the (100) surface (right). [Ref. 363]

is only stable by virtue of the Madelung potential. At surfaces, where this potential is smaller, we might expect a smaller bandgap, or even the stabilization of lower-charge states such as O^-. But even in a perfectly ionic solid, the Madelung potential is not the only factor to consider; polarization and bandwidth terms are also significant.

As a second type of model for MgO, consider the cluster calculation shown in Fig. 4.2.[363,364] The levels of an octahedral $[MgO_6]^{10-}$ cluster were calculated by the discrete-variational (DV-Xα) method. (Some essential aspects of the different computational methods are discussed in § 4.1.5 below.) The empty orbitals labelled $6a_{1g}$ and $6t_{1u}$ are essentially Mg 3s and 3p, respectively, with a small admixture of O 2p. The various levels between –5 and –8 eV are the occupied orbitals, mostly O 2p but with some bonding contribution from Mg. Clearly the full bandwidths cannot be estimated from a small cluster such as this, especially as only one Mg ion is present.

Finally, consider the band-structure calculation shown in Fig. 4.3.[369] This now shows the energies of occupied and empty Bloch states as a function of wave-vector within the Brillouin zone of the fcc lattice.[110,362] Their atomic character is essentially the same as that found in a cluster calculation, but more information is now

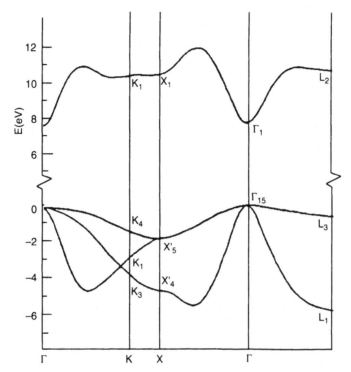

Fig. 4.3 Bulk band structure of MgO calculated by the LCAO (tight binding) method. Only the upper part of the valence band and the lower part of the conduction band (O 2p and Mg 3s contributions, respectively) are shown. [Ref. 369]

conveyed. Not only is the full width of the bands represented, but also their dispersion, which may be important (if the band model is valid here) in understanding various *k*-allowed electronic transitions, for example in angle-resolved photoemission. The dispersion of states in band theory can be interpreted in terms of chemical bonding and the overlap of atomic orbitals, as discussed in Chapter 4 of Ref. 362.

The upper part of the valence band in Fig. 4.3 consists of states with very little bonding character, whereas the states near the bottom of the band have been stabilized by the bonding interaction with Mg 3s and 3p orbitals. Many oxides have UPS spectra that show a rough division of the valence-band density-of-states into two regions. Sometimes these are referred to as 'bonding' and 'non-bonding' parts of the valence band, but the labels 'σ' and 'π' are also used in this context. The latter nomenclature implies division of the band into σ-bonding states which are more strongly bonding and at lower energy than the π-bonding ones. The division of the valence band this way is sometimes useful but is not at all rigorous; in the rocksalt structure in particular, the O 2p orbitals cannot be divided into ones with σ and π overlap, as this notation suggests.

More details of the electronic structure will be presented later in the discussion of individual compounds. Many of the essential features, however, are similar to those found for MgO. Post-transition-metal oxides are different in two important respects. Firstly, the bandgaps are smaller. This can be understood in the ionic model from the higher ionization energies of these atoms: compare Zn (second ionization energy = 18.0 eV) with Mg (15.0 eV). But the smaller bandgap also has the consequence that there is less ionic character, as the metal- and O-based orbitals are closer in energy and so hybridize more efficiently. Another difference is the presence of the filled d levels, Zn 3d and Sn 4d. From a chemical point of view these can normally be regarded as core orbitals, not involved in bonding. In ZnO this is only marginally true, and both calculations and UPS measurements show that Zn 3d overlaps in energy with the lower part of the O 2p valence band.[365]

Another crucial difference with some post-transition-metal oxides is the existence of a stable lower oxidation state.[39] The only example among the oxides treated here is Sn, which has a stable SnO oxide as well as SnO_2. In an ionic picture, Sn^{2+} has a $5s^2$ electron configuration. However, the stabilization of 5s relative to its normal position as a conduction band level is always associated with a change in geometry of the Sn; as noted in Chapter 2, the coordination of s^2 post-transition-metal ions is generally distorted. Such a distortion leads to some mixing between Sn 5s and 5p levels, so that the non-bonding orbitals are best regarded as hybrid states (in the chemical sense). The relative energies of these filled Sn orbitals and the O 2p valence states are quite significant in understanding the surface electronic properties of SnO_2 (see § 4.6). The evidence from both experimental and theoretical studies is that these levels have similar energies.[366]

4.1.2 Bulk defects and doping

The fact that all solids contain defects – and indeed that these are *required* in equilibrium by thermodynamic arguments – was mentioned in § 2.1.2, where some different types of structural defects were also described briefly. Defects are important because they generally give rise to electronic levels within the bandgap of the solid.[315,362] The electronic properties associated with defects depend upon both the energies of these levels relative to the band edges of the perfect crystal and the number of electrons occupying the defect levels. Particularly significant in the present context are defects which are associated with electronic carriers (extra electrons or holes).

Two simple types of structural defect that have been studied in MgO are the O vacancy or **F center** and the cation vacancy or **V center**. A neutral O vacancy must have two electrons remaining, which may be trapped by the unbalanced

Coulomb potential associated with the vacancy.[41] Removing one of these electrons gives the **F⁺ center**. F and F⁺ centers may be characterized by a variety of spectroscopic methods, including their optical absorption bands at 5.01 and 4.95 eV, respectively.[42] With an F⁺ center the other electron must be somewhere in the solid in order to preserve overall charge neutrality. Indeed, in a solid containing just one F⁺ center, we would expect this electron to be held fairly close by. That would normally be at the site of another defect, for example at a V⁻ center described below. One of the difficulties with all defect studies, however, is that the types of defect that compensate the particular defect under study are generally unknown.

The **V⁻ center** is a Mg^{2+} vacancy, with one associated electronic hole. The hole may be regarded as an O^- ion and is especially interesting since similar defects are probably common at surfaces. Spectroscopic studies of the V⁻ center show that the electronic hole is not delocalized over the oxide ions surrounding the vacancy, as might be deduced from symmetry arguments; on the contrary, in the ground state it is trapped at a single O^- site.[41] This localized configuration is stabilized by a distortion and polarization of the surrounding lattice, a situation analogous to the small polarons described in Chapter 5. Structural relaxation or rearrangements are important with all types of defect and are one of the many factors that complicate their study.

Stable defects in pre-transition-metal oxides generally give rise to deep impurity levels having energies at least 1 eV away from the band edges. They are important in spectroscopy but have very little influence on transport properties at room temperature. In contrast, similar defects in post-transition-metal oxides often have quite small carrier binding energies, and so thermal carrier ionization leads to semiconducting properties. This is the case with ZnO and SnO_2, which in their nominally 'pure' state are n-type semiconductors. SnO_2 has a proportion of O vacancies present, and the same may be true of ZnO, although there is a great deal of complexity in the defect chemistry there.[260] Experimentally it is established from many studies that the true stoichiometry should really be written as $Zn_{1+\delta}O$, with δ having a value ranging up to 10^{-3}, depending upon the conditions of preparation. This composition could be consistent with the presence of either O vacancies or Zn interstitials. The n-type conductivity arises because of the extra electrons, which are bound in the ground state but with a low donor ionization energy of around 0.04 eV. However, there is disagreement about the majority defect type, and it may be that both Zn interstitials and O vacancies are present. There are also indications from transport measurements of the presence of *compensating* defects in the form of either (or both ?) O interstitials and Zn vacancies, although at smaller concentrations.

Whatever the defects present, it is crucial that in both ZnO and SnO_2 n-type semiconductivity is associated with oxygen deficiency. The equilibrium defect concentration, and hence the number of carriers, therefore depend not only on the temperature, but also decrease with the ambient oxygen partial pressure.[41,260] The defect and carrier concentrations are frequently fitted to an equation of the form

$$ x \;=\; A\,p(O_2)^{-\alpha}\,e^{-E_a/RT}, \qquad (4.1) $$

where $p(O_2)$ is the oxygen partial pressure, α a parameter depending on details of the defect equilibria, and E_a an energy term relating to defect formation and/or ionization energies. A similar equation may be used for surface defect concentrations, but it is important to remember that the conditions used to anneal samples in preparation for surface studies will also influence the *bulk* of these materials.

Defect levels and carriers in oxides may also result from impurities. The most interesting cases arise where a substitutional impurity changes the ionic charge at a lattice site. For example, replacing Mg^{2+} in MgO with Li^+ or Na^+ removes one electron from the neutral solid, thus creating a hole. In the ground state, this hole is in the form of O^- at a site neighboring the impurity. In n-type ZnO, substitutional Cu is probably present in the form of Cu^+, so that it acts as a compensating defect and removes some of the electrons normally present. In the case of SnO_2, additional n-type doping may be achieved by replacing some Sn by Sb (which behaves effectively as Sb^{5+}) or some O^{2-} by F^-. At concentrations above 5×10^{18} cm^{-3} the impurity levels in SnO_2 overlap sufficiently to give rise to a **semiconductor-to-metal transition**.[367] The electrons introduced by doping can now be regarded as occupying a free-electron-like conduction band, based on the Sn 5s atomic orbitals. As discussed in the following section, the conduction electrons give rise to a plasma excitation which, unusual for a metal, lies in the infrared. This fact, combined with a bandgap (3.6 eV) also lying outside the visible spectrum, makes doped SnO_2 a transparent metal and leads to various important applications.[368]

4.1.3 Electronic excitations

The most familiar types of electronic excitation in all non-metallic solids are those associated with defects and bandgaps. Both types of excitation have been studied extensively in oxides, and indeed optical absorption, over a photon energy range from the IR into the UV, is the most important source of information about the associated energy levels.[41] Most of the defects mentioned in the previous section have been studied in this way, and there is a particularly large literature on the spectroscopic transitions of various impurity ions.[375]

When defect concentrations are very low, one expects the electronic excitation of lowest energy to be related to the bandgap: i.e., the transition of one electron from the top of the valence band to the bottom of the conduction band. A list of

Table 4.1. *Bulk bandgaps in non-transition-metal oxides*

Compound	Bandgap (eV)
MgO	7.7
CaO	6.9
SrO	5.3
BaO	4.4
ZnO	3.4
SnO_2	3.6

Refs: 367 (SnO_2); 375 (others)

optical bandgaps of non-transition-metal oxides is given in Table 4.1. There are, however, a number of complications in these supposedly 'simple' bandgap transitions. In the first place, bandgaps may be classified as **direct** or **indirect** according to whether the lowest energy transition is allowed by the electronic *k* selection rule for electrons in a crystal.[362] It appears that the compounds listed in Table 4.1 have direct gaps, although the bandgap excitation in SnO_2 is forbidden because of the local symmetry of the highest-occupied and lowest-unoccupied states.[370] A more serious problem is the appearance of **excitons** near the absorption edge. These are essentially bound states of the electron and hole 'orbiting' in each other's Coulomb field. They are therefore manifestations of the breakdown of the independent-electron model. Tightly bound **Frenkel excitons** occur when the Coulomb field is strong compared to the bandwidth; the crystal-field excitations of some transition-metal oxides are of this type (see Chapter 5). In the compounds discussed in this chapter, the presence of relatively high dielectric constants and wide bands leads to weakly bound **Wannier excitons**, with binding energies of the order of 0.1 eV.[375] These have been studied in the alkaline-earth oxides and in ZnO. Low temperatures are generally required to study such weakly bound excitons, because they are obscured by thermal broadening of the band edge even at room temperature.

There is an important difference between spectroscopically observed energy levels and those appropriate to calculating the thermal distribution of electrons.[40,41] This is because spectroscopic transitions occur rapidly, and, according to the Franck–Condon principle, no structural relaxation can take place. Electronic polarization terms do contribute, but the excited state corresponds to one which is 'frozen' geometrically. In thermal equilibrium, however, all states will relax to their equilibrium geometries. This additional relaxation may contribute to both defect and bandgap transitions.[371] For example, defect-lattice calculations for MgO suggest a value of 6.1 eV for the bulk thermal bandgap of MgO, compared to the optical value of 7.8 eV.[254]

Transitions at much higher energies than the bandgap have been studied by absorption and reflection spectroscopy using synchrotron radiation.[371,372] At very high energies, transitions from various core levels may be observed, but the intermediate energy region (10–100 eV, say) seems to be particularly difficult to interpret. At a one-electron level, the absorption coefficient should be related to the **joint density-of-states function**, which can be calculated from the band structure. But poor agreement is generally found,[372] suggesting that electron correlation effects may be important. Another significant observation is that at higher energies the spectra seem to reflect less and less the electronic structure of the solid, becoming more like those expected for free ions.[371] These difficulties in interpretation are important in the context of surface studies because of the common use of ELS, as discussed below. That technique can easily be used to study excitations in the 10–100 eV range, and it is frequently unclear how the observed transitions should be assigned.

One type of excitation that is particularly strong in energy-loss studies is the **plasmon**.[362,374] All of the electronic excitations discussed so far involve only a single electron or a small number of electrons. There are also collective excitation modes that involve essentially all of the electrons in a particular band moving in unison relative to the nuclei. These plasma excitations are accompanied by macroscopic dipole moments that couple very strongly to electrons, and so they can appear in ELS spectra or as satellites accompanying core-level excitations in XPS.

The occurrence of plasmons, and the distinction between bulk and surface excitations, is best discussed in terms of simple dielectric functions (see Chapter 3).[110] In materials with mobile electrons that can be described by an itinerant-electron picture, one may write as a first approximation

$$\varepsilon(\omega) = \varepsilon(\infty) - (\omega_p/\omega)^2 . \tag{4.2}$$

This is Eqn 3.10 with the imaginary damping term omitted. $\varepsilon(\infty)$ is the background electronic dielectric function and ω_p is the unscreened plasma frequency (see Eqn 3.11), given in terms of the conduction-electron concentration (n) and effective mass ($m*$) by

$$\omega_p^2 = \frac{ne^2}{\varepsilon_0 m *} . \tag{4.3}$$

Bulk excitations correspond to the condition $\varepsilon(\omega) = 0$, which is true at a frequency ω_b given by

$$\omega_b^2 = \frac{\omega_p^2}{\varepsilon(\infty)} . \tag{4.4}$$

On the other hand, with low- or medium-energy electrons that do not penetrate a sample but are reflected from the surface, the excitation condition is given instead by $\varepsilon(\omega) = -1$ (see the discussion on ELS in § 3.2.2). Thus one observes a **surface plasma frequency** given by

$$\omega_s^2 = \frac{\omega_p^2}{[\varepsilon(\infty)+1]}.$$ (4.5)

The peaks appearing in an energy-loss experiment correspond to quantized excitations of these plasma frequencies, with an energy of $\hbar\omega$. The example of Sb-doped metallic SnO_2 is discussed later (see Fig. 4.31, where the surface plasmon corresponds to the loss feature at 0.6 eV). The presence of the background dielectric term $\varepsilon(\infty)$ in the above equations is quite important, since for SnO_2 it has a value ≈ 5. It considerably lowers the observed plasma frequency compared with the unscreened prediction (Eqn 4.3); also, the difference between bulk and surface values is much less than the ratio $\omega_b/\omega_s = \sqrt{2}$ often quoted.[376]

The existence of conduction-band plasma excitations is uncontroversial. There have, however, been suggestions that loss peaks at higher energies in MgO and ZnO might be associated with plasma oscillations of *valence-band* electrons. To see how these might arise, consider the following approximation to the dielectric function including an average electronic (bandgap) excitation frequency ω_g

$$\varepsilon(\omega) = \varepsilon(\infty) - \frac{\omega_p^2}{[\omega^2 - \omega_g^2]}.$$ (4.6)

The condition for a bulk plasmon $\varepsilon(\omega) = 0$ now gives

$$\omega_b^2 = \frac{\omega_p^2}{\varepsilon(\infty)} + \omega_g^2,$$ (4.7)

with a similar equation for the corresponding surface value. If the plasmon term on the right-hand side of Eqn 4.7 is considerably larger than the bandgap, plasma excitations may occur at energies not very much higher than those predicted without any bandgap.

The above discussion is based on a very approximate representation of the dielectric properties. Whether the conclusions are correct must depend on details of the electronic structure, such as the width of the bands (which determines the effective mass and hence the 'free' plasma frequency), the bandgap, and the real distribution of oscillator strength in one-electron transitions, assumed above to be concentrated at a single frequency ω_g. A loss peak apparent at 22.2 eV in the ELS spectra of MgO (shown in Fig. 4.11) has been assigned as a bulk valence-band plasmon.[377] The observed transition energy is very close to the value of 22.7 eV calculated in the free-electron approximation without any bandgap. Similar losses at 13.5–16 eV in the ELS of ZnO have been assigned to the analogous surface plasmons.[378–381] But another group has attributed excitations in this energy region to an interband 3d-to-conduction-band transition.[382] Thus the status of valence-band plasmons is uncertain; clearly it is bound up in the other difficulties involved in interpreting electronic excitations in the energy region above 10 eV.

4.1.4 Surface electronic structure

Our discussion of bulk electronic structure suggests a number of reasons why the electronic structure of surfaces might be different from that of the bulk. The coordination number of surface ions is smaller, leading to reduced Madelung potentials, smaller bandwidths, and the possibility of 'dangling bond' states in covalent materials. These effects could lead to reduced bandgaps, 'surface enhanced covalency' and various types of surface states. In addition, the types and concentrations of defects might be quite different than those of the bulk.

The importance of the Madelung potential to the bandgap of an ionic material was emphasized in § 4.1.1. Compared with the bulk rocksalt value of 1.747 65 for the **Madelung constant**, the value for an unrelaxed (100) surface is only slightly less: 1.681 55. For less stable surfaces with lower coordinations, much lower values can be obtained, and this has led naturally to suggestions that bandgaps might be lower on such surfaces.[383,384] Our discussion emphasized, however, that polarization and bandwidth terms are also crucial, and both of these effects will be smaller at surfaces. Thus, even in the ionic model one cannot easily guess the effect of lower coordination numbers. The existence of relaxations and reconstructions may also be important, as these may often act to stabilize filled levels and destabilize empty levels, as is found at many semiconductor surfaces.[385]

The later sections of this chapter will review in detail what is known about the electronic structure of perfect as well as defective surfaces of non-transition-metal oxides. In this section we will look briefly at two calculations for MgO surfaces, using the same methods as described in § 4.1.1 for the bulk solid. The techniques employed in these calculations are discussed below in § 4.1.5.

One study, which used Green's function and LCAO formalisms, calculated the one-electron surface energy bands for both MgO (100) and (110).[369] These are shown in Fig. 4.4, where the energy dispersion in the two-dimensional Brillouin zone appropriate to the surface lattice is superimposed on a projection of the three-dimensional band structure. For the (100) surface, Fig. 4.4 (a), a surface-state band is seen to extend about 0.5 eV above the top of the bulk O 2p valence band, while the Mg 3s surface band lies entirely above the bulk conduction-band minimum.

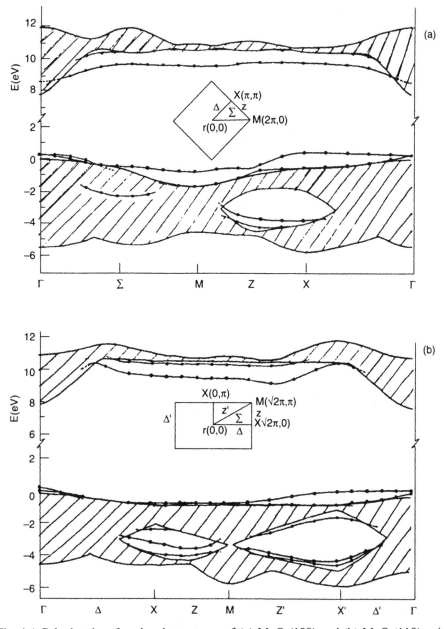

Fig. 4.4 Calculated surface band structures of (a) MgO (100) and (b) MgO (110), with a projection of the bulk bands shown as the shaded regions. The surface Brillouin zone is shown in each case. [Ref. 369]

No direct transitions involving surface states are predicted for energies less than 7.6 eV, which is close to the bulk optical bandgap of 7.8 eV; no indirect transitions exist for energies less than 7.3 eV.

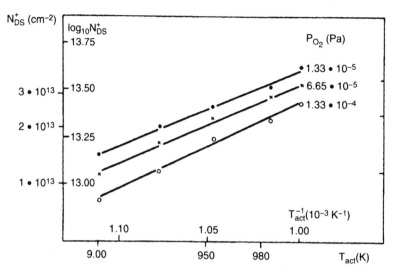

Fig. 4.5 Surface O-vacancy defect density on ZnO ($10\bar{1}0$) as a function of temperature and O_2 partial pressure. [Ref. 260]

The DV-Xα calculations for an MgO cluster were discussed above.[363,364] Figure 4.2 also shows the molecular-orbital energies for a cluster with one oxygen removed, intended as a simulation of the (100) surface. Note that the bulk bandgap in this calculation, defined as the difference in energy between the $3e_g$ and $6a_{1g}$ orbitals, is about 9.7 eV, 2 eV larger than the experimentally determined value; this is not uncommon for cluster calculations. Comparison of the bulk and surface calculations may nevertheless give an estimate for the narrowing of the gap at the surface. The surface bandgap (i.e., the difference between the $4b_1$ and $11a_1$ orbitals) is about 2.5 eV less than that of the bulk, primarily because of lowering of the lowest unfilled (predominantly Mg 3s) level. This bandgap narrowing at the surface comes from a combination of the reduction of the Madelung potential near the surface, differences between cation–anion charge transfer for surface and bulk ions, and a polarization of the wavefunction of the surface ions due to the potential gradient at the surface.[363]

Calculations such as these, and the various experimental measurements we shall describe later, generally indicate that the occupied electronic levels of nearly perfect surfaces of oxides are only slightly perturbed from those of the bulk. Thus with electronic structure (as with chemistry) the most interesting surface features are determined by the presence of defects. There are various reasons why defect concentrations on metal-oxide surfaces may be much higher than in the bulk. In the first place, defect formation energies may be lower on surfaces, so that their *equilibrium* concentrations are expected to be larger. As with bulk solids, such

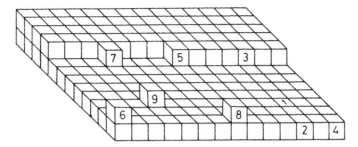

Fig. 4.6 Different defect positions on MgO (100) terraces and steps. See text for the surface Madelung potentials at the labelled sites. [Ref. 386]

concentrations will depend on the temperature and other variables such as oxygen pressure. An example of the dependence of surface defect density on temperature is shown in Fig. 4.5 for O-vacancy point defects on ZnO (10$\bar{1}$0).[260] These results may be fitted to an equation like (4.1), although the parameters for surface defects will be different. Note that the O-vacancy concentration is quite high, corresponding to surface coverages up to 0.03. But it is also important to realize that many surfaces may have many *more* defects than predicted by such equilibrium considerations. Preparation methods such as fracture, cleaving, etching, etc., can leave high concentrations of steps, corners, and point defects, and although annealing conditions are designed to remove these, they may be only partially successful.

The electronic structure of defects on oxide surfaces can be, and generally is, very different from that of perfect terraces. One simple way to see the origin of the effect is to consider the potential that an ion sees at various types of surface defect sites. The Madelung constant has been calculated for the sites on unrelaxed rock-salt (100) indicated by numbers in Fig. 4.6.[383,386] (Each small cube in this model represents a single ion, so adjacent cubes are ions with opposite charge.) As mentioned above, the Madelung constant for a five-fold coordinated terrace site (1), 1.681 55, is only slightly smaller than the bulk value. But it decreases rapidly as the ligand coordination of the surface site decreases. Sites (3), (5), (7) and (9) with four, three, two and one nearest neighbors respectively, have Madelung constants 1.566 93, 0.873 78, 0.180 63 and 0.066 01. These are values for an unrelaxed surface; the relaxation that occurs at surface defects will bring the Madelung constants closer to the bulk value. Furthermore, it has been emphasized above that Madelung potentials are not the only important factors in determining electronic levels, even in 'perfectly' ionic crystals. But these values serve to show that ions at defect sites on oxide surfaces see a very different environment from those on nearly perfect surfaces. One result may be to stabilize charge configurations that are not normally found in the bulk: for example, O$^-$ may be the usual form of oxygen at some low-coordination sites.

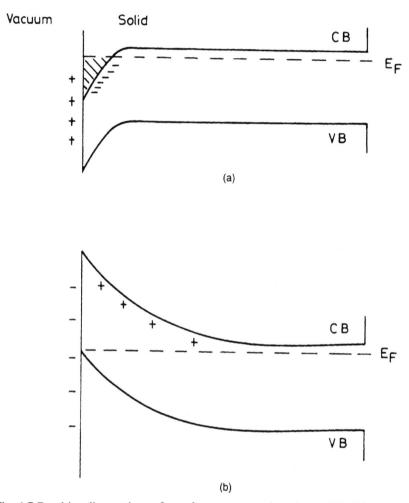

Fig. 4.7 Band-bending at the surface of an n-type semiconductor (a) with an accumulation layer, and (b) with a depletion layer.

Extended defects such as steps do not necessarily give rise to trapped electrons or holes in the same way as point defects do, since both positive and negative ions are removed in creating steps. Little is known either theoretically or experimentally about the electronic structure of steps on metal oxides, although some correlations have been made between the presence of steps and the chemisorption behavior of single-crystal oxides; this will be discussed in Chapters 6 and 7.

As in the bulk solid, the way in which surface defects influence the overall electronic properties depends on the energies of the defect levels with respect to the band edges and on their electronic occupancy. These factors are especially significant with semiconducting materials, where both bulk and surface defects provide easily ionizable carriers. Figure 4.7 illustrates the important phenomenon of **surface band-bending** associated with a change in surface carrier concentration on an n-type semiconductor.[362] An accumulation layer is a surface region where carrier concentration is higher than in the bulk, and it is associated with a downward bending of the bands at the surface, Fig. 4.7 (a). This could be formed by a high concentration of surface O vacancies, leaving extra electrons in the surface region. On the other hand, a depletion layer, which might occur if the surface were covered with adsorbed oxygen (as O_2^- or O^-), gives an upward bending of the bands, as shown in Fig. 4.7 (b). The band-bending in each case is caused by an unequal concentration of (positive) ionized impurity centers and (negative) electrons in the space-charge region. Conventional semiconductor theory predicts quantitative relationships between parameters such as the magnitude of the band-bending, the surface charge density, and the bulk doping level.[362] Unfortunately, the complexity of the defect chemistry of many metal oxides, referred to in § 4.1.2, generally makes it very difficult to apply this theory to oxide surfaces.[260]

Surface carrier concentrations obviously influence the electronic conductivity of surfaces. Even in supposedly 'bulk' studies this may complicate the measurement of conductivity in a polycrystalline ceramic material, as intergrain contacts may have very different conductivities from that of the true bulk solid. Chemisorption of many gases can influence surface conductivity, and this effect leads to the use of some semiconducting oxides as *sensors*.[387] This type of application provides a powerful motivation for fundamental studies of semiconducting oxide surfaces and will be referred to again in Chapter 6.

4.1.5 Computational methods

The experimental methods discussed below in § 4.2 are frequently supplemented by electronic structure calculations. This is not an appropriate place for a detailed review of the great variety of computational methods that have been employed. However, it may be useful to give a brief summary of some of the crucial issues. Details can be found in the references to specific calculations given in the later sections.

Three quite separate issues must be decided when setting up an electronic structure calculation:

(i) How is the appropriate configuration of atoms (crystal structure, defect site, or surface) to be represented?

(ii) What type of mathematical functions are to be used to represent the electronic wavefunction?

(iii) How are the various terms in Schrödinger's equation (potential and kinetic energy, inter-electron repulsion, etc.) to be treated?

In dealing with perfect crystals, conventional solid-state theory uses a single unit cell, repeated to form an infinite three-dimensional array; this, of course, leads to a band structure of the kind illustrated above, Fig 4.3, for MgO. A similar approach can be used for a surface, using only two-dimensional periodicity. Then the solid can be represented as a finite slab, which must be thick enough to assure that bulk and surface states can be truly distinguished. Alternative approaches are possible. For example, one can deal with a semi-infinite solid, starting with the bulk band structure and treating the surface as a perturbation. The appropriate mathematical formulation comes from scattering theory and generally uses Green's functions.[388] A different variant is known as the **transfer-matrix** method, in which the necessary equations for a single layer are solved first, and then a stack of layers is put together to give a quasi-one-dimensional problem.[389] All of these methods end up with a calculation of a two-dimensional surface band structure, and the output may be represented in terms of energy E *vs.* wave-vector k curves, possibly projected onto the bulk bands (e.g., Figs. 4.4, 4.22, 4.23 and 4.27). Alternatively the partial densities-of-states (DOS) in the surface and sub-surface layers may be plotted [see Fig. 4.26 (b)].

Defects can also be treated by scattering theory, using a few atoms embedded within a perfect solid or surface.[40,390] A more common approach, however, is to use an isolated cluster of atoms. The same method can give a crude representation of the ideal bulk solid or surface, as in the calculations illustrated in Fig. 4.2. To get satisfactory results from a cluster calculation, however, it is essential to include the potential from the surrounding lattice. The potential from an array of point charges may be adequate for calculating ground-state properties, but for excited states (including hole states seen in photoemission) it is also desirable to include long-range polarization terms. This has rarely been done, although in accurate calculations it can alter the estimated energy levels by several electron volts.[391]

An especially serious problem in the calculation of surfaces and defects concerns the detailed atomic positions. It is often assumed that atoms remain at the positions appropriate to the ideal bulk lattice, although this may not be a good approximation; defect configurations, surface steps, etc., are undoubtedly associated with significant structural relaxations (see Chapter 2). Only occasionally is the true atomic geometry known from experiment or calculation. Sometimes the electronic structure calculation is used to estimate the total energy, and an energy minimization search is performed to predict the equilibrium geometry. Apart from the very large amount of computing time often required, it is not always obvious that the methods used are very satisfactory for obtaining geometries in this way.

Wavefunctions are nearly always calculated by using some variant of the orbital, or quasi-independent-electron, approximation. This means that inter-electron repulsions are approximated by some average (local or non-local) potential, and that the resulting wavefunction is constructed from a set of occupied one-electron orbitals. These in turn may be represented in various ways. Traditional band theory approaches use a basis of plane waves. Such plane-wave methods may be employed for oxides, so long as valence orbitals of d or f symmetry are not involved.[392] More common in this area, however, are methods based on the **linear combination of atomic orbitals (LCAO)**, or **tight-binding**, method. The atomic orbitals used can range from a large basis set including core, valence and excited orbitals, to a very small one. A 'minimum' basis for ZnO for example might include only the O 2p and Zn 4s functions, contributing to the upper part of the valence band and the bottom of the conduction band, respectively.[365] On the other hand, it has been claimed that more satisfactory results on ZnO require a full set of s and p orbitals on both atoms.[183] Instead of using an explicit basis set of atomic orbitals, these may be generated numerically in the course of the calculation. This is done using the **muffin-tin approximation**, a procedure of spherically averaging the potential around each atomic center, so that Schrödinger's equation can be solved numerically. Muffin-tin methods include the augmented plane-wave (APW) and Korringa–Kohn–Rostoker (KKR) approaches to band theory,[372] and the so-called Xα method for molecular and cluster calculations.[393] The numerical generation of orbitals in this way leads to greater flexibility, but against this advantage must be set the approximations inherent in the muffin-tin approach.

Having chosen a set of atomic positions and suitable basis functions, the computational effort lies in solving Schrödinger's equation at a suitable level of approximation. The choices here are in how to represent the various terms in the Hamiltonian; the methods may range from fully theoretical *ab initio* calculations to a completely empirical parameterization. *Ab initio* approaches aim to represent the Coulomb potential of electrons as accurately as possible, but exchange terms may be treated in different ways. In the **Hartree–Fock** method they are calculated 'exactly' within the independent-electron framework, whereas **local-density** calculations use a representation of the 'exchange-correlation potential' approximated as a local function of the electron density; the latter is done in many band theory calculations, as well as in the Xα method.[394] Both methods can give excellent representations of ground-state properties such as electron densities and total energies, but one must always remember that calculations that accurately describe the ground state are not necessarily directly able to give good values for energies of excited or ionized states. It seems that the local-density method is better at this than Hartree–Fock theory, although local-density calculations tend to underestimate bandgaps. Most experimental methods for investigating electronic structure rely on such excitations or ionizations, and it cannot be assumed that highly 'accurate' *ab initio* calculations will compare more closely with experiment than frankly approximate and semi-empirical ones. The empirical pseudopotential (for plane waves) and empirical tight-binding (using atomic orbital bases) methods treat the matrix elements as empirical quantities, to be determined by fitting some suitable set of experimental results (or sometimes, the results of another calculation). Many of the surface calculations reported later are essentially of this type. Parameters may be optimized for the bulk solid and then applied to defect or surface configurations. In some cases, perturbations such as changes in Madelung potentials or in orbital overlap coming from atomic rearrangements may be included in an approximate way. But one problem with most of these calculations is that they are seldom self-consistent; that is to say, altered atomic charges are not fed back into the calculation of potentials. Nevertheless, they seem to be quite successful and are much easier to carry out than full *ab initio* calculations.

4.2 Experimental methods of surface electronic structure determination

The experimental techniques used to determine the electronic properties of surfaces are, for the most part, very different from those that give geometric structure information. There are several excellent books to which the reader should turn for a discussion of the various surface-sensitive spectroscopies whose results will be utilized here.[4,57,395] But it is necessary to give some discussion of their range of applicability, and especially their limitations, to the study of oxides. It is assumed in this section that the reader is familiar with the techniques under discussion.

4.2.1 Photoemission spectroscopies

In photoelectron spectroscopy, electrons are photoemitted from filled electronic states by absorption of a single photon. Depending upon the energy of the photon, the process is usually classified as either **ultraviolet or x-ray photoelectron spectroscopy** (**UPS** or **XPS**, respectively), although the availability of the entire range of photon energies in synchrotron radiation has resulted in other names being used. Two of the most common gas-discharge atomic lines employed in UPS are He I at 21.22 eV and He II at 40.84 eV. (For both He I and He II discharges there are other, weaker lines in the lamp output that produce 'ghost' spectra shifted in energy form the main spectrum; it is extremely important to correct the spectra for those ghosts when looking for weak emission in a bandgap.)

There is a fundamental problem with photoemission spectroscopies on insulating samples that severely limits the information obtainable from many metal oxides. The incident photons are neutral particles, and the photoemission process involves a net flux of electrons out of the sample surface; thus a positive surface charge will build up unless the sample is sufficiently conducting. The photoemitted electrons are then slowed down due to Coulomb attraction. At best, this gives a shift in the UPS or XPS kinetic-energy scale so that absolute binding energies cannot be measured; at worst, it eliminates the spectra entirely. In principle it is possible to stabilize the surface potential by flooding the surface with a high flux of low-energy electrons from an auxiliary electron source (i.e., using a 'flood gun'), but in practice it is easy to over- or under-compensate the surface charge, and the technique has seldom been used successfully.

Photoelectron spectroscopies, like most of the other techniques discussed below, are not entirely surface-sensitive. The electron mean-free-paths are fairly short, but nevertheless atoms within several atomic planes of the surface may contribute to the measured spectra. The relative weighting of the bulk and surface contributions depends upon electron kinetic energy, angle of electron emission, etc., and these variables are often used to separate the two components, although the procedure is not always straightforward. But the interpretation of experimental data can also be influenced by the goal of the experiment. A good example of this is angle-resolved or resonant photoemission measurements aimed at understanding the bulk electronic structure of materials. When clearly identifiable surface features are not present, the data are often interpreted simply in terms of bulk structure. This is the case in numerous studies of metal oxides, and for the most part it is appropriate, since often the bulk electronic structure is not understood in enough detail to be able to identify deviations from it. But surface electronic structure information is still present in the spectra, particularly when the photoelectrons have kinetic energies in the range of 50–150 eV (and hence mean-free-paths of only 3–4 Å). In some cases we will present such data, even though they have only been interpreted in bulk terms, in the hope that they may be of use to someone in deciphering surface properties.

Although photoelectron spectra are often used simply to give 'binding energy' information, a more sophisticated interpretation of spectra is possible, especially from bulk levels. The first step in the photoionization process can be regarded as the excitation of an electron from a filled to an empty level within the solid, and thus is similar to an absorption process. At a one-electron level, the transition probability depends on the joint density-of-states of filled and empty levels. When tunable synchrotron radiation sources are employed, it is possible in principle to perform measurements in such a way as to separate contributions from the different levels.[372] **Constant final state (CFS)** spectra show the intensity of photoionization from different initial (filled) states as the photon energy is scanned, keeping the measured electron kinetic energy constant. In this way information about occupied densities-of-states is obtained. On the other hand, the measured photoelectron energy may be scanned together with the photon energy, in a **constant initial state (CIS)** experiment, which provides some information about empty levels. In surface studies it is more usual to measure **energy distribution curves** (**EDC**s), corresponding to scans of the photoelectron kinetic energy while the photon energy is kept constant. It is important to remember that these EDCs depend upon the empty as well as the filled levels. At higher photon energies, electrons are excited into high conduction-band levels which closely resemble free-electron states: thus it is often assumed that the structure corresponding to empty levels can be neglected.

While energy is, of course, always conserved in the photoemission process, the crystal momentum of the electron need not be. Excitations in which the momentum of the photoelectron is the same in initial and final states are termed **direct transitions**. **Indirect transitions** involve phonons as well as the photon and the electron, so that the final-state momentum of the electron may not be the same as in the initial state.[362] [The total crystal momentum of the (photon + electron + phonon) system is always conserved.] Both types of transition occur in UPS, and it is not generally possible to separate them. The direction in which a photoelectron is emitted from the surface of a single-crystal sample is related to its final-state momentum, and if direct transitions predominate this also gives information about the initial-state momentum; such measurements are referred to as **angle-resolved photoemission spectroscopy** (**ARPES, ARUPS**, etc.). The component of the photoelectron momentum parallel to the surface is conserved, but this is not the case for the component normal to the surface, as the electron is refracted on passing from the solid into vacuum. By making assumptions about the different potentials seen by the electron in the solid and in vacuum (i.e., the **inner potential**), it is possible to infer the electron's initial-state momentum normal to the surface as well. In an energy-band picture, it is therefore possible in theory to use angle-resolved photoemission to determine the full $E(k)$ dispersion relation for electrons in the solid in their ground state.

A few angle-resolved UPS measurements have been reported on metal oxides, and for the most part they have been used to determine the ground-state $E(k)$ curves. However, the interpretation of this type of experiment depends upon a crucial assumption. In the simplest description of photoemission, all of the photon's energy is transferred to a single electron, so that the kinetic energy with which an electron is emitted from the surface into vacuum allows one to determine the binding energy of the initial state from which the electron came. This assumption is implicit, both in the use of angle-integrated UPS to obtain information about densities-of-states and in the more sophisticated angle-resolved work used for band-mapping. It is important to realize, however, that this 'one-electron' interpretation is a serious approximation. What photoemission is really measuring is the difference in energy between the N-electron ground state of a system and various $(N-1)$-electron ionized states. The removal of an electron causes relaxation and correlation effects that may sometimes completely invalidate the one-electron picture. Whether or not this is a serious problem depends on several factors, including the type of compound being studied, the nature of the levels being ionized, and the type of theoretical calculation with which the experiment is being compared.

The one-electron description generally works well for valence-band photoemission from nearly-free-electron materials for which a delocalized, itinerant-electron energy-band model is appropriate. In other words, it works in cases where the population of a band or an orbital can be changed without seriously perturbing any of the other electrons. Even for free-electron materials, photoemission from a deeper-lying core level causes electrons to move so as to screen the positive charge created. Thus the kinetic energy of the emitted electron is not simply related to the initial state of the electron and the photon energy. For materials whose bonding electrons are partially localized on specific atoms or ions, an itinerant-electron energy-band picture may not be the appropriate model even for the interpretation of valence-band photoemission. This is not always appreciated in work on metal-oxide surfaces, and interpretations of experimental data have sometimes been naive. This is especially true for some of the transition-metal oxides discussed in Chapter 5, where neither solid-state band nor localized molecular-orbital methods are entirely satisfactory. For the valence bands of non-transition-metal oxides, final-state effects do not seem to be so important. Successful band-mapping experiments have been performed by ARPES studies and will be discussed later under MgO and ZnO (see Figs. 4.9 and 4.21). Even here, however, one must realize that some degree of electronic relaxation is present in the final states. As mentioned above in § 4.1.5, the local-density method generally gives band structures more directly comparable with this type of data than do Hartree–Fock calculations.

In addition to electronic final-state effects, another interaction is always present in photoemission experiments. In highly polar solids, electrons and holes interact strongly with lattice phonons, particularly the longitudinal optical ones (see Chapter 3). When localized states are being ionized, this interaction leads to a local relaxation of the atomic geometry surrounding the hole state and manifests itself in the spectra as a Franck–Condon broadening of the photoemission signal. In UPS measurements aimed at determining band dispersions, electron–phonon interactions lead to a breakdown of the normal k-conserving selection rules used in the interpretation of these experiments. Thus only a portion of the total UPS signal (and possibly a rather small one) is expected to show the dispersion effects resulting from electronically direct transitions. Band-mapping experiments of non-metallic solids therefore have a large non-dispersing 'incoherent' background.

4.2.2 Inverse photoelectron spectroscopy

Inverse photoelectron spectroscopy (IPS) is essentially the time reversal of photo-emission, both theoretically and experimentally. A monoenergetic electron incident on a surface enters the sample in an allowed electron energy level, falls to another empty level, and emits a photon, whose energy is measured. For nearly-free-electron materials where electron energies do not depend upon population, a simple orbital energy picture can be used, but this is not always the case for metal oxides. It is then necessary to calculate the energies of both the initial N-electron state and the final $(N+1)$-electron state of the system. Few calculations of this type have been performed to date.

For insulating samples, IPS suffers from surface charging problems similar to those described above for photoemission. Negatively charged particles are incident on the surface, while the emitted particles are neutral photons, so the surface may charge negatively. In this case an electron flood gun is of no use. For this reason a number of IPS studies of metal oxides have been performed on thin oxide films grown on metal substrates, where the surface charge can be neutralized by tunneling to the conducting substrate. One difficulty with this approach is that the very thin films, generally less than 30–50 Å, may not have the structure or stoichiometry of the bulk oxide.

4.2.3 Electron-energy-loss spectroscopy

Electron-energy-loss spectroscopy (ELS) is essentially the same as HREELS except that electronic transitions are excited instead of vibrational ones. The different names are used because the instrumental resolution necessary to study electronic transitions is much less than that required for vibrational spectroscopy. ELS has one advantage over UPS, XPS or IPS for insulating samples; since both the incoming and outgoing particles are electrons, it is often possible to find conditions (i.e., incident-beam energies, angles of incidence) for which the surface potential will only change by a small amount and then become stable. The criterion for this is the same as for LEED: as discussed in § 2.2, the incident-electron kinetic energy must be in the range where the secondary-electron yield of the surface is greater than unity.[50]

As discussed in Chapter 3, the strongest excitations in ELS are ones related to the long-wavelength dielectric function according to

$$P(\omega) \quad \propto \quad \text{Im} \frac{-1}{[\varepsilon(\omega)+1]}. \qquad (4.8)$$

ELS is therefore an excellent method for studying plasmon excitations, although the energy of surface plasmons studied in this way is slightly different from the bulk values given by the maxima of the function Im $[1/\varepsilon(\omega)]$ (see § 4.1.3). ELS is also widely used to study 'one-electron' excitations, including surface bandgaps and defect transitions. A similar *caveat* applies to the interpretation of these results as that discussed above for photoemission. As we have noted in § 4.1.3, it may be quite dangerous to relate electronic transitions, which really measure the energies of N-electron excited states, to simple one-electron energy levels.

4.2.4 Other spectroscopic techniques

There are a many other surface-sensitive spectroscopic techniques that, in one way or another, measure the electronic structure of surfaces. However, they are either rarely used for that purpose or provide more limited or convoluted information. **Auger spectroscopy**, for example, is an atom-specific technique that measures essentially a convolution of three densities-of-states, so little direct information about surface electronic structure can be obtained. There have been specific cases, however, where Auger spectra provide important complementary information. But its primary use is in determining surface composition and cleanliness. In electron-excited Auger spectroscopy it is generally possible to find a range of incident-beam energies for which the surface potential is stable even for highly insulating samples. There is one major drawback, however, in that the high electron energies needed to ionize the core hole (usually > 1 keV) will often break interionic bonds in the sample, leading to electron-stimulated desorption of lattice ions and the creation of surface defects; this will be discussed in § 6.7.

The **scanning tunneling microscope** can explore surface densities-of-states within a few eV of the Fermi level from measurements of the tunneling current versus the voltage between the tip and the sample.[51,52] If the tip is negative with respect to the sample, electrons tunnel from the tip to the sample, and the current is a function of the density of filled states on the tip and the density of empty states on the sample surface; conversely, if the tip is positive, the spectra measure a convolution of empty tip and filled sample states. In principle, the tip densities-of-states can be determined from measurements on samples having known densities-of-states, so the technique has the potential to give rather direct information about sample surface electronic structure. More importantly, the spatial resolution of the technique means that local electronic structure can be determined within regions of a surface that are comparable in area to the size of individual surface atoms, point defects, etc. This potential has been realized for metals and semiconductors, but little work has yet been performed on oxides. STM cannot, of course, be performed on insulating samples.

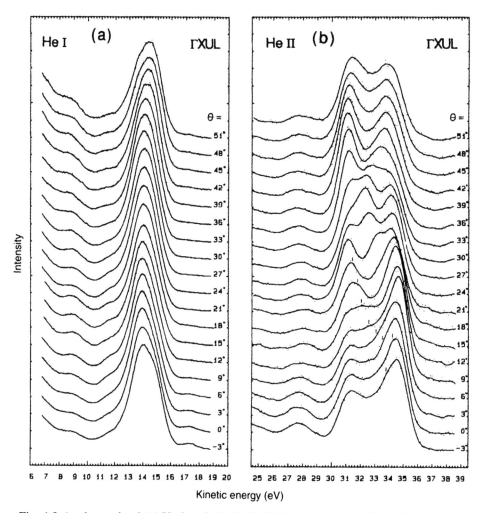

Fig. 4.8 Angle-resolved (a) He I and (b) He II UPS spectra for MgO (100) in the ΓXUL emission plane, normalized to the integrated valence-band intensity. [Ref. 398]

4.3 Alkaline-earth oxides

The most thoroughly studied of the pre-transition-metal oxides are the alkaline-earth oxides, particularly the rocksalt oxides MgO, CaO and BaO. A few UPS measurements of filled electronic states have been performed on single-crystal samples, and lower-resolution XPS measurements have been reported; UPS measurements have also been performed on thin oxide films grown on metal substrates. Auger and ELS measurements have been made on single crystals of most alkaline-earth oxides.

4.3.1 MgO

As discussed in § 4.1.1, experimental measurements and theoretical calculations indicate that the bulk electronic structure of MgO is essentially that given by the purely ionic approximation: $Mg^{2+}(3s^0) O^{2-}(2p^6)$. The effects of reduced coordination at the surface, leading to the possibility of a reduced bandgap and enhanced covalency, have also been considered. The status of these effects on the MgO (100) surface is not entirely clear, although there is consistent experimental evidence for a reduced optical bandgap.

MgO is an excellent insulator, and it cannot be made conducting by doping. Surface charging problems therefore make UPS measurements on well-characterized single-crystal samples difficult. Three UPS measurements have been reported on single-crystal MgO.[396-398] In Ref. 396, no mention was made of surface charging or ways to minimize it, so the results are open to question. In Ref. 397, an electron flood gun was used to reduce charging on polished and annealed MgO (100) and (111) surfaces: the polar (111) surface showed LEED patterns characteristic of (100) facets about 20 Å across. The He II UPS spectra obtained for both surfaces are similar. The O 2p valence band is about 6 eV wide, exhibiting larger emission from the non-bonding region of the valence band. The data were taken with an angle-resolved spectrometer (although neither the measurement angle nor the angular aperture were specified), however, and therefore the relative amplitudes of the bonding and non-bonding components of the valence band may not represent those of the k-averaged density-of-states.

Angle-resolved UPS measurements using both He I and He II sources and an electron flood gun were performed on MgO (100) surfaces that had been cleaved in air and then bombarded with 500 eV Ar^+ ions in the spectrometer;[398] no degradation of the sharp (1×1) LEED patterns was observed as a result of ion bombardment. Representative UPS spectra are shown in Fig. 4.8. The expected two-peaked structure and 6 eV width of the valence band are only observed in the He II spectra, with strong emission from only the non-bonding region of the band visible in He I spectra. The band dispersions were determined by assuming a superposition of angle-integrated and direct bulk transitions, with free-electron final-state wave functions. The resulting experimentally determined bands are shown in Fig. 4.9, superimposed on the (shaded) projection of the theoretically predicted bulk band structure onto the (100) surface. Most of the data are consistent with the expected bulk electronic structure, except that some of the data points at $\overline{M}$ lie outside of the range of the bulk dispersion. It is not clear from the experimental results, however, whether they correspond to surface states.

The effects of surface charging are even more severe in IPS than in UPS, and no IPS measurements on single-crystal MgO have been reported. However, several

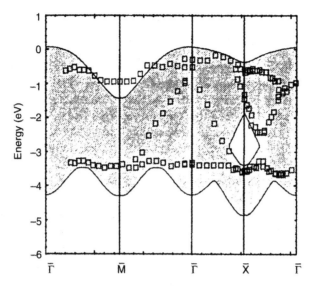

Fig. 4.9 Experimentally determined surface band structure of MgO (100) (squares) super-imposed onto the projection of the bulk energy bands (shaded region). [Ref. 398]

measurements of electronic excitations on nearly perfect MgO (100) surfaces have been obtained by ELS and other techniques. They agree in showing a narrowing of the optical bandgap at the surface.[377,399–401] Figure 4.10 shows an ELS spectrum taken at high resolution (note the surface phonon structure on the elastic peak) for a cleaved and annealed MgO (100) surface;[401] an electron energy of 50 eV was used in order to maximize surface sensitivity. The threshold for loss transitions is seen to be at about 6 eV, almost 2 eV less than the bulk bandgap. That the 6 eV loss is indeed a surface-related feature can be seen by varying the incident-electron energy in an ELS experiment, thereby varying the electron mean-free-path and hence the depth sampled. Figure 4.11 shows first-derivative ELS spectra for cleaved and annealed MgO (100) as a function of incident-electron energy between 100 and 2000 eV; the corresponding mean-free-paths range from about 3 to 20–30 Å.[377] (A peak in the energy-loss spectrum corresponds to the mid-point of a negative slope in these first-derivative spectra.) The loss feature at 6.2 eV dominates the spectrum for surface-sensitive conditions, while it is barely visible for large sampling depths.

The excitations just discussed involve valence-band electrons, but similar effects are seen for intra-ionic transitions of Mg core electrons on MgO (100) sur-faces.[377,399] ELS measurements of transitions from the Mg 2p and 2s core levels were performed with various incident-electron energies so as to change the sam-pling depth. Surface loss features again showed a threshold shifted 2 eV to lower energy relative to the bulk transition.

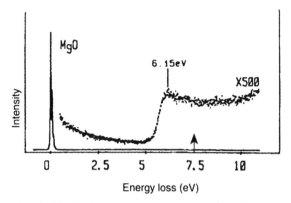

Fig. 4.10 ELS of MgO with 50 eV beam energy measured at high resolution, showing the surface loss at 6.15 eV compared with the bulk bandgap of 7.5 eV (arrow). [Ref. 401]

Transitions at energies less than the bulk optical excitation threshold have also been observed in other experiments. Diffuse reflectance measurements performed on powder samples exhibit several features that were attributed to excitonic surface states.[384] Defects such as steps, edges and corners are present on these samples, but the prevalence of (100) faces (see § 2.3.1.1) suggested that a surface feature at about 6.6 eV was due to an exciton on a (100) terrace. The exact energy of the excitation is uncertain since it was extrapolated beyond the instrumental limit, but it is probably consistent with the exciton energy measured by ELS. Low-energy loss features were also seen in total-electron-reflection measurements from cleaved and annealed single-crystal MgO (100) surfaces.[402] A feature at an incident-electron energy of 5 eV was interpreted as an intrinsic excitation on the nearly perfect surface, although it was attributed to an O 2p 'dangling bond' state rather than a surface-shifted conduction-band state, an interpretation which is different from those discussed above.

A variety of excitations can be seen at higher energy in ELS, as shown in Fig. 4.11. The interpretation of many of these is unclear. The excitation at 22.2 eV has been attributed to a valence-band plasmon: that is, not a true surface excitation, but one resulting from the bulk electronic structure, as discussed in § 4.1.3.[401] Other features in the loss spectra have intensities that are much less dependent on sampling depth, and they correspond to bulk excitations seen by other techniques.[377]

The value of the electron affinity of MgO (and CaO) has been the subject of speculation for many years.[403,404] Since those oxides are such good insulators, it has not been possible to measure it by electron emission from a surface as one would do for a semiconductor. X-ray scattering experiments conclude that the electron affinity is small and positive, with a preferred value of about 1 eV in MgO

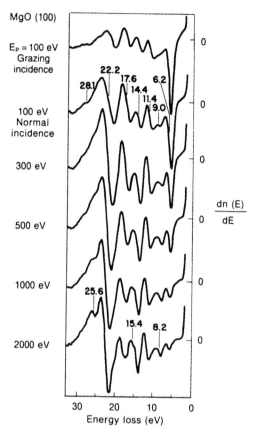

Fig. 4.11 First-derivative ELS spectra for cleaved and annealed MgO (100) as a function
of incident-electron energy (and hence sampling depth). [Ref. 377]

and 0.7 eV in CaO. But theoretical consideration of the stability of Fe^+, Co^+ and
Ni^+ impurities in those oxide hosts has led to the speculation that the electron
affinities of both MgO and CaO may be *negative* (i.e., the vacuum level lies below
the bottom of the cation 3s conduction band) by more than 1.3 eV in MgO and
0.9 eV in CaO. To the best of our knowledge, this discrepancy has not yet been
resolved. However, it is interesting to note that MgO has the highest measured secondary-electron yield of any material. (This quantity measures the number of secondary electrons emitted from a surface when one electron is incident on it).
Values of between 12 and 24 (we prefer the lower value) have been reported for
the total secondary-electron yield of single-crystal MgO for incident-electron
energies of a few hundred electron volts.[405] Such a high yield is consistent with a
negative electron affinity, since electrons that had thermalized near the bottom of
the conduction band, and whose mean-free-path are therefore very long, would
literally 'fall out' of the oxide when they reached the surface. (This is analogous to

cesiated GaAs negative-electron-affinity photocathodes, where electrons photoexcited into the conduction band just roll out of the band into vacuum at the surface.) One interpretation of the surface excitation energy seen in ELS also requires a negative electron affinity for MgO (100), as discussed below.

A number of electronic structure calculations have been performed on MgO (100) surfaces. Band-structure and cluster calculations on unrelaxed surfaces were discussed earlier in § 4.1.4;[363,364,369] see Figs. 4.2 and 4.4. As we have noted, both types of calculation predict some narrowing of the bandgap at the surface, although in a different way and with different magnitudes. As these are ground-state calculations, they may not be directly applicable to the interpretation of spectroscopic data. No calculations of the type necessary for rigorous interpretation of such experimental data have been performed. However, by comparing results of the same type of calculation for bulk and surface configurations, it is often possible to get some idea of the origin of surface features in experimental spectra. The 2 eV lowering in the threshold for surface excitations is more nearly in agreement with the cluster than with the band calculations. Furthermore, the 6 eV transition may give a localized excited state that would be described better by the cluster theory.

A self-consistent tight-binding cluster approach has also been used to address the electronic structure of unrelaxed MgO (100) and (110) surfaces.[406] The change in the Madelung potential and the reduction in the coordination numbers of surface ions results in a narrowing of the bandgap from 7.8 eV in the bulk to 6.7 and 5.6 eV on the (100) and (110) surfaces, respectively. Significant changes in the shape of the O 2p valence-band density-of-states are also predicted, with a progressive reduction in the relative amplitude of the high-binding-energy bonding orbitals in going from the bulk to the (100) and then the (110) surfaces.

One calculation of the electronic structure of relaxed MgO (100) has been performed by using a tight-binding total-energy method.[407] A slight rumpling is predicted, with the surface O ions moving outward and the Mg ions relaxing inward; the amplitude of the rumpling is no more than a few percent, in agreement with the conclusions drawn from LEED and other measurements above. The calculated surface bands, together with the projection of the bulk energy bands on the (100) surface, are shown in Fig. 4.12. These results are in qualitative agreement with the Green's function bands for the unrelaxed surface,[369] except for the absence of two states in the belly gap near X, which are absent in the calculations on the relaxed surface.

Another calculation, which aimed at interpreting the excitations of MgO bulk crystals and (100) surfaces, employed a large cluster for the purpose of calculating electronic levels, and a surrounding lattice of polarizable ions, rather than point charges as more commonly used.[408] The question posed was whether holes in the

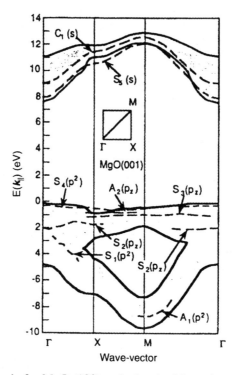

Fig. 4.12 Energy bands for MgO (100), calculated with a relaxed surface. The projection of the bulk bands (shaded) is shown. [Ref. 407]

valence band form a self-trapped state (that is, a small polaron localized by lattice relaxation). The conclusion was that this does not happen in the bulk, but it probably can occur on the (100) surface. It was also suggested from these calculations that the 6.2 eV excitation feature seen on the (100) surface is a self-trapped exciton, although the reason for the difference in surface and bulk excitation energies was not explained in detail.

Less rigorous models have also been used to interpret the surface excitations seen in ELS. The energy lowering in the surface Mg 3s level as predicted by the cluster calculation has been attributed to a Stark effect resulting from the potential gradient at the surface.[377,399] On the other hand, it has been pointed out that surface excitons at energies lower than bulk transitions are only observed in some ionic compounds: for example, the ELS of NaCl and other alkali halides show excitations identical to the bulk transitions measured by UPS. In an attempt to rationalize these differences, it has been suggested that lower-energy surface excitations only appear on ionic solids where the conduction band lies *above* the vacuum level.[401] Surface excitonic states of the type seen in MgO could then correspond to final states in which the electron is localized primarily *outside* of the

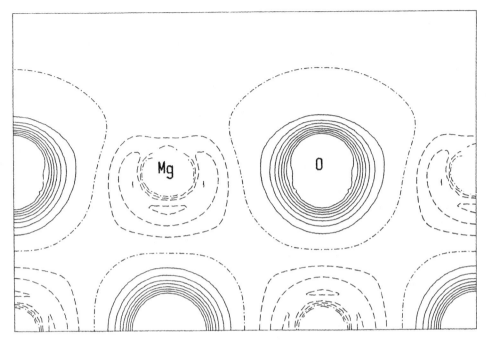

Fig. 4.13 Difference charge densities for MgO (100). The difference between charges calculated by a self-consistent method, and a purely ionic distribution, is plotted as contours in a vertical section containing the surface. [Ref. 87]

crystal. More commonly the bulk excitation has a final-state energy below the vacuum level, and in this case the surface excitation is probably hidden in the loss spectrum. Whether the bulk conduction band in MgO really does lie above the vacuum level is uncertain, since an absolute energy calibration of the band energies cannot be made because of surface charging; this can only be partially overcome by using a flood gun, as in Ref. 389. This question of the electron affinity of MgO was discussed above, and some evidence for a negative value (that is, a conduction-band edge lying above the vacuum level) was given.

Several calculations on relaxed MgO (100) surfaces have estimated surface charge distributions, surface phonon frequencies, ionic polarizabilities, etc., although not electronic energy levels that can be compared with UPS or ELS data.[84,87,90,355,409] All calculations agree that both (100) and (110) have essentially the same ionicity as in the bulk, although the charge distribution on the surface ions is, of course, slightly distorted. Figure 4.13 shows contours of the difference in predicted charge density between a slab LCAO calculation and a superposition of ionic charge distributions.[87] Unfortunately this is not the difference between a surface and a bulk calculation using the same method, and so one must compare the contours of the first- and second-layer ions in order to see the distortion in the

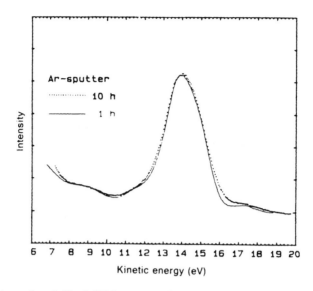

Fig. 4.14 Valence-band He I UPS spectra for MgO (100) as a function of Ar⁺-ion-bombardment time. [Ref. 398]

surface charge densities. The polarizability of the O^{2-} ions on the surface is found to vary only slightly from that of the bulk ions, with the magnitude and sign of the difference depending upon the assumed relaxation or rumpling of the surface.[409]

4.3.2 Defects on MgO

Compared to most other metal oxides, the (100) surface of MgO is remarkably stable against the formation of point defects. Although some defects are formed by ion and electron bombardment, the surface stoichiometry never varies much. The extent to which point defects manifest themselves in UPS spectra is shown in Fig. 4.14. A (100) surface was cleaved in air and then bombarded with 500 eV Ar⁺ ions for various lengths of time; the two curves in Fig. 4.14 are for times of 1 and 10 hours. After even 10 hours of ion bombardment, the (1 × 1) LEED patterns remain clear and sharp; only after 20 hours of bombardment do they begin to degrade. The changes in the UPS spectrum are minuscule.

Point defects on MgO (100) are more evident in ELS measurements on cleaved and annealed samples.[377,400,410,411] Figure 4.15 shows first-derivative spectra for a UHV-cleaved surface measured with two different incident-electron energies, one surface sensitive and the other sampling the bulk (compare with Fig. 4.11).[411] In addition to the surface exciton loss at 6 eV, a feature at about 2 eV increases tremendously in amplitude when the surface is preferentially sampled; measure-

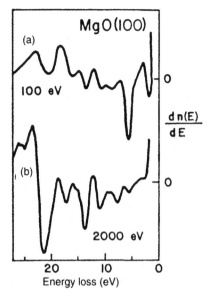

Fig. 4.15 First-derivative ELS spectra for UHV-cleaved MgO (100) for incident-electron energies of (a) 100 eV (surface sensitive) and (b) 2000 eV (bulk sensitive). [Ref. 410]

ments at higher resolution locate that peak more accurately at 2.3 eV.[400] Its amplitude is strongly dependent upon surface treatment. It is sometimes present on freshly cleaved surfaces, as shown in Fig. 4.15, but never on ones that have been annealed in UHV. It appears after bombardment by electrons having a few hundred eV energy, and it can be removed by exposure of the surface to O_2. It was originally attributed to a surface O vacancy, both because that is the dominant type of point defect created on metal-oxide surfaces, and because of the effect of O_2 exposure.[377,410] However, the energy of the loss peak does not agree with the calculated excitation energy (5 eV) of a surface F center,[412,413] and subsequently it was attributed to a surface cation vacancy, or V^- center, whose excitation energy should be close to 2 eV.[400,411] In the experiments on the interaction of Cu atoms with MgO (100) surfaces discussed in § 7.2.1, the disappearance of the defect loss peak with Cu deposition was also taken as evidence for the surface Mg vacancy interpretation. Additional work is necessary before the nature of this defect can be uniquely determined.

Diffuse reflectance spectra from MgO powders show features that were tentatively interpreted in terms of excitons at ions in lower-coordination sites at edges and corners.[384] Structure at 5.8 and 4.6 eV has been attributed to ions with four- and three-fold ligand coordination, respectively: i.e., the sites labelled (3) and (5) in Fig. 4.6.

In addition to this experimental work, various theoretical studies have been performed on defects on MgO, although few of those make predictions that can be directly compared with experiment. The surface O-vacancy defect on MgO (100) has been treated by the DV-Xα technique.[364] The calculation predicts that the defect would create a surface state in the bandgap about 1 eV above the top of the O 2p valence band. Since no UPS or IPS measurements have been performed on single-crystal MgO (100) surfaces with an appreciable density of point defects, it is not known experimentally whether this is correct.

Some attention has been paid to the electronic structure of extended defects on MgO surfaces. The changes in the Madelung potential at step, corner and kink sites have been calculated as a function of the relaxation of the lattice from an ideal termination of the bulk. The most exposed ions at steps in (100) terraces have a Madelung potential of 21.4 eV for an ideal termination, which changes to about 22 eV when relaxation is taken into account.[415] These values should be compared with the bulk Madelung potential of 23.9 eV and that for a relaxed, but perfect, (100) surface (23.0 eV). This shows the importance of relaxation effects in the ionic model and how these effects tend to increase the potentials of exposed surface ions towards the bulk value.

The charge distribution and anion polarizability for ions on MgO (100) terraces and in four- and three-fold coordinated step-edge and corner sites have been calculated using an *ab initio* Hartree–Fock cluster approach.[416] The calculated density maps and moments of charge distributions depend only slightly on the ligand coordination, although the polarizability of the surface O^{2-} ion increases significantly as the coordination number is decreased. The latter effect is plausible since the uncoordinated (i.e., free) O^{2-} ion is unstable. Calculations using a variable-potential *ab initio* method also showed that the charge state of O ions at both the top and bottom of {100} steps on MgO (100) have essentially the same O^{2-} charge state as do bulk and terrace ions.[417]

4.3.3 CaO, SrO and BaO

Although MgO is by far the most exhaustively studied alkaline-earth oxide, some ELS measurements have also been performed on (100) faces of single-crystal CaO, SrO and BaO.[418,419] The results are similar to those obtained on MgO. Surface excitation energies of 4.97 and 4.3 eV were found in CaO and SrO, respectively, which are less than the bulk bandgaps of 6.9 and 5.3 eV. They were attributed to the same type of exciton as that in MgO. No loss peaks at energies less than the bulk bandgap were found for BaO in single-crystal experiments, although ELS measurements on BaO thin films did show a loss peak at 3.7 eV, below the bulk threshold of 4.4 eV.[420] Again this feature was described in terms of a surface exciton. Diffuse reflectance spectra also exhibited low-energy surface features on CaO, SrO and BaO powders.[384] The energies of the features that were attributed to nearly perfect terrace sites were 5.5, 4.6 and 3.5 eV, respectively, which compare well with the ELS values on single crystals.

ELS measurements on (100) faces of single-crystal CaO, SrO and BaO also exhibit low-energy excitations at 1.2, 0.9 and 0.6 eV, respectively, attributable to surface point defects.[418,419] The response of the defect peaks to electron bombardment, O_2 exposure, etc., is the same as for MgO, and they have been interpreted in the same way: as excitations of surface V^- centers. In addition to the low-energy peak, BaO exhibits other changes in ELS spectra upon defect creation. A broad peak appears centered at about 5 eV, which has been attributed to a surface plasma excitation.[419] A surface-related loss peak at about 2 eV was also observed on thin polycrystalline BaO films, although its origin was not firmly established.[420] As the loss peak appeared just at the edge of the elastic peak in this poorly resolved spectrum, its actual energy is probably less than 2 eV.

Diffuse reflectance spectra of CaO, SrO and BaO powders also show results similar to those for MgO, and excitation features were interpreted in terms of ions at four- and three-fold coordinated edge and corner sites.[384]

4.4 Al_2O_3

Alumina, Al_2O_3, is extremely important in ceramics, catalysis, integrated circuit manufacture, etc. There are several polymorphs of Al_2O_3, the most stable being corundum, α-Al_2O_3. Like MgO it is an excellent insulator that cannot be made conducting by doping, and so detailed experimental measurements of its surface properties are limited. There has, however, been some effort to understand its surfaces theoretically.

There have been two reported UPS measurements of the valence-band electronic structure of single-crystal Al_2O_3 (0001),[421,422] although there have been several XPS measurements.[167,421–423] Figure 4.16 shows both He II UPS and $Al_{K\alpha}$ XPS spectra for polished and annealed Al_2O_3 (0001); a high-energy electron flood gun was used to stabilize the surface potential during the measurements. The O 2p valence band has two main peaks and a total bandwidth of about 8 eV, slightly smaller than the value of 10 eV predicted theoretically.[45,169,424] XPS measurements of the valence band have also been performed on single-crystal α-Al_2O_3 ($10\bar{1}2$) surfaces, although the surface treatment used in this work did not yield observable LEED patterns.[423] The valence band exhibited a two-peaked structure very similar to that shown in Fig. 4.16 for the (0001) surface.

ELS measurements have been interpreted in terms of the joint density-of-states between the O 2p conduction band and the Al 3s,p conduction band. For α-Al_2O_3

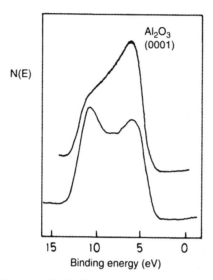

Fig. 4.16 (a) $Al_{K\alpha}$ XPS and (b) He II UPS spectra for polished and annealed Al_2O_3 (0001). A high-energy electron flood gun was used to stabilize the surface potential. [Redrawn from Ref. 421]

(10$\bar{1}$2) surfaces that exhibited fairly good (2 × 1) LEED patterns, most of the observed loss features were consistent with the bulk electronic structure.[423] However, an empty surface state was identified at 4.0 eV below the bottom of the conduction band. A second state, lying 1.0 eV below the conduction-band minimum, was attributed either to an exciton or to a surface state. Figure 4.17 shows ELS measurements of the apparent bandgap on four different Al_2O_3 (0001) surfaces, each displaying one of the reconstructions discussed in § 2.3.4.2.[167] The bandgap is largest for the (1 × 1) surface structure, and the ($\sqrt{31}$ × $\sqrt{31}$) R ± $\tan^{-1}(\sqrt{3}/11)$ surface exhibits transitions throughout the bulk bandgap that are characteristic of a reduced, almost metallic surface. It is interesting that Ar^+-ion bombardment, which destroys all LEED patterns as it disorders the surface, does not reduce the surface to give a metallic surface layer; it merely narrows the bandgap at the surface.

The (0001) surface of α-Al_2O_3 has been treated theoretically by using a self-consistent extended-Hückel tight-binding slab method.[424] The surface Al–O bonding was found to be less ionic than that in the bulk. The resulting energy-band structure for the unrelaxed α-Al_2O_3 (0001) surface is shown in Fig. 4.18, along with the projection of the bulk bands onto that surface. The total density-of-states (DOS) is plotted in the right panel of the figure. The almost dispersionless empty surface band, S_d, at 3 eV above the top of the valence band, is composed primarily of the Al 3s,3p$_z$ 'dangling bonds' of the surface Al ions, with a small admixture of

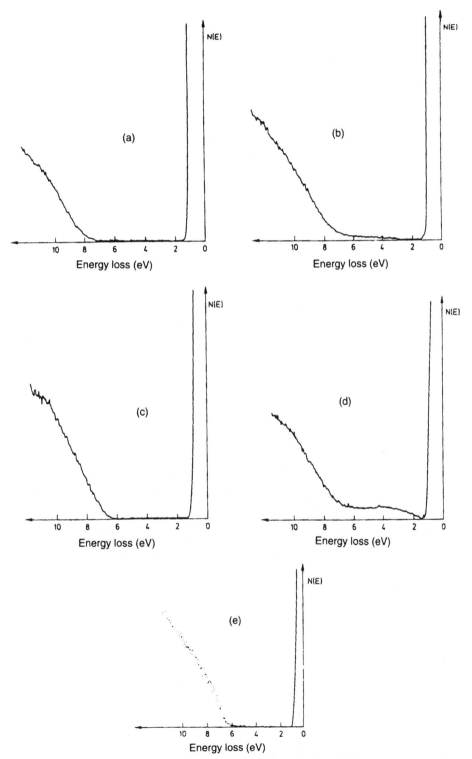

Fig. 4.17 ELS spectra showing the apparent bandgap on Al_2O_3 (0001) surfaces exhibiting
(a) (1 × 1), (b) (2 × 2), (c) (3 $\sqrt{3}$ × 3$\sqrt{3}$) R 30°, (d) ($\sqrt{31}$ × $\sqrt{31}$) R ± $tan^{-1}(\sqrt{3}$ /11)
reconstructions, and (e) the ion-bombarded surface. [Ref. 167]

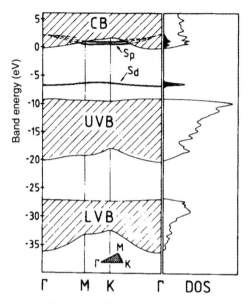

Fig. 4.18 Band structure and density-of-states calculated for the Al_2O_3 (0001) surface. CB, UVB and LVB are projections of the bulk conduction, upper-valence and lower-valence bands, respectively, and S_p and S_d are predicted surface states. [Ref. 424]

O 2p. The predicted energy of this surface state depends strongly on the effective charge of the surface Al ions. A second surface state, S_p, overlaps the bulk conduction band.

A Hartree–Fock slab approach has also been used to calculate the surface electronic structure for α-Al_2O_3 (0001) and (10$\bar{1}$0).[45,169] In these calculations the outermost layer of ions [Al in the case of (0001), O for (10$\bar{1}$0)] was allowed to relax, and the resulting relaxation was found to be as large as 0.4 Å. This relaxation partially restores the ionicity of the Al–O bond at the surface. For the (0001) surface, an empty state is also found in the bulk bandgap, but at a slightly higher energy than in the extended Hückel calculations. The orbital composition of the surface state is similar for both calculations. For α-Al_2O_3 (10$\bar{1}$0), the surface density-of-states is more complicated than for (0001); states of predominantly O 2p character are pulled up from the valence band, while Al 3 s,p states drop below the bottom of the bulk conduction band. However, the slab thickness used in the calculations was too thin to clearly separate surface and bulk states.

The electronic structure of surface cation and anion vacancies on the unrelaxed α-Al_2O_3 (0001) surface has been calculated using extended-Hückel tight-binding methods.[424] For surface O vacancies, three defect surface states are found, having energies 1.3, 2.7 and 8.1 eV below the conduction-band minimum. They arise from redistribution of charge on the Al ions surrounding the defect, although the

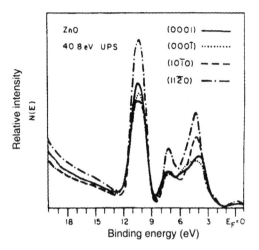

Fig. 4.19 Angle-integrated He II UPS spectra for the non-polar (10$\bar{1}$0) and (11$\bar{2}$0) and the polar (0001)-Zn and (000$\bar{1}$)-O faces of ZnO. [Ref. 193]

state at 8.1 eV has a considerable contribution from O 2p orbitals. Similar calculations for surface Al vacancies did not give any localized surface states in the bulk bandgap. The surface bandgap measured by ELS for an ion-bombarded Al_2O_3 (0001) surface, shown in Fig. 4.17 (e), is slightly larger than that predicted by the tight-binding model, but it is still less than the gap on a stoichiometric surface.

The possible types of surface defect that might be responsible for the catalytic activity of η- and γ-Al_2O_3 have also been considered theoretically,[222] but there is no experimental information on well-characterized surfaces with which they can be compared.

4.5 ZnO

ZnO is the most thoroughly studied post-transition-metal oxide. Its use as a gas sensor, in which the surface conductivity changes in response to adsorbed gases, made it an ideal candidate in the early days of surface science. Several studies have been performed on UHV-cleaved single crystals, and so data are available for direct comparison with theoretical calculations.

4.5.1 Nearly perfect ZnO surfaces

The general features of the electronic structure of ZnO can be seen in Fig. 4.19, which presents angle-integrated UPS spectra measured under highly surface-sensitive condition on the non-polar (10$\bar{1}$0) and (11$\bar{2}$0) faces and the polar (0001)-Zn

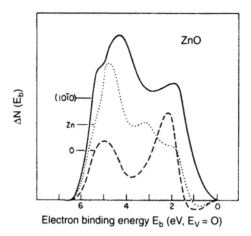

Fig. 4.20 Differences between UPS spectra taken at surface- and bulk-sensitive photon energies for the (10$\bar{1}$0), (0001)-Zn and (000$\bar{1}$)-O surfaces of ZnO, showing the contribution due to surface states and resonances. [Ref. 181]

and (000$\bar{1}$)-O faces.[193] The large peak at about 10 eV binding energy comes from the filled Zn 3d band. The valence band extends from about 3 to 8 eV; theory predicts that the emission from 3 to 5 eV is from the non-bonding O 2p orbitals, and that from 5 to 8 eV is a bonding combination of O 2p and Zn 4s orbitals. In spite of the proximity of the Zn 3d levels to the valence band, the two are not significantly hybridized.[365] The valence-band photoemission has also been separated experimentally into O 2p and Zn 4s contributions by varying the photon energy and hence the relative cross-sections for the two orbitals.[425]

There are large differences in emission between the different surfaces, particularly in the shape of the valence band. The contribution due to surface states and resonances for the (10$\bar{1}$0), (0001) and (000$\bar{1}$) surfaces was estimated from angle-resolved UPS with photon energies in the range 20 to 80 eV;[12,181,426] these contributions are shown in Fig. 4.20 for each of the three faces.[181] Angle-resolved UPS measurements on the (10$\bar{1}$0) face were analyzed, assuming plane-wave final states, to determine the dispersion of the valence bands in the Γ–M direction in reciprocal space; Figure 4.21 shows the resultant band structure compared with theoretical calculations of the bulk bands.[426] Surface states were not identified in this analysis.

There have also been several ELS studies of the (0001), (000$\bar{1}$) and (10$\bar{1}$0) single-crystal faces of ZnO aimed at determining the joint densities-of-states; combined with a knowledge of the filled states from photoemission, this should give information about the empty levels above the Fermi level, E_F.[379–382,427–429] The results have been less than satisfying. The loss spectra are in general agreement with the energy-loss function, $\mathrm{Im}\{-1/[\varepsilon(\omega)+1]\}$, calculated with a dielectric function obtained from optical measurements, but the details of the spectra show

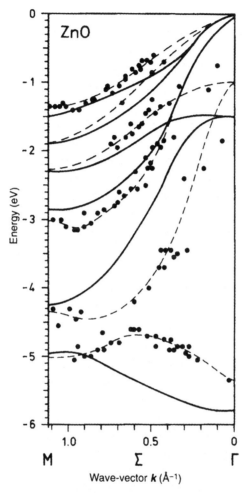

Fig. 4.21 Dispersion of the valence bands in the Γ–M direction for ZnO (10$\bar{1}$0) determined experimentally from angle-resolved UPS measurements (dots and dashed lines) compared with theoretical calculations of the bulk bands (solid lines). [Ref. 426]

significant variation between research groups; the situation has been reviewed in Ref. 381. So far, this work has not provided a definitive model for the empty density-of-states.

Some of the problems involved in interpreting electronic excitations were mentioned in § 4.1.3, where valence-band plasmons were discussed. ELS loss peaks in the energy range of 13.5–16 eV in ZnO have been assigned as surface plasmons of this type.[378–381] However, another group observed a loss peak at 15.3 eV but attributed it to a Zn 3d-to-conduction band transition.[382] The status of valence-band surface plasma excitations is thus still the subject of some controversy.

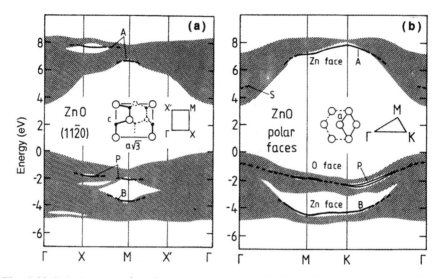

Fig. 4.22 Calculated surface band structures for ZnO, with projection of bulk bands (shaded). (a) shows the (11$\bar{2}$0) surface. For both polar surfaces in (b) the bulk projection is the same; the diagram indicates which surface states appear on the Zn and O faces. [Ref. 365]

The interaction of surface plasmons with Fuchs–Kliewer surface phonon modes in ZnO, and the manifestation of that interaction in HREELS, were discussed in § 3.2.3.

The work function, Φ, of ZnO single-crystal faces has been measured by several techniques on surfaces prepared in various ways, including cleaving in UHV.[5,12,13] The results provide an excellent example of how sensitive the work function is to essentially all possible variables (see § 1.3). For cleaved samples, different cleaves gave different values of Φ for the same crystal plane; these values also depended upon the temperature at which the sample was cleaved. The only things that could be agreed upon were that the work functions for the faces studied followed the trend

$$\Phi(000\bar{1}) > \Phi(10\bar{1}0) > \Phi(0001)$$

and that the work function of cleaved surfaces changed irreversibly upon annealing.

A variety of theoretical approaches have been brought to bear on the surface electronic structure of ZnO, including the tight-binding scattering-theoretical method,[365,430,431] the transfer-matrix method,[389,432] the DV-Xα cluster approach,[364,433–435] an sp^3 tight-binding model,[182–185] and evaluation of surface Madelung potentials.[436] Only one of the approaches predicts surface states in the

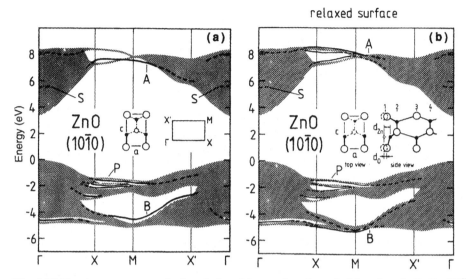

Fig. 4.23 Band structures calculated for (a) unrelaxed and (b) relaxed ZnO (10$\bar{1}$0)
surfaces, with projections of the bulk band structure (shaded). [Ref. 365]

bulk optical bandgap, and the surface states that are predicted lie very close to the
bulk band edges. Figure 4.22 shows the projected bulk band structure and the surface states and resonances for ZnO (11$\bar{2}$0), (0001) and (000$\bar{1}$)) calculated by the
scattering-theoretical method. The surface resonances labelled P and S are entirely
anion-derived and are thus ionic in nature, while states B and A correspond to
more covalent back-bond and anti-back-bond surface states, respectively. The surface ions on ZnO are believed to be somewhat less ionic than the bulk ions.

All of the calculations shown in Fig. 4.22 utilized unrelaxed surface geometries.
The effect of relaxation on the surface electronic structure was explored for the
(10$\bar{1}$0) face (recall that the Zn ions relax inward on this surface by 0.4–0.5 Å), and
Fig. 4.23 shows the calculated band structures for both surfaces. It is found that the
surface resonances P and S are almost unaffected by the relaxation, while the more
covalent back-bond (B) and anti-back-bond (A) surface states change appreciably
in energy.

The results of the transfer-matrix calculations are similar, although they suggest
that all of the surface states contain some Zn 3d character in their wave functions.[389] Surface states were also found to exist in the Zn 3d band itself, but presumably they would be resonances that would be difficult to observe.

The DV-Xα cluster calculations differ significantly from the others in predicting a relatively large density of both anion- and cation-derived surface states in the
gap; this is found for both polar faces and for the (10$\bar{1}$0) face.[364,433–435] One experi-

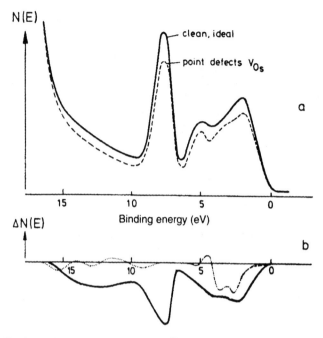

Fig. 4.24 (a) He II UPS spectra for ZnO (10$\bar{1}$0) with (dashed curve) and without (solid curve) point defects. (b) Difference spectra for the data in (a) with (dotted curve) and without (solid curve) normalization of the Zn 3d intensity. [Ref. 258]

mental study also concludes that bandgap surface states exist,[437] although, as we have mentioned, the experimental situation concerning empty states is rather confused.

4.5.2 Defective ZnO surfaces

Point defects on ZnO surfaces are extremely important in gas sensor applications since they produce very large changes in the surface conductivity.[438,439] The dominant surface defects are O vacancies, and their electronic properties and thermodynamics have been studied by a variety of techniques.[258–260] [See Fig. 4.5 above for the equilibrium density of O-vacancy defects on ZnO (10$\bar{1}$0) as a function of temperature and O_2 partial pressure.] The changes in the valence-band electronic structure of ZnO (10$\bar{1}$0) that are caused by such defects are shown in Fig. 4.24. Defects were created by heating to a high temperature and then quenching to room temperature; He II UPS spectra taken before and after this treatment are shown, along with difference spectra that show the changes more clearly.[258]

The creation of O-vacancy surface defects on ZnO does not produce any new filled electronic states in the bandgap. This is one of the important differences between non-transition-metal and transition-metal oxides and is related to the range of stable oxidation states. As discussed in Chapter 5, most transition metals have several stable oxidation states, whereas non-transition metals are limited to one or at most two. In ZnO the electronic charge donated to the surface upon removal of surface O^{2-} ions results in an accumulation layer and an increase in surface conductivity. It is this accumulation layer that is affected by chemisorbed atoms, with the resulting changes in surface conductivity that make ZnO an important material for gas sensors.

4.6 SnO₂

The rutile oxide SnO_2 has several very important applications in transparent conducting coatings, gas sensors and catalytic processes. Most of the surface work on SnO_2 has been concerned with defects and their role in surface conductance and chemisorption; this will be discussed following an account of the electronic properties of nearly perfect single-crystal surfaces.

4.6.1 Nearly perfect SnO₂ surfaces

The most studied SnO_2 surface is (110), since this is the thermodynamically most stable rutile surface (see § 2.3.3.1).[140–145,440–444] The interesting defect properties of this surface arise because the bridging O ions lying above the main surface plane can be removed easily; on the perfect surface they are all present. Figure 4.25 shows UPS spectra for both nearly perfect and defective SnO_2 (110); the zero of binding energy is taken at the Fermi level.[440] The O 2p valence band extends to about 10 eV, with the emission at higher binding energies corresponding to an inelastic background. The shape of the SnO_2 valence band is very different from that for either MgO or Al_2O_3 (compare with Fig. 4.16). The sharp peak at the upper edge of the band comes from non-bonding O 2p orbitals directed perpendicular to the Sn–O axis and forming an almost dispersionless band.[445] The Sn ions on the perfect surface are all in the nominal Sn^{4+} state, as in the bulk. The conduction and valence bands do not appear to be bent at this surface (i.e., the surface is in a 'flat band' state), and the surface has the same resistivity as the bulk.

The unrelaxed SnO_2 (110), (001) and (100) surfaces have all been treated theoretically by the scattering-theoretical method.[445–448] Figure 4.26 shows (a) the projected bulk (shaded) and surface band structures, and (b) the wave-vector-resolved densities-of-states (DOS) at the Γ and M points in the Brillouin zone, compared to the calculated bulk densities-of-states at those points, for an SnO_2 (110) surface *without* the plane of bridging O ions. The B states are back-bonding surface states

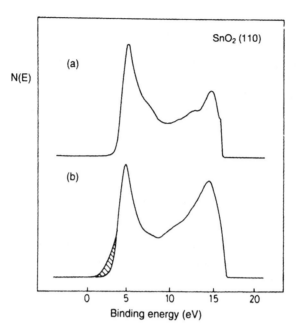

Fig. 4.25 He I UPS spectra for (a) nearly perfect and (b) defective SnO_2 (110). The zero of binding energy is taken at the Fermi level. The surface-defect-induced feature is indicated by hatching. [Ref. 440]

whose dominant orbital component is Sn 5s, with an admixture of O 2p. The P levels are surface resonances having O 2p character. The states S_1 and S_2 are primarily Sn 5s, while the resonance S_3 contains more O 2p than Sn 5s character. Calculations on the perfect surface containing the bridging O ions give similar results except for the absence of the B_1 and S_1 surface states. The projected band structures for the SnO_2 (001) and (100) surfaces are shown in Fig. 4.27 (a) and (b), respectively; however, the geometric model used for (100) does not contain the O ions lying above the main surface plane that would be present on an ideal cleaved surface (see Fig. 2.11). Note that there are no surface states in the optical bandgap for any of these surfaces, even though the models used for the (110) and (100) surfaces do not contain all of the O ions necessary to produce a charge-neutral surface; this is discussed in more detail below.

A tight-binding, total-energy model has also been applied to the stoichiometric SnO_2 (110) surface and to the (110) surface that contains no bridging O ions.[148] In neither case did any electronic states appear in the bulk bandgap, in agreement with both the scattering-theoretical and photoemission results.

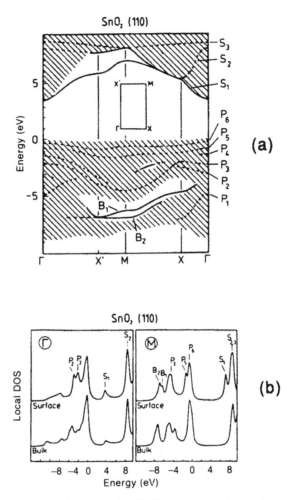

Fig. 4.26 Surface band structures for SnO_2 (110). (a) shows band dispersions compared with a projection of the bulk bands; (b) displays *k*-resolved densities-of-states at two points in the (110) Brillouin zone, comparing surface and bulk contributions. [Ref. 445]

4.6.2 Defective SnO₂ surfaces

Because of its use in gas sensing applications, the properties of surface defects on SnO_2 have been extensively studied.[140–143,145,149,332,354,440–444] Most attention has been paid to the (110) surface, where it is found that the layer of bridging O ions that is present on the perfect surface (see Fig. 2.10) can be easily removed by heating or by particle bombardment. In fact, ISS measurements have shown that a stable surface structure exhibiting (1 × 1) LEED patterns is formed when *all* of the bridging O ions are removed: this is sometimes referred to as the 'compact' surface. It is also possible to replace the bridging O ions with isotopically labelled

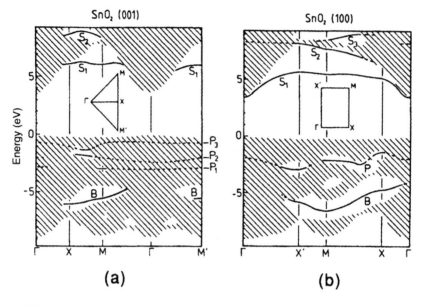

Fig. 4.27 Band structures for SnO_2 (001) and (100) surfaces, presented as in Fig. 4.26 (a). [Ref. 445]

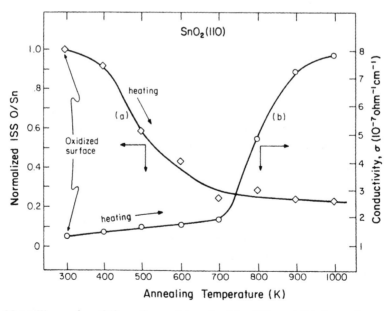

Fig. 4.28 (a) The surface O/Sn ratio, as determined by ISS, and (b) the surface conductivity, as measured by a four-point probe method, for oxidized SnO_2 (110) as a function of annealing temperature. [Ref. 140]

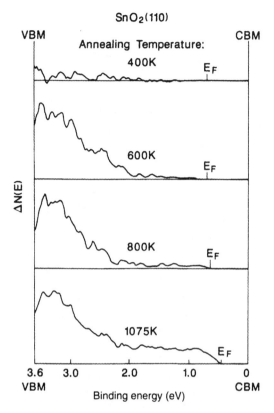

Fig. 4.29 UPS difference spectra showing the increase in occupation of bandgap defect states between the valence-band maximum, VBM, and the conduction-band minimum, CBM, for oxidized SnO_2 (110) as a function of annealing temperature. [Ref. 140]

^{18}O ions selectively, with little mixing with the in-plane O ions.[450] The reason for the stability of the compact surface is that the Sn ion has two stable oxidation states, Sn^{4+} and Sn^{2+}. On the perfect surface all cations are Sn^{4+}. But when surface O^{2-} ions are removed, the two electrons left behind occupy orbitals (a mixture of 5s and 5p) on surface Sn ions, converting them to Sn^{2+}.[332,444]

Measurements of the surface electrical conductivity of SnO_2 (110) as a function of the number of O ions removed show that the Sn^{2+} ions formed still possess a localized electronic structure. This can be seen in the UPS spectra in Fig. 4.25 (b), where the increased density-of-states produced by removing bridging O ions by Ar^+-ion bombardment is localized near the valence-band maximum and not at the Fermi level.[440] Four-point conductivity probe, UPS and ISS measurements on samples whose surface stoichiometry was changed by annealing show that the conductivity of the surface does not increase significantly until after the bridging O ions have been removed.[140] This is shown in Fig. 4.28, which plots both the surface

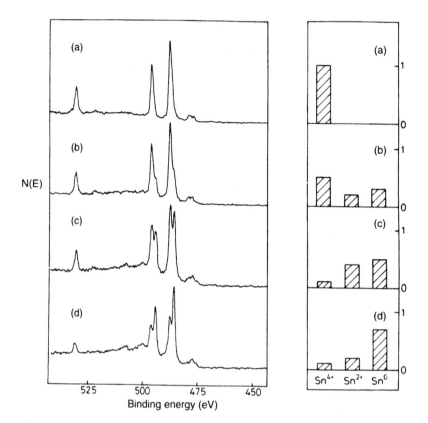

Fig. 4.30 XPS spectra of the Sn 3d core levels for (a) stoichiometric SnO_2 (110) and (b)–(d) that surface after increasing amounts of electron bombardment. The right-hand box shows the corresponding fractional surface concentrations of Sn^{4+}, Sn^{2+} and Sn^0. [Ref. 442]

O/Sn ratio, as determined by ISS, and the surface conductivity as a function of annealing temperature for an initially oxidized sample. The large increase in conductivity between 700 and 800 K is attributed to the removal of O ions from the main Sn–O surface plane or in the near-surface region; the decrease in the ISS O/Sn ratio is due predominantly to removal of bridging O ions. UPS measurements show that the increased surface conductivity is accompanied by the occupation of states in the bulk bandgap, and in particular a measurable density-of-states at E_F; this is shown in Fig. 4.29. In another study, XPS measurements of the Sn 3d core levels as a function of electron bombardment showed that reduction of the SnO_2 (110) surface was accompanied by the creation of both Sn^{2+} and Sn^0 states, as illustrated in Fig. 4.30.[442] Unfortunately, the results of these various experiments cannot be combined to produce a consistent picture of the behavior of reduced SnO_2 (110) surfaces because of the different surface preparation procedures used.

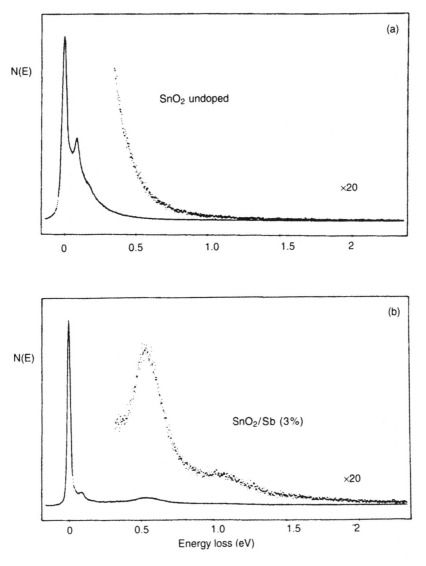

Fig. 4.31 ELS spectra for (a) undoped and (b) 3 at. % Sb-doped SnO_2 ceramic samples, showing the surface plasmon loss in the Sb-doped sample. [Ref. 332]

Several models of surface defects on the unrelaxed SnO_2 (110) surface have been calculated with the scattering-theoretical method.[447,449] The calculated band dispersion for the compact surface has already been presented in Fig. 4.26. Other models included the removal of only one bridging O ion, the removal of either a single or a pair of O ions in the main surface plane (after the removal of all of the bridging O ions), and the removal of an O ion from the next plane below the surface. None of these vacancy systems was predicted to give bound defect states in

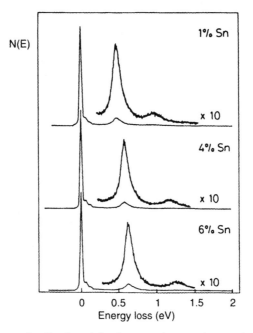

N(E)

1% Sn

x 10

4% Sn

x 10

6% Sn

x 10

0   0.5   1   1.5   2
Energy loss (eV)

Fig. 4.32 ELS spectra for Sn-doped In_2O_3 ceramic samples having different Sn dopings, illustrating the dependence of the surface plasmon frequency on mobile carrier concentration. [Ref. 453]

the bulk optical bandgap, but rather the surface-state features found for the defect-free surfaces were enhanced. This result does not appear to be consistent with the chemical picture discussed above, or with UPS measurements which suggest that the occupied $5s^2$ levels associated with surface Sn^{2+} are located above the valence-band edge.[332]

4.6.3 Doped SnO_2

When SnO_2 is doped with Sb, electrons are introduced into the conduction band, resulting in a concentration of mobile electrons that depends upon the amount of Sb present.[451] ELS spectra for both undoped and 3 at. % Sb-doped SnO_2 ceramic samples are shown in Fig. 4.31.[332,452] The loss peak centered at 0.55 eV, which is only present in the Sb-doped sample, is a surface plasmon excitation. The normally insulating oxide In_2O_3 can also be made conducting by the addition of Sn. ELS spectra shown in Fig. 4.32 for different Sn dopings in In_2O_3 ceramic samples illustrate how the surface plasmon frequency depends on mobile carrier concentration.[453] As the Sn doping level is increased, the surface plasmon energy increases due to the larger conduction-electron density.

The optical reflectivity of materials also depends strongly on the density of mobile electrons present in the conduction band. This phenomenon has been used to determine the presence of carrier-free layers on the surface of metal oxides.[354] Model calculations were made for Sb-doped SnO_2 and compared with infrared reflectance measurements on ceramic samples having various Sb doping levels. The results suggested that, for doping levels greater than about 1 at. %, the samples might be covered with a surface layer having reduced, but not zero, carrier concentration.

5

Electronic structure of transition-metal-oxide surfaces

The surface electronic structures of transition-metal oxides are much more varied than those of the non-transition-metal oxides; this richness of behavior includes other important aspects such as defect formation and chemisorption. It is related to the much wider range of electronic and chemical properties found generally in transition-metal compounds,[39,40] and is the reason for dealing with transition-metal oxides in a separate chapter.

Much of the introductory material in Chapter 4 is still relevant here and need not be repeated. In order to understand surface properties, however, it is essential to have some idea of the different types of electronic behavior found with transition-metal oxides; this is reviewed in § 5.1. A much more extensive account of this subject can be found in Ref. 40, which is recommended to anyone who wants more details. The discussion in § 5.1 below gives some references to relevant sections of that book.

The experimental techniques discussed previously in § 4.2 can, of course, be used also for transition-metal oxides. But, unfortunately, the interpretation of experimental excited-state spectra such as UPS, XPS, IPS, ELS, Auger, etc., is often more complicated than for non-transition-metal oxides. Section 5.1.3 therefore describes some of the peculiar difficulties arising in this area.

Following the introductory survey, compounds will be discussed in detail under the heading of the relevant transition-metal cation.

5.1 Survey of electronic structures of transition-metal oxides

This section will concentrate on the differences between transition-metal oxides and non-transition-metal oxides and will assume the introductory discussion given in Chapter 4 as a starting point. It is largely concerned with bulk properties, as it is important to understand these before the differences that arise at surfaces can be appreciated.

5.1.1 Peculiarities of transition-metal oxides

The common feature of the non-transition-metal oxides discussed in Chapter 4 is that the valence orbitals of the metal atoms are of s and p symmetry. With transition metals, however, the d atomic orbitals assume crucial importance. Many of the complications with transition-metal oxides stem from this difference, because of the different bonding properties associated with d orbitals. These complexities include the existence of variable oxidation states, the frequent failure of the band model, and the crystal-field splitting of the d orbitals.

5.1.1.1 Variable oxidation states and their consequences

The non-transition-metal oxides discussed in Chapter 4 contain elements that, with the exception of Sn, have only one preferred oxidation state. Other states are inaccessible because too much energy is needed to add or remove an electron from the cations when they are coordinated with O^{2-} ligands. Transition-metal oxides behave differently in that the energy difference between a cation d^n configuration and either a d^{n+1} or d^{n-1} configuration is often rather small. The most obvious consequence is that many transition elements have several stable oxides with different compositions. It is also much easier than with non-transition-metal oxides to make defects having different electron configurations. Much of the complexity of both the bulk and surface chemistry of transition-metal oxides results from the presence of high defect concentrations.

The variety inherent in transition-metal oxides is shown in Table 5.1, where the 3d-transition-metal oxides are classified according to their cation d-electron configuration and their crystal structure. One striking point is the existence of isostructural families of compounds. This has facilitated systematic studies, as in some cases it is possible to separate geometric and electronic effects either by comparing the properties of oxides of different metals within a family, or by studying different oxides of the same cation.

Table 5.1 also shows the variety of stable oxidation states found with many transition metals (see § 1.1 in Ref. 40). It includes, for example, the oxides VO, V_2O_3, VO_2, and V_2O_5, where vanadium has the formal charges 2+, 3+, 4+ and 5+, respectively. In fact, this does not exhaust the possibilities, since many phases of intermediate composition are also known, where the oxidation states have mixed, or possibly fractional, values. Sometimes such phases may be of well-defined composition, especially the so-called **Magnéli phases** associated with shear planes, as discussed § 2.1.2. But many transition-metal oxides have phases of variable composition, and the existence of **non-stoichiometry** of this kind is a serious complication in surface as well as bulk studies. Non-stoichiometric compounds may have much higher defect concentrations than in non-transition-metal oxides.

Table 5.1. *d-Electron configuration vs. crystal structure for 3d-transition-metal oxides*

	Bixbyite	Rutile	Corundum	Rocksalt	Spinel	Other
$3d^0$	Sc_2O_3	TiO_2	–	–	–	TiO_2 (anatase and brookite) V_2O_5 (orthorhombic) CrO_3 (orthorhombic)
$3d^1$	–	VO_2 (T ≥ 340 K)	Ti_2O_3	–	–	–
$3d^2$	–	CrO_2	V_2O_3	TiO_x (0.6 ≤ x ≤ 1.28)	–	–
$3d^3$	–	β-MnO_2	Cr_2O_3	VO_x (0.8 ≤ x ≤ 1.3)	–	–
$3d^4$	Mn_2O_3	–	–	–		–
$3d^5$	–	–	α-Fe_2O_3	MnO	Mn_3O_4	–
$3d^6$	–	–	–	FeO	Fe_3O_4	–
$3d^7$	–	–	–	CoO	Co_3O_4	–
$3d^8$	–	–	–	NiO	–	–
$3d^9$	–	–	–	–	–	CuO (monoclinic)
$3d^{10}$	–	–	–	–	–	Cu_2O (cubic) ZnO (wurtzite)

Transition-metal-oxide surfaces may therefore be even more sensitive to preparation conditions than are non-transition-metal oxides. Furthermore, even if one knows the precise bulk composition that is stable under given conditions of temperature and oxygen pressure, it is by no means certain that the surface will have the same equilibrium composition. We encountered this problem earlier in connection with defects in ZnO; it is far more serious with transition-metal compounds.

Associated with the different oxidation states are various electron configurations, also shown in Table 5.1. The principal valence atomic orbitals of the transition-metal cations are of d symmetry. Electron configurations are generally assigned on the basis of a formal ionic charge, which may not, of course, give a realistic indication of the actual charge distribution. Covalent bonding interactions transform the O 2p and the metal d orbitals into bonding and antibonding orbitals of mixed atomic character. The 'd-electron' configuration really represents the number of electrons remaining when all the metal–oxygen bonding levels are filled; it could be thought of (as with the chemical oxidation state) more as a book-keeping device than an indication of the real electron distribution. Nevertheless, it remains an essential concept in understanding the electronic properties of transition-metal oxides.

Trends in the stability of different oxidation states are very important in surface chemistry, as they control the types of defect that may be formed easily and the type of chemisorption that may take place (see Chapter 6). The d^0 configuration represents the highest oxidation state that can ever be attained: thus pure TiO_2, V_2O_5, etc., cannot gain any more oxygen, although they can lose oxygen to form defects or other bulk phases. On the other hand, d^n oxides with $n \geq 1$ are potentially susceptible to oxidation as well as reduction. The stability of high oxidation states declines with increasing atomic number across a given series. For example, the oxides of Ti, Fe and Ni that are normally stable in air are TiO_2, Fe_2O_3 and NiO, respectively. NiO can be oxidized to some extent, forming the semiconducting, slightly Ni-deficient compound $Ni_{1-\delta}O$; the higher oxidation state Ni^{3+} is also present in some ternary compounds such as $LiNiO_2$. By contrast, TiO is instantly oxidized in the presence of O_2. Another trend is apparent in comparing the first (3d) transition series with the lower (4d and 5d) ones in the periodic table: higher oxidation states are generally more stable and resistant to reduction in the lower series. For example, TiO_2 is more easily reduced than ZrO_2, and the 'normal' stable oxide of Ru (which lies below Fe in the periodic table) is RuO_2, in contrast to Fe_2O_3.

5.1.1.2 Competing effects: bandwidth, electron–electron, and electron–phonon interactions

Band structures for transition-metal oxides may be constructed according to the same principles as those for non-transition-metal oxides discussed in § 4.1.1. We expect a filled valence band based predominantly on O 2p orbitals, separated by a gap from a metal-based conduction band (see § 2.3 in Ref. 40). Two obvious differences are that (a) the conduction band is formed from the metal d orbitals, rather than from s orbitals as in the non-transition-metal oxides, and (b) the conduction band is empty only in d^0 compounds; in other oxides it is expected to have some electrons in it. But there is another, less obvious, difference that is crucial to understanding the properties of many transition-metal oxides. The metal d conduction band, especially in the 3d series, may be quite narrow, so that other interactions compete seriously with the band properties of electrons. The repulsion between electrons and the interaction of electrons with lattice phonons are certainly not negligible effects in non-transition-metal oxides and may be important, for example, in the interpretation of spectroscopic transitions. But the independent-electron band model does not fail disastrously for compounds like ZnO. It does appear to fail totally for many transition-metal compounds.

Bandwidth in solids arises from the overlap of orbitals. One important property of d orbitals is that they are quite contracted in size compared with valence s and p orbitals, and so have relatively poor overlap with surrounding atoms. (This is even more true for the 4f orbitals in the rare-earth elements, which hardly overlap at all.) In some oxides of early transition elements, such as TiO, some direct overlap of the 3d orbitals on neighboring Ti ions is possible and can contribute to the width of the d band. But the 3d orbitals contract rapidly with increasing atomic number and with increasing ionic charge, and for later elements such as Ni direct overlap is very small. In most transition-metal oxides, the d bandwidth is largely a consequence of *indirect* bonding interactions via metal–oxygen–metal linkages. It is thus very sensitive to structural and electronic factors, but the important point is that it is often *small*. Typical values may be around 3 eV for titanium oxides, and possibly as small as 1 eV for the oxides of Fe → Cu. (One of the difficulties in finding more precise values is that this bandwidth often does not contribute directly to any of the electronic properties.) For the elements of the 4d and 5d series, d-orbital overlap is larger, and wider bands are normal. The other interactions discussed below are therefore less of a problem.

Band models, like orbital theories of atoms or molecules, assume that the effect of electron–electron repulsion can be incorporated into some time-averaged potential, so that electrons move in a quasi-independent way. This often works well for metals, because long-range Coulomb repulsions are screened by Fermi-surface electrons. But such screening becomes less efficient as the bands become narrower, or as the carrier concentration is reduced. As this happens, a point can be reached where the itinerant properties of the metal break down, and electrons become localized; this is known generally as the **Mott transition**.

An important semi-quantitative approach to understanding the Mott transition in transition-metal oxides is provide by the **Hubbard model** (see § 2.4.1 in Ref. 40). This emphasizes the importance of the repulsion between electrons when they are on the same ion. It is a very approximate theory, in that longer-range Coulomb repulsions are certainly significant, and may be crucial to understanding what happens close to the transition point. Nevertheless, the Hubbard model gives a very useful guide to the type of behavior expected. Consider an oxide containing M^{2+} ions, and imagine the process of moving an electron from one cation to another. Starting from localized electron configurations, this can be written

$$M^{2+} + M^{2+} \rightarrow M^{+} + M^{3+}.$$

Such a process requires energy, the magnitude of which is related to the extra electron repulsion present in M^{+} and is known as the **Hubbard U**. For isolated ions, it represents an energy barrier to the motion of electrons. The competing tendency of electrons to delocalize in a solid is provided by the overlap of orbitals to form bands. Very roughly it is predicted that when the bandwidth W is greater than the Hubbard U a metal should be formed (that is, the bands are wide enough to overcome the electron repulsion), but that electrons remain localized when $W < U$.

Hubbard U values for transition-metal ions in free space may be as large as 20 eV, but in solids they are greatly reduced by screening and polarization effects. In fact, the precise measurement (or even definition!) of Hubbard parameters appropriate to solids is fraught with difficulties, but typical values may be around 3 eV for early 3d elements such as Ti and V, and in the range of 5–7 eV for compounds such as NiO. In the latter case, it is clear that U is much larger than the bandwidth, so that metallic conduction is inhibited by the electron repulsion. NiO is an example of a **magnetic insulator**, an important class of materials discussed in more detail below (§ 5.1.2.3).

The argument given above applies to a stoichiometric compound, with all ions in the same M^{2+} charge state. What happens if the oxygen content is increased, so that some M^{3+} is also present? Electron transfer between M^{2+} and M^{3+} can now take place without incurring any electron repulsion penalty, and we might expect electrons to show delocalized metallic properties, even when this is not possible in pure M^{2+}- or M^{3+}-containing compounds. It is indeed the case that introducing mixed valency of this kind often produces compounds having strikingly higher conductivities than otherwise: an example is Fe_3O_4. But another complication now intervenes: the tendency of electronic carriers to polarize the surrounding lattice. The importance of polarization energies in bandgaps and other properties was emphasized in § 4.1. These terms can have qualitatively different effects in transition-metal oxides because their magnitudes, as with Coulomb interactions, are sometimes similar to the bandwidth.

Polarization energies can be separated into two physically different contributions: that coming from the rapid response of other electrons, and that from the relatively much slower motion of ions or atoms as a whole. When a carrier moves in a band, the electronic polarization follows more-or-less instantaneously, but the slow atomic polarization does not. This latter contribution therefore acts as a 'drag' on the motion of the carrier, and an electron or hole accompanied by its lattice polarization is known as a **polaron** (see § 2.4.5 in Ref. 40). In the **large-polaron** limit the effect is to increase the effective mass of the carrier, but if the polarization energy is sufficiently large compared with the bandwidth, a more localized **small polaron** may result. In this case, the carrier is essentially immobilized by the strong lattice polarization it produces. A chemical interpretation of small-polaron formation can be given by noting that the radius of an ion depends on its charge. Putting an extra electron or hole on one ion will therefore cause a change in size, and hence a local distortion of the lattice. In the small-polaron limit this distortion is sufficient to cause 'valence trapping', with the carrier associated with a particular ion rather than delocalized in the solid. Whether or not this happens depends upon the relative energies associated with the distortion, which tends to localize the carrier, and the bandwidth, which is a measure of the competing tendency towards delocalization.

Under most conditions small polarons move from site to site by a thermally activated 'hopping' process. The activation energy is effectively the energy required to adjust the local distortion enough to allow the electron to interact effectively with a neighboring ion. Activated carrier mobilities associated with small-polaron formation have been found in many semiconducting transition-metal oxides, but the results are rarely completely free of controversy (see § 4.1.2 in Ref. 40). One important conclusion, however, is that the formation of small versus large polarons depends on a rather fine energy balance, which may easily be upset by other factors such as the influence of lattice disorder and the potentials due to impurity centers. Generally all these effects work together to increase the localization of carriers. The implication for surface studies is important: when electrons or holes are introduced by surface defects, it is likely that they will remain localized on transition-metal ions in the surface region. In chemical language, they will be associated with local changes in oxidation state. For example an extra electron may change one Ti^{4+} to Ti^{3+}.

5.1.1.3 Crystal-field splitting

Unlike the non-degenerate s levels that form the lower part of the conduction band in non-transition-metal oxides, the d orbitals of transition metals have five-fold degeneracy in the free ions. However, the different directional properties of the orbitals give rise to different bonding interactions with nearby atoms. The result is known as the **crystal-field** or **ligand-field** splitting and is important in all transition-metal oxides (see § 2.1.1 in Ref. 40).

By far the most common coordination geometry for transition-metal ions in oxides is octahedral (see § 2.1.1). Figure 5.1 shows the orientation of the five d orbitals with respect to the six ligand oxygens. It can be seen that the orbitals fall into two groups according to their symmetry properties (irreducible representation) in the O_h point group. The doubly degenerate e_g orbitals have lobes of maximum density pointing directly towards the ligands, whereas the three t_{2g} orbitals point away. Figure 5.1 also shows the resulting crystal-field splitting Δ between these sets of orbitals. Typical values of Δ for 3d-series ions in oxides are in the range of 1–2 eV. It was originally believed that the splitting is caused by electrostatic repulsion, but modern interpretations emphasize the importance of overlap and bonding interactions. The e_g orbitals can form σ bonds with the appropriately directed O 2p orbitals, while the t_{2g} orbitals form π bonds. What we call the 'd' orbitals in transition-metal compounds are in fact the antibonding counterparts of these bonding orbitals; the predominantly O-based bonding orbitals are fully occupied, even in a d^0 compound, by electrons nominally making up the full $2p^6$ configuration. The ligand-field splitting arises because σ bonding interactions have stronger orbital overlap than π bonds, so that the e_g σ anti-

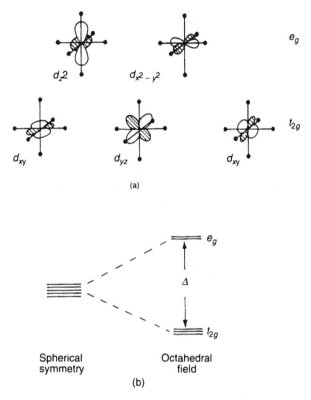

e_g

t_{2g}

d_{z^2} $d_{x^2-y^2}$

d_{xy} d_{yz} d_{xy}

(a)

e_g

Δ

t_{2g}

Spherical symmetry   Octahedral field

(b)

Fig. 5.1 The crystal-field splitting of d orbitals in octahedral coordination: (a) orientation of d orbitals with respect to surrounding oxygens; (b) the resulting orbital energies.

bonding orbitals are higher in energy than the π antibonding t_{2g} (see § 2.2.1 in Ref. 40).

In the band structures for the d^0 and metallic d^n compounds discussed in subsequent sections, ligand-field effects appear as a splitting of the d conduction band into two components of different energy. They are also important in magnetic insulators, where the electronic structure is best represented in terms of localized configurations of d electrons. Consider the example of Mn^{2+}, which has a d^5 configuration. We might expect to obtain the lowest-energy state by putting all five electrons in the more stable t_{2g} orbitals, but there is a competing interaction at work. Electrons with paired spins have more mutual repulsion than ones with parallel spins – this is a manifestation of the **exchange energy**. If the exchange energy terms are larger than the ligand-field splitting, it is better to have as many spins parallel as possible, even if this entails occupying the e_g orbitals. This latter **high-spin** arrangement, giving $(t_{2g})^3 (e_g)^2$ for Mn^{2+}, is in fact the normal ground state for most ions of the 3d series. The alternative **low-spin** state formed by com-

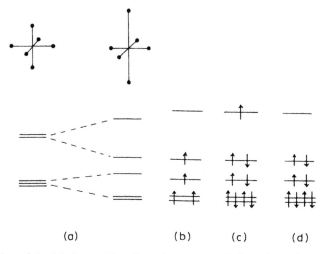

Fig. 5.2 Splitting of d orbitals resulting from the tetragonal distortion of an octahedron: (a) the change in orbital energies; (b)–(d) three electron configurations that commonly exhibit this distortion.

pletely filling the t_{2g} orbitals first is only found as the ground state in a few cases, notably Co^{3+} and Ni^{3+}. The d^8 ion Cu^{3+} can also adopt a diamagnetic low-spin configuration when its ligand coordination is square-planar. In the 4d and 5d series, by contrast, low-spin arrangements are the rule, because the ligand-field splittings are larger and exchange energies smaller.

The detailed pattern of d-orbital energies arising from ligand-field effects obviously depends upon the geometry of the coordinating ligands. In tetrahedral coordination, for example, the arrangement shown for octahedra is reversed, and the magnitude of the splitting is smaller. One important case arises in distortions from regular octahedral geometry which are found particularly with certain electron configurations. High-spin d^4 and d^9 (e.g., Cu^{2+}) ions often occupy tetragonally elongated sites, formed by partially removing two ligands from the octahedron. In the limit, this procedure gives a square-planar coordination, found with low-spin d^8 (e.g., Cu^{3+} and Pd^{2+}). Figure 5.2 shows how the pattern of ligand-field splitting changes during this elongation; it also shows the ground-state arrangements appropriate to the configurations mentioned. Both t_{2g} and e_g sets split in the lower symmetry, but the effect on e_g is much larger because of their stronger interactions with the ligands. It can be seen that the lowering of one orbital is favorable energetically for these particular configurations.

The distortion associated with certain electron configurations is often treated as a manifestation of the **Jahn–Teller effect**. Jahn–Teller distortions might be expected for certain configurations of t_{2g} electrons, but the tendency is generally

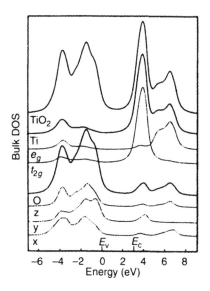

Fig. 5.3 Calculated partial and total densities-of-states (DOS) for TiO_2. The upper-most curve shows the total density-of-states for valence and conduction bands, and the lower curves present the breakdown into different atomic-orbital contributions. E_V is the valence-band maximum and E_C the conduction-band minimum. [Ref. 414]

very weak and obscured by other effects such as crystal packing arrangements (which often give lower symmetries for reasons that have nothing to do with specific electronic interactions) and lattice vibrations.

5.1.2 Types of electronic behavior

The extremely varied electronic properties of transition-metal oxides are a consequence of the different factors discussed in the previous section. Any simple attempt to classify these properties will probably run into some kind of difficulty. However, the scheme adopted below is helpful in understanding the compounds discussed later (see § 1.4 in Ref. 40).

5.1.2.1 d^0 and d^{10} oxides

The d^0 configuration is found in binary transition-metal oxides such as TiO_2, V_2O_5, WO_3, and ternaries such as $SrTiO_3$ and $LiNbO_3$ (see § 3.1 in Ref. 40). These compounds share many features with the non-transition-metal oxides. They have a filled valence band of predominantly O 2p character and a gap between this and an empty conduction band. Typical bandgaps are 3–4 eV. Stoichiometric d^0 oxides are therefore good insulators, diamagnetic, and have no electronic excitations at energies less than the bandgap. The main differences between transition-metal oxides and non-transition-metal oxides are that, for the former, (a) the lower part of the conduction band is based on metal d, rather than s, orbitals, and (b) many transition-metal oxides are relatively easily reduced to form semiconducting or metallic phases (see § 5.1.2.2 and 5.1.2.4 below). The use of d orbitals to form the conduction band gives a more structured density-of-states, as illustrated in Fig. 5.3 for TiO_2.[414] This figure shows the calculated total density-of-states for the valence and conduction bands and its breakdown into contributions from different atomic orbitals. As expected, the valence band is mostly composed of O 2p orbitals, and the conduction band from Ti 3d. There is, however, some covalent mixing or hybridization of these levels, which is significantly greater in the lower 'bonding' part of the valence band than in the upper 'non-bonding' part. The crystal-field splitting of the conduction band into a lower t_{2g} and an upper e_g part can also be seen.

In ternary compounds such as $SrTiO_3$ the pre-transition ion Sr^{2+} contributes its own states based on the empty s and p orbitals. All the evidence suggests that these are at considerably higher energy than the transition-metal d orbitals. Although some details of the band structure may be different because of the different structures of ternary compounds, the major features are essentially similar to those of the binary d^0 oxides.

The post-transition-metal compounds ZnO and SnO_2 discussed in Chapter 4 have cations with the filled d^{10} configuration. The same is found in Cu_2O, although in this compound the Cu 3d states are valence rather than core orbitals. The bandgap of Cu_2O is 2.2 eV, which separates a filled band of Cu 3d character from an empty conduction band formed from Cu 4s orbitals (see § 3.2 in Ref. 40).

5.1.2.2 Metallic oxides

The band model predicts that most oxides having a partially filled d band – that is, for d^n with $0 < n < 10$ – should be metallic. These expectations are frequently not fulfilled because of the intervention of various types of electron–electron and electron–lattice interactions. Nevertheless, 'simple' metallic behavior is found with a number of oxides of elements in the 4d and 5d series, where bandwidths are relatively large (see § 5.1 in Ref. 40). Compounds in this class include ReO_3 ($5d^1$), RuO_2 ($4d^4$), and the tungsten bronze compounds Na_xWO_3 with $x > 0.3$. These oxides have high electronic conductivities that decline with increasing temperature. Valence-level UPS measurements can be interpreted using simple one-electron arguments and are found to agree well with band structure calculations. Other characteristic metallic properties include plasma excitations in ELS. The core-level spectra of metallic oxides do, however, show some complex features; these are discussed in § 5.1.3.3 below.

Some oxides of the 3d series also have high conductivities in the metallic range; examples are Ti_2O_3 and VO_2 at higher temperatures, and the mixed-valency spinel

Fe_3O_4. Superconducting mixed-valency copper oxides such as $YBa_2Cu_3O_{7-x}$ could also be included in this class. Many observations suggest that their behavior is more complex than that of the 4d and 5d metallic oxides, and that the bands are narrow so that one cannot ignore other interactions. Some of these problems will be mentioned later in § 5.1.2.4 and 5.1.2.5.

5.1.2.3 Magnetic insulators

As explained earlier, the narrow bandwidths of many oxides cannot compete with the strong on-site Coulomb interactions. In compounds such as MnO, CoO, NiO and CuO the itinerant behavior of electrons is suppressed, and many of the properties are most simply interpreted in terms of fully localized electron configurations (see § 3.4 in Ref. 40). Thus pure NiO is a good insulator, with a bandgap of about 3.8 eV. Unlike the d^0 compounds, however, it has spectroscopic transitions at energies within the bandgap and magnetic properties indicative of two unpaired electrons per cation. To a first approximation, NiO is remarkably like a very concentrated solution of isolated Ni^{2+} impurities in MgO, or even of complex ions such as $[Ni(H_2O)_6]^{2+}$ in aqueous solution. Small interactions between cations are significant and lead, for example, to antiferromagnetic ordering of the spins below the Néel temperature of 523 K. But the magnitude of such interactions (around 0.1 eV) is at least a factor of 10 smaller than the other important energy terms.

Many attempts have been made to treat magnetic insulators using a band theory formalism. Although some of these have been partially successful in describing the ground state, they fail to describe many of the excited-state properties. Thus the electronic spectra show the ligand-field excitations expected of isolated d^n ions with an appropriate ligand environment, which cannot be interpreted simply using a band model. One may crudely think of transitions between the t_{2g} and e_g levels shown in Fig. 5.1, but ligand field spectroscopy is in fact quite complex. For example, in Ni^{2+} ($3d^8$) the configurations $(t_{2g})^6(e_g)^2$, $(t_{2g})^5(e_g)^3$ and $(t_{2g})^4(e_g)^4$ each give rise to several possible spectroscopic states, the energy of which is determined as much by differential electron repulsion effects within the d shell as by the ligand-field splitting Δ. Ligand-field excitations are best regarded as tightly bound **Frenkel excitons**, with energies depending more on strong Coulomb interactions than anything to do with band structure (see § 2.4.2 in Ref. 40). This is one piece of evidence suggesting that some variant of the Hubbard model, which emphasizes Coulomb interactions, is the most satisfactory theory for magnetic insulators.

Ligand-field excitations may be observed in ELS, and some examples will be given at appropriate points. The photoemission spectra of these compounds is especially complex, and the important **configuration-interaction (CI)** model now used in this area is described in § 5.1.3.2.

5.1.2.4 Defects and semiconduction

Many transition-metal oxides are difficult to prepare without such large defect concentrations that their intrinsic properties are impossible to determine. Even in compounds such as TiO_2 and NiO, where reasonably pure and stoichiometric samples can be made, defects are easily introduced by various preparation procedures. Some important trends in defect chemistry can be related to the stabilities of different oxidation states, as discussed earlier. For example, d^0 oxides, where the cation cannot be oxidized further, can undergo oxygen loss, e.g., to form TiO_{2-x}, where some Ti^{4+} is reduced to Ti^{3+}. On the other hand, the monoxides such as FeO and CoO can take up *excess* oxygen, producing phases containing some M^{3+} ions. Indeed, the extent of defect formation can sometimes be related to the ease of oxidation or reduction of the transition metal. Thus 'FeO' does not exist, but has a range of composition around $Fe_{0.95}O$ (and even so, this wustite phase is thermodynamically stable only at temperatures above 823 K; see § 1.1.2 in Ref. 40). The much greater defect concentration in FeO compared with MnO and CoO is related to the fact that Fe^{2+} is much more easily oxidized to the 3+ state than are the other cations.

As with ZnO and SnO_2, the presence of defects gives rise to semiconducting properties (see § 4.1 in Ref. 40). Loss of oxygen leaves behind extra electrons and produces an n-type semiconductor; extra oxygen (entering the lattice as O^{2-}) creates a deficit of electrons (i.e., it introduces electronic holes), which produces p-type behavior. At low temperatures these carriers will occupy bound states caused by the potential field of the defects, but the binding energies are small enough (typically in the range $0.1 - 0.5$ eV) that significant numbers of carriers may be ionized at room temperature and above. Similar semiconducting properties may be produced by doping with impurities. Thus the insertion of an alkali element or hydrogen into a d^0 oxide may donate electrons and produce an n-type semiconductor such as $Li_xV_2O_5$. On the other hand, the replacement of an M^{2+} cation with Li^+ (which has similar size) effectively removes one electron, so that $Li_xNi_{1-x}O$ is a p-type semiconductor. p-type doping is also possible in d^0 oxides such as TiO_2, by replacing Ti^{4+} with an ion of lower charge such as Al^{3+}.[454]

Although these principles seem fairly straightforward, the detailed properties of semiconducting oxides are often extremely difficult to interpret.[43] The problem of identifying the types of defect present was mentioned in the case of ZnO. The difficulties with transition-metal oxides are often greater, partly because the larger deviations from ideal stoichiometry often make the model of isolated point defects quite inappropriate. A reduced d^0 oxide such as TiO_{2-x} could be expected to have either oxygen vacancies or metal interstitials, but point defects in these compounds are probably only present at very low concentrations, and most of the oxygen deficiency is incorporated by forming **crystallographic shear planes**, as explained in § 2.1.2.[44] Other types of defect interaction take place in the p-type monoxides: for example, $Fe_{1-x}O$ contains defect clusters with a range of sizes that have both Fe^{2+} vacancies and Fe^{3+} interstitials present. These effects clearly make a detailed theoretical understanding of defect electronic structure very difficult.

Another problem concerns the behavior of carriers. Mobilities are often found to be very low, and may be influenced by strong interactions with lattice phonons and with the magnetic moments of other ions present. Small polaron 'hopping' transport is sometimes found, but, as mentioned above, the dividing line between small- and large-polaron behavior is often difficult to establish and may be influenced by many factors.

All of these difficulties must apply to surfaces as well, and in view of the lack of understanding of *bulk* properties of semiconducting oxides, it is not surprising that our knowledge of the surfaces of these materials is still fairly primitive.

5.1.2.5 Metal / non-metal transitions

Several transition-metal oxides show transitions from semiconducting to metallic behavior as a function of temperature, pressure or composition. Some of these transitions are relevant to understanding the surface properties, so it is useful to present a brief summary of them here.

As explained above, insulating oxides may often be made semiconducting by changing the oxygen content or doping in some other way. In more familiar semiconductors such as Si, it is found that the activation energy for conduction disappears above a certain doping level, and metallic behavior follows. The same is true for some, but by no means all, doped transition-metal oxides (see § 4.4 in Ref. 40). An example is in the sodium tungsten bronze series, Na_xWO_3 formed by doping the insulating host WO_3 by insertion of Na. For $x < 0.3$ they are n-type semiconductors, but the temperature dependence of the conductivity decreases with increasing x and has a metallic form for $x > 0.3$. One can think of the impurity levels overlapping to form an impurity band which supports metallic conduction, but many aspects of the transition are not easy to understand in detail. As in this example, quite high doping levels, in the range $0.1 - 0.3$ carriers per transition-metal ion, are usually required to make transition-metal oxides metallic. (Exceptions are $SrTiO_3$ and $KTaO_3$, which show metallic behavior at low temperatures after very slight reduction.) There is evidence that polaronic effects are important in assisting the carrier localization in the semiconducting state, and sometimes these may not entirely disappear until well into the metallic region.

Another type of transition occurs with oxides of early elements in the 3d series. Ti_2O_3, V_2O_3 and VO_2 are all metallic at high temperatures, but undergo transitions to semiconducting phases, accompanied by small structural changes, at lower tem-

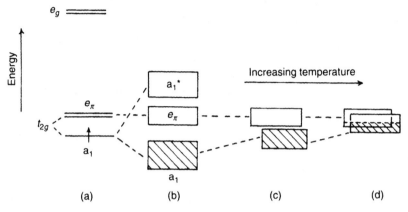

Fig. 5.4 Energy levels of Ti_2O_3. (a) d-orbital energies of a single Ti^{3+} ion; (b) formation of a filled bonding a_1 level (shaded) by interaction between two Ti^{3+} ions; (c) and (d) effect of increasing temperature.

peratures. In the $3d^1$ oxides Ti_2O_3 and VO_2 the low-temperature structures have pairs of metal atoms separated by rather short distances, indicative of some direct metal–metal bonding. In the picture of the corundum structure shown in Fig. 2.2 (b) it can be seen that metal cations occur in pairs of face-sharing octahedra. The electronic levels and the way in which they evolve with increasing temperature are illustrated in Fig. 5.4 (see § 5.3.2 in Ref. 40). Figure 5.4 (a) shows the 3d levels of an individual Ti^{3+} ion, with the large octahedral (e_g – t_{2g}) splitting accompanied by a small additional splitting due to the trigonally distorted site symmetry in the corundum lattice. The lowest (a_1) orbital is oriented so that it can overlap with the similar orbital of the near-neighbor Ti^{3+}. Thus bonding a_1 and antibonding a_1* levels are formed by this direct overlap, as in Fig. 5.4 (b). The lowest a_1 level forms a band which can hold just two electrons per Ti^{3+} pair, and so is full in Ti_2O_3. There is a very small gap (around 0.1 eV) between this and the lowest unoccupied d levels e_π, which are oriented differently so that metal–metal overlap cannot occur. As the temperature increases, lattice vibrations increase the Ti–Ti distance and the overlap weakens; thus the gap also decreases. When the bands start to overlap in energy, metallic conduction begins. At this point some electrons are transferred out of the bonding a_1 level, so that the metal–metal bonding suddenly starts to weaken. The disappearance of metal–metal bonding in this way is shown experimentally by an anomalous expansion of the *c*-axis above the transition temperature.

   Direct overlap and bonding between pairs of Ti^{3+} ions are found in other reduced titanium oxides (see § 5.3.3 in Ref. 40). The same effect is thought to be important in the surface defect properties associated with oxygen vacancies on TiO_2, as mentioned in § 5.2.2 below.

The transitions in V_2O_3 and VO_2 are superficially similar to that in Ti_2O_3 but rather different in detail. The structural changes in V_2O_3 are quite small. The most significant point, however, is that the low-temperature semiconducting phase of V_2O_3 is antiferromagnetic with strong local moments. This is therefore a transition between a metallic phase and a **magnetic insulator**, which is a type of Mott transition (see § 5.2.3 in Ref. 40). The behavior of V_2O_3 is quite sensitive to pressure and to small amounts of substitutional impurities, all of which may act to alter the 3d bandwidth. It seems that the bandwidth is in serious competition with Coulomb repulsion effects in this compound, which is therefore poised on the dividing line between a metal and a magnetic insulator. VO_2 is probably intermediate in behavior between Ti_2O_3 and V_2O_3. The regular rutile structure of the high-temperature metallic phase undergoes a significant distortion, with a pairing of the vanadium atoms. It may be that the transition is driven largely by metal–metal bonding, but although low-temperature VO_2 does not have local moments, both it and the metallic form exhibit anomalous properties which suggest that Coulomb effects are more important than in Ti_2O_3.

It is worth mentioning another type of transition that occurs in some phases based on MoO_3, although no surface electronic studies of these materials have yet been reported. The unusual layered structure of MoO_3 persists in modified form in various reduced phases and in bronzes such as $K_{0.3}MoO_3$. The band structures of these compounds have some low-dimensional aspects, which lead to **Fermi-surface nesting** and charge-density wave anomalies at low temperatures (see § 5.3.1 in Ref. 40).

5.1.2.6 Superconductivity

The best-known oxide superconductors are the 'high-T_c' ones based on copper oxides, although other transition-metal-oxide superconductors are known. These include slightly reduced $SrTiO_3$, with a T_c less than 1 K, and the lithium–titanium spinel $Li_{1+x}Ti_{2-x}O_4$. A large number of surface studies of the high-T_c compounds have been made, and it is appropriate here to summarize briefly some aspects of their bulk properties (see § 5.4 in Ref. 40).

All copper-oxide superconductors appear to have **mixed valency**. Compounds containing only Cu^{2+}, such as La_2CuO_4, are magnetic insulators with strong antiferromagnetic coupling between the local moments of the d^9 ion Cu^{2+}. In most compounds, superconductivity is associated with hole doping, as in $La_{1.85}Sr_{0.15}CuO_4$, which *formally* contains some Cu^{3+}. In the best-known 1:2:3 compound, $YBa_2Cu_3O_{7-x}$, such doping is rather more subtle and is controlled not only by the overall oxygen stoichiometry, but also by the sharing of charge between different types of ions in the structure. Some electron-doped compounds are also known, especially $Nd_{2-x}Ce_xCuO_4$, which contains Ce^{4+} and a corresponding amount of Cu^+. The fact that all of these materials are doped magnetic insula-

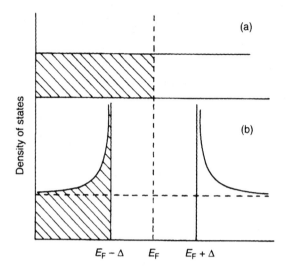

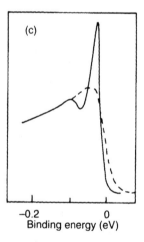

Fig. 5.5 Schematic density-of-states close to the Fermi level for (a) a normal metal, and (b) a superconductor with a BCS energy gap 2Δ. (c) The UPS spectrum of $Bi_2Sr_2CaCu_2O_8$ above and below Tc. [Redrawn from Ref. 456]

tors, with local Cu^{2+} moments probably still present, is one factor that makes the interpretation of their properties very difficult. Spectroscopic measurements, for example, show many of the complex features associated with magnetic insulators, as described below in § 5.1.3.2. One thing that has been clearly established is that the formal chemical assignment of oxidation states gives a misleading idea of the charge distribution. Hole doping, for example, removes electrons more from the O 2p levels than from the Cu 3d.

Of importance to the present account are the unusually short **coherence lengths** in high-T_c superconductors. These mean that the superconducting wavefunction decays rapidly at the surface, and that superconducting contacts between grains in a polycrystalline sample may be very sensitive to surface changes. Such changes are very likely, in fact, because many of these compounds are readily attacked by water and CO_2. This is one reason for the large number of surface studies on high-T_c superconductors. On the other hand, surface techniques have sometimes been found capable of giving information relevant to the bulk materials. An interesting example is the measurement of the superconducting energy gap at the Fermi level, as shown in Fig. 5.5. The upper parts show the predicted density-of-states very close to the Fermi level, E_F, both for (a) a normal metal, and (b) with a superconducting gap as predicted by the BCS theory. Figure 5.5 (c) shows a high-resolution UPS measurement of the filled states in $Bi_2Sr_2CaCu_2O_8$, with T_c about 90 K.[455,456] The change in apparent density-of-states at low temperatures is broadly consistent with theory. However, the BCS *mechanism* for superconduction, based on pairing of electrons through interaction with long-wavelength acoustic phonons, almost certainly does not apply to these materials. As yet there is no consensus on an alternative mechanism.

5.1.3 Photoemission spectroscopy of transition-metal oxides

It was emphasized in Chapter 4 that the simple 'one-electron' interpretation of electronic transitions in solids involves some approximations that may not be valid in many materials. This is particularly true in the case of transition-metal oxides. For example, we have mentioned the occurrence of ligand-field excitations, which are highly localized excitons with Coulomb interactions playing a major role. The photoemission (and inverse photoemission) spectra of transition-metal oxides raise some especially difficult issues, as both core and valence spectra can be dominated by final-state effects. Before discussing these, we shall briefly describe some important resonance effects that occur as a function of photon energy.

5.1.3.1 Resonance effects

Important information on the electronic structure of transition-metal oxides can be obtained by varying the exciting photon energy in UPS. Between the He I (21.2 eV) and He II (40.8 eV) energies the ionization cross-section for metal d orbitals increases relative to the O 2p, so that the changes in the spectra can often be used to distinguish these different levels. Using synchrotron radiation enables the photon energy to be varied continuously over a much wider range, and more dramatic changes in cross-section can occur. (Figures 5.18 and 5.34 below show examples of such changes.) A large resonance in the 3d ionization cross-section occurs at photon energies close to the threshold for cation $3p \rightarrow 3d$ photo-excitation, and has the following interpretation.[457]

The direct photoemission from a $3d^n$ ion can be written

$$3p^6 3d^n + \hbar\omega \rightarrow 3p^6 3d^{n-1} + e^-.$$

When the photon energy is larger than the $3p \rightarrow 3d$ optical absorption threshold, direct photoemission is supplemented by a two-step process. First, there is a $3p \rightarrow 3d$ photoexcitation to an excited state

$$3p^6 3d^n + \hbar\omega \rightarrow \{3p^5 3d^{n+1}\}*.$$

This is followed by a very rapid **super-Coster–Kronig** Auger decay, leading to the same final state as in direct photoemission

$$\{3p^5 3d^{n+1}\}* \rightarrow 3p^6 3d^{n-1} + e^-.$$

The resonance profiles shown in Figs. 5.18 and 5.34 result from interference between the direct and indirect processes.

This model suggests that resonance effects at the $3p \rightarrow 3d$ threshold should be a unique signature to ionization from d levels. However, similar effects of smaller magnitude are found with the valence bands, including those in d^0 compounds such as TiO_2. It is generally assumed that this is due to hybridization between the O 2p and the metal 3d orbitals, and is a sign of the 3d contribution to the valence-band density-of-states, as was illustrated for the calculated levels shown in Fig. 5.1. A number of applications of these resonant effects are illustrated in the detailed discussions of various compounds below.

5.1.3.2 Magnetic insulators: the configuration-interaction model

The photoemission spectra of magnetic insulators are especially susceptible to final-state effects resulting from strong Coulomb interactions. The various types of interpretation that have been used may be illustrated for the example of NiO.

Since the electrons in the ground state of NiO are best described in terms of a local-ized $3d^8$ configuration, one approach to interpreting photoemission spectra is based on the ideas of ligand-field spectroscopy. The configuration $(t_{2g})^6 (e_g)^2$ gives the ground state $^3A_{2g}$. Ionizing an e_g electron gives rise to $(t_{2g})^6 (e_g)^1$ which has only one possible final state, ^2E_g. On the other hand, if the t_{2g} level is ionized, one gets $(t_{2g})^5 (e_g)^2$, where several spectroscopic final states are available. The energies depend upon electron-repulsion parameters as in ligand-field theory, and the intensities of different final states can be predicted by **fractional parentage** arguments.[458] Although these ideas have been applied to the UPS of transition-metal oxides,[156] it is now recognized that a more elaborate interpretation is often required. The supposed 'd' levels appearing at lower binding energy than the valence band in NiO show anomalous behavior under the resonant conditions discussed above. Furthermore, the spectra show strong satellite bands that have no simple interpretation in terms of filled levels. The most satisfactory current model for describing these spectra is that known as the **configuration-interaction (CI)** formalism (see § 2.2.2 in Ref. 40).[459,460]

For NiO the idea is as follows. The ground-state function is written as a linear combination of terms each representing the possible ionic configurations

$$\psi_g = a \, | \, d^8 > + \, b \, | \, d^9 \underline{L} > + \, ... \, . \tag{5.1}$$

Here $\underline{L}$ denotes a hole in the ligand (O 2p) band, so that the second term represents the transfer of charge between 2p and 3d orbitals arising from their overlap and hybridization. a and b are mixing coefficients, and further terms may be added if needed. The same can be done for ionized states with an electron missing

$$\psi_f = \alpha \, | \, d^7 > + \beta \, | \, d^8 \underline{L} > + \gamma \, | \, d^9 \underline{L}^2 > + \, ... \, . \tag{5.2}$$

The mixing coefficients are determined by diagonalizing a matrix in which the diagonal terms represent the energies of the unmixed states, and the off-diagonal terms depend on overlaps. These may be parameterized in a suitable way so as to reproduce both the energies and the intensities of the various states seen in spectroscopic measurements; in principle ligand-field excitations as well as UPS and inverse photoelectron spectroscopy (IPS) information can be included. Figure 5.6 shows how this works for NiO and includes a comparison of the experimental optical absorption and photoemission spectra with the fitted states.[459] In photoemission both the main ionization bands in the region 0–5 eV binding energy and the satellite band around 8 eV are satisfactorily reproduced.

One of the surprising conclusions of the configuration-interaction model is that the first ionization band of many oxides such as NiO does *not* come primarily from ionization of metal 3d orbitals, but has more weight from configurations involving O 2p ionization. This suggests that the bandgap excitation in NiO does not represent an electron transfer between 3d orbitals on different ions, as in the simple Hubbard approach discussed earlier, but is of oxygen $\rightarrow$ metal charge-transfer character. Sawatzky and Allen[461] have suggested that NiO and similar compounds should be called **charge-transfer insulators**, and that the term **Mott insulator** should be reserved for cases where the bandgap is of d $\rightarrow$ d type. Although this may be a useful distinction, it should not obscure the fact that the presence of a d $\rightarrow$ d Coulomb gap of the Hubbard type is still crucial in NiO. The Hubbard U is

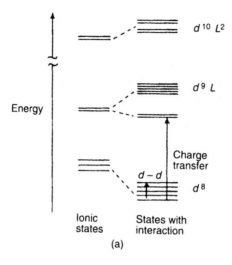

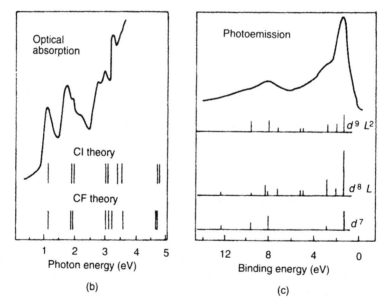

Fig. 5.6 The configuration-interaction (CI) model for NiO: (a) the ionic states, and the results of mixing them; (b) comparison of the CI and crystal-field (CF) models with the optical absorption spectrum; (c) the photoelectron spectrum and breakdown of the ionized states into different configurations. [Redrawn from Ref. 459]

an important parameter in the configuration-interaction model, and without it NiO would have very different properties (see § 3.4.2 in Ref. 40).

An essential feature of the configuration-interaction model is that it enables one to treat the serious *perturbation* of electronic structure that occurs when an elec-

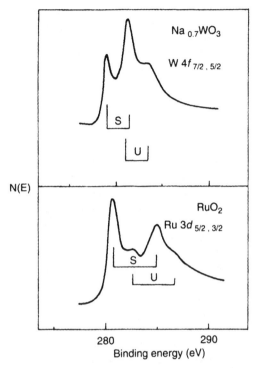

Fig. 5.7 Core-level XPS spectra of $Na_{0.7}WO_3$ and RuO_2 showing satellite structure super-imposed on the expected spin-orbit splitting. See text for explanation of 'S' and 'U'. [Redrawn from Refs. 156 and 463]

tron is ionized; band theory approaches to the spectra of magnetic insulators can-not do this. We shall mention below some angle-resolved band-mapping UPS studies that have been compared with band-theory predictions.[462] We have very serious reservations about this type of interpretation, although at the moment there appears to be no satisfactory alternative model, since the configuration-interaction theory, based on the levels of an *isolated* transition metal ion, is incapable of deal-ing with d-band dispersion.

Strong satellite bands arising from final-state interactions are also a feature of the core-level photoemission spectra of magnetic insulators and may be treated theoretically in a similar way.[460]

5.1.3.3 Core-level spectra of metallic oxides

The valence-level spectra of metallic oxides seem to be relatively simple, and they compare nicely with predicted densities-of-states at the one-electron level. Core-level spectra, however, are often more complicated, and their interpretation has generated a great deal of controversy (see § 5.1.3 in Ref. 40).

Figure 5.7 shows XPS spectra of the W 4f core level in $Na_{0.7}WO_3$ and the Ru 3d

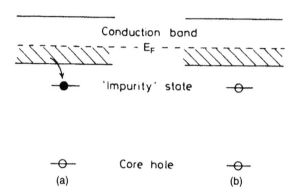

Fig. 5.8 Interpretation of core satellites in XPS of metallic oxides, showing the formation of bound impurity states in the presence of a core hole: (a) screened final state with impurity level occupied; (b) unscreened final state with impurity level empty. [Ref. 40]

level in RuO_2.[156,463] Superimposed on the expected spin-orbit doublet in each case is a more complicated structure that can be resolved into components with at least two separate binding energies. Both spectra have been interpreted by some groups as a sign of mixed valency. Thus for $Na_{0.7}WO_3$ it has been claimed that one is seeing ionization from separate W^{6+} and W^{5+} states present in this compound,[464] and in the case of RuO_2 that there is a surface phase of RuO_3 composition.[465] The interpretation of the tungsten bronze spectrum is contradicted by the fact that this is a metal with delocalized conduction electrons, all tungsten ions appearing structurally equivalent. Also, the intensities of the two core-level signals are not in the correct ratio to correspond to W^{6+} and W^{5+}. The idea of a surface phase on RuO_2 may be harder to refute, but there is no sign of this in the valence-level spectra or the ELS, which are entirely compatible with the electronic structure of the bulk solid.[156]

Complex core-level spectra of the type illustrated seem to be an entirely general phenomenon for metallic transition-metal oxides and are a sign of final-state effects arising from the strong Coulomb interaction between d-band electrons and the core hole produced in ionization.[466,467] This idea is illustrated in Fig. 5.8. The Coulomb effect of the core hole lowers the energy of the valence d orbitals on one ion sufficiently to give a localized impurity state below the conduction band. Different final states may arise depending upon how this is occupied: in the **screened** final state at lower binding energy (labelled 's' in Fig. 5.7) an electron is present, whereas in the **unscreened** final state (labelled 'u') the impurity level is left empty.

The appearance of final-state effects such as this can be a serious complication in surface studies on transition-metal oxides. It does not, of course, *disprove* other possible interpretations such as mixed valency or surface phases of different composition, as these may be present *as well*. But it means that one should be very careful in the interpretation of such spectra and always suspect 'straightforward' interpretations of complex core peaks in terms of different oxidation states.

5.2 The oxides of titanium

The most extensively studied transition-metal-oxide system is that of Ti. The d^0 oxide TiO_2 has become almost the prototypical transition-metal oxide in surface studies for a variety of reasons: it is readily available and inexpensive (TiO_2 is the pigment in most white paints); it can be made conducting in the bulk by simply annealing in UHV, so that electron and ion spectroscopies can be used without any surface charging problems; stoichiometric, well-ordered, nearly perfect surfaces can be produced relatively easily; it is interesting both as a catalyst support (see the discussions in Chapter 7 on metals on oxides and strong metal/support interactions) and in important photocatalytic applications (the discoloring of paint pigment, the photoelectrolysis of water to produce H_2); it is a relatively simple transition-metal oxide to model theoretically; and it has an unusually rich system of reduced oxides (and a correspondingly complicated Ti–O phase diagram!).

5.2.1 Nearly perfect TiO₂ surfaces

The lowest-energy surface of TiO_2 is (110), whose geometric structure is discussed in § 2.3.3.1. It is sufficiently stable that nearly perfect TiO_2 (110) surfaces can be prepared by polishing and annealing; those surfaces exhibit essentially the same electronic structure (as determined by spectroscopies such as UPS, ELS, Auger, etc.) and almost the same chemisorption properties as UHV-fractured surfaces. (As discussed in § 2.3.3.1, TiO_2 does not cleave well, and we will use the term 'fractured'.) Figure 5.9 (a) shows an angle-integrated He I UPS spectrum of UHV-fractured TiO_2 (110).[125] The bulk sample had been reduced, as with all of the TiO_2 samples discussed here, so that the Fermi level is pinned at the bottom of the Ti 3d conduction band (although the density of defect states, and their associated electronic charge, is too low to be observable in UPS). The bulk bandgap region, between 0 and 3.1 eV binding energy, is seen to be relatively free of surface states; those that are present are associated with defects resulting from the fracture process. The emission from the O 2p valence band extends from 3 to 9 eV and rests on a rather high background of inelastically scattered and secondary electrons. (UPS spectra taken with higher photon energies, where the inelastic background is much lower, do not exhibit any additional features in the valence-band region.)

Polished and annealed TiO_2 (110) surfaces exhibit higher quality LEED patterns than do fractured surfaces. They also exhibit sharper structure in UPS spectra of

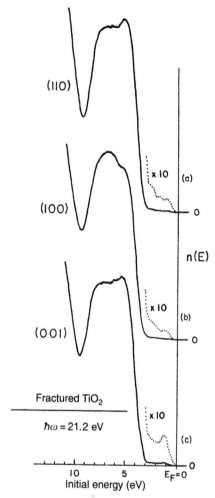

Fig. 5.9 He I UPS spectra for UHV-fractured TiO_2 (110), (100) and (001). All samples
were bulk reduced so that E_F lies at the bottom of the Ti 3d conduction band. [Ref. 125]

of the valence band;[16,126,270,468–470] the solid curve (a) in Fig. 5.10 shows such a
spectrum.[270] ELS spectra for valence-to-conduction band transitions are similar
for the two surfaces, however.

Two other UHV-fractured TiO_2 surfaces, (100) and (001), have been studied by
UPS, as shown in Fig. 5.9 (b) and (c).[125] All three surfaces exhibit essentially the
same spectra, except for different shapes of the valence-band emission (which differ slightly from one fracture to another for the same face anyway) and for a small,
but well-defined, defect peak in the bandgap for the (001) surface [recall that TiO_2
(001) is a highly unstable surface that facets easily (see § 2.3.3.3)]. The important
features to note about the spectra in Fig. 5.9 are that the valence-band width is the

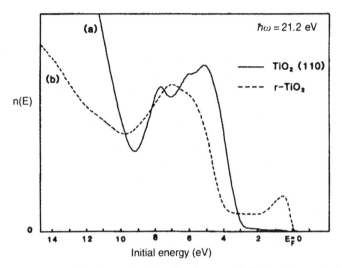

Fig. 5.10 UPS spectra of valence-band region from polished and annealed surfaces of (a) stoichiometric and (b) reduced TiO_2 (110). [Ref. 270]

same for all surfaces, and that no intrinsic surface states are observed in the bulk bandgap for any of the faces. In UPS measurements on polished and annealed TiO_2 (100) and (001) surfaces, the shape of the Ti 3s core-level peak has been interpreted in terms of the sum of bulk and surface-shifted components.[471]

Many ELS spectra for nearly perfect TiO_2 (110) have been reported and interpreted in terms of a joint density-of-states.[16,125,270,271,333,334,468,470,472] Figure 5.11 presents ELS spectra of inter-band transitions on TiO_2 (110) surfaces with and without defects.[468] [Surface phonon losses are also visible at energies below 0.5 eV in Fig. 5.11 (a).] The solid curve is for a nearly perfect annealed surface, and the loss peaks observed at 5.5, 10.2 and 13.7 eV correspond to transitions from the O 2p valence band to empty Ti 3d, 4s and 4p states in the conduction band. There is no evidence in ELS spectra for any empty intrinsic surface states on TiO_2 (110). One IPS measurement of the empty density-of-states of well-characterized single-crystal TiO_2 (110) has been reported.[473] Peaks are found at about 1 and 4 eV above E_F; they are identified respectively as the t_{2g} and e_g components of the Ti 3d-derived bulk conduction band, respectively.

The surface electronic structure of TiO_2 has been treated theoretically using several different approaches. Comparison of the calculated energy levels with experimental results emphasizes some of the difficulties in using one-electron models to describe ionic systems. The earliest calculations used a one-electron LCAO formalism for both $SrTiO_3$ and TiO_2 and predicted a band of surface states, split off from the Ti 3d conduction band, in the bulk bandgap.[474] In order to explain the absence of any bandgap surface states experimentally in these compounds, it was

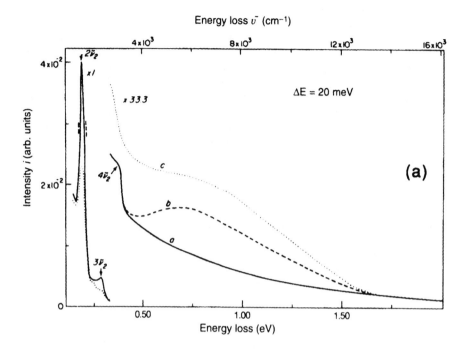

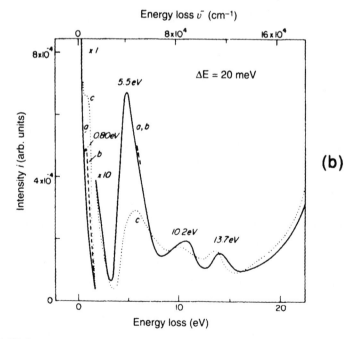

Fig. 5.11 ELS spectra over two energy ranges for TiO_2 (110) surfaces prepared in various ways. Surface (*a*) is stoichiometric, surface (*b*) contains thermally produced defects, and surface (*c*) has been Ar^+ ion bombarded. [Ref. 468]

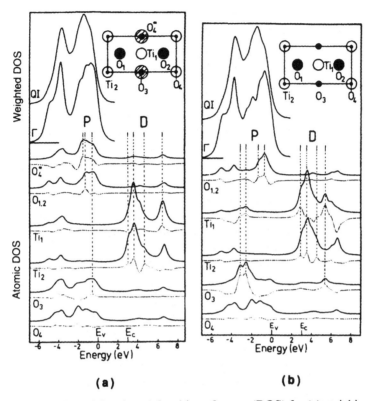

Fig. 5.12 Theoretical partial and total densities-of-states (DOS) for (a) stoichiometric TiO$_2$ (110) and (b) TiO$_2$ (110) with the bridging O ions removed. QI is the k-integrated DOS; Γ is the DOS at the Γ point. [Ref. 479]

necessary to include the effects of Coulomb repulsion between d-electrons on the amount of covalent mixing of O 2p and Ti 3d wavefunctions that occurs at the surface (i.e., **surface-enhanced covalency**).[475]

Calculations of the surface electronic structure of both TiO$_2$ (110) and (001) were also performed using the **linear combination of muffin-tin orbitals (LCMTO)** one-electron method in which relaxation of the surface atoms was included.[476] However, the bandgap on the (110) surface was predicted to be only 1.78 eV, substantially less than the experimental value of 3.1 eV; this resulted from the presence of O 2p-derived surface states in the bulk bandgap. The density-of-states calculated for the (001) surface, in which the surface cations are only four-fold coordinated (see § 2.3.3.3), was even farther from the experimental observations; both O 2p and Ti 3d states were pulled into the gap, resulting in a surface bandgap of only 0.95 eV. DV-Xα cluster calculations were also performed for TiO$_2$ (110).[364,477,478] The surface bandgap predicted by this method was 1.8 eV, in agreement with the LCMTO results, although both O 2p and Ti 3d surface states

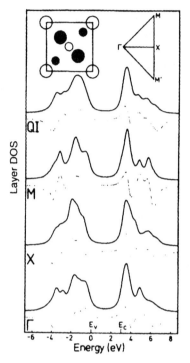

Fig. 5.13 Theoretical wave-vector-resolved (M, X, Γ) and total (QI) densities-of-states for TiO$_2$ (001). [Ref. 479]

were responsible for the narrowing of the bandgap. No cluster calculations were performed for the (001) surface.

Theoretical calculations of the electronic structure of the unrelaxed TiO$_2$ (110), (100) and (001) surfaces that are in agreement with experiment have been performed by the LCAO scattering-theoretical method.[414,448,479] The densities-of-states were calculated not only for single planes of ions, but also on individual ions within a plane. Figure 5.12 shows densities-of-states for layers and individual ions in the surface plane of TiO$_2$ (110), both (a) with and (b) without the layer of bridging O ions; the surface in (a) is the one that we refer to as ideal. Also shown as dotted lines in the figure are the differences between the densities-of-states on individual surface ions and their bulk counterparts. The features labelled P are O 2p surface resonances, and Ti 3d surface resonances in the conduction band are labelled D. These results are similar to those for SnO$_2$ (110) discussed in § 4.6.1; no surface states appear in the bulk bandgap for either the ideal surface model or for the 'compact' surface from which all of the bridging O ions have been removed. The results for the TiO$_2$ (001) surface, presented in a slightly different way, are shown in Fig. 5.13. The solid curves are the densities-of-states for the surface plane at three symmetry points in the Brillouin zone (M, X and Γ), and inte-

integrated over the entire zone (QI). These densities of states are strikingly different from the LCMTO results.

The TiO_2 (110) surface density-of-states has also been calculated by using a tight-binding extended-Hückel method.[480] Once again no surface states are found in the bulk bandgap. The valence-band density-of-states is qualitatively similar to that obtained by the scattering-theoretical method.

5.2.2 Point defects on TiO_2 surfaces

When point defects (predominantly O ion vacancies) are created on TiO_2 surfaces, they cause a dramatic change in electronic structure. In most of the experimental studies on TiO_2, surface defects are created by ion or electron bombardment or by heating to a high temperature and quenching. However, laser irradiation has also been reported to result in the formation of similar surface defects,[481,482] as has the deposition of sub-monolayer amounts of metallic Ti onto TiO_2 (110) and (100).[23,483] The basic physics of defects on transition-metal-oxide surfaces can be seen by considering an idealized O vacancy in the surface plane (see § 2.5). Two electrons are left on the surface upon removal of an O^{2-} ion, and they must occupy some states based on Ti 3d orbitals. From our discussion of bulk electronic structure above, we might expect these electrons to occupy states that are more localized than in most non-transition-metal oxides. In chemical terms, Ti can form oxides with the lower oxidation states Ti^{3+} and Ti^{2+}, and this suggests that one or two electrons could occupy a metal site immediately adjacent to the O vacancy. Most studies of defects on TiO_2 have been interpreted along these lines. The occupied defect state corresponding to partial population of the Ti 3d orbitals lies in the upper half of the bulk bandgap and is readily apparent in UPS spectra.[16,20,23,126,131,269,270,440,469,471,484–489] It can be seen clearly in the dashed spectrum (b) in Fig. 5.10, taken from a TiO_2 (110) surface that had been reduced by bombardment with 500 eV Ar^+ ions.[270] That the state is primarily Ti 3d-derived is corroborated by its resonant behavior in UPS taken across the photon energy range that encompasses the Ti 3p $\rightarrow$ 3d optical excitation threshold.[488] The position and amplitude of the defect surface state depend upon the amount of surface reduction; this will be discussed in detail below. The additional charge on the surface Ti cations is also apparent in XPS spectra of the Ti 2p core levels.[270,468,480,490,491]

Figure 5.10 also shows that the O 2p valence band moves away from E_F when defects are created on the surface. The simplest way in which to interpret this would be to assume a rigid band model, with reduction merely moving E_F up into the conduction band; the occupied defect states would then correspond to originally empty conduction-band states. The situation is more complex, however, since the chemical composition of the reduced surface is different from that of the stoichiometric surface. In other words, a very thin layer of a different compound is formed on the TiO_2 substrate. A surface-sensitive UPS spectrum samples predominantly the surface compound, although contributions from the substrate will also be present. No analysis of how to accurately interpret UPS spectra from such surfaces has been published.

Accompanying the creation of surface defects on reduced TiO_2 is a new feature in the ELS spectra, which can be seen in Fig. 5.11 (a),[334,468] where defects were produced by annealing and quenching (curve *b*). A loss peak appears corresponding to a transition energy of about 1 eV. (In other ELS experiments having lower resolution, and where a larger density of defects was produced by electron or ion bombardment, this loss has been assigned a transition energy of 1.3– 2 eV.[16,269,270,333,335]) The simplest interpretation of this loss peak is that it corresponds to a transition between the occupied Ti 3d defect state and the lowest-lying empty Ti 3d level at the defect site; by analogy with the electronic structure of Ti_2O_3, it has been described in terms of a transition from bonding to antibonding orbitals of paired Ti^{3+} ions.[269] However, as mentioned later, this interpretation of the Ti_2O_3 spectrum may be overly naive. The intensity of the loss peak in reduced TiO_2 suggests that it cannot be a simple d → d transition, but may correspond to the charge-transfer of an electron from Ti^{3+} to a neighboring Ti^{4+}; such a transition may be strongly influenced by polaronic lattice interactions.[333,335]

IPS measurements have been made of the changes that occur in the normally empty density-of-states of TiO_2 (110) upon defect creation.[473] A new peak appears at about 3 eV above E_F, whose location agrees with theoretical calculations based upon removal of bridging and in-plane surface O ions. The effect of point defects on TiO_2 surfaces is apparent in other surface-sensitive spectroscopies, but the information on either geometric or electronic structure is usually more convoluted than in UPS, XPS, or ELS, and we shall not consider those results here.

Detailed experimental studies have been conducted of the changes in surface electronic structure on TiO_2 (110) as a function of the density of defects created by 500 eV Ar^+-ion bombardment.[269] LEED and Auger spectroscopy were used to monitor surface order and composition, and UPS and ELS were used to determine the changes in both filled and empty surface electronic levels. The results are summarized in Fig. 5.14, which plots the relative energies of the Fermi level (E_F), the location of the defect surface-state peak (E_S) and the valence-band maximum (E_V), the change in the sample work function ($\Delta\Phi$), and the amplitude of the defect ELS peak, all as a function of the integrated area (above background) under the UPS defect surface-state peak (α_s). α_s is taken to be proportional to the density of defect states on the surface. The filled symbols correspond to the creation of defects on an annealed, nearly perfect surface, while the open symbols correspond to changes induced by subsequent exposure of the defective surface to O_2; the lat-

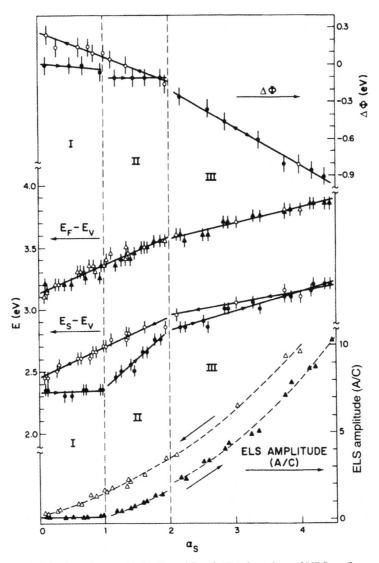

Fig. 5.14 Work function change ($\Delta\Phi$), Fermi level (E_F), location of UPS surface-state peak ($E_s - E_v$) and normalized amplitude of 2 eV ELS peak versus normalized intensity of UPS surface-state peak (α_s) for Ar$^+$-ion bombardment and subsequent O$_2$ exposure of TiO$_2$ (110). Arrows indicate the sequence in which the data were taken. [Ref. 269]

ter will be considered in § 6.4.1. The data are interesting in that three distinct regions, corresponding to different defect phases, are observed. For low defect densities (region I), the amplitude of the peak due to extrinsic surface states increases by more than a factor of ten, but the peak remains centered at 2.3 eV above E_v, suggesting that the defects are isolated and non-interacting. Note, however, that the valence band moves continuously away from E_F as the defect density increases. By $\alpha_s = 1$, the LEED patterns of the annealed surface have virtually dis-appeared, although Auger spectra show almost no loss of O from the surface. The defect ELS peak does not appear in region I, indicating that no well-defined low-lying excited state exists. In region II, the defect surface state moves rapidly closer to the conduction band and away from the valence band, and the ELS peak appears and grows in amplitude. The surface also begins to lose O in this region, and it is at this point that the Ti ions at defect sites appear to interact, as evidenced by the appearance of the low-lying excited state. At $\alpha_s = 2$, a phase transition occurs that is interpreted as the formation of islands of a reduced oxide, similar to Ti_2O_3, on the surface. The work function falls in the region $\alpha_s > 2$, reaching roughly that of bulk Ti_2O_3 for $\alpha_s \approx 4$. The occupied defect state continues to move away from E_V, eventually overlapping the bulk conduction band.

Although the above study is the most detailed one to date of the formation and interaction of defects on oxide surfaces, most of the basic features are found also for defects on other transition-metal oxides. New peaks (or shoulders) in bulk bandgaps are only seen for insulating oxides such as TiO_2, $SrTiO_3$, NiO, CoO, etc. On metallic and narrow-bandgap semiconducting oxides, by contrast, the charge associated with O-vacancy defect sites appears to be less localized, and is seen in UPS as an increase in the amplitude of the emission features near E_F (e.g., Ti_2O_3, V_2O_3). The valence band generally moves away from E_F when surface defects are created.

Resonant photoemission has been used to study the hybridization between O 2p and Ti 3d orbitals on both nearly perfect and reduced TiO_2 (110) surfaces.[488] As the photon energy is varied across the Ti 3p $\rightarrow$ 3d optical absorption threshold near 47 eV, the resonant behavior of the emission from the bonding O 2p region of the valence band (at high binding energy) is different from that for the non-bonding region of the band. Not only is the resonance for the non-bonding orbitals weaker due to the reduced mixing between O and Ti wavefunctions in that region of the valence band, but the shape of the resonance indicates hybridization pre-dominantly with Ti 4s orbitals, while the shape of the resonance for the bonding region of the band shows both Ti 3d and 4s mixing. The hybridization of Ti 4s orbitals with the valence band is enhanced as a result of the reduced symmetry at the surface. When O-vacancy point defects are created on the TiO_2 (110) surface, the shape and amplitude of the resonance profiles suggest that there is less hybridization of the Ti 3d orbitals with the non-bonding O 2p orbitals at surface defects than on the defect-free surface. However, the amplitude of the bonding emission from the valence band increases relative to the non-bonding emission, which suggests that there is more overall covalent mixing between the Ti and O ions for the reduced surface than for the stoichiometric one.

Essentially the same changes in surface electronic structure that occur for TiO_2 (110) upon point defect creation are seen for TiO_2 (100) and (001) as well.[16,23,126,471,489,492] The defect surface-state peak in UPS is often larger on the (001) surface than on the other two, presumably due to the inherent instability of that surface.

Scanning tunneling spectroscopy has been attempted on defective single-crystal TiO_2 surfaces in air.[493–495] While there are still problems with interpretation of the data, and hysteresis is found in many of the spectra, an estimate of the bandgap at the (110) surface of 1.6 eV has been made.[493] This number cannot be compared with the UHV results above, however, due to different surface preparation procedures and the presence of adsorbed molecules. Scanning tunneling spectroscopy has also been performed on the (110) surface of a very heavily reduced TiO_2 sample in UHV.[155] Tunneling spectra showed the bulk conduction-band edge, as well as an electronic state about 0.5 eV below the conduction-band edge. The primary difference between the UPS spectra for defective TiO_2 (110) and the tunneling spectra is that the electronic states 1 eV below the conduction band are occupied in the former case, but they appeared to be unoccupied in the latter. This could be caused by local band-bending at the surface induced by the tunneling measurement itself, or by surface contamination.

Most of the theoretical methods applied to perfect TiO_2 surfaces have also been used to determine the changes in electronic structure produced by surface O-vacancy point defects. The main focus of the calculations has been on the prediction of bandgap surface states. While all calculations predict such states, there are major differences in the type of surface structure found necessary to produce them. DV-Xα cluster calculations treated surface defects on TiO_2 (110) by removing one O ion from the surface of the cluster.[354,478,479] They predict a defect surface state, consisting primarily of 3d wavefunctions on the Ti ions adjacent to the defect, at about 1 eV below the bottom of the bulk conduction band. A tight-binding extended-Hückel calculation reached a similar conclusion for a defect on TiO_2 (110) consisting of the removal of a single bridging O ion;[480] a Ti 3d state appeared in the gap about 0.7 eV below the bottom of the conduction band. *Ab initio* molecular-orbital cluster calculations of a bridging O vacancy have shown that the electronic charge necessary to maintain charge neutrality at an O-vacancy defect site is localized almost entirely on the ions directly adjacent to the vacancy.[496]

The tight-binding LCAO scattering-theoretical method has been used to study the properties of several different surface defect configurations on TiO_2 (110); relaxation of the surface ions to minimize the total energy of the system was included.[449,497] While the calculations do predict that one defect configuration will give rise to a state in the bulk bandgap, they do not appear to be in agreement with experimental observations. In the calculations, removal of bridging O ions does not produce bandgap surface states; in fact, the entire plane of bridging O ions can be removed without creating such a state. This is analogous to the situation for SnO_2 (110) discussed in § 2.6.2. Even removal of an O ion from the main Ti-O surface plane, after all of the bridging O ions have been removed, still does not produce a bandgap state. It is necessary to remove all bridging O ions and then one O ion from the layer *below* the surface plane in order to generate a state in the gap. This is not consistent with experimental results, in which a defect surface state appears in the bulk bandgap *immediately* upon ion or electron bombardment.

5.2.3 Magnéli phases in the Ti–O system

When TiO_2 is reduced in the bulk, a series of interesting phases are formed which consist of slabs of the rutile structure separated by a regular array of shear planes on which the cation density is higher than in TiO_2; these are referred to as **Magnéli phases**.[498] They exist in the composition range between TiO_2 and Ti_2O_3, and their bulk shear plane structure is well-understood. There has been one surface study of a Magnéli phase with a composition close to TiO_2.[153] Single-crystal TiO_2 was heavily reduced by heating in UHV in order that STM imaging and spectroscopy could be performed on it without problems due to charging. The resultant STM images (shown in Fig. 2.21) exhibit a surface structure that corresponded to (121) shear planes intersecting the (110) surface in the [1$\bar{1}$1] direction. Since atomic resolution was not quite achieved, it was not possible to confirm the atomic arrangement on the planes, but the direction of the surface structure can only be explained by such crystallographic shear planes.

5.2.4 Ti_2O_3

When the Ti-O stoichiometry reaches 2:3, a stable, homogeneous bulk phase having the corundum structure results. Each Ti ion in Ti_2O_3 has one 3d orbital occupied, and pairing of the 3d orbitals of nearest-neighbor Ti ions results in a semiconductor having a 0.1 eV bandgap at room temperature. (As mentioned in § 5.1.2.5, there is a transition to metallic behavior over the temperature range 550 – 750 K associated with a weakening of the metal–metal bonding). The material is thus sufficiently conducting that no surface charging occurs. Single crystals of Ti_2O_3 are available, and a number of their surface and defect properties have been studied by using various electron spectroscopies.[26–28,485,499–502] The geometry of the (10$\bar{1}$2) cleavage face is described in § 2.3.4.1.

The electronic structure of the nearly perfect, UHV-cleaved (10$\bar{1}$2) surface of Ti_2O_3 is shown in Fig. 5.15, which presents angle-integrated UPS spectra taken at two different photon energies.[26,173] The emission from 4 to 12 eV corresponds to the predominantly O 2p valence band, while the emission between 0 and 1.5 eV is

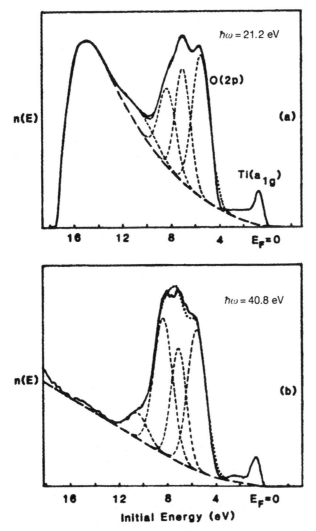

Fig. 5.15 (a) He I and (b) He II UPS spectra for UHV-cleaved Ti$_2$O$_3$ (10$\bar{1}$2). Solid lines are the experimental data, long dashed lines are assumed backgrounds, short dashed curves are Gaussians used to fit the O 2p band, and the dotted curves are the resultant fits. [Ref. 173]

from the Ti 3d a$_1$ band. Since the bandgap in Ti$_2$O$_3$ is only 0.1 eV, the Fermi level, E_F, lies right at the upper edge of the a$_1$ band. The short-dashed curves in Fig. 5.15 are Gaussians used to fit the spectra in order to better interpret the changes in the spectra that result from the different photon energies. (There is no physical significance to either the use of Gaussians in this type of fit or the number and location of Gaussians used; they are merely a convenient functional form that is able to accurately reproduce the measured spectra.) The main difference is an increase in the

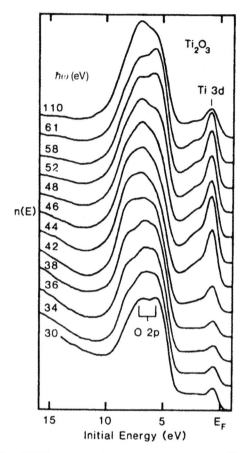

Fig. 5.16 Valence-band UPS spectra of UHV-cleaved Ti_2O_3 ($10\bar{1}2$) as a function of photon energy across the Ti 3p → 3d optical absorption threshold. [Ref. 503]

relative amplitude of the emission near 8 eV as the photon energy is increased from 21.2 to 40.8 eV. Since the photoexcitation cross-section for d states increases relative to that for s and p states as the photon energy is increased, the observed change in amplitude is consistent with the expected admixture of Ti 3d wavefunction into the higher-binding-energy region of the valence band.

The UPS spectra in Fig. 5.15 are in basic agreement with the bulk band structure of Ti_2O_3 as determined from optical absorption measurements, in that there appears to be a gap of about 2.5 eV between the a_1 and O 2p bands. However, there is some question as to whether or not intrinsic surface states are present in that region of the spectrum. UPS spectra taken with photon energies between 40.8 and 110 eV show signals between 1.5 and 4 eV binding energy that resonate with the Ti 3d a_1 band in the vicinity of the Ti 3p → 3d optical absorption threshold, as shown in Fig. 5.16;[503] the bandgap emission is most prominent near $\hbar\omega$ = 46 eV.

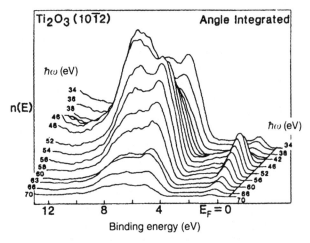

Fig. 5.17 Series of angle-integrated UPS spectra for UHV-cleaved Ti$_2$O$_3$ (10$\bar{1}$2) as a function of photon energy. [Ref. 502]

(The spectra in Fig. 5.16 have been normalized to the intensity of emission from the O 2p band.) This suggests that intrinsic surface states having predominantly Ti 3d character may be present in the bulk bandgap for Ti$_2$O$_3$ (10$\bar{1}$2) surfaces.

A comparison between the spectra in Fig. 5.15 for stoichiometric Ti$_2$O$_3$ with the dashed curve (b) in Fig. 5.10 for defective TiO$_2$ surfaces shows striking similarities. The shapes and energy locations of the bands are almost identical, although there is less structure in the spectra for the reduced and disordered TiO$_2$ surface. It is for this reason that the structure of the heavily reduced TiO$_2$ (110) surface discussed in § 5.2.2 and Fig. 5.14 above was interpreted in terms of the formation of islands having a composition close to that of Ti$_2$O$_3$. The Ti$_2$O$_3$ structure, with pairs of Ti ions sharing faces of the O octahedra, appears to be a relatively stable configuration.

Both angle-integrated and angle-resolved UPS spectra of UHV-cleaved Ti$_2$O$_3$ (10$\bar{1}$2) surfaces have been performed using tunable synchrotron radiation in order to determine the dispersion of the bulk energy bands and the nature of the Ti–O orbital hybridization in the valence band.[500–503] Figure 5.17 presents a series of angle-integrated UPS spectra for UHV-cleaved Ti$_2$O$_3$ (10$\bar{1}$2) for $34 \le \hbar\omega \le 70$ eV; in contrast to the spectra in Fig. 5.16, all spectra in Fig. 5.17 have been normalized to the incident photon flux.[502] Not only are there large, resonant changes in the amplitude of emission from both the O 2p and Ti 3d bands, but the shape of the valence band also changes with photon energy. These changes are similar to those discussed in § 5.2.2 for reduced TiO$_2$ (110) surfaces and result from resonance of the Ti 3d and 4s contributions to the valence band. The resonant behavior of the Ti 3d a$_1$ band near the Ti 3p $\rightarrow$ 3d optical absorption edge depends upon the angle of

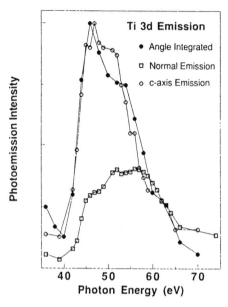

Fig. 5.18 Photon-energy dependence of the Ti 3d UPS emission from UHV-cleaved Ti₂O₃ (10$\bar{1}$2) as measured in the angle-integrated mode, in angle-resolved normal emission, and angle-resolved with the detector set 55° off normal (along the bulk c-axis direction). [Ref. 502]

electron emission in an interesting way, as shown in Fig. 5.18.[502] The total Ti emission in angle-integrated spectra such as Fig. 5.17 rises sharply at the threshold for Ti 3p → 3d excitation, peaking for $\hbar\omega \approx 47$ eV. When only electrons emitted normal to the (10$\bar{1}$2) surface are detected, the peak at 47 eV is absent, although a broad resonance is still observed. The a_1 orbital which gives the Ti 3d emission is composed of atomic d_{z^2} orbitals pointing along the internuclear axis of the two Ti ions that share the face of an O octahedron. Since the pairs of Ti ions are aligned along the c-axis, it was postulated that the emission near 47 eV arose from electrons emitted along the Ti–Ti axis; that axis is oriented 55° away from the normal to the (10$\bar{1}$2) surface, so those electrons would not be detected in normal-emission spectra. This was confirmed by orienting the electron detector along the c-axis, as shown by the open circles in Fig. 5.18. It thus appears as though the photoelectrons ejected from the a_1 orbitals just above resonance are emitted preferentially in the direction of the lobes of charge of the covalent Ti–Ti bond. The broader resonance seen in normal emission, as well as for emission along the c-axis, is believed to correspond to Ti 3p → 4s excitation.

First-derivative ELS spectra for UHV-cleaved Ti₂O₃ (10$\bar{1}$2) are shown in Fig. 5.19.[26] The solid curve is for the clean cleaved surface, while the dotted curve is taken after exposure to O₂ (the effects of O₂ chemisorption on Ti₂O₃ are discussed

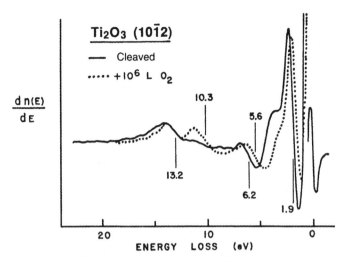

Fig. 5.19 ELS spectra for UHV-cleaved Ti_2O_3 ($10\bar{1}2$) before (solid curve) and after (dotted curve) exposure to 10^6 L O_2. [Ref. 26]

in § 6.5.1 below). The dominant feature in both spectra is the loss peak at about 1.5–1.9 eV, which in a one-electron interpretation must originate from excitation of electrons in the a_1 band to higher states in the conduction band, probably corresponding to a bonding $\rightarrow$ antibonding transition localized on the pairs of Ti ions. However, studies on the bulk electronic structure of Ti_2O_3 suggest that this may be a rather over-simplified picture. Measurements of optical reflectance changes through the semiconductor/metal transition at 393 K suggest that the electronic excitations of Ti_2O_3 must have a considerable *collective electron* character and are better regarded as plasmons, only slightly perturbed by the small bandgap, than as one-electron excitations.[504] The other features in the loss spectra are similar to those seen for either stoichiometric or defective TiO_2 (110); compare with the $N(E)$ spectra in Fig. 5.11.

No theoretical calculations have been reported of the surface electronic structure of Ti_2O_3; there are not even any very complete calculations of its bulk electronic structure. Therefore it has not been possible to separate surface from bulk electronic structure for the nearly perfect Ti_2O_3 ($10\bar{1}2$) surface, although surface features must certainly be present in the UPS and ELS spectra discussed above. The only thing that can be said at present is that the experimental spectra are consistent with what is expected for the bulk electronic structure, and no intrinsic surface states have been definitely identified.

When defects are created on Ti_2O_3 by particle bombardment, distinct changes in the surface electronic structure are observed.[26] As for TiO_2, the predominant defects are O vacancies, since the Auger and XPS data show that the surface is

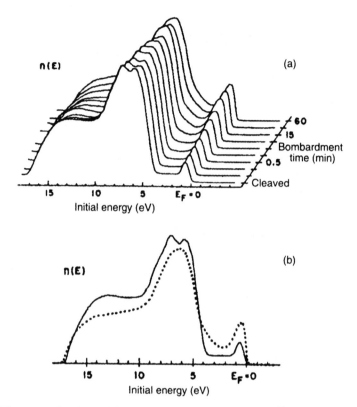

Fig. 5.20 UPS spectra for (a) Ar⁺-ion bombardment of Ti_2O_3 ($10\bar{1}2$) and (b) the cleaved
(solid curve) and steady-state ion-bombarded (dotted curve) surfaces. [Ref. 26]

reduced. The changes in the valence-band density-of-states produced by bombardment with 500 eV Ar⁺ ions are shown in Fig. 5.20. Figure 5.20 (a) presents a series
of UPS spectra as a function of ion-bombardment time for a cleaved ($10\bar{1}2$) surface, while the first and last spectra are replotted in Fig. 5.20 (b). Removal of surface O^{2-} ions deposits more electronic charge into the Ti 3d orbitals, and surface
states are created in the bandgap between the a_1 and O 2p bands. The structure in
the valence band is smeared out, which is consistent with the absence of LEED
patterns for the defective surface.

Efforts have also been made to find a way of preparing nearly perfect Ti_2O_3
($10\bar{1}2$) surfaces by some technique other than cleaving so that the same sample
could be re-used many times. This is not an easy task, as the thermodynamically
most stable oxide in the presence of oxygen is TiO_2. However, a procedure involving ion bombardment followed by rapid annealing and cooling in UHV was found
to give surfaces that are indistinguishable from UHV-cleaved surfaces as determined by LEED, UPS, XPS, Auger spectroscopy and O_2 chemisorption.[28] This

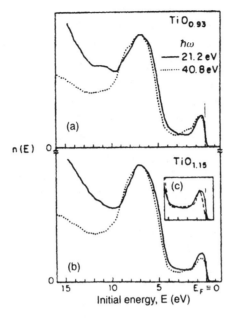

Fig. 5.21 UPS spectra for UHV-fractured (a) $TiO_{0.93}$ and (b) $TiO_{1.15}$. (c) Comparison of UPS spectra for UHV-fractured $TiO_{1.15}$ (solid curve) and Ti_2O_3 (dashed curve). [Ref. 505]

mitted a great many chemisorption measurements to be made on Ti_2O_3 ($10\bar{1}2$) that would have been impossible if cleaved surfaces had to be used; that work is described in § 6.5.1.

5.2.5 $TiO_{x\approx 1}$

The Ti–O system can be further reduced to produce the rocksalt oxide TiO_x. This is not as stable as Ti_2O_3, and a wide stoichiometry range exists, with $0.6 < x \leq 1.28$. For compositions close to $TiO_{1.0}$, the structure distorts to monoclinic, and across the whole stoichiometry range there are substantial densities (up to 15 %) of both cation and anion vacancies in the bulk. Single-crystal samples are not available, but one calculation of the surface band structure of TiO has been performed, as have some surface-sensitive UPS measurements on UHV-fractured polycrystalline samples.

The UPS measurements were performed on samples having compositions in the rocksalt structure range: $TiO_{0.93}$ and $TiO_{1.15}$.[505] Figure 5.21 presents spectra for each surface taken at both He I (21.2 eV) and He II (40.8 eV) photon energies. [While the electronic structure looks very similar to that for Ti_2O_3, note the large density-of-states at E_F that results from the oxide being metallic; the insert (c) directly compares the Ti 3d emission from UHV-fractured $TiO_{1.15}$ with that for

UHV-cleaved Ti_2O_3 ($10\bar{1}2$).] A small shoulder is present on the low-energy side of the Ti 3d band in the He II spectra but not in the He I spectra. This emission may arise from surface states, since the escape depth is smaller for photoelectrons excited by 40.8 eV photons (≈ 5 Å) than for electrons excited by 21.2 eV photons (≈ 10–15 Å). A small peak is also seen below the O 2p band in the surface-sensitive spectra for $TiO_{0.93}$, but not for $TiO_{1.15}$; its origin is unknown.

Theoretical calculations of the electronic structure of the ideal $TiO_{1.0}$ (100) surface have been carried out with the transfer-matrix formalism.[506] Intrinsic surface states and surface resonances were found at various locations within the Brillouin zone. The calculation was highly simplified, however, since the ideal rocksalt structure was assumed, with no cation or anion vacancies. Only Ti 3d and O 2p orbitals were included in the model Hamiltonian, and effects due to electron correlation, surface relaxation and reconstruction were neglected. It is thus doubtful that the results are of much relevance to experiments on real materials.

5.3 Ternary d⁰ oxides

The titanates having the $ATiO_3$ perovskite structure, in which the A cation has a valence of 2+, are close relatives of TiO_2. This is because the highest filled and the lowest empty A orbitals are significantly farther away from E_F than are the O 2p and Ti 3d levels. Thus the electronic structure in the vicinity of E_F and the vacuum level is dominated by the Ti and O ions. The surfaces of two of these perovskite oxides, $SrTiO_3$ and $BaTiO_3$, have been studied on single-crystal samples.

$LiNbO_3$ and $LiTaO_3$ have the ilmenite structure, but they are electronically related to the perovskite titanates in that they are insulators having filled O 2p valence bands and empty Nb 4d or Ta 5d conduction bands. The defect states that appear when the surfaces of these oxides are reduced are also similar to those of $SrTiO_3$ and $BaTiO_3$, so we will consider all four oxides together.

5.3.1 SrTiO₃

There has been a great deal of interest in the surface properties of $SrTiO_3$ both because of its efficiency as a photocatalytic electrode in the photoelectrolysis of water and, more recently, because of its use as a substrate for the growth of high-temperature superconducting oxide films. Several single-crystal studies have been reported,[111–113,410,507–512] and the results are similar to those for TiO_2.

A UPS spectrum for UHV-fractured $SrTiO_3$ (100) is shown in Fig. 5.22.[111] The sample had been bulk reduced by heating in UHV, and E_F is pinned just below the conduction-band minimum. Emission from the predominantly O 2p valence band extends from binding energies of about 3 to 9 eV. Very little emission is seen in the

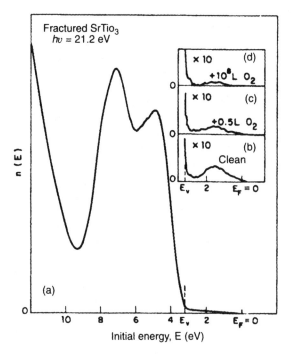

Fig. 5.22 UPS spectrum for UHV-fractured $SrTiO_3$ (100). Insets show depopulation of bandgap defect state upon exposure to O_2. [Ref. 111]

bulk bandgap, and what little is present results from residual defects left by the fracture; they are readily depopulated by exposure to O_2, as shown in the inset. Similar spectra, exhibiting little or no bandgap emission, have been reported for (100) surfaces that were ion bombarded and annealed, and they have been attributed to a stoichiometric surface.[113,508–512] There is some question as to the actual surface composition in that case, however, since studies of ion bombardment and subsequent annealing of UHV-fractured $SrTiO_3$ (100) surfaces showed that bombardment depleted the surface in O and Sr relative to Ti, and that annealing did not restore the surface stoichiometry.[111] It is possible therefore that the surface electronic structure, with no occupied bandgap states, is similar over a range of surface Sr/Ti compositions, provided that the surface is fully oxidized so that the d^0 configuration of surface Ti^{4+} ions is maintained.

As is the case for TiO_2 and Ti_2O_3, the shape of the O 2p valence-band emission in UPS spectra for $SrTiO_3$ is a function of the photon energy used. In one study, the relative increase in amplitude of the lower-binding-energy region of the O 2p band in He I spectra relative to He II was interpreted in terms of the existence of an O-derived intrinsic surface state near the upper edge of the band.[113] An alternative interpretation, however, is that the shape of the two spectra differ because of a

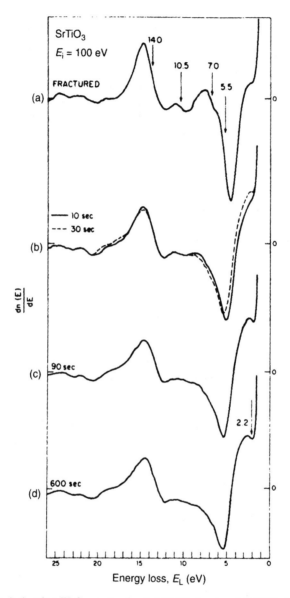

Fig. 5.23 First-derivative ELS spectra for (a) UHV-fractured $SrTiO_3$ (100), and (b)–(d) as defects are created by Ar^+-ion bombardment for the times indicated. [Ref. 111]

change in the cross-section for emission from the Ti 3d orbitals mixed into the higher-binding-energy region of the O 2p band;[173,488,509,511] this effect in Ti_2O_3 $(10\bar{1}2)$ was discussed above in connection with in Fig. 5.15. UPS spectra of the Ti 3s and Sr 3d core levels for both UHV-fractured and polished and annealed $SrTiO_3$ (100) have been interpreted in terms of contributions from bulk and surface ions, with the core levels for the latter shifted by about 2 eV with respect to the former.[513–515] Similar shifts were inferred for emission from the Ti 3d orbitals on SrTiO$_3$ (100) based upon resonant and angle-resolved UPS measurements.[510]

The first-derivative ELS spectrum for UHV-fractured SrTiO$_3$ (100) is shown in Fig. 5.23 (a).[111] It is similar to that for TiO$_2$ (110) in Fig. 5.11 (curve *a*) in that there are no loss features with energies less than the bulk bandgap. However, the location of the loss peaks is slightly different than for TiO$_2$, and there is an additional transition at 7.0 eV. Since the density of filled valence-band states is so similar to that for TiO$_2$, the differences may lie in the empty-state structure above E_F.

One experiment has been performed using IPS to determine the density of empty electronic states in SrTiO$_3$ (100).[516] A polished and annealed sample was used, along with ion-bombarded surfaces that had not been annealed, and surfaces exposed to O$_2$. The IPS spectra were very similar in all cases, indicating that no empty surface states existed above E_F. It should be noted, however, that an n-type sample was used, so that E_F was pinned at the bottom of the bulk conduction band. Therefore any empty surface states, which should have been present on the ion-bombarded surface, would have overlapped the bulk conduction band; they would therefore be surface resonances that would have been very difficult to observe.

The creation of surface point defects on UHV-fractured SrTiO$_3$ (100) surfaces has also been studied, and overall similarities, but detailed differences, between SrTiO$_3$ and TiO$_2$ are found.[111] Figure 5.24 shows the changes produced in He I UPS spectra of fractured SrTiO$_3$ (100) as the surface is bombarded with 500 eV Ar$^+$ ions for different times. The spectra in Fig. 5.24 correspond directly to the ELS spectra in Fig. 5.23. A defect surface state appears in the bandgap as soon as the surface is bombarded, growing in intensity until it reaches saturation (which has occurred by 600 sec of bombardment for the ion flux used in this experiment). The detailed changes in the UPS spectra as a function of the amplitude of the bandgap surface state, α_s, are shown in Fig. 5.25. (The parameters in Fig. 5.25 are the same ones used in Fig. 5.14 above for TiO$_2$.) During the initial stages of defect creation, $0 \leq \alpha_s \leq 1$, LEED and Auger spectroscopy indicate that the surface disorders without any significant change in stoichiometry; this is exactly the same behavior as for TiO$_2$ (110). For $\alpha_s > 1$, the surface becomes deficient in both O and Sr relative to Ti. For SrTiO$_3$ the location of the surface-state peak moves away from the valence band for all values of α_s, and the valence band no longer moves away from E_F for $\alpha_s > 1$; this behavior is different from that for TiO$_2$ (110).

The bandgap emission on defective SrTiO$_3$ (100) surfaces has also been studied on polished and annealed samples for different surface treatments.[508] Different ion-bombardment and annealing procedures and O$_2$ chemisorption were used in order to assign various features in the bandgap to different types of defects. A state centered 0.7 eV below the conduction-band minimum was ascribed to bulk Ti - O

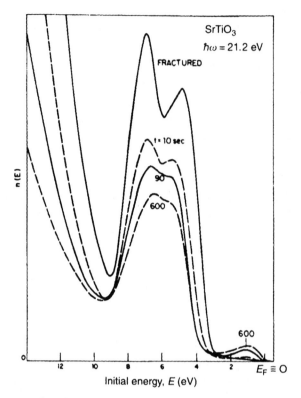

Fig. 5.24 UPS spectra for UHV-fractured $SrTiO_3$ (100) for various Ar^+-ion-bombardment
times. [Ref. 111]

vacancy complexes since it was not completely suppressed by O_2 adsorption.
Emission centered 1.3 eV below the conduction band was assigned to surface O
vacancies. A broad band of emission covering the entire bandgap was also
observed, but no assignment of its origin was made.

The changes that occur in the first-derivative ELS spectra for UHV-fractured
$SrTiO_3$ (100) upon Ar^+-ion bombardment are shown in Fig. 5.23 (b)–(d). A feature
appears at a loss energy of about 2 eV that is similar to, but smaller than, that
found for TiO_2 (110). Undifferentiated loss spectra for polished and annealed
$SrTiO_3$ (100) surfaces show intensities for the defect-derived loss feature at
1–2 eV that are comparable in magnitude to those found for TiO_2 (110) (see
Fig. 5.11).[339,468] No definite conclusions have yet been drawn about the atomic-
scale nature of defective $SrTiO_3$ (100) surfaces.

One study has compared the UPS spectra for polished and annealed $SrTiO_3$
(100) surfaces with those for stepped surfaces oriented slightly off (100);[509,511] the
stepped surface was prepared by fracturing in UHV, the (100) surface by polishing

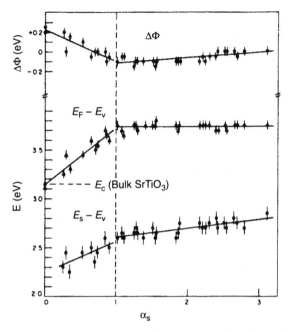

Fig. 5.25 Band structure changes for $SrTiO_3$ (100) as a function of the amplitude of the bandgap surface state. Compare with Fig. 5.14. [Ref. 111]

and annealing. Only slight differences in peak widths and intensities between the planar and stepped surfaces were seen in angle-resolved UPS spectra taken at several different photon energies, indicating that the electronic structure in the vicinity of steps on $SrTiO_3$ is similar to that on flat terraces. This is in agreement with qualitative observations for UHV-fractured and for annealed $SrTiO_3$ (100) and TiO_2 (110) surfaces based upon angle-integrated UPS spectra and the quality of LEED patterns.[517]

Single-crystal studies of the polar (and much less stable) $SrTiO_3$ (111) surface have also been performed;[25,112,518-520] all measurements were made on polished and annealed surfaces. The dependence of surface stoichiometry on preparation procedure was monitored, and it was found that annealing did not remove the bandgap surface defect levels in UPS spectra [similar to those for $SrTiO_3$ (100)] created by ion bombardment.[25] The intensity of the defect emission varied reversibly with temperature, decreasing as the sample was heated, until it completely disappeared at 873 K; the emission then reappeared when the sample was cooled to room temperature. Attempts at bombarding the sample at high temperature produced an irreversible loss of Sr from the surface. ELS spectra of defective $SrTiO_3$ (111) differ from those for the (100) surface primarily in that defect loss features are larger for the (111) surface.

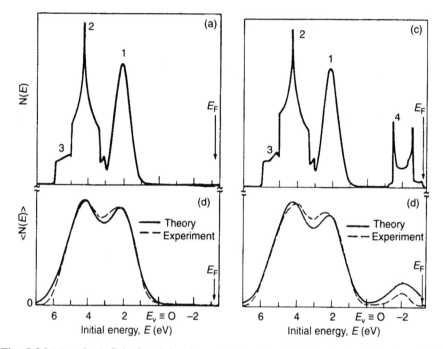

Fig. 5.26 (a) and (c) Calculated densities-of-states for stoichiometric and reduced $SrTiO_3$ (100), respectively. (b) and (d) Comparison of broadened density-of-states (solid curve) with experimental UPS spectra (dashed curves). [Ref. 524]

Theoretical calculations of the electronic structure of both perfect and reduced $SrTiO_3$ (100) surfaces have been performed by several methods. The first calculations to address these surfaces used a simple, analytical LCAO model that was generally applicable to d-band perovskites.[474,521] That theory predicted that the perfect (100) surface would posses a high density of intrinsic surface states in the upper half of the bulk bandgap. Since subsequent experiments did not find intrinsic bandgap surface states, the theory was modified to include, in an approximately self-consistent manner, the additional Coulomb repulsion that would result from occupation of the surface d-orbitals, the effects of all of the fourteen primary bulk bands, and correlation effects.[522-525] With those effects included, the perfect surface exhibited only surface resonances overlapping the bulk conduction and valence bands; however, creation of surface O-vacancy defects (modeled in a phenomenological way by changing the surface potential) pulled a Ti 3d-derived surface state below the bulk conduction band, resulting in a defect state in the upper half of the bandgap. This is in general agreement with experimental observations, as shown in Fig. 5.26.[524] Figure 5.26 (a) and (c) present the densities-of-states from the LCAO calculations for stoichiometric and reduced $SrTiO_3$ (100), respectively, and Fig. 5.26 (b) and (d) compare the theoretical densities-of-states, convo-

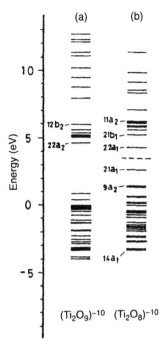

Fig. 5.27 Molecular-orbital levels for cluster calculations corresponding to (a) stoichi-
metric SrTiO$_3$ (100) and (b) the same surface after removal of one O ion. [Ref. 364]

luted with an instrumental broadening function, with experimental UPS spectra.
While the energy of the defect surface band is adequately modelled in the calculation, its amplitude is larger than observed; this discrepancy could, however, be due
to the conditions of measurement.

DV-Xα cluster calculations have also been performed for both the perfect
SrTiO$_3$ (100)-BO$_2$ surface and for the unrelaxed BO$_2$ plane after removal of one O
ion.[364,526] The density-of-states for the defect-free surface exhibits a slight narrowing of the bandgap at the surface, but no states lying deep in the gap. When an O
ion is removed from the BO$_2$ surface, filled states occur in the bandgap; this is
shown in Fig. 5.27, which presents the cluster levels for both the perfect SrTiO$_3$
(100)-BO$_2$ surface and for that surface with one O-ion vacancy.[326] The calculated
charge-density contours for the highest occupied and lowest unoccupied molecular orbitals (HOMO and LUMO) at the defect site are shown in Fig. 5.28. The
orbital composition of the HOMO, 21a$_1$, is about 16 % Ti 3d, 41 % Ti 4p, and
43 % Ti 4s. It has a broad maximum in the vacancy site and extends somewhat
into the vacuum.

The effects of both surface O-ion vacancies and Ti adatoms on SrTiO$_3$ (100)
surfaces have been considered by using a Green's function method coupled with a

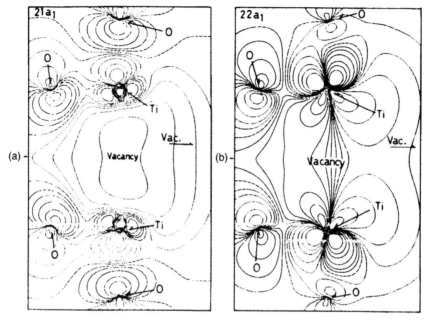

Fig. 5.28 Contours of constant charge density for the (a) $21a_1$ (HOMO) and (b) $22a_1$ (LUMO) orbitals of $SrTiO_3$ (100). [Ref. 364]

tight-binding parameterization of the bulk $SrTiO_3$ band structure.[527,528] The results are somewhat different than those obtained by the $X\alpha$ cluster calculations. Removing one O ion from the (100) plane produced only a slight narrowing of the surface bandgap because of a surface resonance near the bottom of the conduction band. Even removal of a sub-surface O ion, or allowing the Ti ions adjacent to the defect to relax, did little to change the defect state. It was necessary to add a Ti ion to the surface plane, or at a step site, in order to produce the deep gap states seen experimentally. Presumably some of the differences between these results for $SrTiO_3$ and the results of other studies of TiO_2 arise from the larger Ti–Ti distances in $SrTiO_3$. However, neither the Ti 4s nor 4p levels were included in the tight-binding calculations, whereas the cluster calculations found large Ti 4s and 4p components in the defect-state wavefunctions.

5.3.2 BaTiO₃

There have been several experimental studies of the surface electronic structure of $BaTiO_3$ (100).[113,512,529,530] In Refs. 113, 512 and 530, polished and annealed surfaces were studied by UPS, and the results are similar to those found for $SrTiO_3$ discussed above.[113,530] The width of the valence band was found to be the same in both materials, as was the overall shape of the band. Some emission was found in the bulk bandgap for all surface treatments used, but it was largest for samples that were ion bombarded and not annealed, where a distinct peak appeared about 0.9 eV below the conduction-band minimum; the emission was interpreted as arising from Ti^{3+} ions at surface O-vacancy defect sites. The emission from the lower-binding-energy half of the valence band was larger relative to that in the higher-binding-energy region for He I spectra than for He II; as for $SrTiO_3$, this was interpreted as the presence of an O-derived surface state. The same alternative interpretation in terms of enhanced Ti 3d emission at the higher photon energy that was presented above for $SrTiO_3$ applies here also. The other study concentrated on the shallow Ba 5s and 5p core levels;[529] it was found that the $BaTiO_3$ surface was readily reduced by ion bombardment, and that subsequent annealing did not necessarily result in the restoration of a stoichiometric surface.

Although the detailed theories of the surface electronic structure of perovskite oxides have been worked out for $SrTiO_3$, the conclusions should be largely valid for $BaTiO_3$ as well, since no Ba energy levels lie close to either the O 2p or Ti 3d bands.

5.3.3 LiNbO₃ and LiTaO₃

These two oxides have the ilmenite structure, based on corundum. The Li^+ ions have a closed-shell configuration, and Nb^{5+} and Ta^{5+} the d^0 configuration. The bulk optical bandgaps are 3.5 eV and about 4 eV for $LiNbO_3$ and $LiTaO_3$, respectively. UPS experiments have been performed on single-crystal samples that were either scraped or fractured in UHV;[531,532] the results were similar for both surface preparations. Experiments have also been performed on polished and annealed $LiNbO_3$ (0001) surfaces.[512,530] Figure 5.29 presents He II UPS spectra for UHV-fractured $LiNbO_3$ (10$\bar{1}$2) before and after Ar^+-ion bombardment, and for exposure of the ion-bombarded surface to O_2;[532] spectra for $LiTaO_3$ (10$\bar{1}$2) surfaces were very similar. Only emission from the O 2p valence band is seen for fractured surfaces, with no emission in the bulk bandgap region. A band of defect surface states does appear in the bandgap upon either ion or electron bombardment; similar bandgap surface states were found on polished and annealed $LiNbO_3$ samples.[533] That they are associated with reduced surfaces is shown clearly in Fig. 5.30, which presents Nb 3d core-level XPS spectra for $LiNbO_3$ as a function of electron bombardment time.[531] Both Nb^{4+} and Nb^{3+} peaks become visible with increased electron bombardment. (As we noted in § 5.1.3.3, one must be careful in this kind of interpretation of core-level spectra, in view of the possibility of final-state effects in XPS. In this case, however, the location of the peaks that appear upon bombardment agree well with Nb^{4+} and Nb^{3+} spectra for reference compounds.) The

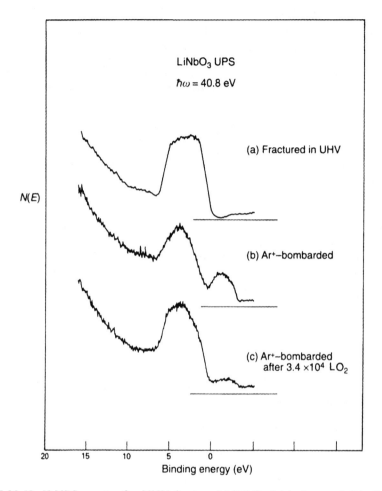

Fig. 5.29 He II UPS spectra for UHV-fractured $LiNbO_3$ (a) before and (b) after Ar^+-ion bombardment, and (c) after subsequent exposure to O_2. [Ref. 532]

bandgap surface states presumably correspond to electrons occupying normally empty Nb 4d or Ta 5d orbitals in the vicinity of surface O vacancies.

ELS spectra have also been taken for $LiNbO_3$ and $LiTaO_3$,[532,534] and transitions associated with surface O-vacancy defects are observed. Ar^+-ion bombardment results in a loss peak at 2.5–3.0 eV in $LiNbO_3$ and 3.1–3.5 eV in $LiTaO_3$; the initial state for those transitions is presumably the occupied bandgap surface state. The same loss peaks, but having reduced amplitude, are also present on UHV-fractured surfaces, leading to the conclusion that a small density of O-vacancy point defects is present on fractured surfaces. Transitions associated with a second type of surface defect, thought to correspond to Li^+-ion vacancies, were also seen in ELS. Exposure to O_2 does not reduce the amplitude of those loss features, as it does the

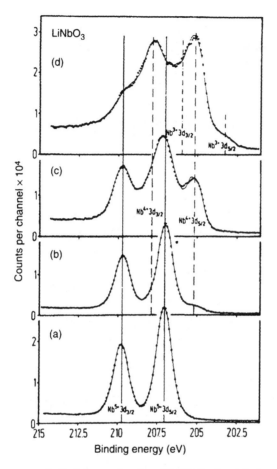

Fig. 5.30 XPS spectra of Nb 3d core levels of LiNbO$_3$ for (a) the clean, stoichiometric surface, and (b)–(d) for increasing amounts of electron bombardment. [Ref. 531]

ones associated with O vacancies, but they can be removed by annealing the sample so that Li ions can diffuse to the surface from the bulk.

5.4 The oxides of vanadium

The V–O system is an even richer and more complex one than Ti–O. Since vanadium can have oxidation states of 2+, 3+, 4+ or 5+, a wide range of stable bulk phases exists. The Magnéli phases discussed above for Ti–O also occur for V–O, so that the number of possible V oxides is very large; as no surface work has been performed on any of the Magnéli phases, however, we will not consider them further here. The surface properties of vanadium oxides are extremely important since V$_2$O$_5$ is an active catalyst for many reactions, including the oxidation and reduction of hydrocarbons, and the oxidation of SO_2 to SO_3, which is used in the manufacture of sulfuric acid.[39] However, the amount of effort that has been devoted to understanding the surface electronic structure of V oxides is far smaller than that for Ti oxides. As for titanium, we will discuss the oxides of vanadium in terms of their bulk cation oxidation state.

There have been several photoemission studies of single-crystal vanadium oxides which, although they did not address surface properties, are worth noting. In the most comprehensive study, the bulk electronic structure of V_2O_3, VO_2, V_6O_{13} and V_2O_5 was addressed through a series of resonant UPS experiments on UHV-cleaved single crystals.[535] The photon energies used were in the most surface-sensitive range, so surface electronic structure information is no doubt present in the spectra. However, no experiments were performed to try to separate bulk and surface properties, and the data were interpreted entirely in terms of bulk electronic structure. XPS and Auger spectroscopy, which are, of course, much less surface sensitive, were used in a study of the electronic structure of VO_2 and V_2O_3, and XPS spectra of the valence band above and below the semiconductor/metal transitions were presented.[536] Qualitative LEED measurements have been reported on V_2O_5 (001) and V_6O_{13} (001) surfaces, along with studies of the transition from the former to the latter under electron bombardment; XPS valence-band spectra of V_2O_5 were also reported.[219] ISS and LEED measurements have also been reported on V_6O_{13} (001) layers that were grown topotactically on V_2O_5 (001) single-crystal surfaces;[537] some measurements were also performed on single crystal V_6O_{13} (001).[306]

5.4.1 V_2O_5

The maximal valency oxide of V is V_2O_5. Its layered orthorhombic structure has only one stable surface face, the basal (001) plane. Single crystals cleaved along that plane in UHV exhibit LEED patterns characteristic of a termination of the bulk lattice.[218,219] However, even the low electron energies used in LEED reduce the surface rapidly, and the LEED patterns change to ones characteristic of V_6O_{13}. XPS spectra of the valence-band region of V_2O_5 exhibit a 6 eV wide O 2p band with no bandgap emission from V 3d states;[218] however, the type of sample used was not described. UPS measurements have been performed both on annealed and on UHV-cleaved V_2O_5 (001).[538] The electronic structure of the stoichiometric surface is similar to that of the bulk. Reduced (001) surfaces exhibit a large peak in the bulk bandgap due to partial population of the 3d orbitals on reduced V cations at O-vacancy defect sites.

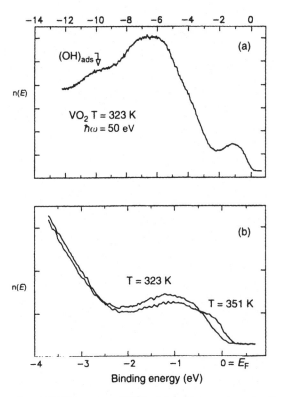

Fig. 5.31 Valence-band UPS spectra of VO_2 (a) in the semiconducting region, and (b) the V 3d peak above and below the semiconductor/metal transition. [Ref. 539]

5.4.2 VO₂

The rutile oxide VO_2 is one of the most interesting of the vanadias both because it exhibits a strong semiconductor/metal transition as a function of temperature and since actual vanadia catalysts usually begin as V_2O_5, but are reduced to lower oxides under reaction conditions. Unfortunately, single crystals of VO_2 large enough to be useful for surface analysis are very difficult to grow. There has been only one report of surface-science measurements of the electronic properties of VO_2, and that was performed on bundles of small needle-like single crystals bound together with indium metal or epoxy.[539] Atomically clean surfaces were then prepared by fracturing or scraping in UHV, although the exact orientation of the surfaces so obtained was unknown. Figure 5.31 shows very surface-sensitive UPS spectra for such samples both above and below the semiconductor/metal phase transition at 340 K. The difference in electron density at the Fermi level is clearly visible in Fig. 5.31 (b). (Similar observations of the phase transition have also been reported in XPS measurements on single-crystal VO_2[536] and in UPS and XPS spectra from powder samples.[535,540])

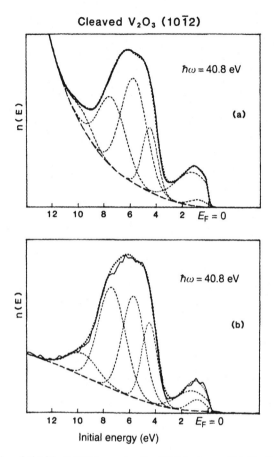

Fig. 5.32 (a) He I and (b) He II UPS spectra for UHV-cleaved V_2O_3 ($10\bar{1}2$). Compare with Fig. 5.15. [Ref. 29]

There have been no theoretical calculations of either the surface geometric or electronic properties of VO_2.

5.4.3 V_2O_3

Several experiments have addressed the electronic structure of the V_2O_3 ($10\bar{1}2$) cleavage face by using a variety of techniques.[29,30,485,500–502,541] The main difficulty is that the bulk electronic structure is not very well understood (see § 5.2.3 in Ref. 40). V_2O_3 is metallic at room temperature, but with quite anomalous electrical and magnetic properties. Furthermore, at around 150 K it undergoes a transition to a magnetic insulator. Electron correlation effects therefore appear to be very important in V_2O_3, and although many of the experiments reported below have been interpreted in terms of one-electron energy levels, this may not be a good approximation. Because of these difficulties it has not yet been possible to determine the surface electronic properties in much detail, except for the influence of defects.

The He I and He II UPS spectra for UHV-cleaved V_2O_3 ($10\bar{1}2$) are shown in Fig. 5.32.[29] The emission from about 3 to 12 eV is from the predominantly O 2p valence band, while the emission between E_F and 3 eV arises from the partially filled V 3d a_1 and e_π bands (see the energy-level scheme in Fig. 5.4 for Ti_2O_3, which has the same crystal structure, although one fewer 3d electron per metal atom). Note the large density-of-states at E_F, similar to that for TiO_x in Fig. 5.21, and characteristic of a metallic oxide. As was discussed in connection with Fig. 5.15 for Ti_2O_3, the difference in the shape of the two spectra is a result of the change in ionization cross-section of the V 3d orbital hybridized with the higher-binding-energy region of the valence band.

The UPS spectra for cleaved V_2O_3 ($10\bar{1}2$) do not correspond to the density-of-states expected for bulk V_2O_3. Based upon optical measurements, there should be a gap of about 1.5 eV between the O 2p and V 3d bands. While there is a minimum in the density-of-states between those bands, any bandgap that might exist is much smaller than 1.5 eV. (Compare this with Fig. 5.15 for Ti_2O_3, where the flat region between the Ti 3d a_1 and O 2p bands was in good agreement with the expected bulk bandgap.) Chemisorption of O_2 or H_2O on the cleaved V_2O_3 ($10\bar{1}2$) surface does not significantly alter the emission in that region of the spectrum, so it probably does not arise from defect surface states resulting from cleaving. Intrinsic surface states could be present, but there have been no UPS measurements as a function of electron emission angle to confirm that. We have also mentioned previously that discussions based on simple one-electron levels may be quite misleading here.

UPS measurements using tunable synchrotron radiation have been made on UHV-cleaved V_2O_3 ($10\bar{1}2$) surfaces in both angle-integrated and angle-resolved modes.[500–502,541] Figure 5.33 shows a series of spectra, all normalized to the incident photon flux, for photon energies in the vicinity of the V 3p → 3d optical excitation threshold; similar measurements were discussed for Ti_2O_3 in § 5.2.4 above. The resonant behavior of the emission across the threshold is clearly visible, as are changes in the valence band. The resonant behavior of the integrated intensity of the O 2p band in V_2O_3 is different than that in either Ti_2O_3 or TiO_2 in that the onset of the resonance occurs at about 8 eV lower photon energy than that for the V 3d emission. The reason for this behavior is not understood, but it is presumably of bulk origin. Figure 5.34 shows the photon-energy dependence of the intensity of the V 3d emission for V_2O_3 ($10\bar{1}2$) measured in both the angle-integrated and normal-emission angle-resolved modes.[502] The results are strikingly similar to those

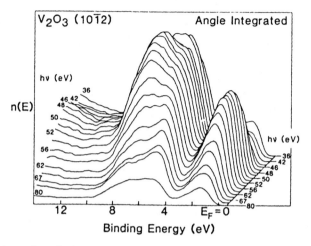

Fig. 5.33 Series of angle-integrated UPS spectra for UHV-cleaved V_2O_3 (10$\bar{1}$2) as a function of photon energy. [Ref. 502]

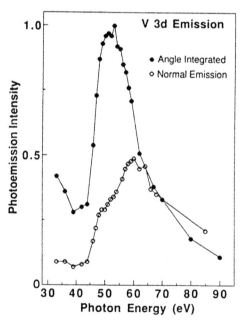

Fig. 5.34 Photon-energy dependence of the V 3d UPS emission from UHV-cleaved V_2O_3 (10$\bar{1}$2) as measured in the angle-integrated and angle-resolved normal-emission modes. [Ref. 502]

for Ti_2O_3 (see Fig. 5.18) in that the angle-integrated emission exhibits a sharper resonance at a lower photon energy than does the normal-emission intensity. This is probably also due to directed emission of photoelectrons from a_1 orbitals just above threshold, although no measurements were made for emission along the c-axis for V_2O_3. The situation here should be more complicated than for Ti_2O_3, however, since the occupied V 3d levels consist of e_π orbitals as well as a_{1g} orbitals.

When defects are created on V_2O_3 (10$\bar{1}$2) by Ar^+-ion bombardment, the UPS spectra change in the same way as do the spectra for Ti_2O_3 (10$\bar{1}$2) shown in Fig. 5.20, with an increase in emission from the V 3d bands and a decrease in amplitude of the O 2p valence band. LEED patterns show that the surface disorders upon ion bombardment, and Auger spectra indicate surface reduction. The density-of-states at E_F increases compared to that for the cleaved surface, showing that the reduced surface layer is even more metallic than the bulk.

The bulk electrical properties of V_2O_3 are drastically changed by the addition of small amounts of Cr. A new metal/insulator transition appears whose transition temperature depends strongly on Cr concentration. Single crystals of Cr-doped V_2O_3 have been studied by UPS using surface-sensitive photon energies; the changes in shape and position of the V 3d band have been observed across two of the metal/insulator transitions as a function of temperature, and one transition as a function of Cr doping level.[542] But the results are preliminary and have been interpreted only in terms of bulk band-structure changes. Earlier XPS measurements also observed the changes in the valence-band region across the semiconductor/metal transition.[536]

To date, no theoretical calculations have been reported for the electronic structure of V_2O_3 surfaces.

5.4.4 $VO_{x\approx1}$

The situation for VO_x is exactly the opposite of that for V_2O_3; no experimental measurements have been performed to determine any of its surface properties, but an idealized model of the VO surface has been considered theoretically. Calculations of the electronic structure of a defect-free rocksalt $VO_{1.0}$ (100) surface have been performed by using the transfer-matrix formalism.[506,543] Cation or anion vacancies were neglected (although VO actually contains a high density of both types of vacancy), as was distortion of the bulk crystal structure from rocksalt; only V 3d and O 2p orbitals were considered, and effects due to electron correlation were neglected. A narrow intrinsic surface state, arising from the hybridization of V 3d and O 2p orbitals, was predicted slightly above the Fermi level in the V 3d band, as well as other less striking changes in the surface density-of-states.

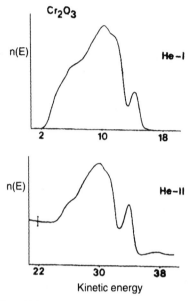

Fig. 5.35 He I and He II UPS spectra for Cr_2O_3 powder. [Ref. 544]

5.5 Chromium oxides

The chromium oxide system is the least studied in terms of surface electronic properties among all of the 3d series elements. The most stable bulk oxide of Cr is Cr_2O_3, a magnetic insulator having the corundum structure. Single crystals of Cr_2O_3 and not readily available, and no surface electronic structure studies using single-crystal samples have been published to date. UPS measurements have been reported for polycrystalline Cr_2O_3, however, and the He I and He II spectra are included as Fig. 5.35 for completeness;[544] similar spectra have been obtained on oxidized Cr films using photon energies of 40 and 50 eV.[545,546] The electronic configuration of Cr^{3+} is $3d^3$, with all three d electrons occupying the majority spin t_{2g} orbital. Since the electrons in Cr_2O_3 behave in a localized manner, there is one narrow UPS peak, labelled 'c', that corresponds to emission from the t_{2g} orbital. The remainder of the emission is from the O 2p band. Cr_2O_3 is a semiconductor, so the Fermi level lies at a slightly higher kinetic energy than the Cr 3d peak in the spectra in Fig. 5.35.

No theories have been published for any of the surface properties of any Cr oxide.

5.6 Manganese oxides: MnO

As can be seen in Table 5.1, several stable bulk oxides of Mn exist; in fact, Mn

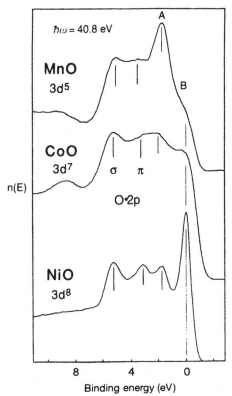

Fig. 5.36 He II UPS spectra of UHV-cleaved (100) surfaces of MnO, CoO and NiO. The spectra have been aligned at the uppermost emission feature B.

exhibits stable oxidation states of 2+, 3+, 4+, 6+ and 7+. The most studied oxide of Mn is MnO, a magnetic insulator having the rocksalt structure. Several UPS and XPS studies of single-crystal MnO (100) have been conducted,[99,503,547–551] but the main goal of most of them was to understand the bulk electronic structure of MnO. Thus little effort was made to identify any surface features in the data. Only in connection with point defects have clearly identifiable surface states been observed.

The angle-integrated He II UPS spectra for UHV-cleaved (100) surfaces of MnO and two other antiferromagnetic rocksalt transition-metal monoxides – CoO and NiO – are presented in Fig. 5.36. The electronic structure of the valence bands in these oxides is complex, with both cation and anion character extending across the entire occupied band. The features labelled π and σ correspond to the regions of the spectra that are dominated by direct emission from the bonding and non-bonding parts of the O 2p band, respectively. (See the discussion of π and σ labels in § 4.1.1.) The emission features A and B correspond to cation emission that is screened by charge transfer from O^{2-} ligands. Spectra taken at comparable photon energies on UHV-scraped single crystals and FeO-doped polycrystalline MnO are in agreement with those in Fig. 5.36, except for the amplitude and location of the small peak near 9 eV;[549] this peak will be discussed in detail below. The spectra can be explained in terms of the bulk MnO electronic structure, and no intrinsic surface states have been identified.

There is a problem with the cleaved MnO (100) surface, however. Recall (§ 2.3.1.1) that only diffuse LEED patterns are obtained from that surface; it is necessary to use polished and annealed MnO (100) surfaces in order to obtain good LEED patterns. Experiments on the interaction of UHV-cleaved MnO (100) with adsorbed molecules show the surface to be very active for chemisorption (this will be discussed in § 6.5.4); this is distinctly different behavior from that observed for virtually any other transition-metal oxide. It is therefore probable that a high density of surface defects is created in the cleavage process for MnO. No controlled study on a cleaved MnO (100) surface has yet been made to determine whether or not annealing retains the stoichiometry of the surface, how effectively it removes surface defects, etc. In studies on ion-bombarded and annealed MnO (100) surfaces, however, it was observed that the surface was quite inert, at least when maintained in UHV,[550,551] so there are probably significant differences between cleaved and annealed surfaces.

The small peak near 9 eV in the MnO spectrum in Fig. 5.36 resembles the so-called 'satellite' peaks that occur in UPS valence-band spectra for many transition metals and their compounds. The origin of these peaks is different for the light 3d-transition-metal oxides ($Z < 25$) than it is for the heavy 3d oxides ($Z > 25$). Since Mn lies at the boundary between those two regimes, its satellite structure is particularly interesting. The 9 eV peak has been studied in detail for UHV-cleaved MnO (100) samples, and seems to be a surface-related feature;[552] in the higher-Z oxides similar peaks correspond to bulk electronic structure. Resonant UPS data show that the peak is associated with the Mn 3d electrons, although its behavior cannot be explained in terms of bulk electronic structure. Its amplitude is extremely sensitive to surface treatment, including ion bombardment, annealing or O_2 exposure, and it is not observed in less surface-sensitive XPS spectra. It is believed that the feature is associated with point defects on the surface, probably Mn vacancies, but further work is necessary in order to clarify its origin.

When the MnO (100) surface is bombarded with Ar^+ ions, it becomes slightly reduced as a result of the net removal of O from the surface. No new features appear in the UPS spectra, but emission in the region of feature A increases slightly, presumably due to the additional electrons on the Mn ions adjacent to surface defects, and there is a reduction in the amplitude of emission from the bonding O 2p orbital.

A very interesting experiment involving spin-polarized photoelectron diffraction has been performed on polished and annealed MnO (100) surfaces.[550,551]

While most of the information obtained concerned the bulk electronic structure of MnO, one surface-structure-related conclusion was drawn. The Mn 3s core level in MnO is split into a doublet in photoemission spectra, with a separation of 6 eV between the high-binding-energy ^5S and the low-binding-energy ^7S peaks. The doublet arises from intra-atomic final-state exchange splittings. Although MnO is an antiferromagnet, with alternate cations having opposite net spins, the Mn 3s doublet can be used as a source of polarized electrons for local magnetic studies, since, with respect to the spin direction of the emitting ion, the ^5S component is 100 % majority-spin polarized and the ^7S component is 71 % minority-spin polarized. The energies of the two spin states are easily distinguishable in photoemission experiments, and it has been possible to use them to study local magnetic order in MnO as a function of temperature. The measurements have identified a new type of bulk short-range magnetic-order transition at 530 K, which is more than four times the bulk Néel temperature of 120 K. In the course of the experiments it was necessary to eliminate any other phenomena as possible causes for the observed transition. As the experiment involved diffraction of the spin-polarized electrons from atoms in the surface plane, one potential complication would be a structural transition taking place on the surface as the short-range-order transition occurred. No evidence was found for any such surface structural changes at or near 530 K.

There has been one published report of ELS spectra of UHV-fractured MnO surfaces.[553] However, all of the features in the loss spectrum were interpreted in terms of bulk transitions, and no experimental attempts were made to specifically identify surface excitations. ELS spectra on MnO powders have been compared with those from other rocksalt transition-metal oxides, and the intensity of the $3d \rightarrow 3d$ dipole- and spin-forbidden transitions in the bulk bandgap was found to be anomalously large;[554] here also, however, interpretation was purely in terms of bulk electronic structure.

No theoretical calculations have been reported for the surface electronic structure of MnO or any other Mn oxide.

5.7 The oxides of iron

The solid-state chemistry of iron is very different from that of V or Mn. The only two stable oxidation states of Fe in binary oxides are 2+ and 3+, and only three stable bulk oxide phases exist: Fe_xO, Fe_3O_4 and Fe_2O_3. Both the chemistry and the bulk electronic properties of these compounds are quite complex, and they have received a great deal of attention both theoretically and experimentally. Fe_xO is a p-type semiconductor which is highly non-stoichiometric and very easily oxidized. Fe_2O_3 is a magnetic insulator with the $3d^5$ configuration. The mixed-

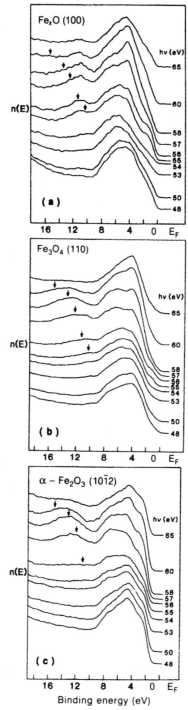

Fig. 5.37 Angle-integrated UPS spectra as a function of photon energy for UHV-cleaved (a) Fe$_x$O (100), (b) Fe$_3$O$_4$ (110) and (c) α-Fe$_2$O$_3$ (10$\bar{1}$2). The predicted location of the M$_{23}$M$_{45}$M$_{45}$ Auger emission is indicated by the arrows. [Ref. 565]

valence inverse spinel Fe_3O_4 shows high conductivity at room temperature and above, but at 120 K undergoes the so-called **Verwey transition** to a much less conducting phase: in spite of much work, the details of this transition are still rather obscure (see § 5.3.3 in Ref. 40).

Much of the work on the electronic properties of iron oxides has involved the use of surface-sensitive techniques such as UPS, and there are numerous published spectra that no doubt contain a wealth of information about the surface electronic structure. [31,170,172,537,555–566] Such information has yet to be extracted, however, since the data were interpreted almost exclusively in terms of bulk properties, with the exception of some work on point defects. Nonetheless, we will present a brief review of that work here, partly because it is necessary for an understanding of the chemisorption properties of Fe oxides to be discussed in § 6.5.5.

Angle-integrated UPS spectra for UHV-fractured surfaces of Fe_xO, Fe_3O_4 and α-Fe_2O_3 are presented in Fig. 5.37.[564,565] The single-crystal surfaces studied were those that fractured most easily: Fe_xO (100), Fe_3O_4 (110) and α-Fe_2O_3 (10$\bar{1}$2). The valence band extends over a binding-energy range of about 10 eV, with Fe and O emission intimately intertwined. The photon-energy dependence of the photoemission cross-sections for Fe and O has been used in an effort to separate the O 2p and Fe 3d contributions to the spectra. Figure 5.38 (a) shows spectra for the three oxides taken at $\hbar\omega = 30$ eV, where emission from the O 2p orbitals is emphasized. To separate the Fe 3d contributions, the resonant enhancement of the Fe emission at the Fe 3p $\rightarrow$ 3d optical absorption threshold was used. Figure 5.38 (b) presents differences between spectra taken for photon energies just above and just below the resonance, so that the O 2p emission should have been removed. As expected, the O 2p contribution is nearly the same for all oxides, but this is not so for the Fe 3d emission. This has to do both with the electron configurations of the Fe ions and their ligand environment. Fe_xO contains mostly Fe^{2+} with the $3d^6$ configuration, while α-Fe_2O_3 has the d^5 ion Fe^{3+}; in both cases, however, the Fe is octahedrally coordinated. Fe_3O_4 is more complicated in that two-thirds of the cations are Fe^{3+} and one-third are Fe^{2+}, while two-thirds of the cations are octahedrally coordinated and one-third of them are tetrahedral. With the help of configuration-interaction calculations for the final states in photoemission, such as the one shown in Fig. 5.38 (b) for α-Fe_2O_3, it is possible to assign the features in UPS or XPS spectra to various combinations of emission from screened and unscreened cation configurations and from O 2p orbitals; however, these assignments are entirely in terms of the bulk rather than surface electronic structure.

Although none of the studies conducted to date has succeeded in separating the bulk and surface contributions for stoichiometric UHV-fractured surfaces, extrinsic defect surface states have been identified on both α-Fe_2O_3 (0001) and (10$\bar{1}$2)

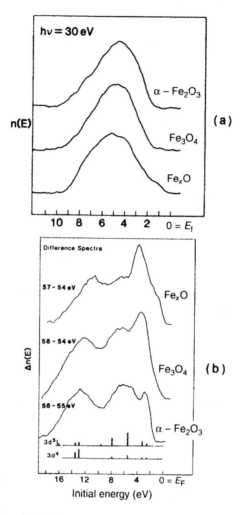

Fig. 5.38 (a) UPS spectra for UHV-cleaved Fe_xO (100), Fe_3O_4 (110) and Fe_2O_3 ($10\bar{1}2$) for $\hbar\omega = 30$ eV, which emphasizes the O 2p emission. (b) UPS difference spectra taken for photon energies just above and just below the Fe 3p → 3d resonance, which emphasizes the Fe 3d-derived final states. The vertical lines in (b) represent the relative intensities of the final states calculated in a configuration-interaction approximation. An inelastic background has been removed from each spectrum. [Refs. 564 and 565]

surfaces.[31,170,172,562,565] Most of the work has been performed on the (0001) face, and Fig. 5.39 compares UPS spectra for Ar^+-ion-bombarded α-Fe_2O_3 (0001) with that for the same surface after annealing to produce the most nearly stoichiometric surface.[31] As for most other metal oxides, ion bombardment preferentially removes O ions, thus reducing the surface. Comparison of Fig. 5.39 (c) with the 48 eV spectrum of UHV-fractured α Fe_2O_3 ($10\bar{1}2$) in Fig. 5.37 (c) shows that the

α – Fe₂O₃

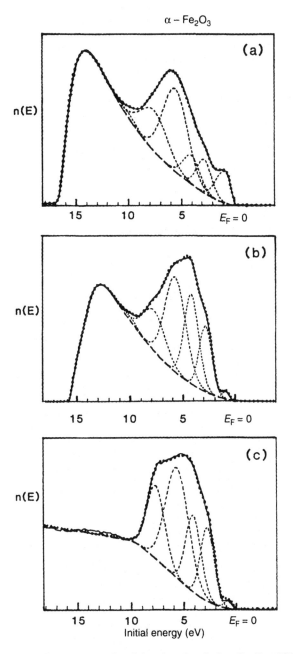

Fig. 5.39 (a) He I UPS spectrum of Ar⁺-ion-bombarded α-Fe₂O₃ (0001); (b) He I and (c) He II spectra for that surface after annealing at 1100 K. [Ref. 31]

primary difference between the two is the presence of a small amount of emission just below E_F for the annealed surface. There is strong evidence that this feature is a cation-derived surface state, probably associated with Fe^{2+} ions (XPS spectra show no indication of Fe^0 on this surface). Its amplitude is extremely sensitive to adsorbates, completely disappearing upon exposure to O_2 (see Fig. 6.46). Ion bombardment of the (0001) surface reduces a significant fraction of the Fe^{3+} ions in the surface region to Fe^{2+} and Fe^0. This produces a number of changes in the valence-band density-of-states, as shown in Fig. 5.39 (a). The surface becomes metallic, as evidenced by the finite emission at E_F. However, it has not yet been possible to associate any of the features in the valence band of the reduced surface with specific ionic configurations.

Although the bulk electronic structure of Fe oxides has been considered theoretically in great detail, no calculations have been reported of the surface electronic structure of any Fe oxide.

5.8 Cobalt oxide

The most stable cobalt oxide is CoO, a magnetic insulator in which Co^{2+} has the $3d^7$ configuration, and it is the only one on which surface-spectroscopic measurements have been performed. As for the iron oxides, however, most studies on well-characterized single-crystal samples have been aimed at understanding the bulk electronic structure of CoO.[462,503,537,553,567,568] Only two studies directly addressed its surface electronic structure, and they were both concerned with defects.[274,569] No theoretical calculations have been performed of the surface electronic structure of CoO.

An angle-integrated He II UPS spectrum of UHV-cleaved CoO (100) is shown in Fig. 5.36; other published UPS spectra at comparable photon energies are similar. The valence band contains a mixture of O 2p and Co 3d character throughout its entire width, so a simple identification of UPS peaks as arising from either O or Co orbitals is not appropriate. Peak A consists largely of electrons emitted from the Co 3d majority-spin t_{2g} orbitals, and peak B contains contributions from electrons originating from both the majority-spin e_g and the minority-spin t_{2g} levels; both of these final states are screened by charge transfer from the O^{2-} ligands. Spectra such as these have only been used to date to interpret the bulk CoO electronic structure.

When CoO (100) is bombarded with Ar^+ ions, its surface becomes slightly reduced, and the angle-integrated UPS spectra change as shown in Fig. 5.40.[274] The three spectra correspond to the cleaved surface before ion bombardment and after bombardment by 500 eV or 5 keV ions. Auger spectra indicate that the 500-eV-bombarded surface contains about 4 % of a monolayer of O vacancies, while the 5-keV-bombarded surface has about 16 % surface O vacancies. The 500-eV-

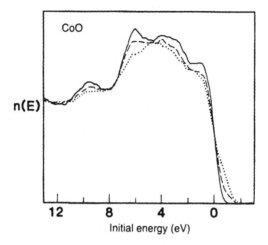

Fig. 5.40 He II UPS spectra for UHV-cleaved CoO (100) (solid curve); after bombardment with 500 eV Ar$^+$ ions (dashed curve); and after subsequent bombardment with 5 keV Ar$^+$ ions (dotted curve). [Ref. 274]

bombarded surface still exhibits LEED patterns, although the beams are broader and the background more intense than for the cleaved surface. The most significant change in the UPS spectra upon defect creation is the appearance of a distinct bandgap surface state near the valence-band maximum; this state is thought to correspond to the increased occupation of Co 3d orbitals on cations adjacent to defects. With only a small density of surface defects, the bandgap emission consists merely of a broadening of the upper edge of the valence band. Since in the ground state the minority-spin t_{2g} orbital on the stoichiometric surface contains only two out of a possible three 3d electrons, the broadening presumably arises from increased occupation of that level at defect sites. For larger defect densities, where adjacent defects begin to interact, more than one additional electron may be associated with a surface Co ion, so partial population of the minority-spin e_g levels probably also occurs; this gives rise to the appearance of a distinct shoulder on the upper edge of the valence band. The resonant photoemission behavior of the bandgap emission shows that holes in the cation d orbitals are screened by transfer of charge from O^{2-} ligands.[569] For high defect densities, of course, many types of defect will exist on the surface, including reduced Co^0, so no unique assignment of the origin of the shoulder can be made.

Additional evidence for population of the remaining t_{2g} orbitals upon defect creation is provided by the behavior of the surface conductivity of CoO as a function of defect density.[274] For most other oxides, including NiO, creation of surface O-vacancy point defects leads to a more conducting surface due to the greater mobility of the excess electrons at defect sites. For CoO (100), however, the 500-eV

Ar$^+$-ion-bombarded surface is found to be *less* conducting than the UHV-cleaved surface; only after 5 keV bombardment does the surface become more conducting than the cleaved one. In addition to CoO, this behavior has only been seen on high-T_c oxide superconductors. It may occur because the additional electronic charge on Co ions adjacent to O-vacancy defects gives rise to the more stable 3d^8 electronic configuration (which is the same as that for bulk NiO), in which the minority-spin t$_{2g}$ orbital is filled. The energy of the slightly defective CoO (100) surface may thus be *lower* than that of the perfect surface.

The changes in resonant photoemission on CoO (100) on forming surface O-vacancy defects have been interpreted in terms of the decrease in the ionic component of cation–anion bonding expected from adding 3d electrons to cations adjacent to the defect.[569] The idea behind the method comes from bulk electronic structure investigations.[459] In the configuration-interaction model, the ground state electronic configuration of CoO can be expressed as

$$\psi_g = a \,|\, 3d^7 > + \, b \,|\, 3d^8\underline{L}> \qquad (5.3)$$

where the 3d^7 term corresponds to the ionic contribution to the Co–O bond and 3d$^8\underline{L}$ represents the charge-transfer state, i.e., the ground-state covalency. The final-state wavefunction, after one electron has been emitted, can be written as

$$\psi_f = \alpha \,|\, 3d^6 > + \, \beta \,|\, 3d^7\underline{L}> \qquad (5.4)$$

(Higher order terms such as d$^9\underline{L}^2$ and d$^8\underline{L}^2$ are neglected because they either lie at much higher energies or contribute negligible amplitude to the wavefunctions.) The 3d^6 final-state term can *only* be derived from the 3d^7 ground-state, while the 3d$^7\underline{L}$ term can be obtained from either the 3d^7 or 3d$^8\underline{L}$ initial states, depending upon whether the charge transfer occurs in the initial or final state. It has been demonstrated for a number of transition-metal compounds that the intensity of 3d^{n-1} final-state features resonates more strongly near the cation 3p $\rightarrow$ 3d threshold than do the final states involving electron transfer from the ligands to the cations. Resonant photoemission can therefore be used to determine the 3d^6 contribution to the final state in CoO, which is proportional to the 3d^7 initial-state amplitude.

The changes in cation electron configuration that occur at defect sites are not specifically included in Eqns 5.3 and 5.4 since no ligand hole is present. But if it is assumed that 3d^{n-1} final states resonate more strongly than charge-transfer states in this case also, then the amplitude of the observed resonance is still a measure of the ionic contribution to cation–anion bonding. UPS spectra for UHV-cleaved CoO (100) showed that a reduction of the amount of ionic bonding on the surface did indeed occur when surface O-vacancy defects were present.[569]

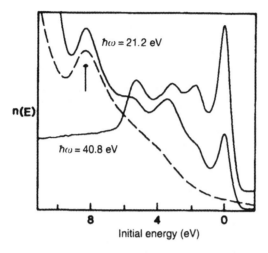

Fig. 5.41 Angle-integrated He I and He II UPS spectra of the valence levels (solid curves) and an electron-beam-induced secondary-electron spectrum (dashed curve; $E_i = 100$ eV) for UHV-cleaved NiO (100). [Ref. 191]

5.9 Nickel oxide

The most thoroughly studied rocksalt transition-metal oxide is NiO. Several experimental studies have used very surface-sensitive electron spectroscopies such as UPS and ELS applied to UHV-cleaved single-crystal NiO (100) surfaces.[32,97,191,461,462,547,553,570–572] The primary interest in most of those studies was understanding the bulk electronic properties, and the data were thus interpreted entirely in bulk terms. Figure 5.36 shows an angle-integrated He II UPS spectrum for the valence-band region of UHV-cleaved NiO (100). The Ni^{2+} ion has a $3d^8$ ground-state configuration, and emission features A and B correspond to the screened $3d^7$ ligand-field final states. As discussed in § 5.1.3.2, however, it is better to use the configuration-interaction model (see Fig. 5.6). There is no indication in this or any other surface-sensitive UPS spectrum from cleaved NiO (100) that intrinsic surface states (i.e., features not explicable in terms of the bulk electronic structure) are present. Measurements of both the filled and empty densities-of-states near E_F do not show any evidence of surface states in the bulk bandgap.[461] Angle-resolved UPS measurements on NiO (100) were compared with bulk band calculations; although we do not believe that this is a satisfactory procedure, there is no indication in these spectra of any intrinsic surface states.[462]

One surface-related feature has been seen in UPS and secondary-electron spectra from UHV-cleaved NiO (100).[32,191] It is shown in Fig. 5.41, which presents He I and He II UPS spectra (solid curves) and an electron-beam-induced secondary-electron spectrum (dashed curve); the binding energies are referenced to the main

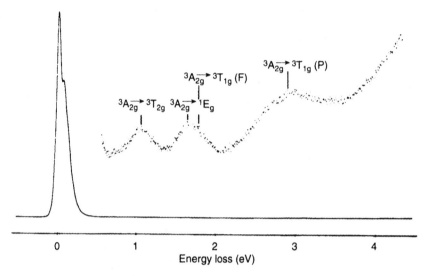

Fig. 5.42 Specular ELS spectrum ($E_i = 25$ eV) from polished and annealed NiO (100). [Ref. 97]

Ni 3d peak at the upper edge of the valence band. A peak occurs in the He I spectrum at an apparent initial-state energy of 8 eV which is absent in the He II spectrum. The feature actually appears at fixed final-state kinetic energy, and it is present whether electrons in the sample are excited by photons or electrons. (Its primary-electron energy dependence rules out the possibility that it results from an Auger transition.) Such peaks arise when excited electrons lose energy by inelastic collisions and drop into lower-lying allowed states in the band structure. If strong maxima occur in the empty density-of-states, electrons will preferentially pile up in those levels, giving rise to increased emission at the corresponding energies in secondary-electron spectra. The peak in NiO occurs at an energy 6.9 eV above the vacuum level, which is close to the position of the peak attributed to the bulk Ni 4p level in Bremstrahlung isochromat spectra.[461] What is surprising about the peak is that it is more surface-sensitive than the rest of the photoelectron spectrum. A maximum in the emission of the peak is observed for emission 45° from the sample normal, and the intensity reduces to zero for either normal or grazing emission. Its intensity also decreases by one-half after five minutes of Ar⁺-ion bombardment, while few changes are observed in the emission from filled levels after that treatment. Unfortunately, no efforts after the initial observation of this unoccupied surface feature have been made to determine its origin.

Several ELS measurements have been performed on cleaved NiO (100) surfaces in very surface-sensitive energy regimes, but all of the features have been correlated with the bulk joint density-of-states.[97,553,570] The loss spectra do show very weak transitions at energies less than the bandgap, as shown in Fig. 5.42, but

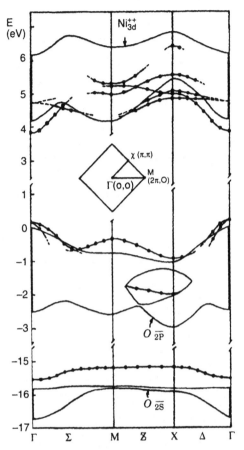

Fig. 5.43 Calculated LCAO surface bands and bulk continuum structure for NiO (100).
[Ref. 573]

they correspond to localized optically forbidden Ni 3d → 3d transitions and not to intrinsic surface states.[97,554]

For the most part, theorists are still trying to understand the bulk electronic structure of NiO (see § 3.4.2 in Ref. 40). A non-magnetic LCAO band structure performed for NiO (100) is the only published surface electronic structure calculation.[573] As shown in Fig. 5.43, surface states are predicted close to the bulk band edges; however, except for a slight narrowing of the bandgap at the surface that would be difficult to observe experimentally, none of them lie in the bandgap. Angle-resolved UPS or IPS measurements would be required to identify those states.

One aspect of the surface magnetic properties of NiO has received both experimental and theoretical attention. In some LEED investigations of UHV-cleaved NiO (100) surfaces, half-order spots were observed that were attributed to

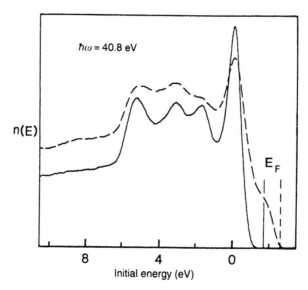

Fig. 5.44 He II UPS spectra of UHV-cleaved NiO (100) (solid curve) and the steady-state
500 eV Ar⁺-ion-bombarded surface (dashed curve). [Ref. 32]

exchange scattering of the incident electrons by the spins on the Ni^{2+} ions.[574–577]
Theoretical efforts to explain the effect quantitatively included the derivation of
the properties of surface magnons (i.e., low-energy antiferromagnetic surface
spin-wave states that propagate parallel to the surface but are attenuated with
increasing depth into the crystal),[578] multiple-scattering LEED calculations
including exchange potentials,[579] and calculations of the cross-sections for inelastic scattering of electrons by surface magnons on antiferromagnets.[580]
Unfortunately, no more quantitative experimental investigation of the phenomenon has been reported to date.

The changes in UPS spectra of NiO (100) as surface point defects are created by
ion bombardment at room temperature are similar to those observed for MnO and
CoO. Figure 5.44 presents He II UPS spectra for UHV-cleaved NiO (100) before
and after bombardment with 500-eV Ar⁺ ions.[32] The surface becomes only slightly
reduced, with a relatively small concentration of O vacancies. (Surface reduction
of rocksalt transition-metal oxides under ion bombardment is highly dependent
upon sample temperature, and severe reduction, including metal island formation,
can occur for bombardment at elevated temperatures.[273]) Based upon Auger spectroscopy measurements of the surface stoichiometry, the ion-bombarded surface in
Fig. 5.44 contains roughly 15 % O vacancies. Most of the vacancies appear to be
in the surface plane, since LEED patterns are still present after ion bombardment.
Detailed differences between the electronic structure of defective NiO (100) surfaces and those of MnO and CoO are apparent, however. Since the minority-spin

t_{2g} orbitals are filled in the ground state for stoichiometric NiO, any additional electrons residing at surface O-vacancy defect sites must begin populating the normally empty minority-spin e_g levels. This gives rise to a distinct surface state in the bulk bandgap, even for very low surface defect densities. (Recall that for $3d^7$ CoO, a shoulder only appeared when the defect density was high enough that adjacent defects interacted.) Measurements of the photoemission threshold for annealed NiO (100) surfaces also indicate filled surface states lying just above the bulk valence-band maximum.[33,35] The finite density of electrons at E_F shows that the surface has become metallic; the increased electron population at E_F is also apparent in an increased asymmetry of core levels in XPS spectra.

Surface O vacancies are also produced on NiO (100) by annealing in H_2.[96,581,582] The kinetics of reduction are controlled by the rate of removal of lattice O at the surface and by the diffusion of subsurface O to the surface. For severe reduction, Ni islands have been observed to grow epitaxially on the NiO substrate.

The equilibrium surface defects on p-type oxides such as NiO are not O vacancies, however. When NiO (100) is annealed at high temperatures in UHV and then quenched to room temperature, Ni vacancies are observed.[583] This is in contrast to similar experiments on the n-type oxides TiO_2 and ZnO, where the surface defects in equilibrium at high temperatures (studied by quenching samples to room temperature to freeze in the high-temperature distribution) consist of O vacancies. Detailed studies of the types of surface defects produced by high-temperature treatment of NiO have not been reported.

5.10 Copper oxides and high-T_c superconductors

Surface-science experiments on single crystals of Cu oxides have begun only recently, spurred in part by the advent of Cu-oxide high-T_c superconductors. UPS, XPS and LEED measurements have been performed on (111) and (100) faces of polished and annealed single-crystal Cu_2O.[208] The (111) surface is non-polar, consisting of a plane of Cu ions sandwiched between two planes of O ions. The (100) surface, on the other hand, is polar; ideal, atomically flat surfaces could be terminated by either Cu or O ions. Not surprisingly, (111) surfaces that were stable and exhibited good quality LEED patterns were easier to prepare than were (100) surfaces (see § 2.3.7). Figure 5.45 presents He II UPS spectra for Cu_2O (111)-(1 × 1) and (100)-(3√2 × √2) R 45° surfaces. Cu^+ has a closed-shell $3d^{10}$ electronic configuration, and 3d orbitals constitute the upper part of the valence band. They appear in the 1–4 eV binding energy region of the valence band, whereas the emission from 5–8 eV corresponds to the predominantly O 2p levels.[584]

The details of the UPS spectra for Cu_2O depend upon the history of surface preparation.[208] The spectra for the (111) surface are always very similar to that

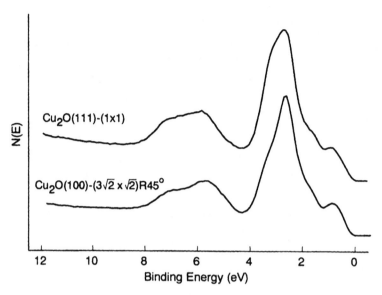

Fig. 5.45 He II UPS spectra for Cu$_2$O (111) and (100) surfaces reconstructed as indicated. [Ref. 617]

shown in Fig. 5.45. But for Cu$_2$O (100), large changes were observed both in the region of the O 2p emission and at about 9.4 eV binding energy (referenced to the top of the valence band as in Fig. 5.45). No definitive assignment could be made for the 9.4 eV emission, but it is believed to correspond to some non-incorporated form of adsorbed oxygen. It had not been reported on previous UPS studies of polycrystalline Cu$_2$O prepared by oxidizing Cu.[584]

Resonant photoemission measurements have also been reported on CuO; while most of the experiments have been performed on oxidized Cu surfaces,[585,586] UHV-cleaved CuO single crystals were used in one study.[587] Figure 5.46 compares UPS spectra from two studies of CuO on oxidized Cu [curves (1) and (2) in Fig. 5.46 (a)] with single-crystal data [curve(3)]; the face exposed on the cleaved single crystal was not specified. Al Kα XPS data for the single crystal are presented in Fig. 5.46 (b), along with the results of a calculation of the Cu 3d spectral weight for a [CuO$_4$]$^{6-}$ cluster.[588] In the ground state CuO has a 3d^9 electronic configuration and is an antiferromagnetic semiconductor with a bandgap of about 1.4 eV.[584] The resonant photoemission measurements established that the unscreened 3d^8 final states dominate the region of the spectrum between 8 and 16 eV, while the structure closer to the bandgap involves final states in which charge has been transferred to the Cu cations from the O ligands. Thus the bandgap in CuO is identified as being of charge-transfer origin rather than arising from d → d transitions. No specific surface states were identified in any of the spectra, but changes were

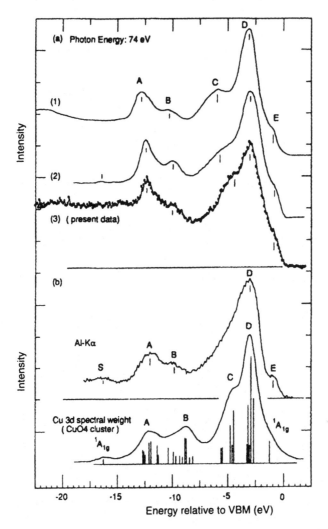

Fig. 5.46 (a) Valence-band UPS spectra taken with $\hbar\omega = 74$ eV for (1) and (2) CuO on oxidized Cu from Refs. 585 and 586, respectively, and (3) UHV-cleaved single-crystal CuO (face unspecified); and (b) Al Kα XPS for single-crystal CuO compared with the results of a calculation of the Cu 3d spectral weight for a $[CuO_4]^{6-}$ cluster from Ref. 588. [Ref. 587]

observed in the single-crystal spectra as a function of time (at room temperature) after cleaving in UHV that were interpreted in terms of oxygen loss from the sample.

The surface properties of Cu-oxide-based high-T$_c$ superconductors are important in two contexts. Some of the compounds lose oxygen easily, and O-deficient surface regions can form when they are used in ambients with low O$_2$ partial pressure. This non-stoichiometric surface layer does not have the bulk superconducting properties, which adversely affects the behavior of polycrystalline or thin-film samples. The surface non-stoichiometry has made it very difficult to use inherently surface-sensitive analytical techniques such as photoemission to study the electronic properties of these materials.[589,590] The extent of the problem varies for different compositions, with the largest O mobility for the $YBa_2Cu_3O_{7-x}$ oxides. On the other hand, the $Bi_2Sr_2CaCu_2O_{8-x}$ compounds exhibit almost no loss of O at the surface, although they are sensitive to electron-beam reduction of the surface for incident-electron energies between 1 and 2 keV.[591]

Another area where the surface properties of superconducting oxides become important is in the response to adsorbed atoms or molecules. Although one normally does not think in terms of the 'corrosion' of oxides, that is precisely what happens when many of these materials are exposed to ambient gases. It is a severe problem for many practical applications and has caused countless problems in determining the true bulk properties of ceramic and thin-film samples. These interactions will be addressed in § 6.6 below. The interaction between high-T_c superconductors and metal and semiconductor atoms is very important for the formation of electrical contacts to the materials and for their compatibility with other materials in device structures; this will be discussed in § 7.4.

Almost all of the experimental studies of the electronic structure of high-T_c superconductors performed to date on well-characterized single-crystal samples, even those that utilized surface-sensitive techniques such as UPS, have addressed the bulk properties and interpreted their data in terms of them.[275] Surface information may be present in the spectra, but it has not yet been extracted in a systematic way. An example of the UPS spectra obtained on single crystals of $YBa_2Cu_3O_{6.9}$ using a spectrometer having the highest resolution currently available in both energy and angle is shown in Fig. 5.47.[592] The sample was cleaved in UHV at 20 K exposing the (001) plane, and the spectra were taken at 20 K. Spectra are shown for different symmetry points in the Brillouin zone as indicated, and for two different incident photon polarizations at $\bar{S}$. Interest has concentrated around the features at E_F in order to characterize the superconducting gap; in the study shown here it was concluded that the gap is less than 10 meV in $YBa_2Cu_3O_{6.9}$, and significant emission is seen at E_F in certain directions in reciprocal space. In many of the published studies of both annealed ceramic pellets and UHV-cleaved single crystals of the high-T_c Cu-oxide superconductors, UPS spectra exhibit a very low density of occupied electronic states just below E_F.[590,593] It is possible, however, that the experimental results may have been affected by a loss of O from the surface. UPS measurements on $EuBa_2Cu_3O_{7-x}$ single-crystals cleaved below 20 K do exhibit significant emission at E_F; that emission decreases as the samples are warmed to room temperature.[589] For the 80 K phase of Bi-Sr-Ca-Cu-O, however, XPS measurements on single crystals that were either cleaved in vacuum or

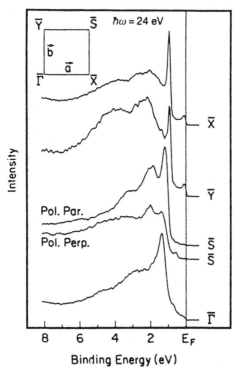

Fig. 5.47 Angle-resolved UPS spectra ($\hbar\omega = 24$ eV) for the (001) face of UHV-cleaved $YBa_2Cu_3O_{7-x}$ at the indicated points in the Brillouin zone. Data were taken at 20 K. [Ref. 592]

cleaved in air and then annealed at 600 K in vacuum indicated that the chemical state of the ions in the surface region was nearly the same for either preparation procedure.[594]

There has been one theoretical calculation of the effect that the surface would have on the valence-band density-of-states for the (001) surface of $YBa_2Cu_3O_7$.[595] Three different possible terminations of the lattice were considered, all of them giving a stoichiometric surface. The one-hole final-state spectra that would be seen in UPS were calculated, and small differences between the bulk and surface contributions were found. However, the differences were sufficiently subtle that it would be difficult to identify them in experimental spectra.

5.11 Molybdenum oxides

Problems of surface instability have made it difficult to study the oxides of Mo. The d^0 binary oxide, and the only one that has been studied in single-crystal form, is MoO_3, an insulator having a layered orthorhombic structure.[194,596] The most

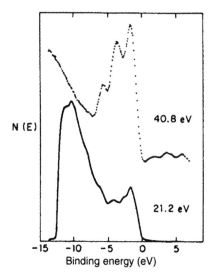

Fig. 5.48 He I and He II UPS valence-band spectra for MoO_3 (010). Binding energy is referenced to the valence-band maximum. [Ref. 194]

stable, basal plane is (010), on which only O ions are exposed. At room temperature that surface is quite inert chemically, and surfaces prepared by cleaving in air give the same UPS, XPS and LEED results as surfaces scraped in UHV. Figure 5.48 shows angle-integrated He I and He II UPS spectra for stoichiometric MoO_3 (010); both spectra are referenced to the valence-band maximum. There is negligible emission in the bulk bandgap region (above 0 eV) in the He I spectrum. (The apparent bandgap structure in the He II spectrum is valence-band structure excited by He II_β photons at 48.4 eV.)

However, when the MoO_3 (010) surface is exposed to electron or ion beams, even of relatively low energy, or to near-bandgap ultraviolet photons,[597] the surface becomes reduced. The ease of reduction no doubt occurs because of the many reduced oxides of Mo that are stable in the bulk; for example, between MoO_3 and MoO_2 stable Magnéli phases of $Mo_{18}O_{52}$, $Mo_{17}O_{47}$, Mo_9O_{26}, Mo_8O_{23}, Mo_7O_{20}, Mo_6O_{17}, Mo_5O_{14} and Mo_4O_{11} all exist.[598] Bombardment of MoO_3 (010) with 500 eV Ar^+ ions has been found to reduce the surface to $MoO_{1.6}$. The reduced surface exhibits significant emission in the bulk bandgap, as shown in Fig. 5.49 for various types of surface treatment. The bandgap emission must correspond to states having predominantly Mo 4d character. Only after annealing in a high pressure of O_2 does the surface re-oxidize to MoO_3. The bandgap defect states have been studied by using angle-resolved UPS and a range of photon energies.[596] While the valence-band emission features were found to disperse with $k_\parallel$ but not with $k_\perp$, as expected for bulk bands, the bandgap emission was dispersionless with $k_\parallel$, but dis-

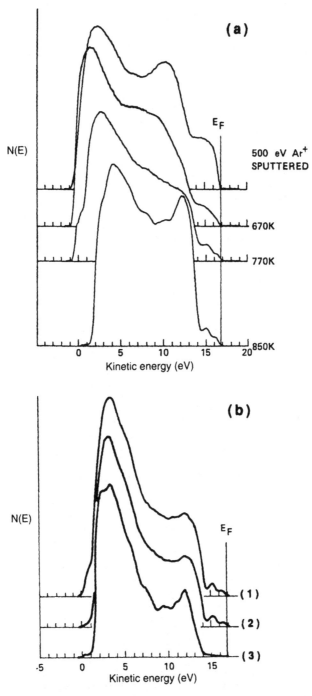

Fig. 5.49 He I UPS spectra of (a) Ar⁺-ion-bombarded MoO_3 (010) before and after anneal-ing in UHV at the temperatures shown, and (b) partially annealed in UHV at 850 K [curve (1)], annealed at 770 K in 0.8 Pa O_2 [curve (2)], and annealed at 770 K in 100 Pa O_2 [curve (3)]. [Ref. 194]

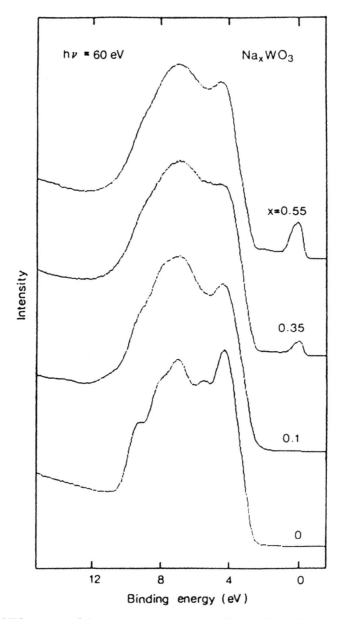

Fig. 5.50 UPS spectra of the valence-band region of Na_xWO_3 (100) as a function of x.
[Ref. 604]

persed over 1 eV with $k_\perp$. The electronic defect levels are therefore quite unlike those associated with point defects, but are extended ones having electron delocalization perpendicular to the MoO_3 layers. Thus the 'surface' defect structure of MoO_3 is different from that for most other transition-metal oxides, presumably because of the shear-plane structure of the reduced phases.

5.12 Tungsten and rhenium oxides, and Na$_x$WO$_3$

The 5d-transition-metal oxides WO_3, Na_xWO_3 and ReO_3 are similar in electronic structure to the early 3d oxides. They have very nearly the ABO_3 perovskite crystal structure, with the W or Re ions occupying the octahedral B cation sites. In WO_3 and ReO_3 the A cation site is vacant, while in Na_xWO_3 the Na^+ ions occupy that site. Stoichiometric WO_3 is an insulator, since the W 5d band is empty, while ReO_3, in which Re^{6+} has the $5d^1$ configuration, is metallic. When Na ions are added to WO_3, they donate their 3s electron to the W 5d band, resulting in bulk metallic behavior for $x \geq 0.3$ (see § 5.1 in Ref. 40). There is disagreement in the literature as to whether or not the surface region of Na_xWO_3 is depleted of Na compared to the bulk, and it is difficult to judge since different surface preparation procedures and electron beam exposures are used in each report.[121,124,599–603]

5.12.1 WO$_3$ and Na$_x$WO$_3$

Several surface-sensitive UPS experiments have been reported for UHV-cleaved WO_3 (100) and Na_xWO_3 (100) and (110) surfaces.[124,599,602,604] Figure 5.50 presents angle-integrated spectra for WO_3 (100) and three different Na concentrations of Na_xWO_3 (100).[604] In addition to changes in the shape of the O 2p valence band, emission from the W 5d orbital is seen to increase with increasing Na concentration. A 2 eV bandgap separates the W 5d and O 2p bands, as expected for the bulk electronic structure. However, intrinsic surface states may be present in that bandgap. Figure 5.51 shows two sets of angle-resolved UPS spectra for UHV-cleaved $Na_{0.85}WO_3$ (100); the W 5d and bandgap emission is examined in more detail along the right side of each panel.[599] The bandgap peak disperses from 2.1 eV for $k_\parallel = 0$ to 1.2 eV for $k_\parallel$ along $\Gamma \rightarrow M$. This state was interpreted in Ref. 599 in terms of the surface-state theory for 3d perovskites. In work by the same authors on pure WO_3, no evidence was found for intrinsic surface states in the bandgap.[605] Similar, although much weaker, bandgap emission was seen on the polished and annealed $Na_{0.64}WO_3$ (110) surface, but it was attributed to a surface plasmon loss associated with the W 5d emission and not to a surface state.[124] An alternative possibility for the origin of the structure in the bandgap of Na_xWO_3 might be a valence-band satellite.[606]

ELS measurements have also been made on Na_xWO_3 samples, but no surface information other than surface composition was obtained.[124,600,601] The main emphasis of the work was on understanding the properties of conduction electrons in the bulk and demonstrating that the bulk electron density persists up to the surface. An IPS measurement on $Na_{0.64}WO_3$ (110) was likewise interpreted solely in bulk terms.[607]

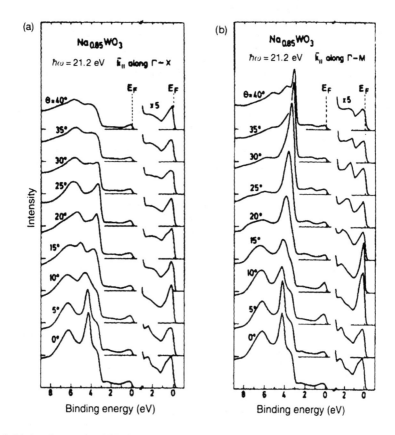

Fig. 5.51 Angle-resolved He I UPS spectra for $Na_{0.85}WO_3$ (100) for $k_{||}$ along (a) Γ–X and (b) Γ–M. Each spectrum is for a fixed value of the polar angle θ. [Ref. 599]

Theoretical calculations of surface electronic structure have been performed for model WO_3 (100) surfaces,[608] but not for Na_xWO_3. No surface states were found in the bulk WO_3 bandgap, in agreement with the experimental observations mentioned above.

The nature of surface defects on WO_3 and Na_xWO_3 parallels that of the early 3d-transition-metal oxides. Surface O vacancies are produced when samples are bombarded with ions, electrons or photons, or by exposure to atomic H.[118,121,597,605,609–611] The effect of O-vacancy defects on the electronic structure of WO_3 (100) can be seen in the UPS spectra in Fig. 5.52, which shows the effects of both electron and Ar^+-ion bombardment.[609] A broad band of defect states is created in the bulk bandgap, the detailed shape of which depends upon what type of ion is used for bombardment, whether the surface is subsequently annealed, etc. In

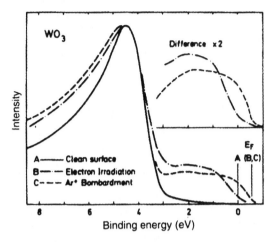

Fig. 5.52 Normal-emission He I UPS spectra for WO$_3$ (100) subjected to the surface treatments shown. The spectra have been aligned in amplitude and energy at the strong valence-band peak. The difference spectra (B–A) and (C–A) are shown on an expanded scale. [Ref. 609]

ELS spectra, a loss peak corresponding to a transition energy of 20 eV was found for both WO$_3$ and Na$_x$WO$_3$; it was attributed to surface O vacancies.[611] The nature of defect surface states on these materials has not been addressed theoretically.

A very different experiment has been performed on the polished (but not annealed) (100) surface of Na$_x$WO$_3$;[352] model calculations were also used to interpret the data.[354] Infrared reflectance measurements performed in air exhibited minima near the highest bulk longitudinal optical phonon frequency. Comparison of the spectra with a model of a depletion layer on the surface gave satisfactory agreement with experiment if the layer was taken to be 10 Å thick. Similar experiments on better-characterized surfaces prepared in UHV have not been reported.

5.12.2 ReO$_3$

Two single-crystal photoemission studies have been performed on ReO$_3$.[612,613] Figure 5.53 shows a UPS spectrum for UHV-cleaved ReO$_3$ (100), together with similar spectra for Na$_{0.83}$WO$_3$ and Na$_{0.40}$WO$_3$ for comparison.[613] The major difference between ReO$_3$ and the others is the significantly wider valence band and the absence of a gap between the Re 5d and O 2p bands; only a minimum in the density-of-states occurs for ReO$_3$. The data in both studies were interpreted solely in terms of bulk parameters, and yet the data in Fig. 5.53 are taken in the most surface-sensitive region of photon energy. Thus there may well be surface states present in the spectra, although they have yet to be identified.

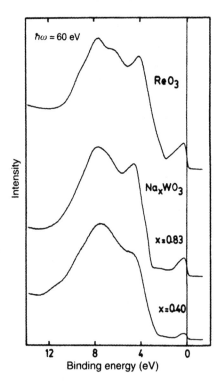

Fig. 5.53 UPS spectra of UHV-cleaved (100) surfaces of ReO_3, $Na_{0.83}WO_3$ and $Na_{0.4}WO_3$.
[Ref. 613]

Theory is slightly ahead of experiment for ReO_3. DV-Xα cluster calculations
have been performed for both ideal and O-deficient ReO_3 (100) surfaces.[364,614]
Two models for the stoichiometric surface were used, since it is not clear what the
geometry of a real (100) surface would be. The atomically flat Re–O plane has the
composition ReO_2 and hence is not charge neutral. Charge neutrality can be
achieved by adding 0.5 monolayer of O ions over surface Re sites, but the geometric arrangement of the additional O ions is not defined. The electronic structure
predicted for these two model surfaces differs in the location of filled Re 5d surface states, since for the flat surface electrons must move from the bulk into the
surface region to restore charge neutrality. However, no comparison has been
made between theory and the experiments on UHV-cleaved ReO_3 surfaces.
Calculations of the electronic structure of an O vacancy on the flat surface predicted a narrowing of the bandgap between the O 2p and Re 5d bands relative to
the defect-free flat surface, but again no comparison with experiment has been
made.

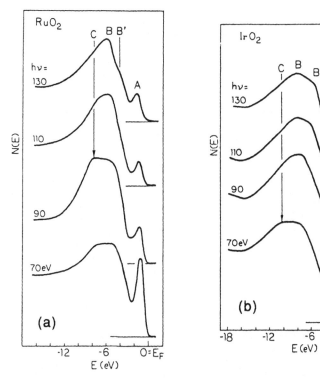

Fig. 5.54 Angle-integrated UPS spectra for UHV-scraped single-crystals of (a) RuO_2 and (b) IrO_2 as a function of photon energy [Ref. 615]

5.13 Other oxides

Only a few other transition-metal oxides have been studied using sufficiently well-characterized surfaces that unique interpretation of the data can be made. Single crystals of **RuO_2** have been studied by LEED, XPS and UPS, and the surfaces were found to be rather complicated.[150,151,615] The only measurements on clearly stoichiometric surfaces were performed on single crystals whose surfaces were scraped in UHV. Angle-integrated UPS measurements were made on those surfaces (whose orientation was unknown), and Fig. 5.54 (a) presents the spectra for four different photon energies.[615] The electron configuration of Ru in bulk RuO_2 should be Ru^{4+} $4d^4$, and peak A is attributed to emission from those 4d orbitals. The remainder of the emission is attributed to the predominantly O 2p orbitals. While these spectra are extremely surface-sensitive, no consideration has been given to surface electronic structure, and there are no theories of surface properties with which to compare the data.

As-grown single-crystal (110) and (100) RuO_2 surfaces have been studied by LEED, XPS and Auger spectroscopy, and the results suggest highly non-stoichio-

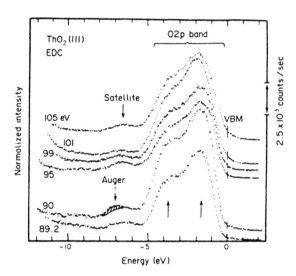

Fig. 5.55 UPS valence-band spectra of ThO$_2$ (111) for photon energies in the vicinity of the Th 5d → 5f edge. VBM = valence-band maximum. [Ref. 616]

metric surfaces.[150,151] Nothing has been determined about the electronic structure of those surfaces, but evidence for the presence of RuO$_3$ in the surface region was found, and it was hypothesized that RuO$_3$ played a role in the stability of the RuO$_2$ surfaces. The electronic structure of such surfaces would surely be extremely complex. On the other hand, it appears from UPS and ELS measurements on polycrystalline samples of RuO$_2$ that no differences from the expected bulk electronic structure can be seen.[156] The surface stoichiometry and properties of this important oxide must therefore be regarded as an open question.

Figure 5.54 (b) shows analogous spectra for UHV-scraped single-crystal **IrO$_2$**,[615] but again no consideration was given to surface properties. The peak labelled A should correspond in this case to emission from the Ir^{4+} 5d^5 orbitals.

UPS spectra have also been taken for (111) surfaces of **ThO$_2$**.[616] ThO$_2$ has the same crystal structure as UO$_2$, which was discussed in § 2.3.7; the (111) surface is charge neutral and non-polar. Polished and annealed surfaces showing sharp (1 × 1) LEED patterns were used. Spectra were taken for photon energies from 89 to 105 eV in order to study resonant effects on transitions originating with the Th 5d level, as shown in Fig. 5.55. Although the data were interpreted in terms of bulk electronic structure, it was noted that no emission in the bulk bandgap, which would indicate occupied surface states, was observed. The shape of the O 2p valence band in these very surface-sensitive UPS spectra is similar to the O 2p band in most other transition-metal oxides.

6

Molecular adsorption on oxides

This chapter will discuss the interaction between metal-oxide surfaces and atoms and molecules that are normally gases or liquids at room temperature; this includes all molecules that are of catalytic importance. The only adsorbates excluded are metals and other metal oxides, which will be considered in Chapter 7.

The logical starting point in the study of molecular adsorption is to consider how molecules interact with defect-free surfaces. This is easy to do theoretically, and most calculations of adsorption on oxides have assumed a perfect surface structure. Unfortunately, no truly defect-free surfaces can be prepared in practice, although the density of defects on surfaces can sometimes be made so small that it is not possible to detect them by using surface-spectroscopic techniques. However, attempts to correlate the catalytic behavior of single-crystal surfaces with surface-spectroscopic measurements often fail because surface defects having a density too small to be observed directly may well constitute the active sites for catalysis. This is much more important for ionic materials such as oxides than it is for metals, since point defects on oxides possess very different electronic structure than does the perfect surface.

It is still useful to prepare the most nearly perfect, defect-free oxide surfaces possible, and many measurements have been performed on such surfaces. We shall refer to those surfaces as 'stoichiometric', since the density of defects is sufficiently small that the surface composition is for all intents and purposes the same as that of the bulk. However, it must be borne in mind that the separation of surfaces into 'stoichiometric' and 'defect' categories is only a matter of degree. And some of the results that are attributed to perfect surfaces – no matter how careful one may be in the preparation of the surface – may really be due to the presence of defects.

A word of caution is in order concerning the comparison of chemisorption studies with catalysis. Virtually all chemisorption measurements on well-characterized surfaces are performed at very low pressures under essentially UHV conditions.

Commercial catalytic reactions, on the other hand, are almost always run at well above atmospheric pressure. There are two reasons why effects may occur in catalysis that cannot be observed in chemisorption measurements. The fundamental difference results from the finite residence times of species on surfaces. At high pressures the surface generally has a much higher steady-state coverage of adsorbed species than at low pressure. Not only can more reactions between adsorbates, or between impinging molecules and adsorbates, take place per unit time at high pressure, but at high surface coverages the way in which a moiety is adsorbed may well be different than at low coverage. Therefore, types of reaction may take place that are simply not possible at low pressures and coverages. The second reason that catalytic reactions may not be reproducible at low pressures is merely one of degree. If a catalytic reaction takes place at an 'active site' whose surface density is extremely small, the techniques used in chemisorption measurements on single crystals may not be sensitive enough to observe adsorption at such sites. At higher pressures, however, a catalytic reaction will be seen to occur, even though only a small fraction of the reactant species impinging on the surface reacts per collision. This inability to accurately mimic high-pressure reactions under low-pressure conditions is often referred to as the **pressure gap**. A number of experimental approaches are currently being explored in order to find ways to bridge that gap.

In § 6.1 the experimental methods and theoretical approaches to adsorption on metal oxides will be described. The types of adsorption that can occur on oxide surfaces will be outlined in § 6.2, both in general terms and with regard to the behavior of specific molecules. Specific adsorption systems will then be treated in two ways. The results of both experimental and theoretical studies of molecular adsorption on oxide surfaces are compiled in Tables 6.1 through 6.7, and references to the relevant literature are included therein. More detailed discussions of adsorption on non-transition-metal oxides, d^0 transition-metal oxides, and d^n transition-metal oxides are given in § 6.3, 6.4 and 6.5, respectively. Adsorption on high-T_c oxide superconductors is considered in § 6.6. The desorption of atoms, ions and molecules from metal-oxide surfaces by electron and photon excitation is discussed in § 6.7.

6.1 Experimental and theoretical methods in chemisorption

The study of molecules and atoms adsorbed on surfaces poses some experimental and theoretical challenges that do not occur for atomically clean surfaces. In this section we will briefly outline how various techniques are applied to the study of adsorption on single-crystal surfaces and what complications arise in their application to metal oxides.

6.1.1 Experimental methods

The experimental methods that are used to study chemisorption on solid surfaces fall into two categories: those that determine what is on the surface, and those that determine what comes off. Obviously the most complete characterization of a chemisorption system will be obtained if both types of technique are used, and that has been done in a number of the experiments discussed below. But there are still many studies that use only one or the other type of technique, so the experimental information available is often less than adequate to completely determine how a molecule interacts with a surface. It may thus be necessary to infer some steps in the process based upon knowledge of the behavior of related systems.

The major experimental techniques that are used to determine what is on a surface have already been discussed in § 2.2, 3.2.5 and 4.2. In this section we will only consider the aspects that are important in studying adsorbed molecules. **Auger spectroscopy** and **XPS** are used in a straightforward way to determine what atoms are present on the surface and, where possible, their valence states. The potential problems that can be encountered in using them to study adsorbates on oxides are the possibility of electron- or photon-stimulated desorption (ESD or PSD) of species from the surface and the electron-beam damage to the surface itself mentioned in § 4.2. Those problems are generally not severe with XPS, since the cross-section for PSD is very small at x-ray energies. However, the problem of ESD induced by the incident electron beam used in Auger is much greater for molecules adsorbed on metal oxides than it is for the same molecules adsorbed on metals. The problem of lattice dissociation discussed previously is also present, of course. Thus Auger spectroscopy is rarely used in adsorption studies except for initial characterization of the surface before adsorption.

UPS is a powerful tool for observing adsorption-induced changes in the electronic structure of a surface in the vicinity of the Fermi level. The cross-section for PSD of adsorbates is small, although one should still check experimentally in specific cases to be certain that is the case. The most useful application of UPS is to take **difference spectra**, which is the name given to the process of subtracting one UPS spectrum from another. Such spectra were discussed in Chapters 4 and 5, primarily for spectra taken from the same surface but for different photon energies. In the case of adsorption, one usually subtracts the spectrum of an atomically clean surface from the spectrum for the same surface after adsorption, with both spectra taken at the same photon energy. In the limit of small adsorbate/substrate interaction, such difference spectra will consist of electrons emitted from the molecular orbitals of the adsorbate. In reality the difference spectra also contain changes in the electronic structure of the substrate, but the technique has still proven to be extremely useful in identifying adsorbed species.

One precaution must be exercised when working with anything other than metallic samples, however, or UPS difference spectra will be meaningless. For example, consider Fig. 6.41 in § 6.5.2 below, which plots UPS spectra for a UHV-cleaved Cr-doped V_2O_3 ($10\bar{1}2$) surface before and after exposure to various amounts of SO_2. The energy scales for all of the spectra are aligned at E_F, which is the normal way in which data are obtained. Notice that, while there are only small changes in the shape of the O 2p valence band upon adsorption, the entire valence band moves closer to E_F. This is a result of the partial depopulation of the V 3d band just below E_F and corresponds to a rigid shift of the O 2p band relative to E_F. If one were to take differences between an adsorbate-covered surface and the clean surface with the spectra aligned in energy as shown, the difference spectra would be dominated by large peaks that resulted simply from the rigid shift of the two spectra; they would contain no real physics, and the meaningful information would be either distorted or destroyed. It is thus necessary to shift the two spectra relative to one another by the amount of the rigid-band shift before taking differences; this has been done for the difference spectra shown in Fig. 6.42. It is not always straightforward to determine the correct amount to shift the spectra, and this is a possible source of error. In many cases that is not a severe problem. However, if the adsorbate/substrate interaction is strong, then the entire concept of difference spectra breaks down; this often occurs for adsorption on defective oxide surfaces.

Useful information can often be obtained by comparing UPS difference spectra obtained on chemisorption of a particular molecule with the gas-phase spectrum of the same molecule. The electrostatic polarizability of the nearby surface gives rise to an **extra-molecular relaxation-polarization shift**, ΔE_R, appearing as a (generally) uniform shift of all of the molecular orbitals to lower binding energies relative to the free molecule. In addition to ΔE_R, those orbitals involved in bonding to the surface will be shifted toward higher binding energy by an amount ΔE_B, so that the net shift in the binding energy of bonding orbitals relative to their gas-phase counterparts is ($\Delta E_B - \Delta E_R$). (See also the discussion of the adsorption of Xe on surfaces in § 6.2.3.6 below.)

An example of the use of UPS difference spectra to measure ΔE_R and ΔE_B is presented in Fig. 6.1 for benzene, C_6H_6, adsorption on ZnO ($10\bar{1}0$). The dashed curve in (a) is the He II UPS spectrum for the clean ZnO surface, and the solid curve is the spectrum in a 10^{-7} Torr C_6H_6 ambient. The difference spectrum in (b) is the adsorbate-covered spectrum minus one-half of the clean surface spectrum, where the factor of one-half is to compensate for the attenuation of the substrate photoemission features by the adsorbed C_6H_6 layer. Panel (c) is the gas-phase C_6H_6 UPS spectrum measured in a similar spectrometer, shifted to lower binding energy by $\Delta E_R = 1.3$ eV, which aligns all but the three lowest-binding-energy mol-

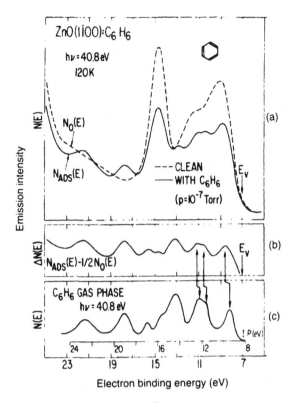

Fig. 6.1 (a) He II UPS spectra for ZnO (10$\bar{1}$0) at 120 K, clean (dashed curve) and in the presence of a 10^{-7} Torr C_6H_6 ambient; (b) difference spectrum, as described in the text; and (c) gas-phase C_6H_6 UPS spectrum. [Ref. 763]

ecular orbital peaks. These peaks are shifted to higher binding energy by 0.2–0.4 eV relative to the other orbitals, which represents the energy stabilization coming from the interaction of the molecular orbitals with the surface. In some of the figures presented later (e.g., Figs. 6.37 and 6.48), gas-phase spectra presented for comparison have been shifted to take account of the polarization shift ΔE_R.

The vibrational spectroscopies, **infrared reflection/absorption (IRRAS or RAIRS)**, and **high-resolution electron-energy loss (HREELS)**, were discussed in § 3.2 in connection with the measurement of surface phonon modes on ionic materials. The most powerful use of these techniques in surface science is for measuring the vibrational modes of molecules adsorbed on surfaces. In HREELS the low-energy incident-electron beam couples to time-varying electric fields at the surface, and it is most sensitive to dipole fields. Thus the incident-electron beam can stimulate the adsorbate to absorb (or emit) a quantum of vibrational energy, which then appears as a corresponding loss (or gain) in the energy of the electron. The information obtained is thus similar to that in IRRAS, although HREELS is more surface-sensitive. It is also possible to relax the usual selection rule, which for IR spectroscopy requires that the vibrational mode must have a dipole component normal to the surface if it is to appear in the spectrum. The same is true for the strong dipole scattering appearing in specular HREELS. But in the off-specular direction, the dominant loss come from **impact scattering**, which has less stringent selection rules. By measuring both dipole and non-dipole spectra, increased information can be obtained.

There are problems with the use of both HREELS and IRRAS on semiconducting or insulating oxides. The appearance of strong intrinsic phonon spectra discussed in Chapter 3 can make it very difficult to see adsorbate spectra. In IRRAS, it is possible to study adsorbate vibrations at significantly higher frequency than the optical phonon modes of the oxide; although many experimental results have been reported, e.g., for CO and H_2O adsorption on polycrystalline samples, the technique has not yet been exploited for single-crystal oxides. In HREELS, not only is the fundamental optical phonon frequency excited, but strong overtones as well, which may lie in the same energy-loss region as adsorbate vibrations. The phonon losses can be much larger than the adsorbate features and may obscure or distort them. However, methods have been developed to deconvolute phonon overtones from HREELS spectra so that adsorbate losses can be observed (see § 3.2.5).

The geometric structure techniques of **LEED** and **RHEED** described in Chapter 2 are also useful in chemisorption studies. Molecules often either adsorb in an ordered array on single-crystal surfaces or induce a reconstruction in the substrate, both of which give rise to changes in LEED or RHEED patterns. RHEED is also useful for monitoring the topography of thin-film growth on single-crystal surfaces, as long as the growth is epitaxial. A few recent studies have utilized **SEXAFS** and **NEXAFS (near-edge extended x-ray absorption fine structure)** in studies of atoms and molecules adsorbed on oxide surfaces. The former technique gives distances between the atom being probed and its ligands, while the latter is sensitive primarily to the electronic structure of the ligands.

Two indirect techniques for studying adsorption are **work-function** measurements (either using a Kelvin probe or measuring the low-energy cut-off of secondary-electron spectra) and **surface conductivity** measurements. Both are extremely sensitive to adsorbates that transfer charge to or from the surface, and effects can be seen in those measurements before any evidence of adsorption is apparent in other spectroscopies. They provide information on the charge state of the surface, including surface dipole moments, that is complementary to that of most other techniques.

The most powerful technique for measuring what comes off a surface after adsorption is **thermal desorption spectroscopy (TDS)**.[4] After adsorption at low temperature, the sample temperature is increased at a (usually) constant rate, and the species that desorb from the surface are monitored as a function of temperature with a mass spectrometer. Not only can the desorbing species be determined, but analysis of the shape of the desorption spectra provides information about the kinetics of desorption; from this information it is often possible to deduce what species were adsorbed at various temperatures and even the path of a surface reaction. Of course, the information obtained is much more useful if direct determination of the adsorbed species is conducted simultaneously.

It is often difficult to differentiate O-containing adsorbates from the oxygen of the metal-oxide lattice. In TDS and HREELS this can sometimes be overcome by the use of isotopes. Adsorbed species containing ^{18}O have easily distinguishable vibrational frequencies and mass/charge ratios and can thus be separated from lattice O. Isotopically labeled molecules have also been used to determine whether or not O exchange occurs between the adsorbate and the oxide surface.

Three other desorption techniques can be used on single-crystal surfaces, but they have not found widespread application to metal oxides. **Secondary-ion mass spectroscopy (SIMS)** uses an incident ion beam to desorb surface species;[4] it has been used in a few studies of oxides. **Electron- and photon-stimulated desorption (ESD** and **PSD)** have also been used in a few instances, and that work is discussed in § 6.7 below.

One area where experimental information on adsorbed species is seriously lacking for well-characterized metal-oxide surfaces is **surface diffusion**. One group has attempted measurements of the diffusion of CO_2 and NH_3 on single-crystal MgO (100) surfaces by means of laser-induced thermal desorption.[618,619] The experiments were performed by adsorbing a uniform layer of molecules onto the surface, desorbing all of the molecules from a small region of the surface by means of a high-power laser pulse, and then probing that region with another laser pulse at later times to determine the rate at which molecules diffuse back into the clean region from the surrounding layer. The single-crystal MgO surfaces were cleaved in air and then cleaned in the vacuum chamber by heating to 950 K in O_2; they exhibited sharp (1 × 1) LEED patterns. Since the MgO (100) surface is so inert, it was necessary to cool the samples to below 100 K for CO_2 and 165 K for NH_3 in order for the adsorbed layer to be stable. At those temperatures no diffusion was detected, even for times as long as one hour between the desorbing and the probe laser pulses. An upper limit on the diffusion coefficients of CO_2 and NH_3 of 1×10^{-9} cm^2/s at those temperatures was determined. It was speculated that surface diffusion might be inhibited by either the magnitude of the surface corrugation or the presence of steps.

No measurements of surface diffusion on single-crystal metal oxides by any other techniques have been reported. Field-ion microscopy has proven useful in studying the details of atomic diffusion on single-crystal metal surfaces; perhaps it could also be applied to thin oxide layers on metal tips in some specialized cases.

6.1.2 Theoretical techniques

Theoretical calculations have been applied extensively to adsorbates on oxides, partly to complement experimental techniques which often give inadequate information, and partly as a guide to systems that have not been studied experimentally. Most commonly, the aim of a calculation is to predict the likely structure of an adsorbate and its mode and energy of bonding to the surface. An additional, or sometimes alternative, goal is to calculate quantities such as orbital energies which can be directly compared with spectroscopic measurements.

The discussion of electronic structure calculations in § 4.1.5 is also relevant to adsorbates and need not be repeated. It is worth noting, however, that nearly all calculations on adsorbates deal with a finite cluster of atoms: that is, the adsorbate molecule itself, together with a small part of the substrate. Band-structure methods can be used for an ordered array of adsorbates, but in practice this has only been done for the simplest systems such as hydrogen atoms. Most cluster calculations use some variant of the molecular-orbital method, with a suitable basis of atomic orbitals. The main differences are (a) the size of the basis set, which may range from a very extensive one down to a minimum basis of the essential valence orbitals, and (b) the way in which they compute the different integrals which make up the Hamiltonian and overlap matrices. So many different types of approximation are employed, and the approximations are justified in so many different ways, that it is very hard to judge a priori whether a method is likely to give reliable results. Ultimately, the test of a theoretical method must come from comparison with experiment. The detailed discussion in following sections will show examples where calculations apparently agree with experiment, as well as many where they do not; but most numerous are cases where no direct comparison is possible. The most urgent need in this area therefore is for a larger data base of directly comparable calculations and experiments. There are various reasons why this comparison is rather difficult. One problem is that there is very little detailed experimental information about adsorbate structures. Another serious difficulty is that calculations generally aim to show the relative stability of different *static* configurations of atoms, but have little to say about the *kinetic* routes by which these can be achieved. It may be that many adsorption processes are controlled by kinetic limitations, rather than the ultimate stability of the adsorbed species. Thus, for many reasons, the species treated by calculations are sometimes frankly unrealistic; they must be considered more as models of adsorbate bonding than as representations of real adsorbate systems.

6.2 Types of chemisorption

Chemisorption on oxide surfaces follows significantly different patterns from that on metals. In the first place, the relatively ionic nature of the solids leads to a predominance of acid/base, or donor/acceptor, interactions. Cation sites are Lewis acids, and may interact with donor molecules such as H_2O through a combination of electrostatics (ion–dipole attraction) and orbital overlap. Oxide ions similarly act as basic sites and can interact with acceptors such as H^+. One of the most common dissociative reactions is indeed the deprotonation of an adsorbate to produce surface hydroxyl groups, which are almost universally present on polycrystalline oxides. Other types of heterolytic dissociation may be favored, even with molecules which would normally be expected to dissociate homolytically (that is, into neutral radicals) on metals.

Another important feature is the involvement of lattice oxygens. Towards acidic molecules such as CO_2 they are essentially basic, forming surface CO_3^{2-} for example. But in addition to their acid/base properties, many oxides are known for their ability to perform selective oxidation reactions. In these, oxygen may be added to an adsorbate, not as the oxide ion O^{2-}, but in the form of a neutral O atom. Oxidation of an adsorbate must be accompanied by a corresponding reduction of the substrate, in the form of electrons which might be free carriers, but more often it will lead to a localized decrease in oxidation state of metal atoms at the surface. In a catalytic cycle, re-oxidation takes place by filling of surface oxygen vacancies; but before this happens the electrons involved must be accommodated somewhere, a principle which often seems to be ignored in both experimental and theoretical adsorption studies.

Stoichiometric low-index faces of oxides are often rather unreactive, and chemisorption, especially dissociative, is promoted by surface defects and steps. The difficulty in seeing small concentrations of such features by surface-science techniques leads to considerable difficulties in the interpretation of adsorption studies. Is the proposed adsorption mechanism really taking place on the perfect crystal surface, or at unrecognized defect sites, or with some other undetected adsorbate? This question is often hard to answer.

An important concept in the reactivity of oxide surfaces is that of **coordinative unsaturation**. This implies the existence of surface sites where the coordination of metal and oxygen is less than optimal. Coordinatively unsaturated ions are obviously able to act as acidic or basic adsorption sites, but also they are bonded to the lattice less strongly than coordinatively saturated ones, so that they may be more easily removed, for example in oxidation reactions. The degree of coordinative unsaturation differs greatly between different surfaces, but in general it is rather slight for the most stable low-index faces, a fact that explains both their stability and their general lack of reactivity. Step and defect sites clearly have a greater degree of coordinative unsaturation. This increases their acidic and basic properties and leads to greater reactivity. Such sites may also have different electronic properties; for example, oxygen may exist as O^- and metal ions may have additional electrons present. Thus defects can act as 'sources' or 'sinks' for electrons, and thereby promote oxidation or reduction reactions.

6.2.1 Adsorption mechanisms

With the above ideas in mind, the following classification of adsorption mechanisms is useful. Firstly, there is the obvious distinction between

   I non-dissociative or molecular

and

   II dissociative.

Secondly, one should distinguish between the type of interaction with the surface:

   a weak electrostatic or dispersion

implies physisorption only, of which only a few examples are considered here;

   b acid/base or donor/acceptor

was mentioned above. Donors such as H_2O and NH_3 have lone-pair electrons that may interact with acidic (cation) surface sites, to give either non-dissociative [*type I (b)*] or dissociative [*type II (b)*] chemisorption. The latter implies heterolytic cleavage, with transfer of H^+ to a basic oxide site. The normal dissociative interaction of water may thus be written

$$H_2O + O^{2-}{}_{lattice} \rightarrow OH^-{}_{lattice} + OH^-{}_{surface} \qquad\qquad type\ II\ (b)$$

where the new OH^-, being a basic species, will coordinate to a surface metal ion.

   An essential feature of acid/base interactions is that they do not directly involve electronic carriers. Adsorbates may be classified as 'electron donors' or 'acceptors' according to the predominate direction of charge transfer; but the electrons concerned are never in any sense 'free'. This is discussed in more detail in § 6.2.2 below.

   c oxidation/reduction with electron transfer

implies the change in oxidation state of an adsorbate, with the release or capture of electrons. Atomic examples are the reaction of H to form H^+, and that of Cl to give

Cl⁻. The dissociative adsorption of H_2 or Cl_2 might be expected to proceed similarly. Because electronic carriers are involved, such reactions would not be expected on wide-bandgap insulators such as MgO, unless surface defects such as O⁻ or vacancies containing electrons were present. On semiconducting and metallic oxides, this type of adsorption may directly influence the surface carrier concentration.

d oxidation/reduction with oxygen transfer

is an alternative mechanism for oxidation. An example is the reaction

$$CO + O^{2-} \rightarrow CO_2 + 2\,e^-. \qquad\qquad type\ I\ (d)$$

As indicated, the oxidation of the adsorbate implies that the surface must be reduced, and the consequences are similar to that in *type (c)*. This type of reaction is common with organic molecules, for example in the oxidation of aldehydes RCHO to carboxylic acids RCOOH. It may not always be possible to distinguish it from *type (c)*. For example, the oxidation of an alcohol RCH_2OH to an aldehyde RCHO is essentially a **dehydrogenation**, and can be written

$$RCH_2OH + 2\,O^{2-} \rightarrow RCHO + 2\,OH^- + 2\,e^-. \qquad type\ II\ (c)$$

However, the reaction is very often accompanied by the loss of water, with lattice oxygen involved

$$RCH_2OH + O^{2-} \rightarrow RCHO + H_2O + 2\,e^-. \qquad type\ II\ (d)$$

The distinction between acid/base and oxidation/reduction (**redox**) reactions is fundamental from a chemical point of view and does not depend on a precise assignment of ionic charges as one might think from the above discussion. The reaction of CO to form CO_2 or surface carbonate is unambiguously an oxidation, and must be compensated by a corresponding reduction of the surface. This point is not always clearly recognized in the literature on chemisorption, and there may be various reasons for this. Experimentalists are naturally reluctant to assign definite charge states to surface species that may not be very well-characterized. In theoretical studies (for example, of the cluster molecular-orbital type) atoms are generally found to have fractional charges, less than the full 'ionic' value, so that again a definite assignment of ionic charges may not seem appropriate. The difficulties become greater when surface defects are involved, with electron configurations that may be different from normal. Surface oxidations may sometimes involve O⁻ ions, for example, so that electronic charges need not be left behind on the surface.

A particular problem often arises in dissociative chemisorption: should one regard it as **homolytic**, AB giving neutral A and B radicals, or as **heterolytic**, to A^+ and B^-? It could be argued that such a distinction is only real in the context of a completely ionic viewpoint, but that is not true. Even if A and A^+ do not represent 'real' charges, they correspond to systems with different electronic configurations, which will therefore bond to the surface in different ways. It might be hard in practice to decide between the two, especially if the sites of bonding are not well-characterized. There remains, however, an important difference in principle. Ultimately this type of question (and the general distinction between acid/base and redox interactions) has to do with the proper accounting for the electrons present. If this cannot be done in a given situation, then it follows that the adsorption process is not fully understood.

6.2.2 Electronic interactions in chemisorption

Changes in the electronic structure of the surface are frequently important in chemisorption. The band-bending and alterations in surface conductivity occurring with some semiconducting oxides form the basis of gas sensor operation. In addition, measurements of occupied electronic levels by UPS may frequently be the most important source of information about chemisorbed systems. It is therefore important to have some idea of how these changes are related to the chemical models of chemisorption just discussed.

Donor/acceptor interactions result from the overlap of a filled orbital on the donor with an empty one on the acceptor. This leads to the stabilization of the filled level, which is now shared to some extent between the two halves of the combined system. The empty orbital acquires some corresponding antibonding character and is raised in energy; this is illustrated in Fig. 6.2. In the more usual case where the adsorbed molecule is the donor, the occupied orbital in this figure represents one of the molecular orbitals of this adsorbate. Stabilization of an orbital in this way can sometimes be seen by comparing UPS difference spectra with those for free molecules, after correction for the extra-molecular relaxation–polarization shift, as discussed above in § 6.1.1. An example can be seen in Fig. 6.37 in § 6.5.1 below, which shows the increase in binding energy of the a_1 orbital (the 'lone-pair') of H_2O on adsorption on Ti_2O_3. It should be noted that the uppermost filled orbital of water, b_1, does not show a similar shift because it is oriented so that it cannot overlap strongly with orbitals in the substrate. On the other hand, the shifts discussed previously for benzene on ZnO (see Fig. 6.1) show that the upper three occupied orbitals of benzene can interact with empty orbitals in the substrate.

In the cases just discussed, the empty orbitals involved come from cations in the surface; they should really be represented in a more elaborate way in Fig. 6.2, as

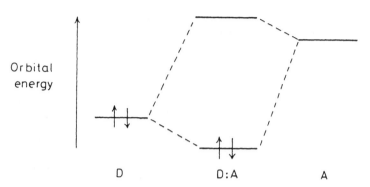

Fig. 6.2 Interaction between the filled orbital of a donor (D) and the empty orbital of an acceptor (A), giving the molecular orbitals of the combined system (D:A).

part of the empty density-of-states in the surface layer. A shift in energy of such empty levels should not appear in UPS, but it could be detectable either in IPS or in excitation experiments such as ELS; very little information is in fact available to date from those techniques. In some cases, however, changes in *occupied* levels have been seen in UPS from chemisorbed systems where a donor/acceptor interaction is expected. This might occur with a reduced surface, or a d^n oxide, where cation orbitals at the surface are not empty as in the picture just described. However, one does not always expect to see shifts in the d levels of the substrate, and Fig. 6.3 illustrates what might happen when a donor absorbs at the surface of a d^n oxide. The d orbitals of surface cations can be divided roughly into 't_{2g}-like' and 'e_g-like' ones (see § 5.1.1.3); one of each type is illustrated. Clearly the surface e_g orbital is oriented so as to achieve strong σ-type overlap with the occupied orbital of a donor adsorbate, whereas the t_{2g} orbitals will have weaker overlap. Orbital interaction will therefore raise the energy of the e_g orbital, but may hardly affect t_{2g}. With transition-metal cations having relatively few d electrons (including reduced d^0 oxides such as TiO_2), only the t_{2g} orbitals are occupied, and relatively little interaction should be seen between these and adsorbates. With transition metals further along in the series, e_g orbitals may also be occupied, and some shift in the energies of these may be seen in UPS when adsorption takes place.

The situation with semiconducting non-transition-metal oxides such as ZnO is not so clear. The empty cation orbitals are of s and p symmetry and should overlap strongly with adsorbate orbitals. Because of the larger conduction-band widths found in these oxides, however, a picture of localized electrons on particular cations is not so appropriate; rather we should think normally of free carriers. As mentioned previously, donor/acceptor interactions cannot directly alter the number of free carriers, and would not be expected to change the surface conductivity.

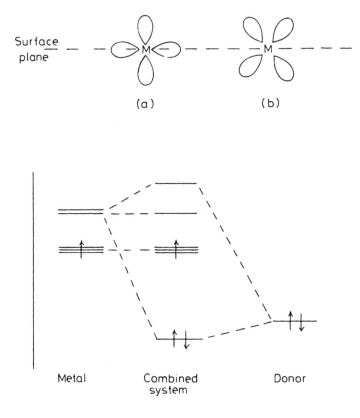

(c)

Fig. 6.3 The orientation of (a) e_g-type and (b) t_{2g}-type d orbitals of a transition-metal ion on the (100) surface of a perovskite oxide. (c) The resulting perturbation in orbital energies arising from the adsorption of a σ donor molecule onto a d^1 ion at the surface.

Since the adsorption of molecules such as H_2O does appear to change the surface conductivity of ZnO and SnO_2, their interaction with carriers must be indirect. One way in which this might happen is when electrons are bound in surface states below the Fermi level on the clean substrate. Interaction with the filled level of a donor might then raise the energy of these states above E_F, so that the electrons are released. An alternative possibility is that pre-adsorbed species such as O_2, which interact by a redox mechanism that *does* involve free carriers, may be displaced by the new adsorbate.

In contrast to the donor/acceptor type of interaction, the redox mechanism directly alters the electron configuration of the surface. In transition-metal oxides, reduction (or oxidation) of the surface by an adsorbate is expected to show up clearly as a change in population of the d levels. In ZnO and SnO_2, band-bending and changes in surface conductivity are expected. The transfer of charge to or from the surface also creates a substantial dipole, and large changes of work function can occur.

Band-bending effects at semiconductor surfaces were discussed in § 4.1.4 and illustrated in Fig. 4.7. The release of electrons by an adsorbate causes a downward band-bending (to higher binding energy), and the uptake of electrons a shift in the opposite direction, as shown in that figure. On the n-type material illustrated, extra electrons can be accommodated in the conduction band to give an **accumulation layer**. On the other hand, the electrons available for withdrawal by an adsorbate are those provided by the donor levels. A **depletion layer** is therefore formed, where band-bending is caused by the unbalanced charge of the ionized donors. In general, the density of conduction-band states is expected to be much higher than the concentration of donors, so that a given amount of charge can be concentrated much closer to the surface in an accumulation layer than is possible in a depletion layer. One consequence is that much more band-bending is predicted in the latter situation. On n-type oxides therefore the largest effects are expected with electron-withdrawing (oxidizing) adsorbates such as O_2 and Cl_2, which can form negative ions on the surface. With p-type oxides, larger effects would be expected with electron-donating (reducing) adsorbates such as H_2, although band-bending effects on these oxides have been less studied than on n-type ones. In fact, as mentioned above, the influence of adsorbates on band-bending and surface conductivity is quite complex: the problems of co-adsorption and reaction on the surface, and the details of surface defect chemistry which are often not well understood, are complicating factors that make a detailed understanding difficult to achieve.

6.2.3 Chemisorption behavior of specific molecules

Before discussing individual oxides in detail, it is useful to give some idea about the typical ways in which different molecules interact with surfaces. Tables 6.1 through 6.7 summarize the literature reports of chemisorption on oxide single-crystal surfaces. These studies use a wide variety of techniques, both experimental and theoretical. Sometimes specific identifications of the adsorbed species are claimed, but often these have only been inferred from the desorption products or some other measurement. In later sections a number of these results are discussed in detail.

6.2.3.1 H_2 and H

The H_2 molecule is non-polar, has a very low polarizability, and very weak donor or acceptor properties. Thus molecular adsorption is confined to very weak physisorption. Furthermore, the H–H bond is very strong, and dissociation is only possible when the resulting atoms can bond sufficiently strongly to the surface to

Table 6.1. H_2 and H chemisorption on metal oxides

Oxide	Face	Surface preparation	Type of adsorption		Experiment	Theory
MgO	100	P&A	H_2: none at RT	–	397	
		–	H_2 does not adsorb	–		255,262, 620–622
		–	H^+ bonds to O^{2-}, partially covalent	c		406
		Defect	H → OH^- at O_s^-, V_0 and V_S^- centers	c		255,262, 621,622
		Defect	H_2 → OH^- at O_s^-, V_0 and V_S^- centers	IIc		255,262, 620,621
		–	H_2 → Mg-H^+ and O–H^-, or H_2 → $2OH^-$	IIb IIc		623
ZnO	$10\bar{1}0$	P&A	H_2: none at RT, weak interaction at 200 K	–	624	
	0001 $000\bar{1}$ $10\bar{1}0$	P&A	H: strong interaction, H is donor, increases conductivity	c	13,625, 626	
	$000\bar{1}$	–	H is e^- donor, bonding to O (H + O^{2-} → OH^- + e^-)	c		433,434
	0001	–	H is e^- acceptor	c		433
	$10\bar{1}0$	–	H_2 → Strong O–H bonds, weaker Zn–H bonds	IIb		627
	$10\bar{1}0$	–	H is e^- donor	c		628
SnO_2	110	P&A	Surface conductivity increases, H_2 → $2H^+$ + e^-	IIc	441, 629–631	
TiO_2	110	UHV fractured	H_2 does not react at RT	–	125	
		P&A	H_2 is e^- donor	IIc	632	
		P&A	H_2 → OH^-, also Ti–H at defects	IIb or c	334,470, 626,633	
		–	Generalized chemisorption concepts	–		524,525, 634
	100	P&A stoichiometric	H_2 is e^- donor	IIc	23	
		P&A, slightly reduced	H_2 is e^- acceptor at Ti^{3+} sites	IIc	23	
	001	P&A	Ti–H bonds (ESD; not an adsorption experiment)	c ?	635,636	
$SrTiO_3$	100	P&A	Sr–H and Ti–H bonds (ESD; not an adsorption experiment)	c ?	635,636	
		–	H does not adsorb on perfect surface; need a hole to make H_2 → H^+	c		364
	111	Reduced	Some interaction, → OH^-	IIc	518–520	
VO_2	?	UHV scraped	H reduces surface → OH^-; impedes semiconductor/metal phase transition	IIc	539	
MoO_3	010	P&A	Inert to H_2 for T < 550 K; surface reduces to MoO_2 for T > 550 K	IId	195	
		–	H adsorbs at O^{2-} → OH_x, MoO_3 bronze	IIc		637

Table 6.1. (*cont.*)

Oxide	Face	Surface preparation	Type of adsorption		Experiment	Theory
WO$_3$	100	UHV scraped	No H$_2$ adsorption at RT; H$_x$WO$_3$ bronze	*IIc*	610	
α-Fe$_2$O$_3$	0001	P&A	H$_2 \rightarrow$ OH$^-$, no depopulation of 3d states	*IIc*	31	
NiO	100	P&A and UHV cleaved	H$_2$ and H reduce surface for T > 400 K	*IId*	581,638–641	
		–	H weakly bound to stoichiometric surface; adsorbs to surface O$^- \rightarrow$ OH$^-$; $\rightarrow$ OH at defects	*IIc*		642,643

overcome this. Although dissociative chemisorption takes place on many metals, this only seems possible on the most reactive oxide surfaces. A number of studies have overcome this problem by the direct interaction of H *atoms* with a surface. Both theoretical and experimental work indicates that the normal mode of bonding is with surface oxygen to make hydroxide

$$H + O^{2-} \rightarrow OH^- + e^-$$

that is, *type (c)* according to the scheme above. Theoretical work suggests that on some surfaces holes (O$^-$) or other defects may be required to promote this reaction.

There is also evidence that H may bond to metal atoms, either on some reduced surfaces (TiO$_2$) or on the (0001)-Zn face of ZnO. One can think of this as a hydride. For example

$$H + Zn^{2+} + e^- \rightarrow ZnH^+ . \qquad\qquad type\ (c)$$

There are therefore a number of possibilities for dissociative H$_2$ chemisorption. Homolytic dissociation could result either in two OH$^-$ or (less frequently) in two metal-hydrogen species

$$H_2 + 2\,O^{2-} \rightarrow 2\,OH^- + 2\,e^-$$

$$\qquad\qquad\qquad\qquad\qquad\qquad\qquad both\ type\ II\ (c)$$

$$H_2 + M^{n+} + 2\,e^- \rightarrow 2\,MH^{(n-1)+} .$$

Heterolytic dissociation might also occur

$$H_2 + M^{n+} + O^{2-} \rightarrow MH^{(n-1)+} + OH^- . \qquad type\ II\ (b)$$

Table 6.2. O_2 chemisorption on metal oxides

Oxide	Face	Surface preparation	Type of adsorption		Experiment	Theory
MgO	100	P&A	None at RT	–	397	
SnO_2	110	Reduced	e⁻ acceptor, probably dissociates	*IIc*	140,142,441, 629,631	
		Stoichiometric	None at RT	–	140,142,441, 629,631	
ZnO	0001 0001̄ 101̄0	UHV cleaved	Some interaction at RT (ELS)	*Ic*	644	
	101̄0	P&A, stoichiometric	$O_2 \rightarrow O_2^-$ at Zn ions, weak	*Ic*	258,624,628, 645–647	
	101̄0	P&A, O-vacancy	$O_2 \rightarrow O$ or O^-	*IIb* or *c*	645,646	
	0001	P&A	Adsorption at RT, but type unknown (ELS)	?	381	
	101̄0 0001 404̄1 505̄1	P&A	Weak, probably at O vacancies; more on stepped surface	*Ic* or *IIc*	236,667	
	0001	–	O does not adsorb	–		648
	0001̄	–	O adsorbs over surface O atoms	*c*		648
TiO_2	110	P&A, stoichiometric	Weak at RT, probably at defects	?	269,470,626, 633,649–651	524,525
		Reduced	1st phase: $O_2 \rightarrow O^{2-}$, depopulates Ti states; 2nd phase: probably molecular	*IIc; Ic*	269,470,626 633,649–651	524,525
	100	P&A, stoichiometric	Weak at RT, probably at defects	*IIc* ?	23	
		Reduced	1st phase: $O_2 \rightarrow O^{2-}$, depopulates Ti states; 2nd phase: probably molecular	*IIc; Ic*	23	
$SrTiO_3$	100	UHV fractured	$O_2 \rightarrow O^{2-}$; some molecular adsorption at high exposures (O_2^{2-} ?)	*IIIc; (Ic?)*	649	
		P&A, stoichiometric	Weak $O_2 \rightarrow O^{2-}$; no second phase	*IIc*	111,649,652	524,525
		Reduced	$O_2 \rightarrow O^{2-}$, depopulates Ti states; then some molecular adsorption	*IIc; Ic*	111,649,652	524,525
	111	P&A, stoichiometric	None at RT	–	519,520	
		Reduced	1st phase: $O_2 \rightarrow O^{2-}$, depopulates Ti states; 2nd phase: probably molecular	*IIc; Ic*	519,520,653	
MoO_3	010	Stoichiometric	Inert	–	194	
$Mo_{18}O_{52}$	100	?	Dissociative adsorption at shear planes, followed by bulk oxidation	?	197,654	
Ti_2O_3	101̄2	UHV cleaved	$O_2 \rightarrow O^{2-}$ (?), depopulates Ti 3d	*IIc* ?	26,28,655	
		Reduced	$O_2 \rightarrow O^{2-}$ (?), depopulates Ti 3d	*IIc* ?	26,28,655	
TiO_x	Poly-cryst.	UHV fractured	Sample oxidizes to TiO_2	*IIc*	505	

Table 6.2. (*cont.*)

Oxide	Face	Surface preparation	Type of adsorption		Experiment	Theory
V_2O_5	010	Reduced	Oxidizes surface	*IIc* ?	538	
V_2O_3	$10\bar{1}2$	UHV cleaved	$O_2 \rightarrow O^{2-}$, depopulates V 3d	*IIc*	29, 655	
Cr_2O_3	'0001'	Epitaxial on metal	Dissociative and molecular at RT	*IIc* and *Ic*	656	
MnO	100	UHV cleaved	Strong dissociative adsorption, depopulates Mn 3d	*IIc*	548	
		–	O adsorbs atop Mn; O_2 assumed parallel to surface, adsorption atop Mn	*c*		661
α-Fe_2O_3	0001	P&A, stoichiometric	No interaction at RT until 10^3 L, then depopulates Fe state	*IIc* ?	31,655	
		Reduced	$O_2 \rightarrow O^{2-}$, depopulates Fe states	*IIc*	31	
Fe_3O_4	110	UHV cleaved	Strong interaction with Fe^{2+} ions (depopulation); little interaction with Fe^{3+}	*IIc*	564	
FeO	100	–	O adsorbs atop Fe ions	*c*		648
CoO	100	UHV cleaved	None at RT	–	274	
		Reduced	$O_2 \rightarrow O^{2-}$; then some molecular	*IIc; Ic*	274	
NiO	100	UHV cleaved	None at RT	–	32,657	658,659
		Reduced	$O_2 \rightarrow O^{2-}$; then some molecular $(O_2^{2-}$?)	*IIc; Ic*	32,657	
		Reduced	O adsorbs at defects	*c*		658,659
		–	O_2 adsorbs molecularly to surface Ni	*Ic*		660
		–	O adsorbs atop Ni ions	*c*		648
Cu_2O	111	P&A, stoichiometric	Molecular (O_2^{2-} ?)	*Ic*	208	
		Reduced	Dissociates $\rightarrow O^{2-}_{latt.}$ and $O_{ads.}$	*IId*	208	
	100	P&A, probably defective	Dissociates $\rightarrow O^{2-}_{latt.}$ and $O_{ads.}$	*IId*	208	
YBa_2Cu_3 O_{7-x}	Poly-cryst.	UHV scraped	Very weak interaction	–	662,663	
$La_{2-x}Sr_x$ CuO_4	Poly cryst.	UHV scraped	Very weak interaction	–	662,664	
Bi_2Sr_2 $CaCu_2O_{8-x}$	001	Air cleaved	No interaction	–	665	
$Tl_2Ba_2Ca_2$ Cu_3O_{10}	Poly-cryst.	UHV scraped	No interaction	–	666	

The reactions forming hydroxide are undoubtedly more common, and are likely to be promoted by defects such as O⁻. At elevated temperatures water may desorb, leading to net oxygen loss, either just from the surface, or even within the bulk

$$H_2 + MO \rightarrow H_2O + M.$$ *type II (d)*

Since hydrogen is quite mobile, another bulk reaction is possible: the formation of reduced **hydrogen bronze** compounds, for example with MoO_3 and WO_3

$$\frac{x}{2} H_2 + MoO_3 \rightarrow H_x MoO_3.$$ *type II (c)*

6.2.3.2 *O₂ and O*

The interaction of O_2 with metal oxides is important in many catalytic applications, but it is difficult to study experimentally by means of surface-science techniques because of the extremely large background of lattice O present. Oxygen is a powerful electron acceptor and can be reduced in several steps:

$$O_2 + e^- \rightarrow O_2^-$$

$$O_2 + 2\,e^- \rightarrow O_2^{2-}$$ *both type I (c)*

$$O_2 + 2\,e^- \rightarrow 2\,O^-$$

$$O_2 + 4\,e^- \rightarrow 2\,O^{2-}.$$ *both type II (c)*

There is electron-spin resonance (ESR) evidence in many polycrystalline studies for the existence of the paramagnetic O_2^- (superoxide) and O⁻ species. On single crystals the adsorbed species may sometimes be inferred from UPS, but are frequently unknown. Which type of adsorption predominates probably depends very much on the type and state of the surface: molecular adsorption, *type I (c)*, is generally assumed on stoichiometric surfaces of n-type semiconductors (ZnO and SnO_2), dissociation being promoted by defects such as oxygen vacancies on etched or reduced surfaces.

6.2.3.3 *H₂O and OH*

Water has a large dipole moment and lone-pair electrons, and thus is a good donor. Molecular adsorption occurs by acid/base, *type I (b)*, interaction with surface metal ions, although this may be quite weak on defect-free low-index

Table 6.3. *H₂O (and OH) chemisorption on metal oxides*

Oxide	Face	Surface preparation	Type of adsorption		Experiment	Theory
MgO	100	P&A	$H_2O + O^{2-} \rightarrow 2OH^-$ at RT (used e^- flood gun)	IIb	397	
	'111'	P&A, faceted to (100)	Same as (100), but more adsorption [so probably adsorbs at defects on (100)]	IIb	397	
	100	–	H_2O dissoc.$\rightarrow$ OH$^-$ and H$_2$ at corner defect sites	IIb and IId		668
		–	Weak OH$^-$ adsorption at Mg^{2+} site	Ib		406
		P&A, stoichiometric	None at RT; $\rightarrow$ OH$^-$ at 200 K	IIb	669,670	
		Ion bombarded	RT dissociation; also at 200 K	IIb	669,670	
		Smoke cubes	Dissociates $\rightarrow$ OH$^-$ at step and kink sites	IIb	671	
BaO	Poly-cryst.	–	$H_2O + BaO \rightarrow Ba(OH)_2$	IIb	672,673	
Al₂O₃	0001	?	OH adsorption creates new (O 2p + H 1s) bond in bulk bandgap	b		674,675
ZnO	0001 0001̄ 101̄0	P&A	None for T > 130 K; molecular at lower T; strongest bonding to Zn sites	Ib	676, 677	
	0001 101̄0	–	OH is e^- acceptor; bonds to Zn as OH$^-$	b		678
	0001	–	H_2O adsorbs via O atom to Zn site; e^- donor	Ib		679
	101̄0	–	Assume $H_2O \rightarrow OH^- + H^+$ (or $O^{2-} \rightarrow OH^-$)	IIb		435
SnO₂	110	P&A	H_2O is e^- donor	Ib or IIb	146,441, 629,631	
TiO₂	110 100	Stoichiometric	See § 6.4.1 in text. Probably inert at RT		20,23,469, 471,487,489, 636,650,651, 680–687	
	110 100	Reduced	See § 6.4.1 in text, Probably dissociates $\rightarrow$ OH$^-$, little interaction with 3d electrons		20,23,469, 471,487,489, 636,650,651, 680–687	
	110	–	OH$^-$ adsorbs at 4-fold Ti	b		634,688
		–	Both H_2O and OH$^-$ bond to 5-fold Ti	b		689
SrTiO₃	100	P&A, stoichiometric and reduced	See § 6.4.2 in text		20,248, 339,341, 346,487, 690	
		UHV cleaved, stepped	$H_2O \rightarrow OH^-$ at step sites	IIb	513–515, 691,692	
		–	OH$^-$ is stable on 5-fold Ti	c		364
	111	Stoichiometric and reduced	See § 6.4.2 in text		518–520, 636	
Na₀.₇ WO₃	100	Stoichiometric	None at RT; molecular at low T	Ib	343,346, 693	

Table 6.3. (*cont.*)

Oxide	Face	Surface preparation	Type of adsorption		Experiment	Theory
Ti_2O_3	$10\bar{1}2$	UHV cleaved	Molecular at RT, < 1 ML, weak Ti 3d interaction	*Ib*	173,655	
		Reduced	$H_2O \rightarrow OH^-$ at low exposure; molecular for higher exposure	*IIb; Ib*	173,655	
V_2O_3	$10\bar{1}2$	UHV cleaved	Probably dissociates $\rightarrow OH^-$, perhaps at steps	*IIb*	29,655	
		Reduced	$H_2O \rightarrow OH^-$	*IIb*	29,655	
MnO	100	UHV cleaved	$H_2O \rightarrow OH^-$	*IIb*	548	
α-Fe_2O_3	0001	P&A, stoichiometric	$H_2O \rightarrow OH^-$; weak 3d interaction	*IIb*	31,655	
		P&A, reduced	$H_2O \rightarrow OH^-$; weak 3d interaction	*IIb*	31,562,655	
		P&A, stoichiometric	None at RT; molecular for T < 200 K	*Ib*	562	
FeO	100	–	$H_2O \rightarrow OH^-$	*IIb*		694
CoO	100	UHV cleaved	Inert at RT	–	274	
		Reduced	$H_2O \rightarrow OH^-$ (?)	*IIb* ?	274	
NiO	100	UHV cleaved	None at RT	–	32,657	
		Reduced	Dissociates *only* with pre-adsorbed $O \rightarrow OH^-$; no molecular adsorption	*IIb*	32,657	
		P&A	Inert at RT; some interaction at 500 K (OH$^-$?)	*IIb* ?	695	
Cu_2O	100	P&A	$H_2O \rightarrow OH^-$ at RT; some dissociation at 110 K	*IIb*	696	
YBa_2Cu_3 O_{7-x}	Poly-cryst.	–	Strong corrosion $\rightarrow Ba(OH)_2$, CuO, Y_2BaCuO_5	*IIb*	662,663, 697–704	
YBa_4Cu_2 O_{8-x}	Poly-cryst.	–	Strong corrosion	*IIb*	700	
$La_{2-x}Sr_x$ $Cu O_4$	Poly-cryst.	UHV scraped	Corrosion $\rightarrow$ La and Sr hydroxides	*IIb*	662,664	
$GdBa_2$ Cu_3O_{7-x}	Poly-cryst.	–	Corrosion	*IIb*	705	
$Bi_2Sr_{2-x}Ca_{1+x}$ Cu_2O_{8+y}	100	UHV cleaved	Only weak physisorption at 90 K	*Ia*	706	
	Poly-cryst.	–	Weak corrosion $\rightarrow$ CuO, $Sr(OH)_2$, $Ca(OH)_2$, $CuBi_2O_8$	*IIb*	663,707	
$Bi_{1.4}Pb_{0.6}Sr_2$ $Ca_2Cu_{3.6}O_x$	Poly-cryst.	–	Slow decomposition into CuO, $Sr(OH)_2$, $Ca(OH)_2$ and $CuBi_2O_8$	*IIb*	708	

Table 6.3. (*cont.*)

Oxide	Face	Surface preparation	Type of adsorption		Experiment	Theory
$Tl_2Ba_2Ca_2$ Cu_3O_{10}	Poly- cryst.	UHV scraped	→ hydroxide	*IIb*	666	

surfaces. Stronger adsorption occurs at steps and defects and is often dissociative

$$H_2O + O^{2-} \rightarrow 2\,OH^-. \qquad\qquad type\ II\ (b)$$

This reaction is common on polycrystalline surfaces so that they are frequently covered with hydroxyl groups. On prolonged exposure, especially with very basic oxides such as BaO, hydroxide formation may proceed into the bulk of the solid.

The conditions which may give rise to either dissociative or non-dissociative adsorption of water have been investigated in special detail on some transition-metal-oxide surfaces such as TiO_2 and $SrTiO_3$. Although some of the results, discussed in detail below, are conflicting, it seems on balance that dissociation is promoted by particular *structural* features, such as the presence of defects on the surface, and is not so sensitive to electronic aspects such as the presence or absence of d electrons.

6.2.3.4 CO

CO has only a small dipole moment, and in spite of its lone-pair electrons is a very weak donor. Its bonding to transition-metal surfaces is strongly enhanced by the 'back-donation' of d electrons into the antibonding π orbital; that is, it acts simultaneously as a donor and acceptor. Since the strength of bonding depends upon the availability of d electrons at the surface, adsorption will be favored by low oxidation states. Some weak molecular adsorption is reported on ZnO and some transition-metal oxides, but frequently CO seems to react with pre-adsorbed or lattice oxygen to give CO_2 (which may sometimes further react to form carbonate)

$$CO + O^{2-} \rightarrow CO_2 + 2\,e^-. \qquad\qquad type\ II\ (d)$$

In many studies the electrons liberated by this oxidation are not accounted for: in some cases the reaction may be with pre-adsorbed oxygen. As with H_2 and other reducing adsorbates, bulk reaction may occur at higher temperatures.

On more reactive substrates (as with early transition-metal surfaces) dissociation to C and O^{2-} is possible.

Table 6.4. *CO chemisorption on metal oxides*

Oxide	Face	Surface preparation	Type of adsorption		Experiment	Theory
MgO	100	P&A	None at RT	–	397	
		–	Molecular, weak electrostatic	*Ia*		709
		–	Molecular via quadruple moment, C to Mg^{2+}	*Ia*		710,711
		–	Mg–C bond, C–O axis normal to surface	*Ia*		255,262,621, 712,713
		Steps, edges, corners	Strongest at corners; edges similar to terraces	*Ia*		255
		–	Assume molecular, calculate potential surfaces	*Ia*		714
	110	–	Either Mg–C or Mg–O bonding possible	*Ia*		715
Al_2O_3	0001	–	Molecular, C–Al	*Ib*		716
ZnO	0001 $000\bar{1}$ $10\bar{1}0$	UHV cleaved	Very weak interaction	*Ia*	644	
	$000\bar{1}$	Sputtered & annealed growth face	$CO + O^{2-} \rightarrow CO_3^{2-}$ on stoichiometric; no adsorption at 120 K on defect surface	*Id*	717,718	
	$10\bar{1}0$	P&A	$CO + O^{2-} \rightarrow CO_3^{gas}$ on stoichiometric surface	*Id*	624,628 719	
	$10\bar{1}0$ $11\bar{2}0$	P&A	Molecular to Zn dangling bond	*Ib*	9,193,625, 720	
	0001	P&A	Molecular to surface Zn via C atom	*Ib*	9,193	
	$000\bar{1}$	P&A	Molecular at steps, to exposed Zn	*Ib*	9,193	
	$10\bar{1}0$	P&A, stoichiometric	$CO + O^{2-} \rightarrow CO_2^{gas}$	*Id*	667	
	$10\bar{1}0$	P&A, defect	None at RT	–	667	
	$40\bar{4}1$ $50\bar{5}1$	P&A, stepped	Mostly molecular; some $CO + O^{2-} \rightarrow CO_2$	*Ib* (some *Id*)	677	
	0001	P&A	Molecular, probably at steps	*Ib*	667	
	$10\bar{1}0$	P&A	Mixture of $CO + H_2$: no reaction at 120 K and low pressure	–	721	
	0001	–	Weak molecular, atop Zn via C; CO accepts e^-	*Ib*		722
	$000\bar{1}$	–	CO interacts with pre-adsorbed O to give CO_2; also abstracts lattice O to form CO_2	*Id*		648
	$10\bar{1}0$	–	Strong molecular to Zn; CO donates e^-	*Ib*		723
	0001	–	Strong molecular atop Zn; CO donates e^-	*Ib*		723
	$10\bar{1}0$	–	Molecular to Zn via C; CO is e^- donor	*Ib*		724
	$10\bar{1}0$ $000\bar{1}$	–	Pre-adsorbed O or $O^- \rightarrow CO_2$	*Id*		725
TiO_2	110	P&A	Only bonds at O vacancies, to adjacent $O \rightarrow CO_2$	*Id*	470,622, 623,626	
		–	Weak molecular at 5-fold Ti, or stronger bond to Ti adjacent to O vacancy	*Ib*		496

Table 6.4. (*cont.*)

Oxide	Face	Surface preparation	Type of adsorption		Experiment	Theory
Ti_2O_3	$10\bar{1}2$	P&A	Weak, molecular or dissociative?	?	27	
V_2O_5	010	P&A, stoichiometric	Reduces surface; $\rightarrow CO_2$ (?)	*Id* ?	538	
		Reduced	Inert at RT	–	538	
Fe_3O_4	110	P&A	CO dissociates at RT $\rightarrow$ C + O (?) (poorly characterized surface)	*IId* ?	726	
CoO	100	UHV cleaved	None at RT	–	274	
		Reduced	$CO + O_{latt.} \rightarrow CO_2$ (?) for small defect densities	*Id*	274	
NiO	100	UHV cleaved	None	–	640	
		UHV cleaved	None $< 10^8$ L; then $CO + O_{latt.} \rightarrow CO_2$ (?)	*Id*	657	
		Reduced	Molecular at RT	*Ib*	657	
		P&A, stoichiometric	None for T < 300 K; $CO + O_{latt.} \rightarrow CO_2^{gas}$ for T > 500 K	*Id*	695	
		Reduced	None for T < 300 K; $CO + O_{latt.} \rightarrow CO_2^{gas}$ for T > 500 K	*Id*	695	
		–	Molecular, weak, mostly electrostatic, some covalent	*Ia, b*		709
MnO	100	UHV cleaved	Weak molecular	*Id*	548	
		–	Molecular, atop Mn via C	*Ia*		661
Cu_2O	100	P&A	Molecular at 120 K; e^- transerred from CO to surface Cu^+ 4p orbitals	*Ib*	727	
$YBa_2Cu_3O_{7-x}$	Poly-cryst.	UHV scraped	Some carbonate formation; much less active than CO_2	*IIb*	662	
$La_{2-x}Sr_xCuO_4$	Poly-cryst.	UHV scraped	Some carbonate formation; much less active than CO_2	*IIb*	662,664	
$Bi_2Sr_2Ca\,Cu_2O_{8-x}$	001	UHV cleaved	Molecular at 26 K; desorbs by 300 K, no loss of surface O	*Ib*	591	
		Air cleaved	Some interaction, nature unknown	?	665	

6.2.3.5 CO_2

CO_2 can act as a weak donor or acceptor, calculations supporting chemical intuition in suggesting that acceptor character increases when the normally linear molecule becomes bent. In fact, the common reaction is the acid/base interaction to give surface carbonate

$$CO_2 + O^{2-} \rightarrow CO_3^{2-}. \qquad\qquad \textit{type I (b)}$$

Table 6.5. *CO$_2$ chemisorption on metal oxides*

Oxide	Face	Surface preparation	Type of adsorption		Experiment	Theory
MgO	100	P&A	$CO_2 + O^2 \rightarrow CO_3^{2-}$ (e⁻ flood gun used)	*Ib*	397	
	'111'	P&A, faceted to (100)	Same as above for (100), only stronger [so probably absorbs at defects on (100)]	*Ib*	397	
BaO	Poly-cryst.	–	$CO_2 + BaO \rightarrow BaCO_3$	*Ib*	672,673	
ZnO	10$\bar{1}$0	P&A stoichiometric	$CO_2 + O^{2-} \rightarrow CO_3^{2-}$; less than on sputtered surface	*Ib*	258,626, 628,647, 719	
	10$\bar{1}$0	Reduced	Molecular at O-vacancy sites → $(ZnCO_2)^-$, decrease in surface conductivity, more adsorption than on stoichiometric surface	*Ic*	258,626, 628,647, 719	
		P&A	$CO_2 + O^{2-} \rightarrow CO_3^{2-}$	*Ib*	728	
		–	$CO_2 + O^{2-} \rightarrow CO_3^{2-}$	*Ib*		628
		–	$\rightarrow CO_3^{2-}$	*Ib*		725
	000$\bar{1}$	P&A, stoichiometric	None for T > 100 K (did not go lower)	–	717	
	000$\bar{1}$	Reduced	$CO_2 + O^{2-} \rightarrow CO_3^{2-}$ at O-vacancy defect sites	*Ib*	717	
	000$\bar{1}$	–	$CO_2 + O^{2-} \rightarrow CO_3^{2-}$	*Ib*		725
	10$\bar{1}$0	P&A	Molecular	*Ib*	236,667	
	40$\bar{4}$1 50$\bar{5}$1	P&A	Molecular; similar to (10$\bar{1}$0)	*Ib*	236,667	
	10$\bar{1}$0	P&A	Mixture of CO_2 + H_2: no co-adsorption at low pressure; → formate at 1 atm.		728	
	0001	–	Molecular atop Zn; e⁻ acceptor when bent, donor when linear	*Ib*		678
	0001	–	$\rightarrow CO_2^-$	*Ic*		725
TiO$_2$	110	Stoichiometric	$CO_2 + O^{2-} \rightarrow CO_3^{2-}$	*Ib*	626,633	
MnO	100	–	Weak molecular, atop Mn, linear, vertical	*Ib*		661
NiO	100	UHV cleaved	Some molecular below 473 K	*Ib*	640	
YBa$_2$Cu$_3$O$_{7-x}$	Poly-cryst.	–	Strong corrosion → $BaCO_3$, CuO, Y_2BaCuO_5	*IIb*	662,663, 697,704, 705	
La$_{1-x}$Sr$_x$CuO$_4$	Poly-cryst.	UHV scraped	Corrosion → carbonate	*IIb*	662,664	
Bi$_2$Sr$_2$CaCu$_2$O$_{8-x}$	001	Air cleaved	No interaction	–	665	
	Poly-cryst.	–	→ carbonate in moist atmosphere	*IIb*	663	
Tl$_2$Ba$_2$Ca$_2$Cu$_3$O$_{10}$	Poly-cryst.	UHV scraped	→ carbonate	*IIb*	666	

On BaO this may proceed to yield bulk carbonate; indeed the reaction is so favorable that complex oxides containing barium (such as $YBa_2Cu_3O_{7-x}$ superconductors) may be covered with a layer of $BaCO_3$.

Occasionally it appears that CO_2 may accept a single electron, forming the bent CO_2^- species coordinated to a metal

$$CO_2 + e^- \rightarrow CO_2^-. \qquad\qquad \textit{type I (c)}$$

6.2.3.6 Miscellaneous inorganic molecules

N_2 and N_2O: Although N_2O might react with oxygen vacancies to give N_2, no experimental studies have reported adsorption of these rather unreactive molecules on single-crystal oxides.

NO: NO is more reactive than N_2 or N_2O and adsorbs on many metallic surfaces. The non-dissociative interaction may perhaps be represented as

$$NO \rightarrow NO^+ + e^-. \qquad\qquad \textit{type I (c)}$$

NO^+ is isoelectronic with CO. Only one report (on NiO) suggests molecular adsorption on an oxide single crystal. Dissociative adsorption is also possible, and a common reaction at low temperatures (found on ZnO) is

$$2NO \rightarrow N_2O + O. \qquad\qquad \textit{type II (d)}$$

NH_3: Ammonia is a strong donor, and molecular adsorption similar to that for water, *type I (b)*, can occur. Dissociation to adsorbed H^+ and NH_2^- species is also possible in principle, and has been reported on faceted TiO_2.

PF_3: Phosphines react with transition metals in a similar way to CO, with the donor properties of the lone pair combined with a π acceptor character: in the strongly electronegative PF_3, the acceptor properties are likely to dominate, as suggested by calculations. But there are no reports of experimental studies on oxides.

H_2S: Similar to water, H_2S may adsorb either as a molecule or by proton loss (OH^- formation), that is through acid/base type interaction. Higher temperatures often lead to sulfur deposition, either through direct decomposition of H_2S, or more likely by replacement of oxygen

$$H_2S + O^{2-} \rightarrow H_2O + S^{2-}. \qquad\qquad \textit{type II (b)}$$

Table 6.6. *Miscellaneous inorganic molecule chemisorption on metal oxides*

Oxide	Face	Surface preparation	Molecule	Type of adsorption		Experiment	Theory
MgO	100	P&A	N_2O	None at RT	–	397	
		–	N_2^-	Assume molecular, calculate potential surface	*Ia*		714
		–	NH_3	Assume molecular, calculate potential surface	*Ia*		714
		–	Xe	Calculate potential surface	*Ia*		714
CaO	100	Sputtered & annealed, no LEED	SO_2	$SO_2 + 2O^{2-} \rightarrow SO_4^{2-} + 2e^-$ (?)	*Id*	729,730	
ZnO	0001 000$\bar{1}$ 10$\bar{1}$0 11$\bar{2}$0	P&A	NH_3	Molecular, strong bond	*Ib*	193	
	0001	P&A	Cl_2	$Cl_2 \rightarrow 2Cl^-$, strong	*IIc*	731	
	000$\bar{1}$ 10$\bar{1}$0	P&A	Cl_2	$Cl_2 \rightarrow 2Cl^-$, but less reaction than for (0001)	*IIc*	732	
	0001 10$\bar{1}$0	–	Cl	Strong Cl–Zn bond; Cl $\rightarrow$ Cl$^-$	*c*		733
	0001	–	NH_3	Molecular, N–Zn bond, e$^-$ donor	*Ib*		734
	10$\bar{1}$0	P&A	NO	Molecular at 90 K; some N_2O formed also	?	735	
	0001	–	PF_3	Molecular atop Zn via P, e$^-$ acceptor	*Ib*	733	
TiO_2	110	P&A, stoichiometric	SO_2	Inert at RT	–	18,484, 485,499	
		Reduced	SO_2	$\rightarrow TiO_2 + TiS_2$ at defects only	*IId*	18,484, 485,499	
		Stoichiometric	SO_2	Molecular at 100 K; heat $\rightarrow$ SO_4^{2-} at RT	*Id*	692,736,737	
		P&A	SO_2	Adsorbs as SO_3^{2-} at RT	*Ib*	18	
	441	Stepped	SO_2	$\rightarrow SO_3^{2-}$ and S^{2-} at RT	*Ib* and *IId*	18	
	110	P&A	H_2S	Initial dissociation $\rightarrow$ $H^+ + S^{2-}$, weak d-electron interaction; then molecular at higher exposure	*IIb*; *Ib*	486	
	001	P&A, faceted	NH_3	Molecular at 300 K; some dissociation to NH_2 and OH$^-$ at 340 K	*Ib*; *IIb*	738	
MoO_3	010	P&A	H_2S	Dissociation $\rightarrow$ S adlayer at 570 K; reduces surface for longer exposure	*IId*	739, 740	
V_2O_5	010	Reduced	SO_2	Oxidizes surface; both S–O and S–V species formed	*IId*; *I* ?	538	
		–	NH_3	Molecular, N–V bond	*Ib*		741
Ti_2O_3	10$\bar{1}$2	UHV cleaved	SO_2	Violent dissociative adsorption at RT $\rightarrow TiO_2 + TiS_2$	*IId*	27, 173, 485	

Table 6.6. (*cont.*)

Oxide	Face	Surface preparation	Molecule	Type of adsorption		Experiment	Theory
V_2O_3	$10\bar{1}2$	UHV cleaved	SO_2	Weak interaction, some V 3d depopulation, S–O and S–V bonds	*Ib* and *IId* ?	30, 485	
α-Fe_2O_3	0001	P&A	SO_2	Weak at RT, SO_4^{2-} (?)	*Id* ?	31	
Cr_2O_3	0001	Epitaxial on metal	Cl	Adsorbs at RT, state unknown	*c* ?	656	
NiO	100	UHV cleaved	NO	Weak molecular for T < 250 K	*Ic* ?	742	
		UHV cleaved	H_2S	Reduces NiO → Ni + adsorbed S at 570 K; weak interaction at RT	*IId*	692, 736, 743, 744	
		–	SO_2	Molecular, S–Ni bond	*Ib*		660
		–	SO_3^-	Molecular, S–Ni bond slightly stronger than S–$O_{surf.}$ bond	*Ib*		660
CoO	100	UHV cleaved	H_2S	Reduces surface at 370 K, Co + $S_{ads.}$	*IId*	745	

Removal of oxygen might be thought of as surface reduction, but in fact the oxidation states are not changing, and as written this is an acid/base, not a redox, reaction.

H_2S does nasty things to the inside of UHV systems, which is why its adsorption is rarely studied by surface-science techniques.

SO_2: This molecule is capable of a variety of interactions: like CO_2 (except that it is bent) it can act as a donor or acceptor, but it can also be oxidized or reduced. Molecular adsorption is possible with either S or O atoms bonding to a metal; interaction with oxide leads to sulfite

$$SO_2 + O^{2-} \rightarrow SO_3^{2-} . \qquad\qquad type\ I\ (b)$$

In some cases the formation of sulfate SO_4^{2-} has been reported

$$SO_2 + 2\,O^{2-} \rightarrow SO_4^{2-} + 2\,e^- . \qquad\qquad type\ I\ (d)$$

As with the formation of CO_2 or carbonate from CO, some corresponding reduction of other species should occur, a problem not discussed in the studies reported.

On some surfaces sulfide is produced; with a reduced substrate such as Ti_2O_3 the reaction might be

$$SO_2 + 2\,e^- \rightarrow S^{2-} + O_2 , \qquad\qquad type\ II\ (c)$$

but other possibilities, involving oxide ions, could also be considered.

Cl_2:   Along with the other halogens (which have not been studied) this is a strong electron acceptor, and the expected reaction is

$$Cl_2 + 2\,e^- \rightarrow 2\,Cl^-.$$ *type II (c)*

Xe: As one of the noble gases, xenon is chemically inert in most situations and only undergoes physisorption. The adsorption of Xe at low temperatures has been used to study steps, defects, evaporated metal islands, etc., on single-crystal metal surfaces; the technique has been given the name **PAX (photoemission of adsorbed xenon)**.[778] The binding energy of the shallow $5p_{1/2}$ core level, which is measured by photoemission, is found to be referenced to the vacuum level of the surface rather than E_F since the diameter of the Xe atom is so large that its center is located outside of the strongly varying electrostatic surface potential. Thus changes in the measured binding energy, referenced to E_F, measure the local work function of the surface. The Xe $5p_{1/2}$ binding energy is found to decrease as the number of metal ligands increases for different adsorption sites because of increased screening of the core hole. Even for defect-free metal surfaces, the $5p_{1/2}$ binding energy is larger for the first adsorbed Xe layer than for succeeding layers for the same reason. ZnO is the only oxide on which Xe adsorption has been studied experimentally, and core-level shifts are apparently *not* seen.

6.2.3.7 Organic molecules

Organic molecules are, for the most part, far more complex than the inorganic molecules considered above. Their complexity makes it more difficult to obtain information from experimental spectroscopies that is as detailed as that for molecules such as H_2O or CO_2. The techniques used to study the interaction of organic molecules with metal-oxide surfaces are thus somewhat different. Thermal desorption spectroscopy (TDS) has been widely used to measure the species that desorb after a particular molecule has been adsorbed onto a surface at a lower temperature. This technique does not provide any direct measure of the species actually adsorbed on the surface, but it is often possible to infer those species from the distribution of desorption products. In some cases XPS and UPS have been used to monitor the adsorbed species, but the determination is often not unique due to the complexity of the species. We will therefore have much less to say in detail about organic molecule adsorption, although general trends are usually straightforward to determine.

A general feature of oxide surfaces (which is important in selective oxidation catalysis) is the rather facile breaking of C–H, but *not* of C–C, bonds.

Table 6.7. *Organic molecule chemisorption on metal oxides*

Oxide	Face	Surface preparation	Molecule	Type of adsorption		Experiment	Theory
MgO	100	P&A	CH_4	None at RT	–	397	
		–	CH_4	Assume molecular, calculate potential surface	Ia		714
		Li doped	CH_4	Weak bond to O^-, $\rightarrow CH_3$–Mg^+ + H–O^-	IIb ?		746–748
		–	CH_3F	Assume molecular, calculate potential surface	Ia		714
		P&A	C_2H_4	None at RT	–	397	
		P&A	$1,3$-C_4H_6	None at RT	–	397	
		P&A, stoichiometric	HCCH	None at 170 K	–	670	
		Ion bombarded	HCCH	None at RT; dissociates $\rightarrow$ [CCH]$^-$ at 170 K	IIb	670	
		P&A	CH_3OH	$\rightarrow CH_3O^- + OH^-$ [adsorbs at defects on (100), more on faceted (111)]	IIb	397	
		P&A, stoichiometric	CH_3OH	None at RT; dissociates $\rightarrow$ CH_3O^- at 180 K	IIb	670	
		Ion bombarded	CH_3OH	Dissociates $\rightarrow CH_3O^-$ at RT and 180 K	IIb	670	
		P&A	HCHO	Dissociates $\rightarrow CH_3O^-$ and $HCOO^-$ at 170 K	IId	749	
		P&A	HCOOH	$\rightarrow HCOO^-$ [same amount on (100) and faceted (111)]	IIb	397	
		P&A, stoichiometric; and ion bombarded	HCOOH	Dissociates $\rightarrow HCOO^-$ (on Mg^{2+}) + H^+ (on O^{2-}) at RT and 180 K	IIb	670,750 751	
		P&A, stoichiometric; and ion bombarded	CH_3COOH	Dissociates $CH_3COO^- + H^+$ at RT and 180 K	IIb	670,750, 751	
		P&A	$HCOOCH_3$	$\rightarrow HCOO^-$ and CH_3O^- [adsorbs at defects on (100), more on faceted (111)]	IIb	397	
		–	HCO	O–Mg bond preferred	b		713
		–	HOC (not yet observed)	O–Mg bond	b		713
		–	CH_3O	Mg–O bond	b		713
		–	HCHO	Mg–O bond	b		713
		–	HCOH	Mg–C bond	b		713
		–	CH_2OH	Mg–C bond	b		713
ZnO	0001	P&A	C_2H_4	Molecular	Ib	718	
	$10\bar{1}0$	P&A	C_2H_4	None	–	752	
	0001	P&A	HCCH	RT dissociation $\rightarrow$ [HCC]$^-$	IIb	753,754	
	$000\bar{1}$	P&A	HCCH	Molecular	Ib	754	
	$10\bar{1}0$	P&A	HCCH	Molecular at 120 K	Ib	752	
	0001	P&A	CH_3CCH	RT dissociation $\rightarrow$ [CH_3CC]$^-$ and [CH_2CCH]$^-$	IIb	753,754	
	$000\bar{1}$	P&A	CH_3CCH	Molecular	Ib	754	
	0001	P&A	C_6H_5CCH	RT dissociation $\rightarrow C_6H_5CC^-$	IIb	754	
	$000\bar{1}$	P&A	C_6H_5CCH	Molecular	Ib	754	

Table 6.7. (*cont.*)

Oxide	Face	Surface preparation	Molecule	Type of adsorption		Experiment	Theory
ZnO	0001	P&A	CH$_3$OH	Dissociates at RT and 105 K → CH$_3$O$^-$ and HCOO$^-$	IIb & IId	677,718, 755–760	
	000$\bar{1}$	P&A	CH$_3$OH	Molecular	Ib	677,758	
	000$\bar{1}$	Annealing of growth face	CH$_3$OH	Dissociates → CH$_3$O$^-$ and some CH$_2$O at 100 K	IIb & IIc	761	
	000$\bar{1}$	P&A	CH$_3$OH	Oxidation → HCOO$^-$	IId	762	
	10$\bar{1}$0	Annealing of growth face	CH$_3$OH	Both molecular and dissociates → CH$_3$O$^-$ at 100 K	Ib & IIb	761	
	10$\bar{1}$0	P&A	CH$_3$OH	Molecular at 120 K	Ib	752,763	
	10$\bar{1}$0 5051 4041	P&A	CH$_3$OH	Needs defects for reaction: dissociates → CH$_3$O$^-$ and HCOO$^-$	IIb & IId	236,755 756	
	0001	P&A	C$_2$H$_5$OH	RT dehydrogenation → CH$_3$CHO	IIc	764	
	000$\bar{1}$	P&A	C$_2$H$_5$OH	Molecular for T < 300 K	Ib	764	
	000$\bar{1}$ 10$\bar{1}$0	Air cleaved and annealed	C$_2$H$_5$OH	RT adsorption and TDS → CH$_3$CHO	IIc	765	
	0001	P&A	C$_3$H$_7$OH (1-propanol)	RT dehydrogenation → C$_2$H$_5$CHO	IIc	764	
	000$\bar{1}$	P&A	C$_3$H$_7$OH (1-propanol)	Molecular for 165–300 K	Ib	764	
	0001	P&A	i-C$_3$H$_7$OH (2-propanol)	RT dissociation (CH$_3$)$_2$HCO$^-$	IIb	677,766	
	000$\bar{1}$	P&A	i-C$_3$H$_7$OH (2-propanol)	Molecular 150–300 K	Ib	677,766	
	0001 5051 000$\bar{1}$	P&A	i-C$_3$H$_7$OH	Decomposes → CH$_3$COCH$_3$ and CH$_3$CH=CH$_2$	IId	767	
	0001	P&A	C$_6$H$_5$OH	RT dissociation → C$_6$H$_5$O$^-$	IIb	768	
	000$\bar{1}$	P&A	C$_6$H$_5$CH$_2$OH	RT oxidation → C$_6$H$_5$COO$^-$	IId	768	
	0001	P&A	HCHO	RT dehydrogenation → HCOO$^-$	IIb	756,758	
	0001	–	HCHO	Molecular via O–Zn, e$^-$ donor	Ib		734
	000$\bar{1}$	P&A	HCHO	Molecular	Ib	758	
	000$\bar{1}$	P&A	HCHO	Dehydrogenation → HCOO$^-$	IIc	762	
	10$\bar{1}$0	P&A	HCHO	Molecular at 120 K	Ib	769	
	10$\bar{1}$0 5051	P&A	HCHO	Dehydrogenation → HCOO$^-$	IIc	756	
	0001	P&A	CH$_3$CHO	RT ocidation → miscellaneous oxidation products	IId	770,771	
	000$\bar{1}$	Air cleaved and annealed	CH$_3$CHO	Molecular	Ib	756	
	10$\bar{1}$0	Air cleaved and annealed	CH$_3$CHO	Oxidation → CH$_3$COOH	IId	756	
	10$\bar{1}$0	P&A	CH$_3$CHO	Molecular at 120 K	Ib	769	
	0001	P&A	C$_2$H$_5$CHO	RT oxidation → miscellaneous oxidation products	IId	770,771	

Table 6.7. (*cont.*)

Oxide	Face	Surface preparation	Molecule	Type of adsorption		Experiment	Theory
ZnO	0001 0001̄	P&A	C_6H_5CHO	RT oxidation → $C_6H_5COO^-$	*IId*	768	
	0001	P&A	$CH_2=CHCHO$	RT oxidation $CH_2=CHCOO^-$ and other carboxylates	*IId*	770	
	0001	P&A	$(CH_3)_2CO$	RT dissociation → $[CH_3COCH_2]^-$ intermediates	*IIb*	766	
	0001̄	P&A	$(CH_3)_2CO$	Molecular 150–300 K	*Ib*	766	
	101̄0	P&A	$(CH_3)_2CO$	Molecular at 120 K	*Ib*	752,769, 772	
	0001	P&A	HCOOH	RT dissociation → $HCOO^-$ and OH^-	*IIb*	360,758, 756,773 728	
	0001̄	P&A	HCOOH	Molecular	*Ib*	758	
	0001̄	P&A	HCOOH	Dissociation → $HCOO^-$	*IIb*	762	
	101̄0	P&A	HCOOH	Oxidation at 120 K and RT → CO + O	*IId*	721	
	101̄0 505̄1	P&A	HCOOH	Dissociates → $HCOO^-$	*IIb*	756	
	0001	P&A	CH_3COOH	RT dissociation → CH_3COO^-	*IIb*	774	
	0001̄	P&A	CH_3COOH	Molecular	*Ib*	774	
	0001	P&A	C_2H_5COOH	RT dissociation $C_2H_5COO^-$	*IIb*	770,774	
	0001̄	P&A	C_2H_5COOH	Molecular	*Ib*	774	
	0001 0001̄	P&A	C_6H_5COOH	RT dissociation $C_6H_5COO^-$	*IIb*	768	
	0001	P&A	$CH_2=C=CH_2$	RT dissociation → $[CH_2CCH]^-$	*IIb*	753	
	0001 0001̄	P&A	$(C_2H_5)_2Zn$	Molecular for T < 450 K	*Ib*	775	
	0001	P&A	$HCOOCH_3$	RT dissociation → $HCOO^-$ and CH_3O^-	*IIb*	773	
	101̄0	P&A	$(CH_3)_2SO$	Molecular at 120 K	*Ib*	725,769 772	
	101̄0	P&A	C_2H_4O	Molecular at 120 K	*Ib*	752,763	
	101̄0	P&A	C_6H_6	Molecular	*Ib*	752,763, 776	
	101̄0	P&A	C_5H_5N	Molecular at 300 K	*Ib*	772	
	0001	–	C_5H_5N	Molecular via N–Zn, e^- donor	*Ib*		734
	0001	–	CH_3	Adsorbs at Zn sites, strong e^- acceptor	*c*		733
	0001	–	CCH	Adsorbs at Zn sites, strong e^- acceptor	*c*		733
	0001	–	HCOO	Molecular via O_2–Zn, e^- acceptor	*c*		734
	0001̄	–	CH_3O	Molecular via O–Zn, e^- acceptor	*c*		734

Table 6.7. (*cont.*)

Oxide	Face	Surface preparation	Molecule	Type of adsorption		Experiment	Theory
SnO$_2$	110	P&A	CH$_4$	Weak interaction at 673 K, increase in surface conductivity ?	?	630	
	101	Annealing of growth face	CH$_4$	Reacts at RT, probably to CH$_3$ + H with involvement of lattice O	IIb or IId	631,777	
	110	Annealing of growth face	C$_2$H$_5$OH	Some RT dissociation → C$_2$H$_5$O$^-$; some increase in surface conductivity, which is attributed to creation of O vacancies ?	IIb or IIc	779	
		Annealing of growth face	CH$_3$CHO	Some RT dissociation → C$_2$H$_5$O$^-$; some increase in surface conductivity, which is attributed to creation of O vacancies ?	IIc	779	
		Annealing of growth face	CH$_3$CHO OH	Molecular or → CH$_3$COO$^-$	Ib or IIb	631	
	101	Annealing of growth face	CH$_3$CHO OH	RT dissociation → HCOO$^-$ (?)	IId	777	
TiO$_2$	110 441	P&A	CH$_3$OH	Molecular at RT	Ib	18	
	001	Faceted	CH$_3$OH	Some dissociation → CH$_3$O$^-$ at 200 K, more at 300 K	IIb	780,781	
		(011)-faceted	C$_2$H$_5$OH	RT dissociation → C$_2$H$_5$O$^-$; both molecular and dissociative at 200 K	Ib, IIb	782	
		(011)-faceted	C$_3$H$_7$OH	RT dissociation → C$_3$H$_7$OH$^-$; both molecular and dissociative at 200 K	Ib, IIb	782	
		(011)-faceted	i-C$_3$H$_7$OH	RT dissociation → i-C$_3$H$_7$OH$^-$; both molecular and dissociative at 200 K	Ib, IIc	782	
		Reduced	CH$_3$CHO	Coupling → CH$_3$CH=CHCH$_3$	IId	783	
		Reduced	CH$_2$=CHC HO	Coupling → CH$_2$=CHCH =CHCH=CH$_2$	IId	783	
		Reduced	C$_6$H$_5$CHO	Coupling → C$_6$H$_5$CH=CHC$_6$H$_5$	IId	783	
	110 441	P&A	HCOOH	Molecular at RT	Ib	18	
	001	Faceted to (011) and (114)	HCOOH	RT dissociation → HCOO$^-$	IIb	784	
			CH$_3$COOH	RT dissociation → CH$_3$COO$^-$	IIb	785	
		Faceted to (144)	CH$_3$COOH	Dissociates → CH$_3$COO$^-$ + OH$^-$ at RT; → CH$_3$COCH$_3$ at higher T	IIb, IId	786	
		Faceted	C$_2$H$_5$COOH	Dissociates → C$_2$H$_5$COO$^-$; other products upon heating	IIb; IId	786	
		Faceted	CH$_2$=CHC OOH	RT dissociation → CH$_2$=CHCOO$^-$	IIb	785	
	110 001	Polished, not annealed	C$_5$H$_5$N	Molecular, via N atom to Ti	Ib	787	
	001	Faceted	Rh(allyl)$_3$	→ Rh(allyl)$_2$ bound to O ions	IIb	685,778–790	209

Table 6.7. (*cont.*)

Oxide	Face	Surface preparation	Molecule	Type of adsorption	Experiment	Theory
V_2O_5	001	UHV cleaved	$CH_3CH=CH_2$	None at RT	–	219
		Reduced	$CH_3CH=CH_2$	Oxidation → ?	*IId?*	219
NiO	100	P&A, stoichiometric	C_2H_4	Only weak molecular at low T	*Ib*	695,791
		Reduced	C_2H_4	Molecular < 300 K; reduces NiO surface > 500 K	*Ib; IId*	695,791
		Air cleaved	C_5H_5N	Molecular, via N atom to Ni	*Ib*	787
MoO_3	010	Stoichiometric	CH_3OH	None at RT	–	792,793
		Reduced	CH_3OH	RT dissociation → CH_3O^-	*IIb*	792,793
	?	–	CH_3O	Bonds to unsaturated Mo	*b*	794
	100	–	CH_4	Reacts with O^- hole centers → $MoCH_3 + OH^-$	*IIb* or *IIc*	795
Cu_2O	111	P&A	$CH_2=CHCH_3$	Primarily molecular at 100 K; some dissociation at O-vacancy defects → C_3H_5 (?)	*Ib, IIb*	796
	100	P&A	$CH_2=CHCH_3$	Primarily molecular at 100 K	*Ib*	796
	111	P&A	CH_3OH	Dissociates at 90 K → CH_3O^-	*IIb*	797
	100	P&A	CH_3OH	Dissociates at 90 K → CH_3O^-	*IIb*	797
α-Bi_2O_3	–	–	CH_4	Dissociates	*IIb* and *IId*	798
	–	–	$H_2C=CHCH_3$	Dissociation and oxidation	*IIb* and *IId*	798
$Bi_2Mo_3O_{12}$	–	–	$H_2C=CHCH_3$	Dissociation and oxidation	*IIb* and *IId*	799
$Bi_2Sr_2CaCu_2O_{8-x}$	001	UHV cleaved	CH_4	No interaction	–	665

Nevertheless, in many of the studies listed, a proportion of 'deep oxidation' products (CO and CO_2) and other species formed by breaking the carbon framework have been reported.

Saturated hydrocarbons (alkanes): Molecular adsorption probably involves only weak dispersion forces. The possibility of C–H bond breaking is interesting, as it is the first step in the potentially important methane coupling reaction, leading to C_2 and higher products *via* methyl radical intermediates. One or two experimental and theoretical studies suggest that, as expected, this is promoted by surface O^-

$$CH_4 + O^- \rightarrow CH_3{}^{\cdot} + OH^-.$$ *type II (c)*

The methyl species may bond to surface metal ions, where they are probably best regarded as $CH_3{}^-$.

Unsaturated hydrocarbons (alkenes and aromatics): The presence of π-bonding systems now leads to the possibility of stronger molecular adsorption, through their interaction with metal ions. The C–H bond in an alkene is slightly more acidic than in a saturated molecule, and heterolytic dissociation also becomes more likely, especially when the resulting anion can be stabilized by resonance or by π interaction with the surface

$$RCH{=}CH_2 + O^{2-} \rightarrow RCH{=}CH^- + OH^-.$$ *type II (b)*

This reaction (or a similar homolytic process) is probably the first step in the selective oxidation of alkenes with catalysts such as bismuth molybdate; no experimental single-crystal studies exist, although there has been some theoretical modelling.

Alkynes (acetylenes): As with alkenes, molecular adsorption of alkynes, $RC{\equiv}CH$, is possible by π interaction. However the acetylenic C–H is quite appreciably acidic, and heterolytic dissociation, *type II (b)* as with alkenes above, is common.

Alcohols: In a very similar way to water, ROH may undergo molecular adsorption through the oxygen lone pair, or dissociate by deprotonation

$$ROH + O^{2-} \rightarrow RO^- + OH^-.$$ *type II (b)*

On some surfaces dehydrogenation occurs, primary alcohols giving aldehydes, and secondary alcohols giving ketones

$$RCH_2OH + 2\,O^{2-} \rightarrow RCHO + 2\,OH^- + 2\,e^-$$

                                                              both type II (c)
$$R_1R_2CHOH + 2\,O^{2-} \rightarrow R_1R_2CO + 2\,OH^- + 2\,e^-.$$

As mentioned previously, loss of water may occur, making this reaction effectively *type II (d)*. Under conditions where oxidation of an alcohol occurs, further oxidation of the resulting aldehyde (but not of a ketone) is likely.

Another possible reaction is dehydration to an alkene, e.g., with 2-propanol (i-propyl alcohol)

$$(CH_3)_2CHOH \rightarrow CH_3CH=CH_2 + H_2O .$$

Aldehydes: Molecular chemisorption is possible, presumably with the oxygen acting as a donor to a surface acidic (cation) site. Aldehydes also undergo facile oxidation to carboxylates

$$RCHO + 2\,O^{2-} \rightarrow RCOO^- + OH^- + 2\,e^- . \qquad \textit{type II (d)}$$

On highly reduced surfaces, the reductive coupling to form alkenes also seems possible

$$2\,RCHO + 4\,e^- \rightarrow RCH=CHR + 2\,O^{2-} . \qquad \textit{type II (d)}$$

Ketones: The oxygen may act as a donor to surface acid sites in a similar way to aldehydes, but ketones are much more resistant to oxidation. The C–H adjacent to the carbonyl group can, however, dissociate heterolytically to give an enolate; e.g., with acetone

$$CH_3COCH_3 + O^{2-} \rightarrow [CH_2=COCH_3]^- + OH^- . \qquad \textit{type II (b)}$$

Carboxylic acids: As with aldehydes and ketones, molecular chemisorption is possible, although here there may also be the possibility of hydrogen bonding of the positively charged carboxyl H to surface oxygens. Because of the acidity of this hydrogen, dissociation to carboxylate is very likely

$$RCOOH + O^{2-} \rightarrow RCOO^- + OH^- . \qquad \textit{type II (b)}$$

Surface carboxylate is a common product of adsorption of primary alcohols and aldehydes; as already mentioned, further products (including CO and CO_2) may sometimes be generated by C–C bond breaking reactions.

Other compounds: Only three other molecules are listed in Table 6.7, and may be discussed very briefly. **Pyridine**, C_5H_5N, and **dimethyl sulfoxide**, $(CH_3)_2SO$, can undergo molecular chemisorption through lone pairs on N and O respectively; the former compound is an especially strong base. **Methyl fluoride**, CH_3F, is quite unreactive and only physisorbs.

6.2.4 Behavior of different oxide surfaces

This section will consider briefly the chemisorption behavior expected for different surfaces, based upon the above chemical ideas.

6.2.4.1 Acid/base properties

The extensive literature on the acid/base properties of polycrystalline oxide surfaces refers almost exclusively to **Brønsted acidity**; that is, to the availability of H$^+$ ions from water and OH groups already adsorbed on the surface. The comparative behavior of different oxides can be rationalized to some extent in terms of electrostatic and chemical bonding models.[358] We cannot expect these correlations to be *directly* applicable to the Lewis acid/base properties of clean surfaces, but it is nevertheless possible to apply the same ideas and to predict qualitatively analogous results.

Lewis acidity should depend on the existence of exposed metal cations at the surface, having empty orbitals and positive charges, that can interact with the filled orbitals and/or negative charges or dipoles of donor molecules. Surfaces such as the fully oxidized TiO$_2$ (110), the polar ZnO ($000\bar{1}$)-O face, or the MoO$_3$ (010) basal plane should not show this; in cases where such surfaces are reported to chemisorb basic molecules (e.g., H$_2$O and NH$_3$), one must suspect some disturbance or removal of surface oxygens. On surfaces where metal ions are exposed, Lewis acid strength is expected to depend on factors such as the ion charge, the degree of coordinative unsaturation, and the availability of empty orbitals (i.e., the bandgap). The stable (100) surfaces of alkaline-earth oxides should show very weak acidity because of the relatively low cation charge and the large bandgap; the latter means that empty orbitals at the surface are too high in energy to interact effectively with donors. Stronger acidity is expected with more highly charged ions such as Ti^{4+}, and also with post-transition-metal ions as in ZnO, where the bandgap is smaller and bonding with empty Zn 4s and 4p orbitals is possible. SnO$_2$ is likely to be rather peculiar in this respect. The fully oxidized (110) surface has a complete complement of bridging oxygens and should not be acidic. On the other hand, removal of these oxygens leads to the formation of surface Sn^{2+} ions, with a pair of electrons occupying the Sn 5s orbitals (or a hybrid derived from these). This lone pair might also be expected to inhibit the acid properties. In fact, H$_2$O does interact with SnO$_2$ (110) and is reported to increase the surface conductivity, a fact not expected from the nature of an acid/base interaction. It might be that water displaces a species such as O$_2^-$, or alternatively that its interaction with the lone-pair electrons leads to the promotion of some of these into the conduction band.

Lewis basicity on oxide surfaces is related to the availability of a pair of 2p electrons associated with oxygen ions. Once again, coordination is expected to be important; however, the charge of the metal cation should now play a role inverse to that expected in acid behavior. Low cation charge and large cation radius lead to weaker bonding and hence more basic O^{2-} ions. Thus the most basic of the oxides considered in this book is BaO, which is easily corroded by acidic adsorbates such as CO$_2$.

These arguments show that not only the compound, but also the particular crystal surface under consideration, are important in determining the acid/base properties. Particularly large differences might be expected with oxides such as ZnO, where a range of polar (only Zn or O exposed) and non-polar (charge neutral Zn–O) faces are available. But the reported behavior does not always follow the simple expectations. One reason is that defects may often be important, even on surfaces supposedly prepared to be defect-free. Surface defects create sites where surrounding ions have lower coordination and thus may be more active as acidic or basic centers. One of the long-standing arguments on surfaces of TiO_2 and $SrTiO_3$ concerns the role of defects in promoting the dissociative chemisorption of H_2O (see § 6.4.1 below). Is it the geometric configuration of the defect that is important, or the change in electron configuration (population of the Ti 3d levels) consequent to removing oxygen ions? According to the model given above, the dissociation of water requires both an acidic and a basic site, and is much more likely to be influenced by the coordinative unsaturation around a defect than by the details of the electronic structure. Some, but not all, of the current evidence discussed below supports this idea.

6.2.4.2 Redox properties

Redox reactions involve an electron transfer, either directly or through the removal or addition of an oxygen atom. We might expect to be able to correlate the behavior of different oxides with their known chemistry, although we should be careful in noting that this knowledge generally comes from a study of bulk properties (thermodynamics, etc.) and might be altered at the surface.

Pre-transition-metal oxides (MgO, etc.) are expected to be quite inert, since they can neither be reduced nor oxidized easily. In terms of electronic structure, this is related to the large bandgap, which means that neither electrons nor holes can easily be formed. A few reports apparently contradict this conclusion, and one must suspect that either surface defects (such as O^-) or unidentified pre-adsorbed or co-adsorbed molecules are involved. On the other hand, the post-transition-metal oxides ZnO and SnO_2, as well as most transition-metal oxides, are active in redox reactions since the electron configurations of the solids may be altered.

Stoichiometric ZnO, SnO_2, and d^0 transition-metal oxides may be reduced (i.e., electrons may be added), but not oxidized. Reaction with oxidizing species such as O_2 is therefore expected only with samples that have been bulk reduced (a situation that is difficult to avoid in some cases) or where the surfaces have been made oxygen deficient. The detailed behavior with reducing molecules may be different for different oxides, as suggested by the discussion of defects in Chapters 4 and 5. On ZnO, reduction leads to the formation of free carriers, which therefore greatly increase the surface conductivity, a fact that is crucial for sensor applications.

Reduction of transition-metal oxides, by contrast, is more likely to lead to local changes in electron configuration, with electrons trapped at specific ions in the surface region. SnO_2 is again more complicated, as we have both the possibility of localized Sn^{2+} formation and of free carriers.

The d^n transition-metal oxides may, in principle, be both reduced and oxidized, so that their redox chemistry is likely to be particularly complex. A comparison of the stabilities of bulk oxide phases enables one to predict some trends. Thus the ease of oxidation of M^{3+} ions (in M_2O_3) and of M^{2+} ions (in MO) follow the sequences:

$$Ti^{3+} > V^{3+} > Mn^{3+} > Cr^{3+} > Fe^{3+},$$

and

$$Ti^{2+} > V^{2+} > Fe^{2+} > Mn^{2+} > Co^{2+} > Ni^{2+}.$$

These sequences follow trends in electronic structure, and they can be related to the order of ionization energies. The ease of oxidation generally falls with increasing atomic number, except for breaks associated with particular electron configurations where either the ligand field stabilization (between d^3 and d^4) or the exchange energy (between d^5 and d^6) falls. Much of the comparative properties of d^n oxides discussed below (see § 6.5.2, for example) does follow these expectations, although it should be noted that many of the surface reactions may be governed by kinetic rather than thermodynamic factors.

6.2.5 Photoelectrolysis

One of the most important reason that UHV surface-science studies on transition-metal oxides were begun in earnest was the discovery by Fujishima and Honda[1] in 1972 that water could be dissociated into H_2 and O_2 by a process referred to as **photoelectrolysis**. They replaced one of the Pt electrodes in a standard electrolysis cell containing an aqueous electrolyte with a TiO_2 electrode, removed the battery and connected the two electrodes together directly, and then shined light having an energy greater than the TiO_2 bandgap (3.1 eV) onto the TiO_2 electrode. They found that O_2 was generated at the TiO_2 electrode and H_2 at the Pt electrode. This discovery generated tremendous interest, particularly in the electrochemistry community, because of the potential of developing passive catalytic generators to produce H_2 as a fuel. One vision was of solar photoelectrolytic panels on the roofs of houses using solar radiation to generate H_2 for use in heating and cooling. The process has not yet been developed to a point where it is competitive with fossil fuels or other methods of generating electricity, but research is continuing on this very attractive alternate energy source.

The photoelectrolytic process involves the photogeneration of charge carriers in the semiconducting oxide electrode and the transfer of those carriers across the electrode/electrolyte interface into solution.[800] The minimum photon energy necessary for the process to occur is 1.23 eV, which is the energy between the redox levels $E^0(H_2/H_2O)$ and $E^0(O_2/H_2O)$ in the electrolyte. In practice the energy must be higher than that due to overvoltages in the system. An additional requirement is that the bottom of the conduction band in the electrode must lie above the $E^0(H_2/H_2O)$ level and the top of the valence band must lie below the $E^0(O_2/H_2O)$ level. For an n-type electrode, the bands should also be bent upwards at the surface. Photoexcited electrons will then be swept away from the electrode/electrolyte interface, eventually entering the metal electrode through the external connection, while the holes will move to the surface where they can enter the electrolyte at the $E^0(O_2/H_2O)$ level, liberating O_2.

The ideal electrode would have a bulk bandgap just enough larger than 1.23 eV to overcome the overvoltages in order that carriers could be generated using the largest possible fraction of the solar spectrum. The bandgap of TiO_2 is clearly too large to make efficient use of the solar spectrum. But experiments with other semiconducting electrodes showed that the smaller the bandgap, the more likely it was for the electrode to dissolve in the electrolyte under illumination. So interest in the use of TiO_2 and $SrTiO_3$, which was found to have a slightly higher efficiency than TiO_2, continued. (Some electrocatalytic experiments even utilized single-crystal TiO_2 and $SrTiO_3$ electrodes.) One of the observations that could not initially be explained was the high quantum efficiency displayed by TiO_2 for greater-than-bandgap illumination, since the 3.1 eV bandgap was too large to provide efficient charge transfer into the $E^0(O_2/H_2O)$ level. This prompted the first experimental work on the electronic properties of TiO_2 and $SrTiO_3$ surfaces using surface-science techniques and single-crystal samples. It was soon determined that the presence of O-vacancy defects on the surface provided filled surface states close to mid-gap that could trap the photoexcited holes at the electrolyte interface at an energy close to the $E^0(O_2/H_2O)$ level in the electrolyte. Chemisorption experiments in UHV showed that defective TiO_2 and $SrTiO_3$ surfaces dissociated H_2O without the necessity of illumination, although the environment of the electrode when immersed in the electrolyte is so different from that in vacuum that the relevance of this observation to photoelectrolysis is questionable.

Surface-science studies of other oxides, including α-Fe_2O_3 and WO_3, were also initiated in the course of evaluating electrodes for use in photoelectrolysis. Other approaches that have been tried include the suspension of very small semiconducting oxide particles, whose surfaces have been partially coated with one or more metals such as Pt, Ru and Rh, in an aqueous electrolyte. Each particle is thus an individual electrolytic cell that generates both O_2 and H_2. This approach has suffered from problems in producing ohmic metal/oxide contacts at the solid/solid interface. Current research includes studies of the nature of the metal/oxide interface.

6.3 Adsorption on non-transition-metal oxides

The chemisorption of molecules on non-transition-metal oxides differs from that on transition-metal oxides in that non-transition-metal cations can usually exist in only one valence state. (The exception is Sn, which, as discussed in § 4.6, can exist as either Sn^{2+} or Sn^{4+} on the surface of SnO_2.) The addition or removal of electrons from those cations requires so much energy that it does not usually occur during chemisorption. As for all types of metal oxides, the lattice O ions have very nearly an $O^{2-} 2p^6$ closed-shell configuration, and since the ionization energy for an O 2p electron is several eV, the transfer of charge from O^{2-} ions to adsorbates is also unlikely. However, surface O ions may form molecular complexes with adsorbates, and in some cases the adsorbate may abstract O ions from the surface (e.g., $CO_{ads} + O_{surface} \rightarrow CO_2$). It is also possible for O^- ions to exist at surface defect sites; such ions are generally very active in chemisorption.

6.3.1 MgO

Because of its structural and electronic simplicity, the MgO (100) surface is popular for both experimental and theoretical studies. Nearly all the work shows that the defect-free surface is quite inert, which should not be surprising given both its stability and the fact that its electronic structure shows little perturbation from that of the bulk solid. On the other hand, chemisorption of some molecules is promoted by the defects which are present on etched and faceted surfaces.

Many calculations have been carried out for adsorbates on MgO, using both cluster and band-structure approaches. Some of the results are helpful in clarifying the mode of adsorption and/or the type of defect configuration necessary to promote bonding. Others are frankly unrealistic and consider systems where no adsorption is found experimentally; these latter calculations can, however, be regarded as models which might be relevant to adsorption on more reactive surfaces.

The interaction of H_2 with MgO has been considered theoretically, although experimentally no interaction has been observed at room temperature. Both *ab initio* cluster calculations and defect-lattice methods have been used to study the interaction of H_2 with defect-free MgO (100).[262,621] It was found that no site on either (100) terraces or on steps or corners interacted strongly enough with H_2 to dissociate it, and that point defects were necessary in order to adsorb H_2 dissociatively. A different *ab initio* cluster calculation reached a somewhat different con- clusion, however.[623] The heterolytic adsorption of H_2 to give both adsorbed O–H and Mg–H species was predicted to proceed on the perfect surface with no activa- tion energy. Calculations of H atom binding at point defects on MgO (100) show that H_2 should dissociatively adsorb at surface cation or anion vacancies and local- ized O^- ions; the resultant adsorbed species is an OH^- ion. Adsorption at all three sites is predicted to be both exothermic and non-activated. O^- ions at step and cor- ner sites, where their nearest-neighbor coordination is reduced, are predicted to have higher binding energies for OH^- than those on terraces, suggesting that H_2 will preferentially dissociate at the lower coordination sites. Self-consistent tight- binding cluster calculations of the interaction of H^+ with MgO (100) predict that the proton will bond strongly (about 8 eV) to surface O^{2-} ions; the bond is in large measure covalent, with about 0.7 electron transferred from the substrate to the adsorbate.

The interaction of **H_2O** with MgO (100) has been studied by using XPS on pol- ished and ion-bombarded single-crystal surfaces both before and after annealing at 900 K. Adsorption was inferred from the appearance of a high-binding-energy feature in the O 1s spectrum, which was attributed to adsorbed OH^- ions. (Molecularly adsorbed H_2O was not detected by this method.) At room tempera- ture, H_2O did not adsorb on annealed surfaces, but the ion-bombarded surface did exhibit features characteristic of OH^-, presumably adsorbed at surface defect sites. (This result is in agreement with earlier infrared absorption measurements on MgO smoke cubes.) At 200 K, both surfaces showed evidence of OH^- after expo- sure to H_2O. Using the angular dependence of the intensity of the OH^- feature, it was estimated that the ion-bombarded surface was covered with one monolayer of OH^- at 200 K. As was noted in § 2.3.1.1, when cleaved MgO (100) surfaces are exposed to air, a high density of point defects is formed due to reaction with water vapor.[79]

There has also been some theoretical consideration of the interaction of H_2O with MgO surfaces. A self-consistent tight-binding approach was used to study the interaction of H^+ and OH^- (but not H_2O) with both perfect and defective MgO (100) surfaces. The proton was found to bond to surface O^{2-} ions with an energy of about 8 eV, while a hydroxyl ion was bound to Mg^{2+} ions with only about 1.56 eV. Some consideration was also given to the adsorption of H^+ and OH^- at step and edge sites on MgO (100). CNDO calculations have also been performed of the interaction of H_2O with both flat and rough MgO (100) surfaces, with the result that H_2O preferentially dissociates to OH^- and H_2 at corner defect sites. The adsorption of H_2O can also result in the formation of a Mg vacancy and $MgOH^+$ or $Mg(OH)_2$. Both calculations thus agree with experiment in that the surface of MgO can become hydroxylated upon exposure to H_2O.

The only alcohol whose chemisorption on MgO has been studied is **CH₃OH**. The experimental results indicate that, like H_2O, methanol adsorbs only at defect sites on MgO (100), undergoing heterolytic dissociation to yield methoxide (CH_3O^-) and OH^-. On the other hand, the two carboxylic acids formic acid, **HCOOH**, and acetic acid, **CH₃COOH**, both appear to dissociate at room temperature and below on (100) terraces, since little difference is seen between adsorption on stoichiometric (100) surfaces and on either ion-bombarded (100) or faceted (111) surfaces. Deprotonation occurs, giving surface hydroxide and carboxylate ($HCOO^-$ and CH_3COO^-) species; presumably the contrast with water and alcohols arises because of the higher acid strength of the carboxylic acids. Formaldehyde, **HCHO**, undergoes a more complex reaction, apparently disproportionating into formate ($HCOO^-$) and methoxide on stoichiometric (100) surfaces at 170 K.

Another molecule that adsorbs only at defect sites on MgO (100) is methyl formate, **HCOOCH₃**. Measurements of room-temperature adsorption onto stoichiometric (100) surfaces and highly defective faceted (111) surfaces indicate that $HCOOCH_3$ dissociates into methoxide and formate on defective surfaces.

One experimental study has reported that **CO₂** adsorbs to the MgO (100) surface at room temperature. Using samples that had been polished and then annealed by electron bombardment, CO_2 adsorption was interpreted in terms of bonding to surface O ions, yielding CO_3^{2-}. The adsorption may well have taken place at surface defect sites, which were no doubt present on a sample that had been subjected to electron bombardment. In addition, an electron flood gun was used during the UPS measurements in order to overcome surface charging, so electron/molecule interactions may also have been present.

Towards most other molecules, MgO is quite inert. No reaction was found between stoichiometric (100) surfaces and **CO**, or between stoichiometric (100) and faceted (111) surfaces and **O₂**. With methane, **CH₄** (the only alkane studied), ethylene, **C₂H₄**, and butadiene, **1,3-C₄H₆**, no room-temperature adsorption occurs on either stoichiometric (100) surfaces or on faceted (111) surfaces that exposed primarily (100) planes containing a high density of defect sites. The same is true of acetylene, **HCCH**, although heterolytic dissociation of this molecule has been observed on defective MgO (100) surfaces at 170 K. In one report of the results of UPS measurements, **N₂O** was found not to adsorb at room temperature onto either polished (100) or (111) surfaces that were annealed by electron bombardment.

Several calculations have addressed the adsorption of CO on MgO, although, as noted above, no adsorption is found experimentally on stoichiometric (100) surfaces. The nature of the interaction of CO with defect-free surfaces has been considered theoretically by using both *ab initio* defect-lattice calculations on small clusters and Hartree–Fock crystalline-orbital calculations on extended two-dimensional structures; similar results are obtained by both methods. CO is predicted to chemisorb molecularly on top of the surface Mg^{2+} ions, with the molecular axis normal to the surface. The bonding is weak, and there is little difference between bonding via the C or the O atom. The calculated binding energy is between 17 and 38 kJ/mol, depending upon the calculational method used. CO does not bond to the O^{2-} ions, regardless of which atom is in contact with the surface. There is very little charge transfer between the molecule and the surface upon adsorption, in contrast to the behavior of CO on many transition-metal oxides. The bonding is essentially electrostatic in nature, but, since there is little charge transfer, and since CO has a small dipole moment, it is thought to involve the quadrupole moment of the CO and the ionic charges of the MgO. There is some increase in the strength of the CO/surface bonding at low-coordination cations at corner or step edge sites, but the nature of the bonding remains the same.

Two-dimensional planar structures have been used to calculate the bonding of CO to the MgO (110) surface, where all surface ions are four-fold coordinated. Bonding occurs only above surface Mg^{2+} ions here also, although the strength of the bonding is about 50 % larger than that for the (100) surface. Adsorption energies are very similar for either the C or O atom in contact with the surface, and the electrostatic nature of the bond is the same for both surfaces.

The effect of substitutional surface cations in the chemisorption of CO at point defects on MgO (100) has also been considered theoretically.[255,262,621,712] Cations with charges from 1+ to 3+ were studied, and, while the calculated binding energies vary from ion to ion, in no case was any major difference found in the adsorption characteristics from those at Mg^{2+} sites on the defect-free surface. No calculations have been performed for CO adsorption at surface cation or anion vacancies.

O^- is thought to be important in promoting the heterolytic dissociation of **CH$_4$** on Li-doped MgO powder catalysts which accomplish the conversion of CH$_4$ to C_2H_6. Calculations have suggested that weak dissociation can indeed occur under these conditions, but this result has not been verified experimentally on well-characterized surfaces. Cluster calculations have also suggested that the O^- ions necessary for H abstraction from CH$_4$ are stabilized at three-fold-coordinated kink sites on MgO (100), but not at four-fold O sites on straight steps.

In a set of experiments not included in Table 6.7, XPS, UPS and TDS were used to study the interaction of **HCOOC, CH$_3$COOH, CH$_3$OH, C$_2$H$_5$OH, HCOOCH$_3$** and **C$_2$H$_2$** with thin MgO films grown on single-crystal Mg (0001).[801] The results obtained were similar to those on single-crystal MgO (100). Although the thin-film MgO surfaces were not well-characterized, the experiments are interesting in that UPS difference spectra were used to identify the adsorbed species.

Ab initio cluster calculations on the adsorption of **HCO, HOC, H$_2$CO, HCOH, CH$_3$O** and **CH$_2$OH** fragments at Mg^{2+} and substitutional Cu$^+$ sites on MgO (100) have been carried out, and the results compared with those for CO chemisorption. The general trends are that the bonding is weak, there is little charge transfer between the adsorbate and the surface, and π back-donation is only observed for adsorption at Cu^+ sites. The necessary ingredient is the presence of surface O^- ions, which are able to accept electrons from the C–H bond. The presence of monovalent Li ions stabilizes O^- ions, but even then the reaction is predicted to be weak and sensitive to the geometry of the surface, i.e., terrace, edge or corner sites.

In addition to the calculations that treat the adsorption of specific molecules, LCAO band-structure calculations have been used to simulate the interaction of generalized hydrogen-like and halogen-like atoms with the MgO (100) surface.[802] The model adatoms had a single non-degenerate energy level that fell inside the bulk MgO bandgap. Coupling between the adatom and the surface turns the atomic level into a surface band, with the exact position and dispersion of the band depending upon the assumed strength of the coupling. Figure 6.4 shows the generalized band structure for the two types of adatom for an intermediate value of the coupling constant. [Compare this with Fig. 4.4 (a), which shows the results of a similar calculation for the clean MgO (100) surface.]

Semiempirical calculations have also been used to study the potential energy surfaces of *physisorbed* molecules, including N_2, NH_3, CH_4, CH_3F and atomic **Xe**, on perfect MgO (100) terraces. The potential surfaces predicted for CH_4 and N_2 are shown in Fig. 6.5 (a) and (b), respectively. [Mg ions lie at positions (0,0), (0,a), (a,0), etc.] The orientation of the molecules was varied to minimize the total energy at each adsorption site. The surfaces are seen to be very different for the two molecules. CH_4 is predicted to adsorb preferentially on top of a Mg ion, with three of its H atoms close to the surface and the fourth directed away from the surface. N_2, on the other hand, prefers to adsorb on a site equidistant between two Mg and two O ions, with its internuclear axis parallel to the surface.

Although we are in general not discussing the vast literature of adsorption on metal-oxide powders in this book, one study of CO adsorption on MgO powder is worth noting.[803] The experiments and theories discussed above indicate that CO either does not adsorb on MgO or adsorbs molecularly. But thermal desorption measurements using isotope-labelled $C^{18}O$ showed that there was a large amount of O exchange between the CO molecule and the MgO surface. *Ab initio* molecular-orbital cluster calculations were performed to determine how such an exchange process might take place. While point defects on flat surfaces are unable to dissociate the CO, it was found that certain three-dimensional structures containing both O vacancies and low-coordinated O ions were able to break the C–O bond through intermediates such as CO_3^{2-}. Such structures are probably present on the high-area powders used as catalyst supports.

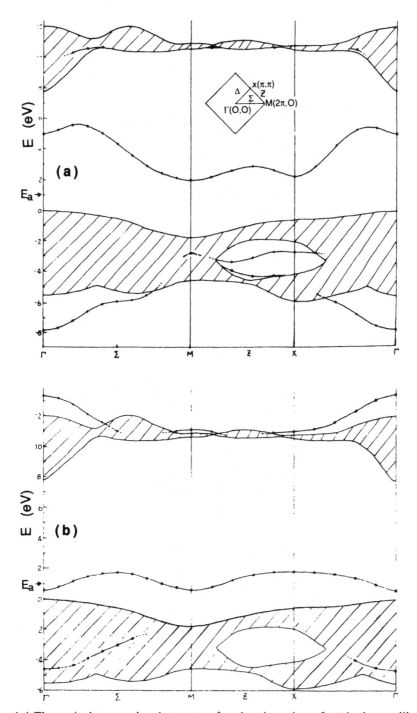

Fig. 6.4 Theoretical energy-band structure for chemisorption of (a) hydrogen-like atoms and (b) halogen-like atoms on MgO (100). [Ref. 802]

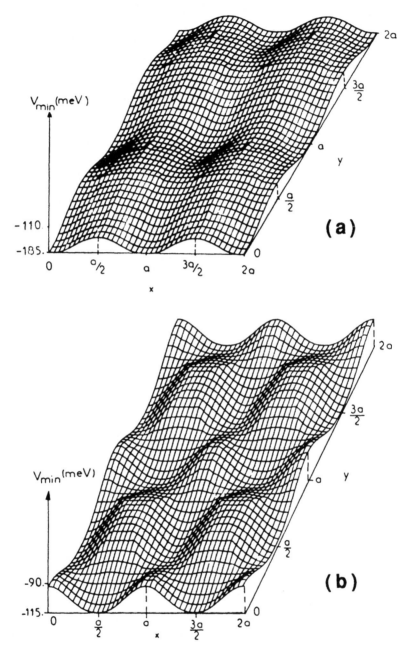

Fig. 6.5 Potential energy surfaces for (a) CH_4 and (b) N_2 adsorbing onto MgO (100).
[Ref. 714]

6.3.2 CaO, SrO and BaO

The geometric and electronic structures of the other alkaline-earth oxides are very similar to those of MgO. Chemically, the oxides of the larger cations are more basic and thus more reactive to H_2O, CO_2, and other acidic molecules. There has, however, been very little work on single-crystal samples of these compounds.

There has been one reported measurement of adsorption on single-crystal **CaO**; the interaction of **SO_2** with ion-bombarded and annealed CaO (100) has been studied by XPS. The structure of the surface was not determined by LEED, nor was its stoichiometry. However, it was atomically clean, and to the extent that CaO behaves like MgO under ion bombardment, the surface was probably nearly stoichiometric. The initial sticking coefficient for SO_2 was found to be 0.4, and the adsorbed species was identified as sulfate, $SO_4{}^{2-}$. This is a somewhat puzzling result, since, as discussed earlier (§ 6.2.3.6), the oxidation of SO_2 to sulfate should liberate two electrons, and it is difficult to see how these can be accommodated in a wide-bandgap material. It is possible that surface defects, or some adventitious O_2, was present in this experiment.

There have been no published reports of chemisorption studies on single-crystal **BaO**. However, the interaction of BaO with components of the atmosphere is extremely important in the stability and operation of BaO-coated thermionic cathodes. Stoichiometric BaO coatings on tungsten cathodes exhibit a work function of about 2 eV, but exposure to the atmosphere results in a significant increase in work function and the necessity of 'reactivating' the cathode before full emission can be obtained. Therefore surface-science studies have been conducted of the interaction of H_2O and CO_2 with polycrystalline BaO layers. H_2O is found to react strongly with the surface, dissociating to form $Ba(OH)_2$. CO_2 similarly combines to form $BaCO_3$, this reaction being catalyzed by the presence of H_2O. The process of reactivating BaO thermionic cathodes consists primarily of decomposing the hydroxide and carbonate to restore the BaO stoichiometry. Reactions similar to those on BaO occur in Ba-containing high-T_c oxide superconductors, as discussed below in § 6.6.

6.3.3 Al_2O_3

Optical studies have been reported of **phenanthrene, pyrene** and **butane** layers adsorbed onto (11$\bar{2}$0) and (0001) surfaces of single-crystal α-Al_2O_3, but the surfaces were not characterized well enough to permit meaningful determination of the nature of the adsorbate/substrate interactions.[804–806] In fact, no experimental reports have been published to date of any adsorption studies on well-characterized single-crystal Al_2O_3 surfaces, which is a pity in view of the importance of alumina in catalysis. However, there have been some calculations of adsorbates on

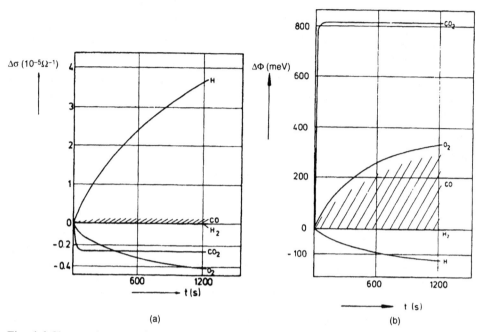

Fig. 6.6 Changes in (a) surface conductivity, $\Delta\sigma$, and (b) work function, $\Delta\Phi$, as a function of time at 300 K after exposure of ZnO $(10\bar{1}0)$ to various gases. [Ref. 626]

the (0001) surface of α-Al$_2$O$_3$. One of these concerns the adsorption of CO. The molecule is predicted to bond to surface Al ions very weakly with an energy of 25.9 kJ/mole, and with very little charge transfer.

The hydroxylation of the α-Al$_2$O$_3$ (0001) surface has been considered theoretically by means of a crystalline-orbital approach. The adsorption of a monolayer of **OH** on top of the surface Al ions and **H** atoms bonded to the surface O ions generates a new band that is shifted a few eV above the top of the valence band in the bulk bandgap. The band is formed primarily from O 2p and H 1s orbitals.

6.3.4 ZnO

Chemisorption on ZnO is important both in catalytic and gas sensor applications, and for this reason it has been the most extensively studied of all the non-transition-metal oxides. Thorough discussions of the chemisorption properties of ZnO as they apply to its use in gas sensing are given in Refs. 807 and 808. Commercial gas sensors utilize powdered ZnO, where the electrical conductivity of the sample changes depending upon the type and amount of gas adsorbed on the surface. The conductivity changes occur at the surface of the ZnO grains as a result of charge transfer and band-bending caused by the adsorbates. The surface chemistry involved is no doubt very complex, and experiments have been performed on single-crystal surfaces in an attempt to determine the important steps in the process. (A good summary of the single-crystal approach and the attendant models of the effect of adsorbed gases on sample conductivity is given in Ref. 626.) Figure 6.6 (a) shows the way in which various gases change the surface conductivity of stoichiometric ZnO $(10\bar{1}0)$;[626] the corresponding changes in work function are plotted in Fig. 6.6 (b).

The only clear conclusion that can be drawn from the work summarized below is that the chemisorption behavior of ZnO is extremely complicated. The variety of polar and non-polar surfaces available, the presence of acidic Zn and basic O sites, the complex and imperfectly understood defect chemistry, and the possibility of electron and oxygen-atom transfer are factors that combine to enable all the types of chemisorption discussed in § 6.2 above. Undoubtedly these features are responsible for the activity of ZnO as a catalyst and a sensor. But they make the elucidation of detailed chemisorption mechanisms very difficult. Many of the results so far obtained conflict with each other and with the expectations of simple models, and much more work needs to be done before a clear understanding can emerge.

Under the low-pressure conditions used in surface-science studies, molecular H_2 essentially does not interact with ZnO surfaces. This is shown by the horizontal lines in Figs. 6.6 (a) and (b), indicating no change in surface conductivity or work function upon exposure. The inability to see any effect of H_2 for low-pressure exposures points out one of the weaknesses of such studies in trying to understand catalysis. Under high-pressure catalytic conditions, H_2 is known to dissociate on ZnO surfaces, which is one of the steps in the catalytic production of methanol from CO and H_2; the nature of the dissociative adsorption site is even known.[809] And yet nothing can be seen in surface-science experiments. This is a manifestation of the so-called **pressure gap** referred to earlier. Atomic H, however, interacts strongly with the ZnO (0001), $(000\bar{1})$ and $(10\bar{1}0)$ surfaces, acting as an electron donor and increasing the carrier concentration in the space charge layer by several orders of magnitude. [Atomic H has also been ion-implanted into the first 10–20 Å of ZnO $(000\bar{1})$ surfaces in order to produce accumulation layers that are stable against exposure to air.[810,811]]

Although H_2 does not chemisorb onto stoichiometric ZnO surfaces, it does physisorb for temperatures below 200 K. However, its heat of adsorption, determined from a Freundlich isotherm, is even lower than that for Xe.

The adsorption of atomic H onto the ZnO $(000\bar{1})$ surface has been treated theoretically by means of cluster calculations. H was assumed to bond in on-top sites directly above surface O ions. The resulting O–H bond is strong, comparable to the O-metal bonding in the bulk oxide lattice, with the H atom donating almost 0.5

electron to the surface O ion. The layer of positively charged H on the surface reduces the work function, as shown in Fig. 6.6 (b), while the electrons donated to the ZnO increase the surface conductivity, as shown in Fig. 6.6 (a). Cluster calculations have also been performed for the adsorption of H on the ZnO ($10\bar{1}0$) surface in order to explain the heterolytic adsorption of H_2 on ZnO. Both strong O–H and fairly strong Zn–H bonds are predicted for the ($10\bar{1}0$) surface, resulting in adjacent Zn^{2+}–H^- and O^{2-}–H^+ sites.

The adsorption of O_2 onto stoichiometric ZnO has been studied for the polar (0001) and (000$\bar{1}$) and the non-polar ($10\bar{1}0$) surfaces, with the ($10\bar{1}0$) receiving most attention; both UHV-cleaved and polished and annealed surfaces have been studied. Molecular oxygen interacts only weakly with all of the surfaces. Below room temperature only physisorption occurs, with no charge transfer between the O_2 molecule and the surface. Between 300 and 650 K, chemisorption involving charge transfer from the surface to the molecule results in adsorbed O_2^-. Although some changes can be seen in the ELS and UPS spectra upon O_2 chemisorption, the actual coverage on the surface has been determined to be less than 2.5×10^{-4} monolayer in the temperature range where chemisorption occurs. The sticking coefficient at low coverages was found to be about 10^{-5}. Adsorption is thus clearly occurring at defect sites. However, even a small coverage of O_2^- is sufficient to change the surface conductivity of undoped ZnO, as shown in Fig. 6.6 (a) for the ($10\bar{1}0$) surface. The decrease in surface conductivity results from the transfer of electrons from the ZnO surface to the adsorbed molecule. The electronic charge associated with surface defects must be responsible for the surface conductivity before O_2 adsorption.

That O_2 adsorption on ZnO surfaces occurs at defects is substantiated by experiments in which O-vacancy point defects were intentionally created on the ($10\bar{1}0$) surface by heating the sample to about 950 K and then quenching to room temperature. The sticking coefficient for O_2 was found to increase exponentially with the density of O-vacancy surface defects, and the changes in surface conductivity were correspondingly larger. O_2 presumably dissociates at defect sites, forming atomic O that at least partially heals the surface defects; O_2^- also adsorbs on defect surfaces.

The interaction of O_2 with surface defects of a different type has also been studied on ZnO ($10\bar{1}0$) by preparing stepped surfaces, where the ligand coordination of the cations at the step edges is less than that of the cations on the ($10\bar{1}0$) terraces. Both ($40\bar{4}1$) and ($50\bar{5}1$) surfaces were prepared and characterized by LEED. More O_2 was observed to adsorb at room temperature on the stepped surfaces than on the ($10\bar{1}0$) surface, and a different adsorption state occurred. This is shown in Fig. 6.7, which plots thermal desorption spectra for O_2 desorbing from both ($10\bar{1}0$) and

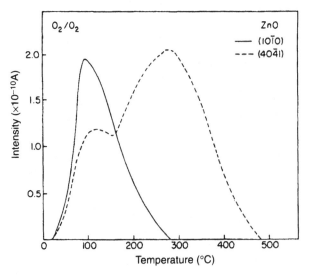

Fig. 6.7 Thermal desorption spectra for O_2 after adsorption of O_2 onto ZnO ($10\bar{1}0$) and stepped ($40\bar{4}1$) surfaces. [Ref. 236]

($40\bar{4}1$) surfaces after room temperature O_2 adsorption. Most of the O_2 is adsorbed in a more tightly bound state on the stepped surface than on the terraces. However, the total surface coverage of adsorbed O_2 is still no more than a few % of a mono-layer, so the O_2–surface interaction is weak even for stepped surfaces.

Atom-superposition electron-delocalization molecular-orbital calculations have been performed for the interaction of atomic **O** with ZnO (0001) and (000$\bar{1}$) surfaces. No adsorption of O on the Zn face is predicted, but O is calculated to adsorb as O^- on top of surface O ions on the (000$\bar{1}$) face with a binding energy of 125 kJ/mol. No calculations were performed for the interaction of O_2 with ZnO surfaces, so the theory cannot be directly compared with experiment.

The interaction of **H_2O** (D_2O was used in thermal desorption measurements for experimental reasons) with ZnO surfaces has been studied on polished and annealed (0001), (000$\bar{1}$) and (10$\bar{1}$0) surfaces. H_2O does not adsorb on those sur-faces for temperatures above 130 K, so the interaction is clearly weak. Adsorption at lower temperature is molecular, with no evidence for dissociation into OH^- ions. After chemisorption at 100 K, thermal desorption spectra showed six differ-ent desorption peaks on each surface, which were interpreted in terms of different adsorption sites. Figure 6.8 compares thermal desorption spectra for the three crystal faces. The 152 K peak is attributed to three-dimensional ice, and the 168 K peak to two-dimensional ice. The peak at 190 K is associated with adsorption of the D_2O molecules at surface O sites, since it is absent on the (0001) face, where

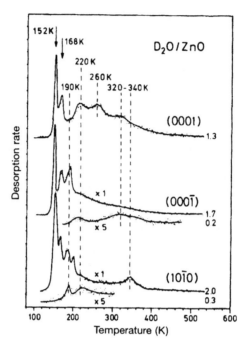

Fig. 6.8 Thermal desorption spectra for D_2O after adsorption of D_2O onto three different faces of ZnO. [Ref. 676]

Zn ions are outermost. The peak at 340 K is thought to correspond to adsorption at surface Zn sites; note that it is virtually absent on the $(000\bar{1})$ surface, where a monolayer of O ions covers the outermost plane of Zn ions. The other desorption peaks are ascribed to two-dimensional D_2O clusters or possibly adsorption at interstitial Zn ions.

UPS was also used in this study of H_2O adsorption, and Fig. 6.9 shows the changes that occur in the UPS spectra as a function of H_2O exposure for both the (0001) and $(000\bar{1})$ surfaces. For submonolayer coverages the spectra for the two surfaces are slightly different, presumably because of the differences in binding to surface Zn and O ions, although no further interpretation of the UPS spectra was made.

The adsorption of H_2O on the (0001)-Zn surface of ZnO has been studied theoretically by means of semi-empirical quantum-chemical calculations (INDO/S) on clusters containing about twenty atoms. The H_2O molecule was assumed to adsorb via its O atom in an on-top site above the surface Zn ions. The O lone-pair orbitals were found to mix primarily with the 3d, 4s and 4p orbitals of the nearest-neighbor Zn, although some interaction with next-nearest-neighbor O ions occurred as well. H_2O donates a net fraction of an electron to the surface. The same technique was also applied to the adsorption of OH at both on-top Zn sites on (0001) and in the

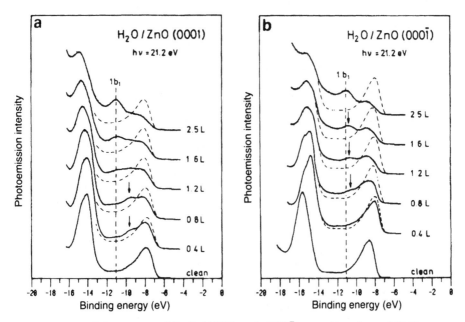

Fig. 6.9 He I UPS spectra of ZnO (0001) and (000$\bar{1}$) as a function of H_2O exposure. The dashed curves are the spectra for the clean surface. [Ref. 676]

direction of the missing O ligand on Zn ions on (10$\bar{1}$0); in both cases OH was found to be an electron acceptor, with the bonding involving σ donation and π acceptance on the OH and predominantly the surface Zn 4s and 4p orbitals.

DV-Xα cluster calculations have also been performed for dissociative adsorption of H_2O onto the ZnO (10$\bar{1}$0) surface, leading to hydroxylation of the surface. The OH$^-$ ions adsorbed above the surface Zn sites, with the other H atom bonding to a surface O ion as H$^+$. The driving force for dissociation in the cluster model is stabilization of the surface O ions relative to the clean surface by the adsorbate.

The adsorption of **CO** on ZnO has been studied in great detail since CO is one of the reactants in methanol production using Cu/ZnO catalysts, and yet there are still reports of conflicting data and different interpretations of the adsorbed species in the literature. Only one study has been performed using UHV-cleaved surfaces. CO was adsorbed onto cleaved (0001), (000$\bar{1}$) and (10$\bar{1}$0) surfaces at room temperature, and small changes in ELS spectra were observed. But the interaction was weak, with equilibrium coverages far below a monolayer. Nothing could be determined about the nature of the adsorbed species.

The results on polished and annealed (0001), (000$\bar{1}$), (11$\bar{2}$0), and (10$\bar{1}$0) surfaces are somewhat different than those on cleaved surfaces. On the stoichiometric (0001) surface, CO adsorbs molecularly and reversibly for temperatures below

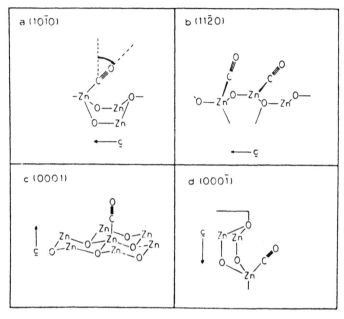

Fig. 6.10 Model of the active sites for adsorption of CO on four different ZnO surfaces. [Ref. 193]

200 K. At 90 K, the coverage was judged to be close to a monolayer. From angle-resolved UPS measurements, CO was determined to bond to the Zn ions with its internuclear axis normal to the surface, as shown in Fig. 6.10. The amount of CO adsorbed as a function of temperature for different pressures was fitted to a Temkin isobar, which yielded a zero-coverage heat of adsorption of 50 kJ/mol. For room temperature CO adsorption, less than 10^{-3} monolayer adsorbed, and it desorbed as CO at about 400 K.

On the (000$\bar{1}$) surface, adsorption of CO at temperatures below 200 K yielded UPS and XPS spectra that were also interpreted in terms of molecular adsorption. The amount of CO adsorbed was less than for the (0001) surface, but more than that reported for the UHV-cleaved (000$\bar{1}$) surface. The polished and annealed (000$\bar{1}$) surface is believed to contain a high density of steps, and the adsorption of CO on that face is thought to occur at Zn sites on the edges of steps, as shown in Fig. 6.10, rather than on the surface O ions on terraces. This interpretation is supported by the fact that the measured heat of adsorption for CO on ZnO (000$\bar{1}$) is the same as that on (0001), indicating a similar adsorption site. However, for temperatures above 200 K, features in the photoemission spectra indicative of a different type of chemisorption were observed. (Evidence of those features was also present in spectra taken at less than 200 K.) The spectra were interpreted in terms of the formation of both CO_2 and a surface carbonate species, CO_3^{2-}. Interestingly,

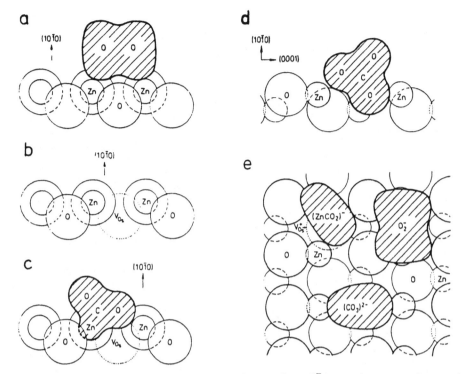

Fig. 6.11 Simplified geometric models for ZnO (10$\bar{1}$0) surface complexes. (a) Chemisorbed 'O_2^-'; (b) O vacancy 'V_{Os}^+' (c) CO_2 associated with V_{Os}^+, '$(ZnCO_2)^-$'; (d) chemisorbed CO_2, 'CO_3^{2-}'; (e) top view. [Ref. 258]

the signatures of the adsorbed species were the same as those for CO_2 adsorbed on defective ZnO (000$\bar{1}$), which will be discussed below.

Defective ZnO (000$\bar{1}$) surfaces prepared by ion bombardment were also exposed to CO, and the results are very different than for the stoichiometric surface. Using UPS and XPS, CO exposure at 120 K did not produce any measurable changes in the spectra. Thus the defect (000$\bar{1}$) surface appears to be essentially inert to CO. This is surprising since the presence of O vacancies on that surface exposes Zn ions, and CO chemisorbs to at least a reasonable fraction of a monolayer onto Zn sites on the (0001) surface at the same temperature.

The ZnO surface onto which CO adsorption has been studied most exhaustively is the non-polar (10$\bar{1}$0). It contains equal numbers of Zn and O ions, and CO is found to chemisorb more strongly on stoichiometric (10$\bar{1}$0) surfaces than on either of the polar faces. Angle-resolved UPS and HREELS measurements indicate that, for temperatures below 200 K, CO adsorbs via its C atom to the coordinatively unsaturated Zn ions, as shown in Fig. 6.10. For adsorption at room temperature and above, CO chemisorbs irreversibly on both stoichiometric and defect surfaces. It adsorbs more readily on the stoichiometric surface, and upon subsequent thermal desorption the only product observed is CO_2, accompanied by the creation of O-vacancy defects on the surface. Both the surface conductivity and the work function increase upon CO adsorption, the former presumably due to the formation of O vacancies. The model for the adsorbed species, shown in Fig 6.11 (c), consists of a $(ZnCO_2)^-$ complex and an associated O vacancy; this is the same surface complex that is believed to be formed when CO_2 adsorbs at an O vacancy on reduced ZnO (10$\bar{1}$0).

The interaction of CO with (40$\bar{4}$1) and (50$\bar{5}$1) stepped ZnO surfaces has also been studied at room temperature. The (40$\bar{4}$1) surface behaved very differently than the (10$\bar{1}$0) surface, which is the terrace orientation on the stepped surfaces. Very little CO_2 desorbed upon subsequent heating of the (40$\bar{4}$1) surface, with the dominant species being CO. This is a puzzling result, since it is difficult to imagine that the presence of steps could so drastically change the adsorption properties of the (10$\bar{1}$0) terraces.

There have been several theoretical cluster calculations of the interaction of CO with ZnO surfaces. Semi-empirical quantum-chemical calculations (INDO) of the adsorption of CO via its C atom in on-top sites on the Zn-terminated (0001) surface indicate that the bond, which is dominated by π back-donation rather than σ donation, is relatively weak.[722] The CO molecule is a net electron acceptor, and the bond is localized to the nearest-neighbor Zn ion. However, tight-binding calculations of the same CO adsorption geometry on ZnO (0001) predict a stronger CO–Zn bond, with CO being a net electron donor.[723] A strong bond is also predicted between CO and Zn ions on the (10$\bar{1}$0) surface. Bonding is again via the C atom, but at this adsorption site CO is an electron donor, transferring charge into the Zn 4s and 4p levels via σ donation. The adsorption energy is computed to be about 50 kJ/mol. Cluster calculations have also been performed for the oxidation of CO by O^- and O on ZnO surfaces. For ZnO (000$\bar{1}$) surfaces onto which atomic O^- had been pre-adsorbed, CO is predicted to react with the O^- to form CO_2.

The adsorption of **CO_2** on ZnO (0001), (000$\bar{1}$), (11$\bar{2}$0), and (10$\bar{1}$0) surfaces has been studied by a variety of techniques on polished and annealed samples. CO_2 adsorbs weakly on stoichiometric (0001) at room temperature, with the maximum coverage estimated to be about 1 % of a monolayer. It is bound fairly tightly, not desorbing until the sample is heated to about 700 K. Considering the small amount adsorbed, it probably occurs at defect sites.

The response of the ZnO (000$\bar{1}$) surface to CO_2 is opposite to that for CO. The stoichiometric polished and annealed surface does not chemisorb a measurable amount of CO_2 even at 100 K, as determined by UPS and XPS. However, on the defective, reduced (000$\bar{1}$) surface CO_2 adsorbs strongly at 100 K, with the maximum coverage measured to be 0.6 monolayer. The adsorbed species was inter-

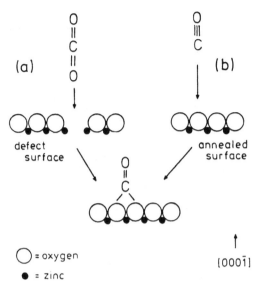

Fig. 6.12 Model of the polar ZnO $(000\bar{1})$-O surface (a) with and (b) without O-vacancy defects, interacting with CO_2 and CO, respectively. [Ref. 717]

preted as a surface carbonate, CO_3^{2-}. The model proposed for the chemisorption of both CO and CO_2 on ZnO $(000\bar{1})$ is shown in Fig. 6.12. The same final state of a surface carbonate is reached via either CO adsorption on the stoichiometric surface or CO_2 adsorbed onto the reduced surface.

For CO_2 chemisorption on defective $(10\bar{1}0)$ surfaces, the surface conductivity decreases while the work function increases; that behavior can be explained by a charge-transfer model in which the adsorbed CO_2 removes an electron from the surface O vacancy upon forming a $(ZnCO_2)^-$ complex such as that shown in Fig 6.11 (c). On stoichiometric $(10\bar{1}0)$ surfaces, CO_2 adsorbs slightly less strongly than it does on defective surfaces, but the conductivity and work function changes are similar. A model for the adsorbed species, which consists of a surface carbonate, CO_3^{2-}, is shown in Fig. 6.11 (d). The zero-coverage sticking coefficient for CO_2 on stoichiometric ZnO $(10\bar{1}0)$ is 0.6, which is more than 10^5 times larger than that for O_2 on the same surface; the work function increases by as much as 1 eV. No differences were found for CO_2 adsorption at room temperature between the stepped $(40\bar{4}1)$ and $(50\bar{5}1)$ surfaces and the $(10\bar{1}0)$ surface.

The adsorption of CO_2 onto ZnO $(10\bar{1}0)$ surfaces has been modelled theoretically by means of a self-consistent semi-empirical cluster method (SINDO1). The results are presented graphically in Fig. 6.13, where the free CO_2 molecule is shown in (a); the relaxed, clean ZnO surface in (b); and the adsorbed complex, which resembles the surface carbonate discussed above, in (c). While the cluster

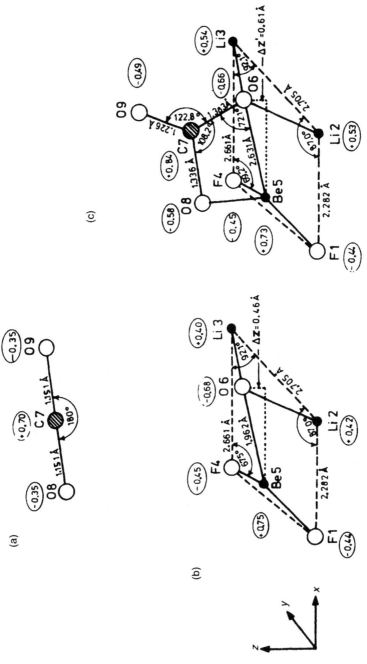

Fig. 6.13 Schematic representation of changes in surface atom positions and local charge distribution upon chemisorption of CO_2 on ZnO (10$\bar{1}$0). (a) Free CO_2 molecule, (b) free surface cluster, and (c) adsorption complex. Filled circles are zinc cations; shaded circles are carbon atoms; open circles are oxygen anions. [Ref. 628]

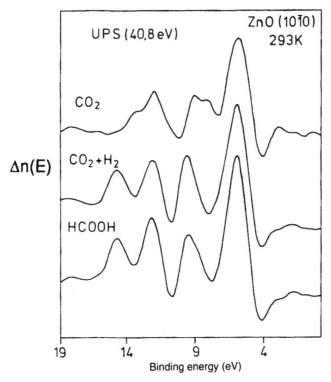

Fig. 6.14 Comparison of UPS difference spectra for CO_2 (150 L), $CO_2 + H_2$ (1 bar, 300 s) and HCOOH (40 L) on ZnO (10$\bar{1}$0) at 293 K. [Ref. 728]

may be too small to accurately represent the surface, the calculations give a general idea of the nature of CO_2 chemisorption on ZnO (10$\bar{1}$0). Atom-superposition electron-delocalization molecular-orbital cluster calculations have also been performed for the bonding of CO_2 to both Zn ions on ZnO (0001) and O ions on ZnO (000$\bar{1}$). In both cases the O–C–O angle is bent upon adsorption, by roughly the same amount as that shown in Fig. 6.13 for adsorption on the (10$\bar{1}$0) face. On the (0001) face, charge is transferred from the surface Zn ions to the $2\pi_u$ orbital of CO_2; the calculated adsorption energy is 1.89 eV. On the (000$\bar{1}$) face, CO_2 bonds to O ions as a surface carbonate, CO_3^{2-}, with a theoretical adsorption energy of 1.07 eV. INDO/S and extended-Hückel cluster calculations have also been performed for various assumed geometries of CO_2 bonded to Zn ions on ZnO (0001). If CO_2 is assumed to retain its linear geometry upon bonding (i.e., if it bonds either perpendicular or parallel to the surface), it acts as an electron donor. However, if its O–C–O bond angle is bent, the adsorbate acts as an electron acceptor. In either event, the substrate orbitals most involved in the bonding are the Zn 4s and 4p.

One reason that chemisorption on ZnO has been so extensively studied is that methanol is produced commercially from CO or CO_2 and H_2 by using a Cu-doped

ZnO catalyst. Some single-crystal experiments have attempted to model the catalytic formation of methanol over ZnO catalysts by exposing well-characterized ZnO surfaces to mixtures of CO_2 or CO and H_2. No evidence for co-adsorption was seen in either UPS or XPS spectra when ZnO ($10\bar{1}0$) was exposed to low pressures ($\leq 10^{-5}$ mbar) of $CO_2 + H_2$ at temperatures between 223 and 373 K. But when the same surface was exposed to $CO_2 + H_2$ at room temperature and atmospheric pressure, the changes in the He II UPS spectra shown in the middle difference spectrum in Fig. 6.14 were observed. They are virtually identical to those for exposure of the surface to formic acid, shown in the bottom difference spectrum. (The top difference spectrum is that due to adsorption of CO_2 alone.) It was thus concluded that in both cases a formate species was being formed on the surface, and that it is one of the intermediate steps in the formation of methanol. When the ZnO ($10\bar{1}0$) surface was exposed to CO + H_2 at 120 K, three peaks appeared in the UPS spectra that correlated with the three highest-binding-energy peaks for adsorbed HCOOH in Fig. 6.14, but the peak at about 6 eV was absent. This led to the interpretation that the lowest-binding-energy feature is associated with the extra O atom available in HCOOH.

Although not performed on what are referred to here as 'well-characterized' surfaces, thermal desorption measurements of ($CO_2 + H_2$)-exposed ZnO powders, in which a significant fraction of the exposed crystal faces consisted of polar surfaces, and single-crystal needles that exposed almost entirely non-polar faces, gave results which indicated that it is predominantly the polar faces that are active for that gas mixture.[812] However, theoretical calculations of various reaction pathways for methanol synthesis on ZnO do not predict that there should be any large differences between the reactivity of the polar (0001) and non-polar ($10\bar{1}0$) surfaces.[723] No experimental single-crystal measurements have been reported for exposure of either CO + H_2 or $CO_2 + H_2$ to polar ZnO surfaces.

The adsorption of **NH_3** has been studied on polished and annealed ZnO (0001), ($000\bar{1}$), ($10\bar{1}0$) and ($11\bar{2}0$) surfaces. It adsorbs molecularly on all four faces and remains on the surface up to 450 K. The strength of the bonding is much greater than that of CO, and NH_3 readily displaces adsorbed CO. For a ($10\bar{1}0$) surface exposed to an ambient of 10^{-6} Torr CO at 90 K, where the CO molecular orbitals are dominant features in the UPS spectra, admission of only 10^{-9} Torr NH_3 results in total displacement of the adsorbed CO by NH_3. This is shown in terms of the UPS difference spectra (where the clean surface spectrum has been subtracted from the adsorbed spectra) in Fig. 6.15. The four times shown correspond to NH_3 exposures of 0, 0.5, 1 and 3.5 Langmuir. Semi-empirical quantum-chemical cluster calculations have been performed for the adsorption of NH_3 on ZnO (0001). Bonding is essentially due to the transfer of electrons from the $3a_1$ orbital of NH_3

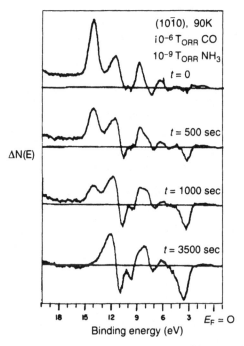

Fig. 6.15 UPS difference spectra showing the competition on the ZnO (10$\bar{1}$0) surface between 10^{-6} Torr of CO and 10^{-9} Torr of NH_3. The length of time of NH_3 exposure is indicated on each curve. [Ref. 193]

to the surface, combined with the electrostatic interaction between the dipole moment of NH_3 and its image charge distribution in the substrate. NH_3 thus acquires a net positive charge upon adsorption.

The interaction of **Cl_2** with the ZnO (0001), (000$\bar{1}$) and (10$\bar{1}$0) surfaces has been studied on polished and annealed samples. Dissociative adsorption was observed on all three faces, as deduced from work-function increases of about 1 eV, although the strength of the interaction varied greatly for the different surfaces. The reaction was strongest for the (0001) surface, where Cl^- ions can presumably bond to the exposed surface Zn ions; the sticking coefficient on that surface was essentially unity. A smaller sticking coefficient was found for the (000$\bar{1}$) surface, and the least reactive surface was the non-polar (10$\bar{1}$0). No UPS measurements were performed, so nothing is known about the details of the bonding to the surface. Semi-empirical quantum-chemical cluster calculations for Cl adsorption on ZnO (0001) and (10$\bar{1}$0) are in agreement with those observations. Cl bonds to surface Zn ions via a highly ionic bond; the computed charge on the adsorbed Cl ion is between -0.72 and -0.78 electron, resulting in an essentially closed-shell configuration.

The interaction of **NO** with polished and annealed ZnO (10$\bar{1}$0) surfaces has been studied by TDS, Auger and UPS. Exposure of the surface at 90 K produces two molecularly adsorbed species, one very weakly bound and the other stable to 420 K. In addition to the molecular adsorption, some N_2O is formed which desorbs at 140 K. Since no O_2 desorption is observed upon N_2O formation, it was hypothesized that the extra O atoms diffuse into the bulk of the crystal.

There has been one study of the physisorption of **Xe** atoms onto ZnO (0001), (000$\bar{1}$) and (10$\bar{1}$0) surfaces.[11] The adsorption of Xe at low temperatures (43 K in this study) has been used to study steps and defects on single-crystal metal surfaces, where the binding energy of the shallow 5p core levels, as measured by photoemission, is found to decrease as the number of metal ligands increases for different adsorption sites because of increased screening of the core hole. The striking changes in 5p binding energy found on metals were *not* observed for Xe adsorption on ZnO. There was little difference between the measured binding energies for the first monolayer of Xe on the three crystal faces, and there was almost no difference between the first and successive monolayers. The largest difference was for adsorption on the (000$\bar{1}$) surface, where all of the nearest-neighbor ligands of the Xe atoms are easily polarizable O^{2-} ions. On the (10$\bar{1}$0) surface, the Xe atoms were found to order one-dimensionally along the troughs in the surface that run parallel to the [$\bar{1}$2$\bar{1}$0] direction [see Fig. 2.15 (a)]. No epitaxial growth was observed on either of the polar faces. The most striking effect of the adsorbed Xe was a redistribution of photoemission intensity in the O 2p valence band. Some of the emission intensity from a region close to the valence-band maximum shifts to higher binding energy by about 1 eV upon Xe adsorption. The extent of the process was dependent on the O content of the top layer of the surface, but no explanation of its origin was given.

The adsorption isotherms for Xe adsorbed onto polished and annealed ZnO (10$\bar{1}$0) have also been measured for temperatures below 130 K.[624] The data were fitted to a Freundlich isotherm, which includes a coverage-dependent isosteric heat of adsorption. The measured values of 22–28 kJ/mol for coverages below 10^{-4} monolayer extrapolated to zero heat of adsorption for coverages close to 0.1 monolayer. For comparable coverages, the adsorption energy for Xe is much lower than that for O_2, CO_2 or CO. This is shown in Figure 6.16, which plots the isosteric heats of adsorption for Xe and several other gases on the stoichiometric ZnO (10$\bar{1}$0) surface.

A great many reports have been published of the interaction of organic molecules on ZnO surfaces, both because Cu/ZnO catalysts are used in methanol synthesis and because ZnO is used in a variety of gas sensing applications. Some of the earliest experiments of chemisorption on well-characterized ZnO surfaces

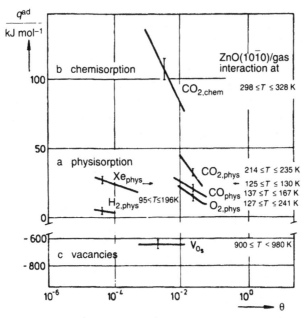

Fig. 6.16 Isosteric heats of adsorption, q^{ad}, as a function of coverage, θ, in monolayers for various gases on ZnO (10$\bar{1}$0). [Ref. 626]

consisted of UPS measurements of hydrocarbon adsorption at 120 and 300 K on stoichiometric ZnO (10$\bar{1}$0). One of the primary goals of that work was to determine whether photoemission difference spectroscopy, which had been used successfully to 'fingerprint' molecules adsorbed on metal surfaces, could be used in the same way on metal oxides. An example of the use of UPS difference spectra for benzene, C_6H_6, adsorption on ZnO (10$\bar{1}$0) was given in Fig. 6.1 and discussed earlier (§ 6.1.1). Another example, which shows how the difference between chemisorbed and physisorbed (condensed) species on ZnO appears in difference spectra, is shown in Fig. 6.17, which compares the gas-phase UPS spectrum of pyridine, C_5H_5N, with the difference spectra for a chemisorbed layer at 300 K and a condensed layer at 120 K. The lowest-binding-energy peaks, which consist of π and N lone-pair orbitals, are shifted about 1 eV to higher binding energy upon chemisorption, indicating that they are the molecular orbitals that are most strongly involved in bonding of pyridine to the ZnO surface. Theoretical cluster calculations predict that pyridine will adsorb at Zn sites via its N lone pair, acting as an electron donor.

Experimental studies of C_2H_4 adsorption on ZnO showed no interaction with the non-polar (10$\bar{1}$0) face and only molecular adsorption on the (0001) face. Three alkynes – **HCCH** (acetylene), **CH$_3$CCH** and **C$_6$H$_5$CCH** – have been studied on several surfaces of ZnO, and the same behavior is observed for all of them; they

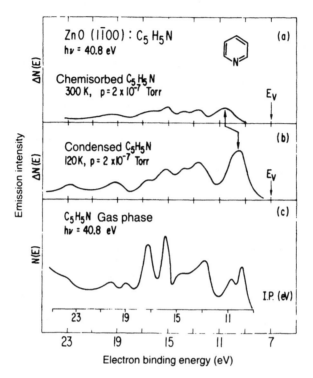

Fig. 6.17 He II UPS spectra for adsorbed and gas-phase C_6H_5N: (a) difference spectrum for chemisorbed C_6H_5N on ZnO (10$\bar{1}$0) at 300 K; (b) difference spectrum for C_6H_5N condensed (physisorbed) at 120 K over the chemisorbed species; and (c) gas-phase UPS spectrum for C_6H_5N. [Ref. 772]

adsorb molecularly on the (10$\bar{1}$0) and (000$\bar{1}$) faces, but dissociate heterolytically to yield adsorbed [CCH]⁻, [CH₃CC]⁻ and [C₆H₅CC]⁻, respectively, on the (0001) face. Apparently the particular type of coordinative unsaturation of the Zn ions on the (0001) face, together with its relaxed geometry, are necessary for alkyne dissociation. Other molecules that adsorb only molecularly on ZnO surfaces include **(CH₃)₂SO, (CH₃)₂CO, C₂H₄O, C₆H₆** and **C₅H₅N**.

Since ZnO is used as a methanol catalyst, there has been a great deal of interest in the adsorption and interaction of alcohols with single-crystal ZnO surfaces. The alcohols that have been studied experimentally on various ZnO faces are **CH₃OH, C₂H₅OH, C₃H₇OH, i-C₃H₇OH, C₆H₅OH** and **C₆H₅CH₂OH**. The interpretations of the results show enough variability that no entirely consistent picture of the nature of the interaction of alcohols with ZnO emerges. This is undoubtedly due to the different sample preparation procedures used by the various groups, and consequently the differences in surface defect type and density. However, some trends are apparent. The (000$\bar{1}$) face appears to be the least reactive, and the reactions that

that have been reported consist primarily of oxidation involving abstraction of a lattice O^{2-} ion. This type of reaction would be expected to occur most readily on the O face due to the availability of coordinatively unsaturated O ions. Due to the propensity for step formation on ZnO $(000\bar{1})$, chemisorption probably occurs preferentially at surface defect sites.

The most reactive surface is the (0001), where the surface Zn ions have been found to catalyze both heterolytic dissociation and dehydrogenation of alcohols. Again a wide range of reactions has been reported, suggesting the importance of defects on this surface also, but the Zn polar surface is clearly more reactive than the O face. [In one TDS study, decomposition of CH_3OH on the (0001) surface was also accompanied by desorption of atomic Zn, presumably resulting from reduction of the ZnO surface during the decomposition reaction.] The non-polar $(10\bar{1}0)$ face appears to be intermediate in activity, and experiments on stepped $(40\bar{4}1)$ and $(50\bar{5}1)$ surfaces indicates that steps play an important role in chemisorption.

Both polar and non-polar ZnO surfaces are fairly reactive towards the aldehydes **HCHO, CH_3CHO, C_2H_5CHO, C_6H_5CHO** and **CH_2=CHCHO**. As for alcohols, the $(000\bar{1})$ and $(10\bar{1}0)$ faces are less reactive than the (0001) surface, but both oxidation and dehydrogenation reactions have been reported on all surfaces. The adsorbed species has usually been identified as a carboxylate, $RCOO^-$. In one TDS study of the interaction of HCHO with ZnO (0001), Zn metal desorption peaks, resulting from reduction of the surface, were also observed.

The only ketone whose interaction with ZnO has been studied is acetone, **$(CH_3)_2CO$**. All crystal faces are less reactive than for alcohols or aldehydes, with the $(000\bar{1})$ and $(10\bar{1}0)$ faces exhibiting only molecular adsorption. On the (0001) face acetone dissociated via deprotonation, leaving enolate, $[CH_3COCH_2]^-$, on the surface. The carboxylic acids **HCOOH, CH_3COOH, C_2H_5COOH** and **C_6H_5COOH** react much more strongly with the (0001) face than with the $(000\bar{1})$ face; the latter exhibits primarily molecular adsorption. Dissociation on the Zn face is heterolytic via deprotonation in all cases.

Theoretical cluster calculations have been performed for the adsorption of **CH_3, CCH, HCOO** and **CH_3O** on the ZnO (0001) surface. All species are found to bond to surface Zn ions and to be electron acceptors, thus resulting in negatively charged adsorbed ions.

6.3.5 SnO$_2$

The other non-transition-metal oxide that is used in gas sensing applications is SnO_2. While much of the adsorption work on SnO_2 has been carried out on powders or thin films, some experiments have been performed on well-characterized single crystals. The vast majority of that work has been on the (110) surface, although some studies of hydrocarbon adsorption have been performed on the (101) surface.

The surface composition of SnO_2 (110), which was discussed in § 2.3.3.1 and 4.6, can easily be changed from stoichiometric, in which one-half of the surface cations are five-fold coordinated with O ions and the other half are six-fold coordinated because of the presence of rows of bridging O ions lying above the main surface plane, to reduced, in which all of the bridging O ions, and perhaps even some of the in-plane O ions, have been removed. As shown in Figs. 4.28 and 4.29, removal of the bridging O ions to form the compact surface does not appreciably increase the surface conductivity; this is due to the localized nature of the Sn^{2+} ions on that surface. However, removal of in-plane surface O ions produces a finite density-of-states at E_F and an n-type accumulation layer in the near-surface region; E_F becomes pinned near the bottom of the bulk conduction band, with the strength of the pinning depending upon the degree of reduction of the surface. (The presence of an accumulation layer does not necessarily lead to higher surface conductivity, however, if the surface is disordered. The electron mobility on disordered surfaces can be quite a bit smaller than that on well-ordered surfaces, thereby resulting in a low value of the surface conductance. This can be seen in Fig. 6.18 below.) On the stoichiometric surface, on the other hand, the bands are generally flat, with the position of the Fermi level determined by the bulk defect density (SnO_2 is always n-type); E_F is thus not pinned at the surface. This difference in surface electronic structure leads to different response of the surfaces to adsorbed molecules. A relatively small density of adsorbed molecules that act as electron donors (acceptors) can easily bend the bands down (up) on stoichiometric, or nearly stoichiometric, surfaces, resulting in an increase (decrease) in the surface conductivity. Providing that the bulk conductivity of the sample is not so large that it shorts out the surface conductivity, the sample resistance is then a measure of the amount and type of gas adsorbed. On heavily reduced surfaces, however, E_F is too firmly pinned to the conduction band for adsorbates to cause a large change in the surface conductivity.

The exposure of SnO_2 (110) surfaces to **H_2** does result in changes in surface conductivity. H_2 probably dissociates at surface defects, with the creation of H^+ ions (which bond to lattice O^{2-} ions as OH^-) and electrons, which bends the bands down at the surface, increasing the surface carrier concentration. On well-ordered surfaces where the electron mobility is high, this results in an increase in surface conductivity. On disordered surfaces, large downward band-bending is observed, but only small conductivity changes are produced because of the low mobility of electrons in the accumulation layer. The effects of H_2 adsorption on SnO_2 (110) can also be seen in changes in the UPS spectra in the region of the bulk bandgap. H_2 adsorption results in a small overall reduction in the bandgap emission,

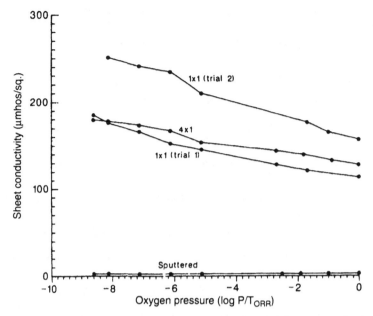

Fig. 6.18 Changes in surface conductivity versus O_2 partial pressure for ion-bombarded ('sputtered') and stoichiometric SnO_2 (110) surfaces. [Ref. 142]

although it produces an increase in emission just below the bottom of the bulk conduction band; this emission is associated with the downward band-bending produced by the extra electrons.

O_2 does not interact with nearly perfect SnO_2, although it does chemisorb on reduced SnO_2 surfaces, with its effect on surface conductivity depending upon the degree of reduction of the surface. Since O_2 is an electron acceptor, it creates a depletion layer and reduces the surface conductivity on nearly stoichiometric surfaces. On sputtered, heavily reduced surfaces, however, exposure to the same amount of O_2 does not change the conductivity. This is shown graphically in Fig. 6.18, which plots the surface conductivity of three different SnO_2 (110) surface structures as a function of O_2 partial pressure. (Note the low value of the conductivity for the disordered, sputtered surface.) The way in which O_2 interacts with the SnO_2 (110) surface on the atomic scale can be better understood by considering the changes that occur in the UPS spectra in the region of the bulk bandgap.[145] On reduced surfaces that exhibit emission from defect surface states in the bandgap, adsorption of O_2 at room temperature results in only a slight reduction in the bandgap emission, leaving most of the defects unaffected. However, exposure of the defective SnO_2 (110) surface to O_2 at higher temperatures (700 K) results in almost complete re-oxidation of the surface, showing that the healing of surface defects is a thermally activated process.

On SnO_2 (110), **H_2O** acts as an electron donor, bending the bands down at the surface and increasing the surface carrier concentration. As for H_2 adsorption on the same surfaces, the increase in surface conductivity caused by H_2O is greater on slightly reduced surfaces than on heavily reduced, disordered ones. In UPS spectra, H_2O adsorption is seen to slightly reduce the emission in the lower part of the bandgap while increasing it just below the bottom of the bulk conduction band as a result of the downward band-bending produced by the added electrons.

The interest in SnO_2 as a gas sensor includes its response to organic molecules. **CH_4** reacts weakly, producing some increase in surface conductivity. The nature of the reaction is not understood, but it most likely involves deprotonation with the involvement of lattice O^{2-} ions. Similar interactions are found for **C_2H_5OH** and **CH_3CHO** on the (110) surface.

As an example of the way in which organic molecules interact with SnO_2, consider acetic acid, **CH_3COOH**, and its effect on the surface conductivity of SnO_2 (101). There are several possible routes for acetic acid decomposition on SnO_2, and a variety of species is observed in thermal desorption spectra after exposure of the surface to CH_3COOH. So when changes in surface conductivity are found as a function of CH_3COOH exposure, it is not clear just what species are responsible. However, some information can be obtained by monitoring the surface conductivity of SnO_2 (101) during cyclic CH_3COOH exposure at different temperatures. Figure 6.19 shows the results of those measurements for surface temperatures of 473 and 673 K. In both cases the conductivity increases during CH_3COOH exposure, but at the lower temperature the changes are reversible, while at the higher temperature an irreversible increase is also present. Based upon the likely mechanisms of decomposition on the surface, it was proposed that the reversible component of the increase in conductivity is due to the presence of adsorbed H, which acts as an electron donor; the irreversible increase is believed to arise from the formation of O-vacancy surface defects, which also increase the conductivity of the surface but are not removed when the flow of CH_3COOH is turned off.

6.4 Adsorption on d^0 transition-metal oxides

Transition-metal oxides exhibit a wider range of adsorption behavior than do non-transition-metal oxides since the transition-metal cations can in general exist in more than one valence state. The energy necessary to change the valence state of the cations is fairly small, so electrons can be added to, or removed from, cation d orbitals during chemisorption. However, in transition-metal oxides whose d-electron orbitals are normally empty, which we refer to as d^0 oxides, there are no cation d electrons available for transfer to adsorbates. Stoichiometric surfaces of d^0 oxides are generally less active for chemisorption than are those of oxides whose d-orbitals

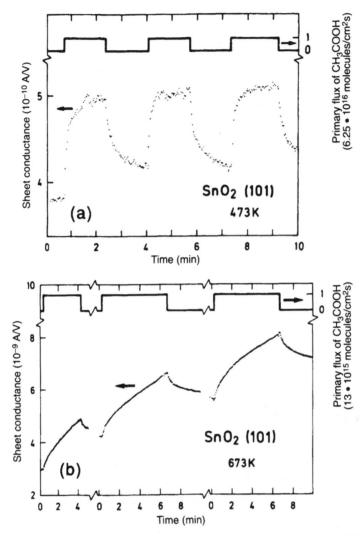

Fig. 6.19 Four-point probe sheet conductance versus time during exposure of SnO_2 (101) to a modulated beam of CH_3COOH at (a) 473 K and (b) 673 K. [Ref. 777]

are partially occupied on the perfect surface. When O-vacancy defects are present on the surfaces of d^0 oxides, however, the d-electron orbitals on adjacent cations can be partially occupied. These reduced surface cations provide the active sites for much of the chemisorption and catalysis that takes place on d^0 transition-metal oxides.

6.4.1 TiO_2

By far the most thoroughly studied d^0 transition-metal oxide is TiO_2. Interest in its chemisorption properties began in earnest in 1972, after the report by Fujishima and Honda[1] that water could be decomposed into H_2 and O_2 in an electrolysis cell having TiO_2 as one of the electrodes, and where energy was supplied by light shining on the TiO_2 electrode instead of from a battery. Soon after that report, it was discovered that $SrTiO_3$ was even more efficient in the photoelectrolysis process than TiO_2, and research on the surface properties of the two materials was conducted in parallel. (Although the properties of TiO_2 and $SrTiO_3$ are quite similar, there are enough differences that they will be discussed separately here.) Adsorption studies on single-crystal samples soon determined that surface defects were a necessary component of the H_2O decomposition reaction, so a great deal of effort has been directed toward understanding the role that they play in chemisorption. (All of the work on single-crystal TiO_2 reported to date has been on rutile, although TiO_2 also exists in the anatase and brookite structures.) The stoichiometric TiO_2 (110) surface, which is the most stable of the low-index faces, is inert to room-temperature chemisorption of any molecule that has been studied to date. However, molecules can be physisorbed at low temperatures, and upon subsequent heating they may dissociate, yielding an adsorbed species that is stable at or above room temperature.

The interaction of **H_2** with TiO_2 is relatively weak and not very well understood. H_2 only chemisorbs at defect sites, and even then the sticking coefficient is only about 10^{-6}. Measurements on the (100)-(1 × 3) surface and the ion-bombarded (100) surface suggested that adsorbed hydrogen (whether it adsorbed as H or H_2 was not determined) behaved as an electron donor on the (1 × 3) surface, but as an acceptor on ion-bombarded surfaces that contained a high density of reduced Ti cations. Measurements on TiO_2 (110) surfaces having different densities of O-vacancy defects were interpreted in terms of dissociative adsorption at defect sites, with the H bonding to Ti cations adjacent to the defects via a hydride bond, Ti^{4+}–H^-; the geometry of that adsorption is shown in Fig. 6.20, which also depicts the adsorption of O_2, CO and CO_2. However, the adsorption of H_2 resulted in an *increase* in surface conductivity on reduced TiO_2 surfaces, not the behavior that would be expected of an acceptor adsorbate. The explanation for this involves diffusion of H atoms into the bulk, producing donor-type bulk defects in which the H atoms bridged two bulk O atoms.[633] [It should be mentioned that O–H stretching vibrations have also been observed on TiO_2 (110) surfaces after exposure to H_2.[334]] In summary, the interaction of H_2 with TiO_2 surfaces is complicated at best, and not enough information is currently available to paint a clear picture of the process.

The generalized way in which H_2 molecules would interact with both perfect and defect TiO_2 (110) surfaces has been considered theoretically in terms of orbital symmetries and charge transfer reactions. H_2 is predicted to dissociate at surface O-vacancy defects, with the atomic H then bonding to lattice O^{2-} ions to

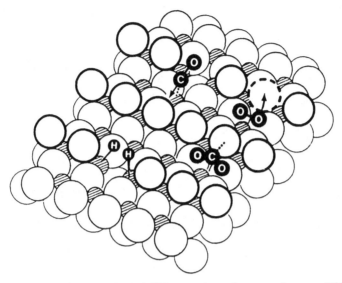

Fig. 6.20 Geometric characterization of different adsorption complexes on TiO_2 (110) containing O-vacancy point defects. Shaded circles are titanium ions; open circles are oxygen ions. [Ref. 626]

form OH^-. This process is predicted to have an activation energy of 1.1 eV in the gas phase.

The interaction of O_2 with TiO_2 (110) and (100) surfaces that were cleaved in UHV, annealed in either UHV or O_2, ion bombarded, or heated and quenched has been studied by several groups. The evidence suggests that O_2 adsorbs only at point defects on TiO_2, and not on either terraces or steps. The argument against adsorption at step sites is indirect, since regular step arrays that could be quantitatively characterized by LEED have not been prepared on TiO_2. However, LEED patterns for UHV-cleaved (110) surfaces exhibit somewhat diffuse spots indicative of coherence areas of such size that perhaps as many as 10 % of the surface ions reside at steps. Little change is seen in UPS spectra upon exposure of those surfaces to as much as 10^8 L O_2;[813] if O_2 did chemisorb at steps, the changes produced in the spectra would be well below the detection limit of the technique.

There is a strong interaction at room temperature between O_2 and point defects (which consist predominantly of O vacancies) on TiO_2. The changes that occur in the He I UPS spectra for ion-bombarded TiO_2 (110) as a function of O_2 exposure are shown in Fig. 6.21. The most significant changes in the spectra are depopulation of the bandgap defect surface states and an increase in the amplitude of the bonding O 2p valence-band orbital relative to the non-bonding component. The valence band also moves toward E_F by almost 1 eV, although that is not apparent

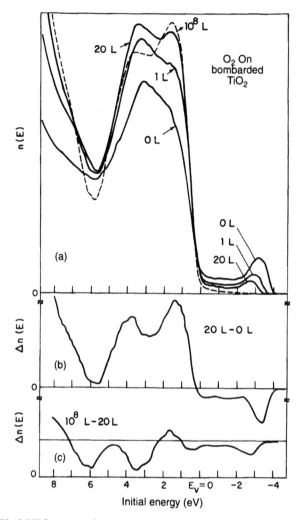

Fig. 6.21 (a) He I UPS spectra for Ar^+-ion-bombarded TiO_2 (110) after various exposures to O_2; (b) UPS difference spectrum for the first adsorbed phase (20 L – 0 L); and (c) UPS difference spectrum for (10^8 L – 20 L). [Ref. 649]

in Fig. 6.21 since all spectra are aligned to the upper edge of the valence band in order that meaningful differences between them may be taken. The valence-band motion can be seen in Fig. 5.14, where the open points correspond to exposure of the ion-bombarded surface to O_2. Also shown in Fig. 5.14 is the observed decrease in the amplitude of the 2 eV ELS peak that corresponds to a bonding-to-antibonding transition on reduced surface cations.

The bottom two panels in Fig. 6.21 show UPS difference spectra for two different O_2 exposure ranges. Figure 6.21 (b) is the difference between the spectra of the ion-bombarded surface before and after exposure to 20 L O_2. The negative region of the spectrum in the bulk bandgap corresponds to depopulation of the 'Ti³⁺' defect state. The region of the difference spectrum where additional intensity appears exactly overlaps the O 2p valence band; the shape of the difference spectrum is also very similar to that of the bulk O 2p band. It is therefore believed that the initial stage of O_2 chemisorption on reduced TiO_2 is dissociative, and that the adsorbed species is O^{2-}, where charge is transferred to the adsorbed O from the surface Ti³⁺ ions. What happens for larger O_2 exposures is not clear. Figure 6.21 (c) shows the difference spectra taken between exposures of 20 and 10^8 L. Some additional depopulation of the bandgap defect states is apparent, but the remainder of the difference spectrum has not been uniquely interpreted. Possibilities for the adsorbed species of O_2^- and O_2^{2-}–have been suggested, but no clear evidence for either species has been presented.

The adsorption of O_2 onto reduced TiO_2 surfaces also lowers the surface conductivity as electrons are transferred from the defect sites to the adsorbed species. Thus TiO_2 is of some interest as a potential gas sensor material. The changes that occur in both the surface conductivity and the work function of reduced TiO_2 (110) upon exposure to O_2 at various temperatures are shown in Fig. 6.22. In those measurements the sample remained in UHV until $t = 2000$ sec, at which time the O_2 partial pressure was increased to the indicated value; the O_2 was then pumped away at $t = 5000$ sec. Due to the O_2 pressure used, the curves correspond to the high-exposure region in Fig. 6.21, and the adsorbed species were assumed to be O_2^-. For the higher temperatures used, the response of the sample involves not only chemisorption but also surface and bulk reaction steps involving annealing of intrinsic defects during O_2 exposure.

The generalized way in which O_2 molecules would interact with both perfect and defect TiO_2 (110) surfaces has been considered theoretically in the same manner as discussed for H_2 above. O_2 is predicted to adsorb molecularly on perfect surfaces (although the strength of the bond was not calculated) and to dissociate at O-vacancy defect sites.

The importance of **H_2O** adsorption in the context of photoelectrolysis on TiO_2 was mentioned above. Even before work on single crystals was begun, the importance of H_2O in determining the surface properties of TiO_2 was well understood.[814] TiO_2 powders that have been exposed to air are always hydroxylated (i.e., covered with a monolayer of OH⁻) unless they have subsequently been heated to at least 775 K in vacuum. Models had been proposed for the adsorption sites of the OH⁻; for example, on the (110) surface it was postulated that H_2O dissociated into OH⁻ ions that adsorbed via their O end on top of five-fold coordinated surface Ti ions, with the other H⁺ ion bonding to a lattice O ion to form a

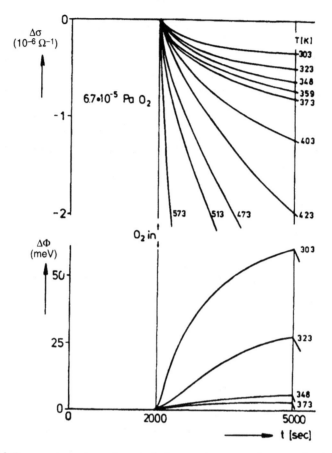

Fig. 6.22 Temperature dependence of changes in the surface conductivity, $\Delta\sigma$, and work function, $\Delta\Phi$, during O_2 exposure to TiO_2 (110). [Ref. 626]

second type of adsorbed OH^- species. The presence of at least two types of adsorbed OH^- was confirmed by infrared spectroscopy.

The experimental work on H_2O adsorption on single-crystal TiO_2 has not resulted in unanimous agreement on the details of adsorption. Several experiments have been performed on TiO_2 surfaces that were prepared so as to be as nearly stoichiometric and perfect as possible, and the conclusions drawn from room-temperature adsorption measurements on those surfaces include:

(i)   the surface is completely inert, chemisorbing neither OH^- nor H_2O;
(ii) H_2O adsorbs only dissociatively, with the surface saturating at 0.1 monolayer of adsorbed OH^-;
(iii) H_2O dissociates for low exposures, but for higher exposures molecular H_2O adsorbs in addition to the OH^-; and
(iv) H_2O adsorbs only molecularly.

Those scenarios cover the entire spectrum of possibilities. One of the reasons that different conclusions were drawn was that most of the experiments involved the use of UPS, and in particular difference spectroscopy, to identify the adsorbed species, and various groups interpreted difference spectra in different ways. In early work, molecular H_2O adsorption was characterized by a single peak occurring below the bottom of the O 2p valence band in the difference spectra and two peaks overlapping the valence band;[20,23] adsorbed OH^- was identified from two peaks, both occurring in the region of the valence band. More recent work on H_2O molecularly adsorbed on both TiO_2 and $SrTiO_3$ surfaces at 130–150 K shows that molecular H_2O gives rise to *two* peaks below the valence band in difference spectra and one peak overlapping the band; see Fig. 6.25 below.[471,487,686] Thus molecular H_2O adsorption should really be characterized by two peaks below the O 2p band rather than a single peak.

Another problem that arises in comparing various studies of 'stoichiometric, nearly perfect' surfaces is that each surface has in reality a different density of step and point defects, regardless of how carefully it was prepared. Since TiO_2 fractures rather than cleaves, even UHV-fractured surfaces (which exhibit a weak interaction with H_2O [813]) have a high density of steps. Attempts to improve the surface quality involve ion bombardment and annealing, which may result in fewer steps (as determined by the improvement in the quality of LEED patterns; see Fig. 2.8), but which probably leave point defects on the surface. Our own appraisal of the situation is that none of the single-crystal TiO_2 surfaces was sufficiently defect free for intrinsic adsorption on perfect terraces to have been measured. Our *guess* is that a truly perfect TiO_2 (110) or (100) surface would be inert to H_2O at room temperature, and that the wide range of results reported is due entirely to either extended or point defects on the surface. [TiO_2 (001) is excluded since we do not have a good feeling for how the four-fold coordinated surface Ti ions on that surface would behave.]

The situation is only slightly better for the case of defective TiO_2 surfaces. However, the weight of the evidence says that H_2O dissociates at room temperature on defective, reduced TiO_2 surfaces. Figure 6.23 shows He II UPS spectra for Ar^+-ion-bombarded TiO_2 (110) before and after H_2O exposure at room temperature. A single peak is seen below the O 2p valence band after exposure, which, as discussed above, is characteristic of dissociative adsorption. Results of a similar experiment by another group are presented in Fig. 6.24; here point defects were created by annealing the sample to 1000 K in UHV and quenching to room temperature. (Note that the two spectra are plotted as mirror images of each other.) In Fig. 6.24 (a) it is clear that only a single peak occurs below the valence band after H_2O exposure, characteristic of dissociation. Figure 6.24 (b) shows difference spectra taken from the data in (a). In addition to the peak at 10.8 eV, a feature is

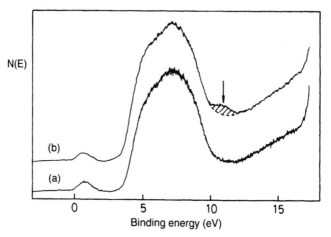

N(E)

(b)

(a)

0   5   10   15

Binding energy (eV)

Fig. 6.23 He II UPS spectra of Ar$^+$-ion-bombarded TiO$_2$ (110): (a) clean surface; (b) after exposure to 6×10^4 L H$_2$O. [Ref. 487]

also found in the region of the valence band. It is, however, difficult to accurately interpret small differences on top of the large O 2p band.

In all of the UPS experiments reported for the interaction of H$_2$O with defective TiO$_2$ surfaces, no changes occur in the bandgap emission from surface defects upon exposure; compare this with the depopulation of the bandgap surface states that occurs on exposure to O$_2$ (Fig. 6.21) or SO$_2$ (Fig. 6.26 below). This observation confirms that the interaction of H$_2$O with the surfaces is of the donor/acceptor type, with no appreciable transfer of charge from reduced surface cations to the adsorbed species.

Although it is generally agreed that H$_2$O dissociates on defective surfaces to give adsorbed OH$^-$ ions and that there is no appreciable interaction between the adsorbed species and the Ti 3d electrons on reduced surfaces, it is still not clear just what features of reduced TiO$_2$ surfaces are responsible for dissociating H$_2$O. The lack of interaction between the adsorbate and Ti 3d electrons at defect sites suggests that the presence of occupied Ti 3d orbitals is not *in itself* sufficient for dissociative adsorption of H$_2$O; this conclusion is supported by the results of H$_2$O adsorption experiments on UHV-cleaved Ti$_2$O$_3$ (10$\bar{1}$2) surfaces, which will be discussed in § 6.5.1 below. We believe that somehow it is the *geometry* more than the electronic structure of O-vacancy point defects that is necessary for dissociation of H$_2$O. The role of steps on stoichiometric surfaces is not clear, but on TiO$_2$ (110) they do not appear to interact as strongly with H$_2$O as do point defects, so the presence of reduced Ti cations probably also plays some role in the dissociation of H$_2$O.

Of course, H$_2$O can always be adsorbed on TiO$_2$ at low temperatures, and disso-

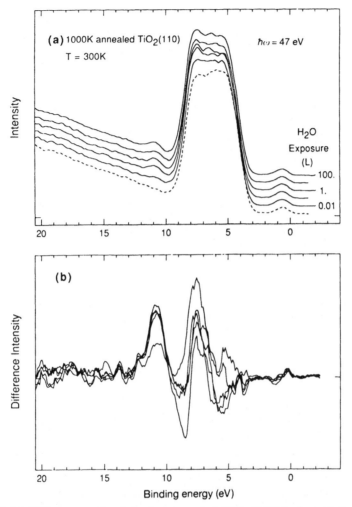

Fig. 6.24 (a) UPS spectra for defective annealed TiO_2 (110) before (dashed curve) and after various H_2O exposures. $\hbar\omega = 47$ eV was used to resonantly enhance the Ti^{3+} emission. (b) Difference spectra obtained from the spectra in (a). [Ref. 469]

ciation may occur when the adsorbed H_2O is heated to room temperature and above. Figure 6.25 shows a series of UPS difference spectra for multilayers of H_2O adsorbed at 130 K onto a well-ordered but slightly reduced TiO_2 (100)-(1 × 3) surface and then heated to the temperatures shown. Below 150 K the $1b_2$, $3a_1$ and $1b_1$ molecular orbitals of H_2O are the only features present. Between 150 and 250 K the spectra change as shown, resulting in a different spectrum which is stable from 250 to 450 K; it has been interpreted as arising from adsorbed OH^-. [No changes in the UPS spectra were observed when the clean (1 × 3) surface was exposed to H_2O at room temperature.] This study also included measurements on

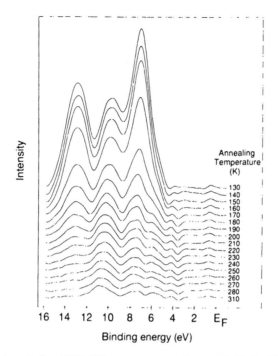

Fig. 6.25 Normal-emission UPS difference spectra ($\hbar\omega = 47$ eV) for TiO$_2$ (110) exposed to H$_2$O at 130 K and then annealed at the temperatures indicated. [Ref. 471]

vicinal (100) surfaces containing steps, and it was concluded that steps did not play any direct role in the adsorption of H$_2$O.

The interaction of H$_2$O and OH$^-$ with TiO$_2$ surfaces has been considered theoretically by means of Xα cluster calculations. Such calculations can determine the hybridization between the molecular orbitals of the adsorbed species and the surface, and the energies at which the molecular orbitals would lie for a given cluster geometry. For the case of OH$^-$ adsorbed over four-fold coordinated Ti ions on TiO$_2$ (110) surfaces from which all of the bridging O ions had been removed, the distance of the adsorbate from the surface was determined by comparing the binding energies of the 3σ and 1π orbitals measured by UPS for OH$^-$ on an ion-bombarded surface with those calculated for different assumed distances. The O end of the OH$^-$ ion was determined to be 2.51 Å above the Ti ion, although the surface on which the experiments were performed contained many more types of defects than the simple one that was calculated. LCAO methods have also been applied to the interaction of OH$^-$ with the surface of TiO$_2$.

The adsorption of **CO** on TiO$_2$ surfaces only occurs in the presence of O-vacancy point defects. At temperatures below about 330 K, exposure of TiO$_2$

(110) surfaces to CO results in a slight decrease of the surface conductivity, while exposure at higher temperatures produces an increase in the conductivity. Thermal desorption measurements on CO-exposed TiO_2 surfaces show small amounts of CO_2 desorbing, which leads to the model of CO chemisorption shown in Fig. 6.20. It is proposed that CO adsorbs only at O-vacancy sites, but that the C end of the molecule bonds to an adjacent O ion. This surface complex acts as a weak electron acceptor, which lowers the surface conductivity. No measurable change is seen in the sample work function during the formation of this complex, however, so the amount of charge transferred is small. At higher temperatures, the CO_2 formed upon CO adsorption desorbs, creating a larger number of surface O vacancies; it is these vacancies that increase the surface conductivity.

Ab initio molecular-orbital cluster calculations have been performed for CO adsorption on both perfect and reduced TiO_2 (110) surfaces. CO would not adsorb at any surface O sites, which was interpreted to preclude the formation of CO_2 or CO_3^{2-} complexes on the surface. However, adsorption of CO, via the C atom, to Ti ions at a bridging O-ion vacancy site was stable, with a calculated binding energy of 125–150 kJ/mol. CO was also found to bond to five-fold coordinated Ti ions on the perfect (110) surface, with a binding energy of 71 kJ/mol, as long as the O end of the molecule pointed away from the surface and there was no steric repulsion by surface O ions. No calculation was performed for the adsorption site shown in Fig. 6.20, which was assumed to be the preferred site in the experimental measurements of CO adsorption on defective TiO_2 surfaces.

The adsorption of CO_2 on TiO_2 (110) was found not to depend upon the density of surface O-vacancy defects, nor does its presence interfere with reactions between O_2 and surface defects. (This behavior is very different than that for CO_2 adsorption on ZnO, which was discussed in § 6.3.4.) It is therefore assumed that CO_2 bonds to surface O ions, as shown schematically in Fig. 6.20. No changes in either the surface conductivity or the work function are observed upon adsorption of CO_2.

One of the most striking examples of the role played by surface O-vacancy defects in chemisorption on TiO_2 is that of SO_2. Stoichiometric TiO_2 (110) surfaces are virtually inert to SO_2 at room temperature for exposures as high as 10^6 L. The reduced TiO_2 (110) surface, on the other hand, reacts very strongly with SO_2, as shown in the UPS spectra of Figure 6.26. The dominant changes in the spectra are the appearance of intensity in the 1.5–4 eV region, which corresponds to emission from the S 3p orbital, and depopulation of the bandgap surface states. The rate of oxidation of the reduced surface cations by SO_2 is even larger than that for exposure to O_2. In fact, the adsorption of SO_2 only occurs to a significant extent if there are reduced Ti cations available. This can be seen more clearly in the series

SO₂ on Defect TiO₂(110)

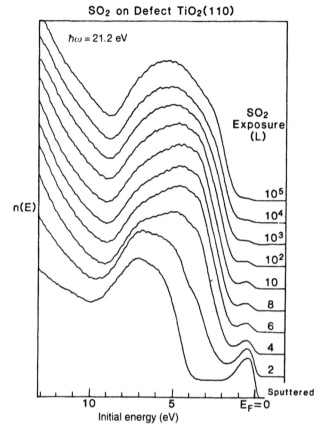

Fig. 6.26 He I UPS spectra for defect TiO₂ (110) exposed to SO₂. [Ref. 484]

of sequential UPS difference spectra in Fig. 6.27. The majority of the adsorption is complete by 10 L (note that the exposures used increase exponentially above 10 L), at which point most of the electrons have been removed from the O-vacancy defects.

The nature of the SO₂/TiO₂ reaction is apparent from the measured binding energy of the S 2p core level in XPS. Studies on a variety of sulfur compounds have shown that the S 2p core level in solids has one of two binding energies, depending upon its ligand environment. When S is bonded to O atoms, its binding energy is about 168 eV, while the S–metal sulfide bond gives a binding energy of about 162 eV. The measured binding energy of the S 2p level for SO₂ on TiO₂ is 163 eV, indicating complete dissociation of the SO₂ and the formation of a Ti sulfide on the surface. This is consistent with the observed strength of the SO₂/Ti interaction at surface defect sites.

A somewhat different SO₂/TiO₂ interaction occurs when SO₂ is physisorbed at low temperatures and then heated. Figure 6.28 shows a series of S K-edge NEX-AFS spectra for adsorption on stoichiometric polished and annealed TiO₂ (110) at

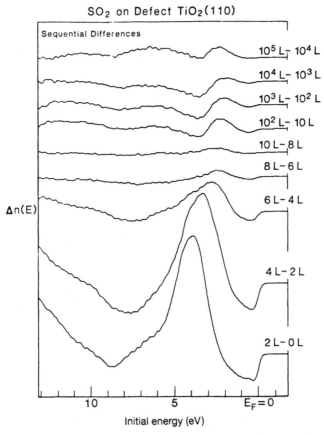

Fig. 6.27 Sequential UPS difference spectra for the data in Fig. 6.26. [Ref. 484]

100 K, followed by annealing at the temperatures shown. The spectrum for condensed SO_2 at 100 K is identical to that for gas-phase SO_2. The spectra after high-temperature annealing have been tentatively interpreted in terms of adsorbed SO_4^{2-}, although the source of the additional oxygen necessary for oxidation of SO_2 is not clear. However, there is clearly a stronger SO_2/surface interaction at room temperature when the SO_2 had been previously adsorbed at low temperature. [The lower panel compares the NEXAFS spectra for reduced TiO_2 (110) exposed to SO_2 at room temperature with that for bulk TiS_2, confirming the interpretation for that system given above.]

There has been one study of SO_2 chemisorption on a stepped TiO_2 (441) surface, and the interpretation of the UPS and XPS spectra is significantly different than that discussed above. S 2p binding energies of both 162.1 and 167.2 eV were observed, indicating both S–cation and S–O bonding. It was concluded that the adsorbed species on (110) terraces was SO_3^{2-}, with S^{2-} adsorbed at step sites. This

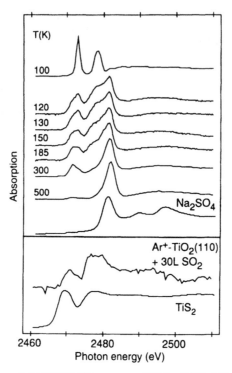

Fig. 6.28 S KLL Auger-yield NEXAFS spectra of TiO$_2$ (110) following SO$_2$ adsorption at 100 K and annealing at the indicated temperatures. Also shown are a spectrum recorded following 300 K adsorption of SO$_2$ on the Ar$^+$-ion-bombarded surface, and fluorescence-yield NEXAFS spectra of TiS$_2$ and Na$_2$SO$_4$. [Ref. 692]

is not consistent with the previous measurements on stoichiometric and reduced TiO$_2$ (110) surfaces, and the origin of the discrepancy is not known.

There has been one brief study of the interaction of **H$_2$S** with defective TiO$_2$ (110) surfaces at room temperature. The interaction is quite weak, as can be seen from the UPS spectra shown in Fig. 6.29. Note the very small interaction between H$_2$S and the 3d electrons on the reduced Ti ions at defects, which is similar to the case of H$_2$O adsorption. Two distinct adsorbed phases are observed: for exposures $\leq$ 1 L, it is proposed that H$_2$S dissociates, with both the H and S atoms bound to surface Ti ions; for exposures > 1 L, molecular H$_2$S is thought to adsorb, although no unambiguous fingerprint was obtained from UPS difference spectra.

Although the H$_2$S/TiO$_2$ interaction is weak, what little does adsorb has a profound inhibiting effect on subsequent O$_2$ adsorption. Figure 6.30 shows UPS spectra for the clean, reduced TiO$_2$ (110) surface before and after exposure to 10^3 L H$_2$S, and the H$_2$S-exposed surface after further exposure to 10^2 L O$_2$. There is virtually no interaction of O$_2$ with the H$_2$S exposed surface, even though a large

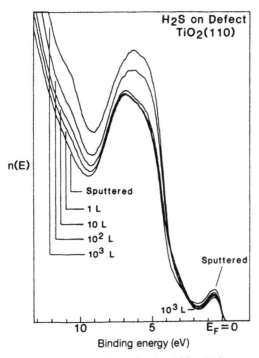

Fig. 6.29 Series of He I UPS spectra for defective TiO_2 (110) exposed to H_2S. [Ref. 486]

amount of charge is still available in the Ti 3d defect states near E_F. This inhibition is presumed to result from site blocking, although further work would be necessary in order to confirm that interpretation.

Polished and annealed TiO_2 (001) surfaces have been exposed to **NH_3**. That surface is severely faceted, and the surfaces used also contained residual surface Ti^{3+} ions as determined by UPS. At 300 K the adsorption was found to be molecular, while at 340 K and above the NH_3 at least partially dissociated. The surface species present at higher temperatures were interpreted as NH_2 and OH^-.

A few experimental studies have considered the interaction of organic molecules with single-crystal TiO_2 surfaces. The alcohols that have been investigated include **CH_3OH, C_2H_5OH, C_3H_7OH** and **i-C_3H_7OH**. Both the (110) surface and the (441) surface, which is stepped and contains (110) terraces, exhibited only molecular adsorption of CH_3OH at room temperature. UPS spectra show bonding-energy shifts, ΔE_B, of about 0.5 eV for the 2a" and 7a' molecular orbitals of CH_3OH relative to the molecular orbitals that were not involved in bonding to the surface. Somewhat different results were found for CH_3OH adsorption on ion-bombarded TiO_2 (001) surfaces and on annealed TiO_2 (001) surfaces that contained either predominantly (011) facets, which contain only five-fold coordinated

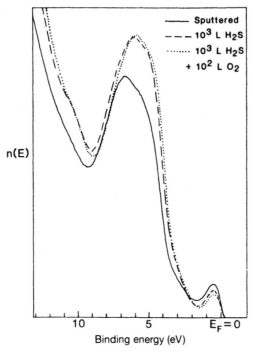

Fig. 6.30 He I UPS spectra for defective TiO_2 (110) exposed to 10^3 L H_2S and subsequently exposed to O_2. [Ref. 486]

Ti ions, or (114) facets, which are predicted to contain four-, five- and six-fold coordinated cations. At 200 K, CH_3OH adsorbed both molecularly and dissociatively, while at 300 K it only adsorbed dissociatively; the adsorbed species after dissociation was determined to be methoxide. In thermal desorption experiments, the three different surfaces were found to have different selectivities for the formation of dimethyl ether, formaldehyde and methane. The adsorption of the higher alcohols has only been studied on (001) surfaces that were prepared so as to contain predominantly (011) facets. In all cases the alcohol dissociated heterolytically via deprotonation at room temperature; at 200K both dissociative and molecular adsorption were observed.

Reduced TiO_2 (001) surfaces have been found to catalyze the coupling of aldehydes to form alkenes. **CH_3CHO, $CH_2=CHCHO$** and **C_6H_5CHO** were all found to couple on surfaces that had been ion bombarded and not annealed. When the TiO_2 (001) surface was annealed at temperatures where (011) facets formed, thus reducing the number of three- and four-fold coordinated Ti ions on the surface, the reactions no longer occurred.

Several carboxylic acids have been adsorbed onto single-crystal surfaces of TiO_2. Formic acid, **HCOOH**, is found to adsorb molecularly on (110) and (441) surfaces at room temperature, with UPS spectra indicating a bonding shift of the 10a and 2a" orbitals relative to the other molecular orbitals of HCOOH. It adsorbed in a regular array on TiO_2 (110), giving rise to a (2 × 1) LEED pattern. In contrast, formic acid dissociated heterolytically at room temperature on (001) surfaces that contained both (011) and (114) facets, yielding adsorbed formate. Acetic acid, **CH₃COOH**, behaved somewhat differently on (001) surfaces that contained predominantly (011) facets than it did on (001) surfaces consisting mainly of (114) facets. On both surfaces deprotonation occurred at room temperature to form adsorbed acetate, CH_3COO^-. However, upon heating to about 600 K, the (011) faceted surface desorbed acetic acid, CO and ketene ($H_2C=CO$), while the (114) faceted surface also desorbed acetone (CH_3COCH_3). Both **C₂H₅COOH** and **CH₂=CHCOOH** dissociate on faceted (001) surfaces via deprotonation.

Surface enhanced Raman scattering was used to monitor the adsorption of pyridine, **C₅H₅N**, on TiO_2 (110) and (001) surfaces at room temperature, and the molecule was found to bond to the surface Ti ions via its N atom, similar to its adsorption on Ti metal. The interaction of **tris(allyl)rhodium** with TiO_2 (001) was studied by UPS in the course of the preparation of model alkene hydrogenation catalysts. The molecule reacts with OH groups on the surface to form a supported di(allyl)rhodium complex. This complex can further react with H_2 to produce the single-crystal analog of the powder [Ti–O]-Rh(allyl)H catalyst. The interaction of tris(allyl)rhodium with model oxide surfaces was studied theoretically by means of extended-Hückel band calculations, and the reducibility of the oxide support through the creation of surface defects was found to be crucial to the molecule–oxide interaction.

There has been one attempt to image an organic molecule on the surface of TiO_2 by means of STM.[14] Phenol, **C₆H₅OH**, was deposited onto an etched TiO_2 surface in air, and rather vague changes in the STM image were seen. It was not possible to uniquely identify the molecule, although it was believed that it lay flat on the surface from the general shape of the contrast in the micrographs. More work has to be done before small organic molecules can be definitively imaged on oxides.

6.4.2 SrTiO₃

The adsorption of H_2 on $SrTiO_3$ has only been studied for the reduced (111) surface. Exposure of the surface to 2000 L H_2 reduces the size of the bandgap defect emission in UPS spectra to about 65 % of its initial value, which is a much smaller reduction than that produced by O_2. No model has been proposed for whether the adsorption is molecular or dissociative, nor has any possible adsorption site been considered. No reports have been published of the interaction of atomic H with $SrTiO_3$ surfaces.

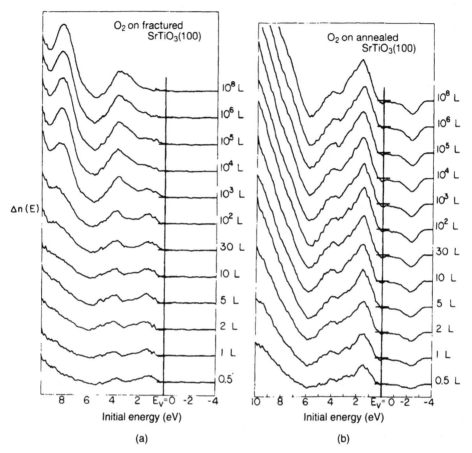

Fig. 6.31 He I UPS difference spectra for (a) UHV-fractured and (b) Ar⁺-ion-bombarded
and annealed SrTiO₃ (100) after successive exposures to O₂. [Ref. 649]

The theoretical considerations of the interaction of H_2 with TiO_2 (110) surfaces
referred to above hold equally well for the 'TiO_2' surface of $SrTiO_3$ (100) [recall
the two types of (100) surface planes in the perovskite crystal structure described
in § 2.3.2.1] due to the symmetry of the surface atomic arrangements chosen.

The interaction of **O_2** with $SrTiO_3$ surfaces prepared by fracturing in UHV, by
ion bombardment, and by subsequent annealing have all been studied. Most studies have been performed on the (100) surface, but the (111) surface has been used
as well. In all cases a strong adsorbate–substrate interaction occurs. On surfaces
that contain O-vacancy defects and associated electronic charge trapped at the
defect site, O_2 exposure rapidly depopulates the electronic states at the defects,
similar to the behavior for TiO_2. However, even for UHV-fractured $SrTiO_3$ (100)
surfaces that exhibit no emission from surface defect states, the O_2 interaction is
strong. Figure 6.31 (a) shows UPS difference spectra for exposure of UHV-frac-

tured $SrTiO_3$ (100) to O_2. For exposures of a few L or less, the difference spectra exhibit a two-peaked structure overlapping the bulk O 2p valence band; this difference spectrum is virtually identical to that for the initial adsorbed phase of O_2 on defective TiO_2 (110) surfaces, which is shown in Fig. 6.21. This adsorbed phase is seen on all $SrTiO_3$ surfaces [Fig. 6.31 (b) presents data for an ion-bombarded and annealed surface, where the depopulation of surface defect states can also be seen] and is attributed to dissociative adsorption of O_2, producing adsorbed O^{2-} ions. The preparation of the surface strongly affects its interaction with O_2 for larger exposures, however. For the annealed surface shown in Fig. 6.31 (b), no second adsorbed phase is observed, while for the fractured surface in (a) a second phase appears to *displace* the initial phase for large exposures. On ion-bombarded $SrTiO_3$ (100) surfaces that were not annealed, a second phase adsorbs *in addition* to the initial one. ELS measurements on UHV-fractured (100) surfaces identified the second adsorbed phase as peroxide, O_2^{2-}. The mechanism proposed in that paper for the formation of the peroxide ions could also account for the disappearance of the O^{2-} phase at high exposures.[652] By analogy with EPR measurements on SrO powder samples, a likely mechanism for the formation of peroxide is via the reaction $O_2 + 2O^{2-} \rightarrow 2O_2^{2-}$, in which the initial adsorbed phase would be used up in the formation of the peroxide ions. As is the case for O_2 adsorption on all transition-metal oxides, however, the interpretation of the species that adsorb for large exposures is still speculative. It is probable, in fact, that several different species are present simultaneously, particularly on defective surfaces.

As was the case for H_2, the generalized theoretical considerations of the interaction of O_2 with TiO_2 (110) surfaces are also valid for the 'TiO_2' surface of $SrTiO_3$ (100).

Many measurements have been performed of **H_2O** adsorption on $SrTiO_3$ surfaces. As for O_2 adsorption, most measurements have been made on the (100) surface, although some work has been done using (111) surfaces. As for TiO_2, differences in assigning peaks in UPS difference spectra have led to differing interpretations of adsorbed species. Only one measurement has been performed on UHV-fractured $SrTiO_3$ (100), and the surface was characterized as having a high density of steps. UPS difference spectra showed that H_2O dissociated at steps, resulting in adsorbed OH^-. Adsorption of OH^- reduced the amplitude of the surface component of both the Ti 3s and Sr 3d core-level spectra. Most, but not all, studies of polished and annealed $SrTiO_3$ (100) surfaces that were nominally stoichiometric showed that H_2O does not adsorb at room temperature. As discussed in § 2.3.2.1, annealed $SrTiO_3$ (100) surfaces do not necessarily have the stoichiometry of the bulk, and the differences in H_2O adsorption results may be due to differences in surface composition. Ion-bombarded, reduced $SrTiO_3$ (100) surfaces give UPS spectra that indicate that both molecular and dissociative adsorption can occur at room temperature. The interaction of H_2O with $SrTiO_3$ is therefore some-what different than that for TiO_2, although in many other respects the two materi-als are very similar.

The fundamental aspects of the interaction of H_2O with the $SrTiO_3$ (100)-'TiO_2' surface should be similar to those for the TiO_2 (110) surface because of the simi-larity in the local atomic arrangements on the two surfaces. So the theoretical work referred to above is also applicable to $SrTiO_3$. DV-Xα calculations have also been performed for unrelaxed clusters simulating the 'TiO_2' surface of $SrTiO_3$ (100) with either OH^- ions adsorbed above surface cations or H^+ ions bonding to surface O^{2-} ions to form surface hydroxyls.

6.4.3 V_2O_5

A small amount of work has been performed on the interaction of gases with sin-gle-crystal V_2O_5 at room temperature, largely aimed at understanding the active sites on vanadia catalysts. Stoichiometric V_2O_5 itself is not useful as a catalyst; surface reduction to create V species having valences of 4+ or lower is necessary in order for the vanadia surface to catalyze chemical reactions. But catalysts are often prepared from V_2O_5, which is then easily reduced under reaction conditions. In one study, V_2O_5 single crystals cleaved along the (001) plane in UHV were exposed to **propene**.[219] Freshly cleaved surfaces were inert, but when the surface was partially reduced, propene oxidation was observed. In another experimental study, stoichiometric V_2O_5 (001) surfaces were prepared by cleaving in air and then annealing in UHV.[538] Lower V valence states were then created on the sur-face by reducing the samples *in situ* either by electron or ion bombardment. The state of surface reduction is easily monitored by UPS, since occupied electronic states appear in the bulk bandgap with the creation of reduced surface V species. Preliminary results show that reduced V_2O_5 surfaces react most strongly with **O_2** and **SO_2**. O_2, and to a lesser extent SO_2, oxidize the reduced V_2O_5 surface. XPS measurements of the S 2p binding energy show that both S–V and S–O species are present on the surface after exposure to SO_2. The behavior of **CO** is very different. It does not appear to react with the reduced surface, but it *reduces* the stoichiomet-ric surface, with UPS, XPS and Auger spectra suggesting that CO_2 is formed and subsequently desorbs.

The adsorption of **NH_3** on the (001) basal plane of V_2O_5 has been considered theoretically by using the extended-Hückel tight-binding method. The most stable adsorption site is predicted to be on top of the exposed V ions, with bonding via the N atom. However, this site does not result in appreciable weakening of the N–H bond, so it is not thought to be the one that is important in catalytic reactions that involve NH_3 bond activation. Other, higher-energy adsorption sites were considered as possible catalytically active sites. Pre-reduction of the surface to partially populate the surface V d-orbitals would also increase the probability of N–H bond activation.

6.4.4 MoO_3 and MoO_{3-x}

The creation of defects on MoO_3 (010) by reaction with H_2 has been studied on single crystals. An H_2 pressure of 3×10^{-3} Torr was maintained, and changes in Auger spectra and RHEED patterns were monitored as a function of time and temperature. For temperatures below 550 K, no changes were observed even for tens of hours of exposure. But for temperatures above 550 K, the surface became reduced. Three-dimensional MoO_2 islands were formed, with nucleation taking place through MoO_2 crystallographic planes in registry with the MoO_3 (010) substrate. A detailed model was presented for the crystallography of the transformation between MoO_3 and MoO_2.

The interaction of atomic **H** with MoO_3 has also been studied theoretically, with H predicted to diffuse into the bulk and form hydrogen bronzes, H_xMoO_3. H should bond most strongly to basal plane O^{2-} ions, while on the edges of the planes it can bond either as H^+ with O^{2-} ions or as H^- to Mo^{6+} ions.

Stoichiometric MoO_3 (010), with its outermost planes of O ions, is inert to O_2. But a continuous range of slightly reduced Mo oxides exist, which consist of regions having the MoO_3 structure separated by crystallographic shear planes similar to those occurring in the Magnéli phases of TiO_{2-x}. The oxidation of one of those phases, $Mo_{18}O_{52}$, has been studied using single-crystal samples. At 670 K under an O_2 pressure of 3×10^3 Pa, the (100) surface was found to oxidize via a three-step process. Initially, O chemisorbs onto crystallographic shear plane boundaries, where the Mo ions do not have the full O coordination that they have in the bulk of MoO_3. Then MoO_3 crystallites nucleate at the shear plane boundaries, a process that involves the mobility of shear planes within the $Mo_{18}O_{52}$ structure. The last step is the epitaxial growth of MoO_3 crystallites across the surface of the reduced oxide. As for the reduction of MoO_3 by H_2, a detailed model is presented for the crystallography of the transformation, including the role played by steps on the surface.

The MoO_3 (010) surface is attacked by H_2S and mixtures of H_2S and H_2 for temperatures above about 500 K. The reaction proceeds in three steps: dissociative adsorption of H_2S to produce a sulfur adlayer on the MoO_3; reduction of the MoO_3 to MoO_2; and an epitaxial overgrowth of highly dispersed bulk-like MoS_2 particles, having their (100) planes parallel to the interface, on the three-dimensional MoO_2 crystallites.

Since MoO_3 is used for the commercially important conversion of methanol to formaldehyde, and since the (010) surface is the thermodynamically most stable one and hence the most prevalent one on powder samples, single-crystal studies to understand catalysis on MoO_3 have concentrated on the (010) face. When stoichiometric, this face exposes only O ions and as such is quite inert. Attempts to adsorb **CH₃OH** on the stoichiometric (010) surface did not succeed. Even exposure of the surface to 0.8 Torr of CH_3OH for 15 min at room temperature did not produce any new features in UPS spectra. It was concluded that no more than 0.01 monolayer was ever present on the surface at room temperature, and that was presumably adsorbed at defects. Reduction of the (010) surface by ion bombardment creates exposed Mo ions, and CH_3OH does chemisorb on such surfaces, presumably as methoxide.

The chemistry of MoO_3 surfaces has been considered theoretically by using a molecular orbital cluster approach. In one calculation C–H bond activation and rupture in **CH₄** was modeled in order to understand catalytic reactions on MoO_3 that involve C–H bond breaking. However, since the basal (010) plane is so inert, it was necessary to consider the (100) surface of MoO_3, which is perpendicular to the basal planes and contains coordinatively unsaturated Mo^{6+} ions, on which the reaction could take place. The mobility on the basal plane of a methyl radical, formed from CH_4, was considered, and it was found to have a very low barrier to diffusion across that O-covered surface. In another calculation, the adsorption of methoxy groups, **CH₃O**, on MoO_3 clusters was considered, again with the aim of understanding C–H bond cleavage. Related calculations have been performed for C–H bond activation in **CH₄** and propylene, **H₂C=CHCH₃**, on α-**Bi₂O₃** and for propylene oxidation to acrolein on **Bi₂Mo₃O₁₂**.

6.4.5 *WO₃ and NaₓWO₃*

Single crystals of WO_3 have been used to study the interaction of **H** with the (100) surface. Exposure to atomic H produces emission from the bulk bandgap that is interpreted in terms of the creation of surface O vacancies by reaction with the H atoms. Consideration was given to the possibility that the H diffused into the WO_3, forming a hydrogen tungsten bronze, H_xWO_3, but the UPS spectra were not consistent with that interpretation. So H either removes lattice O^{2-} ions or adsorbs at some undetermined site on the surface.

The adsorption of **H₂O** on stoichiometric (100) surfaces of $Na_{0.7}WO_3$ has been studied by using HREELS. H_2O does not adsorb at room temperature, but at 150 K losses characteristic of the vibrations of molecular H_2O are seen, as indicated by the arrows in Fig. 6.32. The highest-energy mode corresponds to the H–O–H symmetric stretch mode, although it could also be due to the O–H stretch of adsorbed OH^-. The middle peak, however, corresponds to the H–O–H bend, a mode that can only occur if H_2O adsorbs molecularly. The lowest-energy mode is due to the

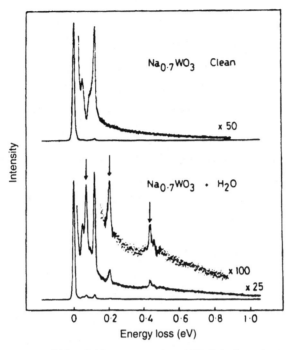

Fig. 6.32 ELS spectra of $Na_{0.7}WO_3$ measured at 150 K before (upper) and after (lower) exposure to 10^{-6} mbar H_2O. The adsorbate-induced loss features are indicated by the arrows. [Ref. 343]

W–H_2O stretch mode of the entire molecule vibrating against the surface. The data therefore suggest that H_2O bonds molecularly via its O lone pair at surface W sites.

6.5 Adsorption on d^n transition-metal oxides

Transition-metal oxides in which the cations have partially filled d orbitals in the ground state, which we refer to as d^n oxides, differ from the d^0 oxides in that at least one of the d electrons can often be removed relatively easily from the cation; i.e., the d^n oxides may be easily oxidized. The d^n oxides are often much more reactive than are d^0 oxides, and they exhibit a wider range of interactions with adsorbed atoms and molecules. The behavior of stoichiometric surfaces of d^n oxides can be similar to that of defective surfaces of the corresponding d^0 oxide, since reduced cations are present at the defect sites. The creation of O-vacancy point defects on d^n oxides results in smaller changes in surface properties than for d^0 oxides, since even the stoichiometric surface contains d electrons that are available to take part in surface chemical reactions.

6.5.1 TiO$_x$ and Ti$_2$O$_3$

Except for the Magnéli phases in TiO$_{2-x}$ (on which no single-crystal adsorption studies have been performed), the first stable reduced oxide of Ti is Ti$_2$O$_3$. While of no real practical importance itself, Ti$_2$O$_3$ is an excellent model oxide for determining the effects of Ti 3d electron population on surface properties, since its bulk cation electronic configuration is 3d^1. Only one group has measured its chemisorption properties on single-crystal samples, but because of its excellent cleavage along the (10$\bar{1}$2) plane, very reproducible results have been obtained.

Since the oxidation of Ti$_2$O$_3$ to TiO$_2$ is a thermodynamically very favorable process, Ti$_2$O$_3$ is expected to interact strongly with **O$_2$**, with the cations oxidizing to Ti^{4+}. Somewhat surprisingly, that interaction is relatively weak at room temperature. Figure 6.33 presents a series of He I UPS spectra for room temperature O$_2$ exposure of the UHV-cleaved (10$\bar{1}$2) surface. [The same data are plotted in (a) and (b), but from different perspectives so that different changes in the spectra are visible.] The decrease in amplitude of the Ti 3d a$_1$ emission indicates clearly that the interaction involves oxidizing the surface Ti ions, as does the disappearance of the shoulder on the high-binding-energy side of the O 2p valence band, which corresponds to the largest Ti 3d admixture to the valence-band wavefunctions. The changes that occur in the surface electronic structure upon O$_2$ chemisorption can be seen more clearly in Fig. 6.34, which plots the change in the work function, $\Delta\Phi$, the position of the Fermi level, $E_F - E_V$, the width of the gap between the O 2p and a$_1$ bands, and the relative intensity of the a$_1$ emission as a function of O$_2$ exposure. There are clearly two phases in the adsorption process. By far the largest changes in the band structure occur during the first 1 L of exposure. The 0.6 eV increase in the work function and the decrease in the intensity of the a$_1$ emission to about 15 % of its clean-surface value indicate that the adsorbed species is negatively charged, having removed electrons form the surface Ti 3d orbitals, and that its sticking coefficient is essentially unity. The valence band also moves toward E_F by 0.6 eV during this phase, exactly tracking the increase in the work function; this indicates that the vacuum level position is fixed relative to the O 2p band.

For exposures above 1 L, the surface band-structure parameters continue to change in the same manner, but at a much lower rate. Even after 10^6 L of O$_2$, Ti^{3+} ions remain in the near-surface region, indicating that chemisorption has effectively stopped. It was not possible to use UPS difference spectroscopy to determine the adsorbed species because of the magnitude of the changes in the substrate band structure. It is almost certain, however, that the first phase of adsorption produces adsorbed O^{2-} ions. Whether that is also the species adsorbing for exposures > 1 L is not clear.

The interaction of O$_2$ with ion-bombarded Ti$_2$O$_3$ (10$\bar{1}$2) surfaces has also been

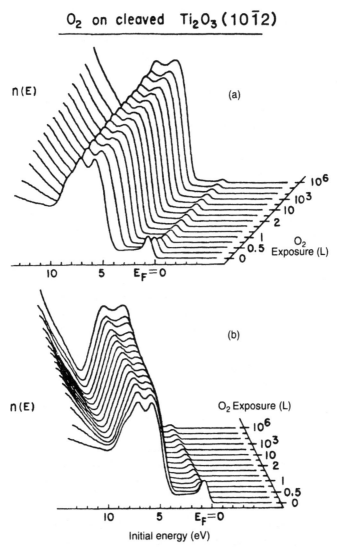

Fig. 6.33 Series of He I UPS spectra for exposure of UHV-cleaved Ti_2O_3 ($10\bar{1}2$) to O_2. The same data are presented in (a) and (b). [Ref. 26]

studied, and the behavior is similar to that for the stoichiometric surface. However, the initial sticking coefficient for O_2 is 2–5 times *smaller* on the reduced surface, even though there are more Ti 3d electrons available on surface cations and at surface defects. This is a surprising result that does not yet have any explanation.

It was mentioned in § 5.2.4 that Ti_2O_3 ($10\bar{1}2$) is one of the few oxide surfaces that can be prepared by ion bombardment and annealing to give surfaces that are essentially indistinguishable from the UHV-cleaved surface. One of the most sen-

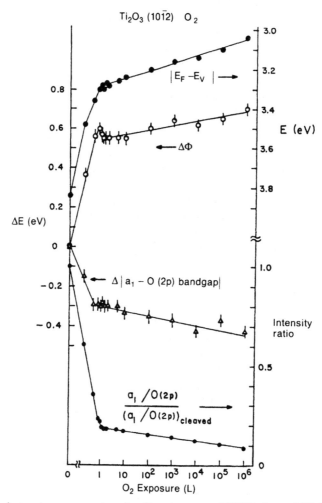

Fig. 6.34 Band-structure parameters for O_2 exposure of UHV-cleaved Ti_2O_3 $(10\bar{1}2)$. See text for details. [Ref. 26]

sitive tests for the quality of the annealed surface involved its response to chemisorption of O_2. Data similar to that in Fig. 6.34 were taken, and virtually every detail of each of the four curves was reproduced quantitatively.[28]

The interaction of O_2 with UHV-fractured polycrystalline samples of $TiO_{x\approx1}$ has been measured as part of an attempt to rationalize some previously published data on the location of the Ti 3d band that did not agree with theory. Published XPS spectra on TiO_x powder samples prepared in air exhibited no emission from the Ti 3d band, although TiO_x is known to be metallic, and surfaces carefully prepared in UHV did exhibit the expected emission (see § 5.2.5 and Fig. 5.21). Samples of

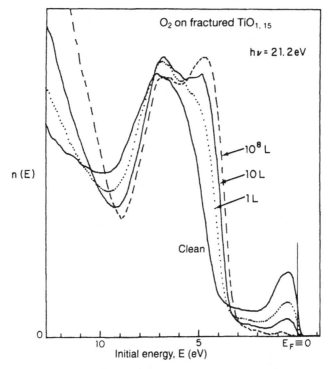

Fig. 6.35 He I UPS spectra for UHV-fractured $TiO_{1.15}$ after various O_2 exposures. Spectra have been aligned at E_F. [Ref. 505]

$TiO_{0.93}$ and $TiO_{1.15}$ were exposed to O_2, and the resulting UPS spectra for the $TiO_{1.15}$ sample are shown in Fig. 6.35. Exposure to O_2 depopulated the Ti 3d band almost completely, far more than it does for Ti_2O_3. The reason for this is believed to be the fact that $TiO_{x\approx1}$ is a highly defective structure; about 15 % of both cation and anion sites are vacant in stoichiometric TiO. The crystal is thus rather like a sponge, and O ions can readily diffuse into the interior of the crystal, oxidizing the Ti ions to Ti^{4+}. This model is consistent with the fact that the UPS spectrum after 10^8 L O_2 exposure is almost indistinguishable from that for TiO_2.

Interest in understanding the photoelectrolytic properties of TiO_2 and $SrTiO_3$ led to the investigation of the interaction of **H₂O** with Ti_2O_3, since that oxide provides the opportunity to study the role of Ti 3d electrons without the necessity of surface disorder. In fact, experimental measurements of the interaction of H_2O with both UHV-cleaved and defective Ti_2O_3 (10$\bar{1}$2) surfaces helped to answer an important question concerning the interaction of H_2O with TiO_2 that could not be answered by experiments on TiO_2 alone. As discussed in § 6.4.1, the interaction of H_2O with stoichiometric TiO_2 (110) at room temperature is weak, and the majority

H₂O on Cleaved Ti₂O₃ (10̄12)

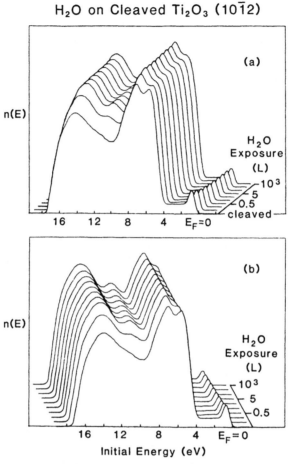

Fig. 6.36 He I UPS spectra for H$_2$O exposure of UHV-cleaved Ti$_2$O$_3$ (10$\bar{1}$2). The same data are presented in (a) and (b). [Ref. 173]

of studies led to the conclusion that what interaction there is probably occurs at residual surface defects. In any event, the perfect TiO$_2$ (110) surface certainly does not dissociate H$_2$O to any appreciable extent. But H$_2$O does dissociate readily at point defects (primarily O vacancies) on reduced surfaces, and such defect sites are believed to be the reason for the relatively high efficiency of TiO$_2$ as a photo-catalytic electrode in photoelectrolysis. O-vacancy point defects are always accompanied by partial population of the 3d orbitals of adjacent Ti cations, and, in spite of the weak interaction seen in UPS spectra between adsorbed OH$^-$ and the Ti 3d electrons at defects, it was assumed for several years (by one of us as well) that the presence of Ti 3d electrons was the necessary and sufficient prerequisite for dissociative adsorption. It was not until experiments were performed on UHV-

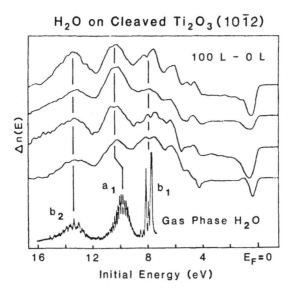

Fig. 6.37 Comparison of UPS difference spectra for four different UHV-cleaved Ti₂O₃ (10Ī2) surfaces exposed to 100 L H₂O with the UPS spectrum for molecular H₂O. (Gas-phase spectrum offset to take account of the polarization shift) [Ref. 173]

cleaved Ti_2O_3 that the true complexity of dissociative adsorption of H_2O was appreciated.

Figure 6.36 shows a series of UPS spectra for exposure of the UHV-cleaved Ti_2O_3 (10$\bar{1}$2) surface to H_2O at room temperature. As for H_2O on reduced TiO_2, there is only a weak perturbation of the Ti 3d electrons in the a_1 band, as expected with a donor–acceptor interaction. The initial sticking coefficient for H_2O on this surface is quite large, as can be seen by the rate at which a new peak in the UPS spectra at about 11 eV appears with exposure in Fig. 6.36 (b); but adsorption effectively stops after 0.3 L exposure. The saturation surface coverage of the adsorbed species is thus less than 0.5 monolayer. There is little band-bending upon chemisorption, and few changes occur in the substrate band structure. It is easy to take UPS difference spectra here, and Fig. 6.37 plots such spectra for H_2O adsorption on four different cleaved surfaces, compared with the UPS spectrum of gas-phase H_2O. The partial depopulation of the a_1 band appears as the negative dip just below E_F, and the features between 4 and 7 eV correspond to other changes in the substrate electronic structure. The features from 7 to 16 eV, however, correspond to emission from the adsorbate molecular orbitals. Comparison of the difference spectra with the H_2O gas-phase UPS spectrum show that adsorption is entirely molecular, and the fact that the energy of the a_1 orbital is shifted by 0.45 eV to higher binding energy relative to the b_1 and b_2 orbitals indicates that bonding to the surface occurs via the a_1 orbital. That orbital consists of the lone-pair electrons on the O atom that point away from the two H atoms, so H_2O bonds to the stoi-chiometric Ti_2O_3 $(10\bar{1}2)$ surface via its O atom, with the H atoms directed away from the surface.

The above results prove that the presence of Ti 3d electrons on the otherwise well-ordered Ti_2O_3 $(10\bar{1}2)$ surface is not sufficient for the dissociation of H_2O (although 3d electrons may well be necessary). This conclusion could not have been reached conclusively from measurements on TiO_2 alone, since there it is not possible to populate the Ti 3d orbitals of surface cations to any appreciable extent without introducing some sort of structural defects. On Ti_2O_3 it is also possible to produce surface structural defects by ion bombardment, and those defects are associated with an even larger number of Ti 3d electrons than on the cleaved sur-face. The nature of the interaction of H_2O with that defective surface is shown in Fig. 6.38, where the UPS spectra are shown in (a) and the corresponding differ-ence spectra are plotted in (b). For exposures below about 5 L, the difference spec-tra consist of only two peaks (the apparent peak at the left edge of the difference spectra corresponds to the change in work function and not to any adsorbate mole-cular-orbital features), which correspond to emission from the 3σ and 1π orbitals of OH^-. Thus H_2O dissociates at defect sites on Ti_2O_3 as well as on TiO_2, showing that some type of surface structural defect is necessary to produce an active site for H_2O dissociation.

Since defective reduced surfaces of both TiO_2 and Ti_2O_3 have Ti 3d electrons associated with the defect sites, the above experiments, while showing that struc-tural defects are necessary, still cannot separate the role played by the d electrons. Some indirect indication that d electrons are also necessary at defect sites in order to dissociate H_2O can be obtained from the fact that UHV-cleaved TiO_2 (110) sur-faces, which contain a large density of steps, interact only very weakly with H_2O at room temperature. To the extent that the local atomic geometry at steps resem-bles that at point defects, it appears that both structural defects and d electrons are necessary in order for Ti oxides to dissociate H_2O.

The adsorption of **CO** with the Ti_2O_3 $(10\bar{1}2)$ surface has been studied on single-crystal samples at room temperature The interaction is weak, with only small changes observed in UPS spectra. For exposures less than 10^5 L, the adsorption is thought to be dissociative, although the evidence for that is not very compelling. Above 10^5 L there is better evidence that CO adsorbs molecularly, since two peaks, located at 7.5 and 10.7 eV below E_F, become evident in difference spectra; they are identified as the $(5\sigma, 1\pi)$ and 4σ molecular orbitals of CO, respectively. Essentially no change in the work function occurs for any exposure.

The strength of the interaction of **SO_2** with Ti 3d electrons at defects on the sur-face of TiO_2 that was discussed in § 6.4.1 suggests that a strong interaction should

H$_2$O on Sputtered Ti$_2$O$_3$ (10$\bar{1}$2)

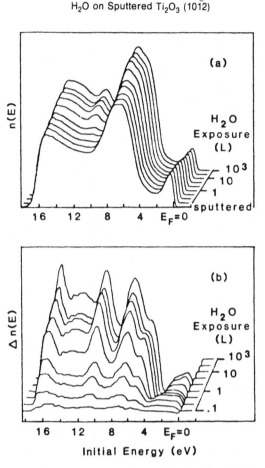

Fig. 6.38 (a) He I UPS spectra for H$_2$O exposure of defective Ti$_2$O$_3$ (10$\bar{1}$2). (b) Difference spectra for the data in (a), aligned to the upper edge of the O 2p valence band. [Ref. 173]

occur with Ti$_2$O$_3$ as well. That is indeed the case. In fact, the interaction of SO$_2$ with Ti$_2$O$_3$ (10$\bar{1}$2) is the strongest adsorbate/substrate interaction that has been observed for any molecule on any metal oxide! The interaction occurs for such low SO$_2$ exposures that Fig. 6.39 shows two different plots of the UPS spectra for SO$_2$ on stoichiometric Ti$_2$O$_3$ (10$\bar{1}$2) at room temperature; (a) shows the first 1 L of exposure, and (b) shows exponential exposures up to 10^4 L. The a$_1$ band is rapidly depopulated throughout the entire depth sampled by He I UPS, as can be seen in Fig. 6.40, which plots both the intensity of the a$_1$ band and the increase in the sample work function. The reaction physically destroys the surface, as indicated by the complete disappearance of all LEED patterns (which are excellent for the clean surface) by 0.5 L exposure. Careful examination of the UPS spectra shows that

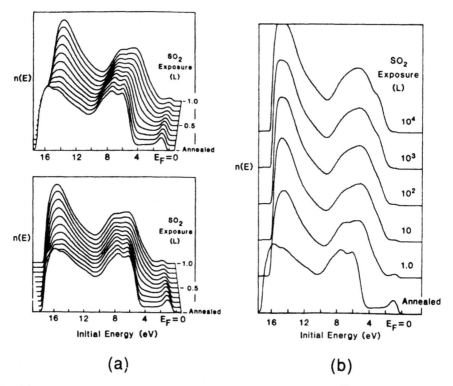

Fig. 6.39 (a) Series of He I UPS spectra for UHV-cleaved Ti_2O_3 $(10\bar{1}2)$ exposed to SO_2 up to 1 L. The same data are presented in the two panels. (b) UPS spectra for UHV-cleaved Ti_2O_3 $(10\bar{1}2)$ exposed to SO_2 up to 10^4 L. [Ref. 27]

changes in the spectra continue to occur up to 10^4 L, the highest exposure used in the study.

The key to understanding the SO_2–Ti_2O_3 interaction comes from XPS measurements of the S 2p binding energy. As discussed in § 6.4.1 above, S bonded to O exhibits a 2p binding energy close to 168 eV, while a S–metal bond has an energy of about 162 eV. The only value for the S 2p binding energy observed for SO_2 on Ti_2O_3 is 162 eV, indicating that SO_2 completely dissociates and bonds to Ti ions in the surface region. From a purely thermodynamic point of view, the most probable reaction is

$$6\,Ti_2O_3 + 2\,SO_2 \rightarrow 11\,TiO_2 + TiS_2\,,$$

which has a net heat of reaction of about –900 kJ/mol. Such a reaction is also consistent with the rapid disappearance of the LEED patterns as a function of SO_2 exposure, since a single SO_2 molecule will affect three Ti_2O_3 units on the surface, as well as with the final value of the work function, which is very close to that of

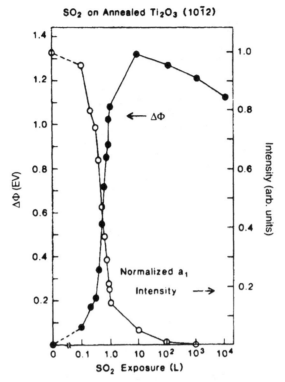

Fig. 6.40 Work function change, $\Delta\Phi$, and normalized Ti a_{1g} intensity, as determined from He I UPS spectra, for annealed Ti_2O_3 $(10\bar{1}2)$ exposed to SO_2. [Ref. 27]

TiO_2. Since changes continue to be observed up to the highest exposures, and since the Ti 3d a_1 band completely disappears from the UPS spectra, the TiO_2 and TiS_2 that are formed must form a microscopic scale that cracks and separates from the bulk, allowing SO_2 continuing access to the Ti_2O_3 below. Thus this reaction is essentially corrosion.

6.5.2 V_2O_3 and VO_2

The corundum oxide V_2O_3 is a close relative of Ti_2O_3, since they have the same crystal structure and cleave along the same $(10\bar{1}2)$ plane (although the quality of cleaves on V_2O_3 is inferior to that for Ti_2O_3). The major difference between them is that the V ions in V_2O_3 have a $3d^2$ electronic configuration, compared to $3d^1$ for Ti_2O_3. The chemisorption behavior of the two oxides has been found to be similar in some respects, but drastically different in others. Similarities occur for the adsorption of O_2 at room temperature, which causes some depopulation of the V 3d band, but less than that seen for Ti_2O_3. The work function increase when either

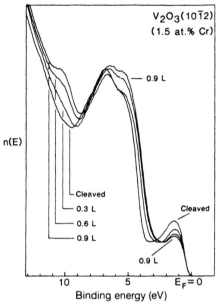

Fig. 6.41 He I UPS spectra for UHV-cleaved Cr-doped V_2O_3 (10$\bar{1}$2) exposed to SO_2 up to 0.9 L. [Ref. 30]

UHV-cleaved or ion-bombarded V_2O_3 surfaces are exposed to O_2 is 3 – 4 times larger than that for Ti_2O_3, increasing by 1.8 eV for exposures over 10^3 L. The interaction between O_2 and V_2O_3 is too strong to permit the use of UPS difference spectroscopy, so no fingerprint of the adsorbed species could be obtained. However, the dramatic increase in the work function is strongly suggestive of adsorbed O^{2-}.

The interaction of **H_2O** with the V_2O_3 (10$\bar{1}$2) surface is similar to, although slightly weaker than, that for Ti_2O_3. Difference spectra for the defective surface show the same two-peaked structure characteristic of OH^- adsorption that those for reduced Ti_2O_3 do. For the cleaved surface, on the other hand, the interpretation of the difference spectra is more ambiguous for V_2O_3. No clear fingerprint of either H_2O or OH^- is apparent; however, the spectra are dominated at low exposures by a two-peaked structure, and it is believed that dissociative adsorption of H_2O predominates. There is some depopulation of the V 3d band upon H_2O adsorption, but it is confined to the upper edge of the band at E_F. Since that region should correspond to the a_1 component of the band, it is probable that chemisorption consists of σ bonding between the substrate cation a_1 and O lone-pair orbitals of OH^-. Since the step density on cleaved V_2O_3 (10$\bar{1}$2) is much larger than that on Ti_2O_3 (10$\bar{1}$2) as determined from the quality of the LEED patterns, it is possible that OH^- adsorption occurs preferentially at step sites.

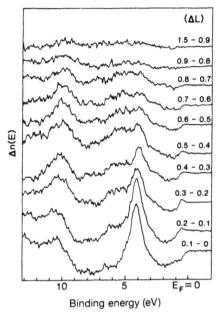

Fig. 6.42 Sequential UPS difference spectra for UHV-cleaved Cr-doped V$_2$O$_3$ (10$\bar{1}$2) exposed to SO$_2$. [Ref. 30]

The similarities between the V$_2$O$_3$ and Ti$_2$O$_3$(10$\bar{1}$2) surfaces that are seen for the chemisorption of O$_2$ and H$_2$O break down completely for the adsorption of **SO$_2$**. Whereas Ti$_2$O$_3$, with its single d electron per cation, interacts violently with SO$_2$, largely via its 3d electrons, V$_2$O$_3$, having twice as many d electrons per cation, hardly reacts at all. Figure 6.41 plots a series of UPS spectra for SO$_2$ exposure of the (10$\bar{1}$2) surface of V$_2$O$_3$ doped with 1.5 atomic % Cr (the Cr doping changes the metal–insulator behavior of V$_2$O$_3$ as a function of temperature, but it has no effect on its chemisorption properties at room temperature). Spectra are shown only for exposures up to 0.9 L, since there are *no changes* in the spectra for larger exposures. The interaction saturates for 1 L, and the surface is inert thereafter. This is shown in more detail in Fig. 6.42, which plots the sequential difference spectra for exposures up to 1.5 L; the majority of the adsorption is complete by 0.5 L. There is some interaction between the V 3d band and SO$_2$, but far less than there is for Ti$_2$O$_3$.

Not only is the degree of interaction of SO$_2$ with V$_2$O$_3$ very different from that for Ti$_2$O$_3$, so is the nature of the interaction. The XPS spectra for the S 2p core level are shown in Fig. 6.43. Peaks having binding energies of both 162.5 and 167.5 eV are observed, indicating that both S–O and S–V bonds are present; the relative amplitudes of the peaks show that the majority of the S is bonded to O,

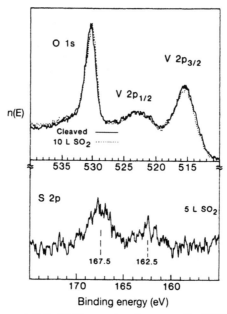

Fig. 6.43 O 1s, V 2p and S 2p XPS core-level spectra for UHV-cleaved Cr-doped V_2O_3 ($10\bar{1}2$) clean and exposed to 10 L SO_2. [Ref. 30]

presumably as sulfate or sulfite. The interaction of SO_2 with V_2O_3 ($10\bar{1}2$) can be blocked completely by pre-exposing the surface to O_2, which bonds much more tightly to the surface. Figure 6.44 shows UPS spectra for a Cr-doped V_2O_3 ($10\bar{1}2$) surface exposed to 10^3 L O_2 and subsequently to 10^3 L SO_2; the differences between the two curves are within measurement error. Contrast this behavior to that shown in Fig. 6.30, where pre-exposure of a TiO_2 surface to H_2S almost completely eliminated its subsequent interaction with O_2.

One study has been reported of the interaction of atomic **H** with UHV-scraped and fractured surfaces of VO_2. The orientation of the exposed faces was not known, but H was found to reduce the surface, forming adsorbed OH^-. Interestingly, the adsorption of H was also found to impede the semiconductor/metal phase transition that occurs in bulk VO_2 at about 340 K (see § 5.4.2).

6.5.3 Cr_2O_3

No studies have been published of chemisorption on well-characterized Cr_2O_3 single crystals, but there has been a report of the interaction of O_2 with Cr_2O_3 (0001) surfaces grown epitaxially on single-crystal Cr(110). Although (1 × 1) LEED patterns indicated a well-ordered surface, the detailed structure and stoichiometry of the surfaces is not known. O_2 adsorbed at room temperature in two distinct phases:

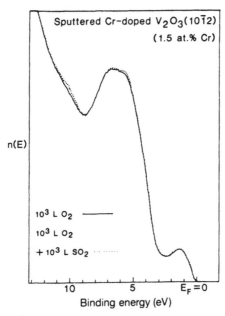

Fig. 6.44 He I UPS spectra for Ar$^+$-ion-bombarded Cr-doped V$_2$O$_3$ (10$\bar{1}$2) exposed to 10^3 L O$_2$ and subsequently exposed to 10^3 L SO$_2$. [Ref. 30]

one phase, attributed to molecular O$_2$, adsorbed linearly in coverage with O$_2$ exposure and desorbed at about 450 K; the second phase, attributed to dissociative adsorption, adsorbed more rapidly for exposures below 10 L than for higher exposures and did not desorb until about 1100 K.

Measurements of **Cl** adsorption on the same epitaxial Cr$_2$O$_3$ surfaces have also been performed. Cl competes for surface sites with the dissociated O phase, and pre-exposure to Cl effectively blocks adsorption of that phase; however, Cl has no effect on adsorption of the low-temperature molecular O$_2$ phase. Because of the nature of the surfaces used, it is not possible to correlate these results with details of surface electronic or geometric structure.

6.5.4 MnO

What little is known about molecular adsorption on single-crystal MnO is very different from what has been found for its close relatives CoO and NiO. One report of the adsorption of **O$_2$** on UHV-cleaved MnO (100) has been published, but there are questions concerning the quality of the cleaved surface. X-ray diffraction analysis of a powder sample made from part of the single crystal revealed that the sample contained approximately 7 at.% α-Mn$_3$O$_4$. Diffuse (100)-(1 × 1) LEED patterns were observed, but their quality was poorer than those from CoO

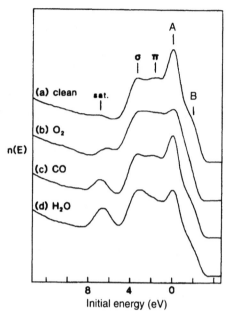

Fig. 6.45 He II UPS valence-band spectra of UHV-cleaved MnO (100) (a) clean, and then exposed to 10^6 L of (b) O_2, or (c) CO, or (d) H_2O. The energy scale is referenced to peak A. [Ref. 548]

and far worse than the beautiful patterns obtained for cleaved NiO, so the surface most likely contained a high density of steps, and perhaps also point defects. The surface used for adsorption is thus not as 'well-characterized' as one would like.

Both the cleaved and the subsequently ion-bombarded MnO (100) surfaces interacted strongly with O_2. The changes in the He II UPS spectra produced by exposure of the cleaved surface to 10^6 L of O_2, CO and H_2O are shown in Fig. 6.45; the changes observed for the ion-bombarded surface are generally similar. The changes produced by O_2 exposure occur rapidly; within the first 10 L, peak A almost disappears, peak B is attenuated, and emission in the O 2p region increases. This is thought to result from dissociative chemisorption, with adsorbed O^{2-} species removing electrons from the Mn^{2+} ions, raising them to higher valence states. Larger O_2 exposures produce the peak at 6 eV, as well as other changes in the valence-band region. No unique fingerprint could be obtained for the adsorbed species, but for high exposures it is probably molecular O_2.

Molecular-orbital cluster calculations have been performed for the interaction of O and O_2 with MnO (100); only perfect surfaces were considered. Atomic O prefers to adsorb on top of surface Mn ions, with a binding energy of 171 kJ/mol. Molecular O_2 adsorbs less strongly than O, but adsorption is still preferred above surface cations.

Both cleaved and ion-bombarded MnO (100) surfaces interact strongly with **H₂O**. The changes in the He II UPS spectra produced by exposure of the cleaved surface to 10^6 L of H_2O are shown in Fig. 6.45; similar changes are observed for the ion-bombarded surface. The initial sticking coefficient for H_2O is comparable to that for O_2, but, as for other transition-metal oxides, H_2O exhibits a much weaker interaction with the Mn 3d electrons than does O_2. UPS difference spectra indicate that the adsorbed species for low exposures is OH^-.

Both stoichiometric and reduced MnO (100) surfaces also interact strongly with **CO**. The changes in the He II UPS spectra produced by exposure of the cleaved surface to 10^6 L of CO are shown in Fig. 6.45; the changes that occur for the ion-bombarded surface are similar. CO interacts with cleaved MnO (100) much less strongly than do O_2 or H_2O. There is even less interaction with the Mn 3d electrons than there is for H_2O, although additional features, such as the peak at 7 eV, appear in the spectra. Difference spectra suggest that the adsorption is molecular, although the rather large separation between the features assigned to the 4σ and (1π, 5σ) CO molecular orbitals leaves that interpretation open to question.

Molecular-orbital cluster calculations have been performed for the interaction of CO with defect-free MnO (100) surfaces. The preferred adsorption site is C end down over surface Mn ions. CO adsorbs more readily in the presence of pre-adsorbed O ions, forming a linear CO_2 complex on the surface. Similar calculations for molecular adsorption of **CO₂** onto MnO (100) determined the lowest energy configuration to be bonding of an O atom above a surface Mn ion, with the O–C–O axis linear and vertical.

6.5.5 FeO, Fe₃O₄ and Fe₂O₃

Very little is known about the interaction of molecules with well-characterized single-crystal Fe oxides. Some measurements have been performed of adsorption on both polished and annealed and ion-bombarded α-Fe_2O_3 (0001), as well as on Fe_3O_4.

There has been a small amount of work on the adsorption of **H₂** on polished and annealed α-Fe_2O_3 (0001) surfaces and on those surfaces after reduction by ion bombardment. Nearly stoichiometric surfaces are relatively inert to H_2 at room temperature; this is probably due to the stability of the half-filled d-band configuration, Fe^{3+} $3d^5$, of α-Fe_2O_3. The measured sticking coefficient for H_2 is 10^{-4}–10^{-5}, so whatever interaction there is must occur at defects. UPS spectra suggest that at room temperature H_2 may dissociate homolytically to produce adsorbed hydroxyl ions, but the results are not definitive. Upon adsorption there is no interaction between H_2 and the 0.9 eV emission feature in the UPS spectra of α-Fe_2O_3 that was discussed in § 5.7 above.

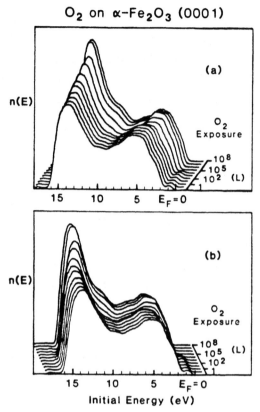

Fig. 6.46 He I UPS spectra for 1093 K-annealed α-Fe$_2$O$_3$ (0001) exposed to O$_2$. [Ref. 31]

Nearly stoichiometric α-Fe$_2$O$_3$ (0001) surfaces are quite inert to O_2, where the measured sticking coefficient for O_2 is 10^{-4}–10^{-5}. This can be seen graphically in Fig. 6.46, which presents a series of He I UPS spectra for the nearly stoichiometric (0001) surface. There are essentially no changes in the valence-band spectra until the O_2 exposure exceeds 10^3 L, at which point the small peak at 0.9 eV disappears, and other changes occur in and below the valence-band region. The nature of the 0.9 eV feature was discussed in § 5.7; it is believed to correspond to a cation-derived surface state associated with Fe^{2+} ions on the α-Fe$_2$O$_3$ surface. The increase in work function indicates a negative adsorbed species, possibly O_2^{2-}.

Adsorption of O_2 at room temperature has also been studied on ion-bombarded, reduced α-Fe$_2$O$_3$ (0001). The changes produced in the UPS spectra are similar to those that occur on the nearly stoichiometric surface, but an observable interaction occurs for exposures as small as 0.5 L, indicating that the reduced surface is several orders of magnitude more reactive than is the nearly stoichiometric one.

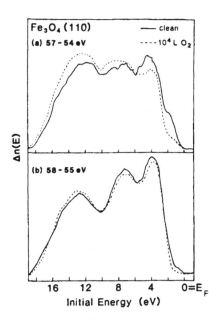

Fig. 6.47 Comparison of the Fe 3d-derived final states measured from the clean Fe_3O_4 (110) surface and following exposure to 10^4 L of O_2: (a) 57 – 54 eV difference spectrum emphasizing Fe^{2+}-derived states, and (b) 58 – 55 eV difference spectrum accentuating the Fe^{3+}-derived states. [Ref. 564]

An extremely interesting case of the interaction of O_2 with Fe oxide surfaces occurs for Fe_3O_4 (110). As discussed in § 5.7, both Fe^{2+} and Fe^{3+} cations should be present on the (110) surface. In a manner similar to that shown in Fig. 5.38 and described in Refs. 564 and 565, differences between UPS valence-band spectra for photon energies above and below the $3p \rightarrow 3d$ resonance were taken for clean and O_2-exposed Fe_3O_4 (110) surfaces in such a way that the contribution to the density-of-states from either Fe^{2+} or Fe^{3+} ions could be accentuated. Figure 6.47 compares such difference spectra for clean and O_2-exposed surfaces taken at energies that enhance the Fe^{2+} emission (a), and that enhance the Fe^{3+} emission (b). The interaction of O_2 with the surface is clearly via the Fe^{2+} ions, with virtually no effect of O_2 on the Fe^{3+} emission. This is understandable since the Fe^{3+} ions have a very stable half-filled shell, $3d^5$ electronic configuration, while Fe^{2+} ions have one additional electron in a minority spin t_{2g} orbital. Note the disappearance of the shoulder in the Fe^{2+} emission near 1 eV; this is analogous to the changes seen in Fig. 6.46 for O_2 exposure of the nearly stoichiometric α-Fe_2O_3 (0001) surface.

The adsorption of **H_2O** onto polished and annealed α-Fe_2O_3 (0001) surfaces and on those surfaces after reduction by ion bombardment has been investigated at room temperature. Nearly stoichiometric surfaces are almost inert to H_2O, presumably for the same reason discussed above for H_2 and O_2. The sticking coefficient was found to be 10^{-5}–10^{-6}. Observable adsorption occurs at room temperature only for exposures even higher than those for O_2; it is thought to be dissociative, with the formation of OH^- ions. H_2O can, of course, be physisorbed on the α-Fe_2O_3 (0001) surface below 200 K, and subsequent heating of the surface also produces chemisorbed species that have been identified as OH^-. There is only a weak interaction between H_2O and the 0.9 eV emission feature, while for H_2 exposure there is no attenuation of that emission. Adsorption on ion-bombarded, reduced α-Fe_2O_3 (0001) surfaces produces changes in the UPS spectra similar to those observed on the nearly stoichiometric surface, except that the effects become apparent at lower exposures. For the defective surface exposed to H_2O, the adsorbed species is identified as OH^-.

While there have been no calculations of the interaction of H_2O with α-Fe_2O_3 surfaces, cluster calculations have been performed for H_2O on the FeO (100) surface. Dissociation is predicted to occur, leaving OH^- on the surface, with an activation energy of 1.6 eV. To the extent that those calculations are relevant to α-Fe_2O_3, the high activation energy may explain the small sticking coefficient for H_2O observed experimentally. H_2O is also predicted to interact strongly with adsorbed O on the FeO surface, yielding two adsorbed OH^- ions. No experiments have been performed to test the theoretical predictions.

There has been one published optical Raman and ellipsometric study of **CO** chemisorption and reaction on single-crystal Fe_3O_4 (110) surfaces, but the surfaces were not sufficiently well-characterized to determine whether or not they were stoichiometric or well-ordered.

A very weak interaction occurs between **SO_2** and the α-Fe_2O_3 (0001) surface. The room-temperature sticking coefficient on the stoichiometric surface is about 10^{-6}. UPS difference spectra obtained for that system are shown in Fig. 6.48. Features only begin to appear in the spectra for exposures of 10^5 L and above. No XPS spectra of the S 2p core-level binding energy were taken, so determination of the adsorbed species was based only on UPS spectra. The adsorbed species was identified as most likely a sulfate complex by analogy with UPS spectra of metal sulfates, as shown in Fig. 6.48 (c), but both SO_3^{2-} and SO_2 also exhibit several peaks in the same region of the UPS spectra, so any assignment based solely on UPS data is open to question.

6.5.6 CoO

The room-temperature adsorption of **O_2** has been studied on three types of CoO (100) surfaces: stoichiometric UHV-cleaved (100); slightly reduced (i.e., bombarded with 500 eV Ar^+ ions); and heavily reduced (i.e., 5 keV ion bombardment). These three surfaces are discussed in detail in § 5.8, and their UPS spectra are

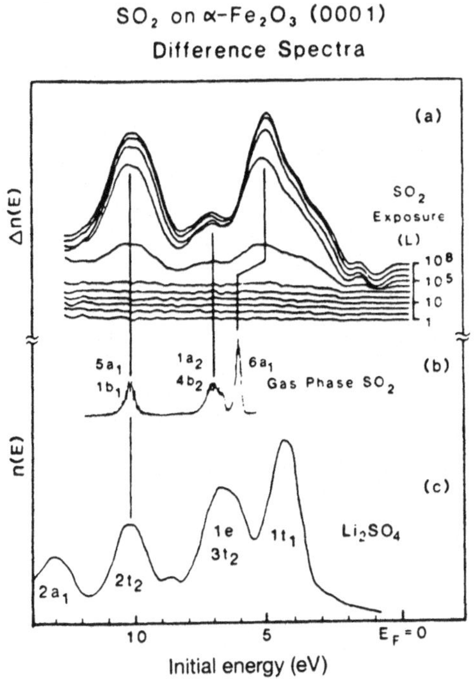

SO$_2$ on α-Fe$_2$O$_3$ (0001)

Difference Spectra

Fig. 6.48 (a) UPS difference spectra for 1093 K-annealed α-Fe$_2$O$_3$ (0001) exposed to SO$_2$; (b) He I UPS spectrum for gas-phase SO$_2$ (offset to take account of polarization shift); and (c) He II UPS spectrum for SO$_4^{2-}$ in Li$_2$SO$_4$. [Ref. 31]

shown in Fig. 5.40. The cleaved surface is virtually inert to O$_2$; what little interaction is observed is attributed to residual surface defects. The defective surfaces interact strongly with O$_2$. Exposure to O$_2$ rapidly depopulates the defect-related features at the upper edge of the valence band, and the adsorption is believed to be dissociative for exposures below about 10^3 L. For larger O$_2$ exposures, an adsorbed phase that is assumed to be molecular is apparent.

The adsorption of **H$_2$O** has been studied on the same three CoO (100) surfaces discussed above. The cleaved surface is virtually inert, with any interaction attributed to defects. The defective surfaces interact strongly with H$_2$O. In common with all other defective transition-metal oxides except NiO, the interaction between H$_2$O and CoO does not strongly involve the 3d electrons at surface O-vacancy defect sites. The UPS difference spectra obtained do not provide a unique fingerprint for the adsorbed species, however, and adsorption could be either dissociative or molecular.

The (100) surface of CoO can be reduced by exposure to **CO**. While the cleaved surface is essentially inert to CO for moderate exposures, slight changes in the UPS spectra suggestive of surface reduction can be seen for very large CO expo-

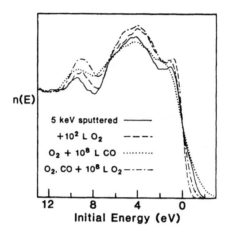

Fig. 6.49 He II UPS spectra of 5 keV Ar⁺-ion-bombarded CoO (100) (solid curve), after exposure to 10^2 L O$_2$ (long dashed curve), and after further exposure to 10^8 L O$_2$ (dot–dashed curve). [Ref. 274]

sures (10^8 L). Similar behavior is found for slightly reduced surfaces, where UPS spectra suggest that CO interacts with lattice O ions, presumably to form CO$_2$, and reduces the surface by the creation of O-vacancy defects. Exposure to CO results in an increased emission at the upper edge of the valence band in UPS spectra, similar to that produced by ion bombardment. Other changes occur in the UPS spectra as well, but nothing that can be correlated with the molecular orbitals of specific adsorbed species. Heavily reduced CoO surfaces also adsorb CO, but no changes in the defect emission at the upper edge of the valence band occur. No specific adsorbed species have been identified in that case either.

An interesting example of the surface reduction produced by adsorbed CO can be seen in Fig. 6.49, which shows He II UPS spectra for sequential exposures of heavily reduced CoO (100) to O$_2$ and CO. The clean reduced surface (solid curve), which exhibits a distinct shoulder of 3d defect emission at the upper edge of the valence band, is first exposed to 10^2 L O$_2$ (dashed curve). The defect states are essentially completely depopulated, with the O$_2$ dissociatively adsorbing as O^{2-} ions. Subsequent exposure to 10^8 L CO results in the dotted curve, which exhibits much more bandgap emission than did the original reduced surface. The adsorbed CO has not only removed the chemisorbed O^{2-}, but some lattice O as well. This removal of O ions can be partially reversed by subsequent exposure of the surface to O$_2$, as shown by the dot–dashed curve.

The interaction of **H$_2$S** with UHV-cleaved CoO (100) surfaces has been studied by LEED, Auger and RHEED. The results are similar to those for NiO discussed below, although the reaction proceeds at lower temperature. No interaction is found at room temperature but, by 373 K, flat, smooth metallic Co islands, having their (100) planes parallel to the substrate, are formed and are covered with a c(2 × 2) array of S atoms (or S^{2-} ions). For higher temperatures and/or longer exposure times, flat, smooth crystallites of Co_9S_8, having their (111) planes parallel to the substrate, are grow on the CoO surface. For temperatures above 623 K, the Co_9S_8 islands roughen.

6.5.7 NiO

Cluster calculations have been performed for the interaction of atomic **H** with both perfect and defect NiO (100) surfaces. H will not bond to perfect (100) surfaces, with either Ni or O vacancies necessary in order for adsorption to occur. At Ni vacancies, H is calculated to bond very strongly to O ions adjacent to the vacancy, with a binding energy of 5–7 eV. At O-vacancy defects, H bonds less strongly to a Ni 4s orbital that is partially populated by removal of the O ion; the bonding energy is about 1 eV. Similar bonding occurs if the Ni or O vacancies lie in the second atomic plane instead of the surface plane. No calculations for H_2 adsorption have been reported. Surface reduction of NiO (100) has been observed experimentally for exposure to H_2 or H at temperatures above about 400 K; for H_2, the kinetics of reduction have been studied as a function of temperature and H_2 pressure, and it is found to be first-order in both surface O-ion concentration and H_2 pressure.

The adsorption of O_2 on UHV-cleaved NiO (100) surfaces has been studied at room temperature. Cleaved (100) surfaces were also ion bombarded to produce surface disorder and a small density of O vacancies, and chemisorption on those reduced surfaces was compared to that on stoichiometric cleaved surfaces. The cleaved NiO (100) surface is inert to O_2. However, O_2 interacts strongly with the defect surface, depopulating the Ni 3d defect structure at the upper edge of the valence band (which is shown in Fig. 5.44) by 100 L exposure. Although the UPS difference spectra do not provide unique identification, the adsorbed phase is thought to be O^{2-}. For exposures above 100 L, a second adsorbed phase is seen, which has been tentatively identified as adsorbed O_2^{2-}. Thus the interaction of NiO with O_2 is essentially the same as that of CoO.

Theoretical cluster calculations have been performed for the interaction of both O and O_2 with NiO surfaces, and different methods reach different conclusions. Hartree–Fock calculations predict the perfect surface to be inert to even atomic O, with the presence of vacancies necessary for adsorption. O_2 is calculated to bond to the defective surface less strongly than O, so dissociative adsorption is postulated. However, atom-superposition electron-delocalization molecular-orbital calculations predict adsorption of both O and O_2 on top of surface Ni sites on the perfect surface, with the preferred geometry for adsorbed O_2 being one in which its internuclear axis is tilted by 45° with respect to the surface normal.

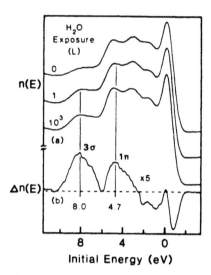

Fig. 6.50 (a) He II UPS spectra for Ar$^+$-ion-bombarded NiO (100) pre-exposed to 10 L O$_2$ as a function of H$_2$O exposure; and (b) (10^3 L – 0 L) difference spectrum. [Ref. 32]

The interaction of **H$_2$O** with NiO (100) surfaces is extremely interesting. The cleaved surface is inert to H$_2$O at room temperature, but, surprisingly, the reduced surface exhibits only very weak dissociative adsorption. However, pre-adsorption of O on the defect surface results in greatly enhanced dissociative adsorption of H$_2$O. This is shown in Fig. 6.50 (a), which plots UPS spectra for reduced NiO (100) that had been pre-exposed to 10 L O$_2$ and then exposed to H$_2$O. The difference spectra in Fig. 6.50 (b) indicate that the adsorbed species is OH$^-$. It thus appears that the presence of non-lattice O is necessary in order to break the O–H bond in H$_2$O and initiate adsorption. Measurements of the interaction of H$_2$O with polished and annealed and with more heavily reduced NiO (100) surfaces also showed that H$_2$O reacts only weakly at room temperature. However, heating the surface to ≥ 450 K greatly enhances dissociative adsorption and surface hydroxylation. [One must be careful in comparing molecular adsorption measurements on single-crystal NiO surfaces prepared in different ways, since the NiO (100) surface obtained by polishing and annealing is *not* the same as a UHV-cleaved surface. Although good (1 × 1) LEED patterns can be obtained on annealed samples, the electronic structure as determined by UPS and other surface-sensitive spectroscopies is significantly different than that of the cleaved surface. There is even some indication that annealed NiO surfaces may be Ni deficient.[583]]

CO is observed to adsorb molecularly on reduced NiO (100) surfaces at room temperature, with UPS spectra exhibiting emission from the 4σ and unresolved (1π, 5σ) molecular orbitals. On UHV-cleaved (100) surfaces, no reaction with CO is seen for small exposures, but for large exposures ($\geq 10^8$ L) CO abstracts lattice O ions from the surface, presumably through the formation of CO_2, which subsequently desorbs; this behavior is the same as that seen for CoO. By reducing the surface, CO molecules generate O-vacancy defect sites at which other CO molecules can adsorb. Thus the O-vacancy site is the active one for the oxidation of CO on NiO (100). Similar behavior is seen on annealed NiO (100) surfaces, where the interaction with CO is weak at room temperature, but heating the sample to ≥ 450 K results in accelerated removal of lattice O ions. *Ab initio* cluster calculations for the bonding of CO via its C atom to surface Ni^{2+} sites on NiO (100) indicate that the bonding is relatively weak, with a heat of adsorption of about 0.25 eV. While the bonding mechanism is primarily electrostatic, there is some covalent component. In contrast, similar calculations for CO adsorbed onto the (100) surface of the d^0 oxide MgO show that the bonding there is purely electrostatic.

There is only one report in the literature of an experimental investigation of the interaction of **CO$_2$** with a d^n transition-metal oxide. UHV-cleaved NiO (100) surfaces exposed to CO_2 between 293 and 473 K exhibited a surface C Auger peak that increased with CO_2 exposure. The adsorption was reversible, but no other determination of the nature of the adsorption was made.

UHV-cleaved (100) surfaces have been used to study the interaction of **NO** with NiO. Adsorption on those surfaces was compared with that on thin NiO layers grown by oxidizing Ni, which were known to contain a large density of defects. NO was found to interact only weakly with both NiO surfaces, and then only at low temperatures; all NO desorbed by about 250 K. The similarity of the behavior on the two surfaces indicates that defects do not play a role in NO adsorption on NiO.

The interaction of **H$_2$S** with UHV-cleaved NiO (100) has been studied by using LEED, RHEED, Auger and SEXAFS. At 570 K, H$_2$S reacts dissociatively with NiO, removing the O atoms from the surface and replacing them with S atoms. The resulting geometric structure is believed to consist of a square array of Ni atoms with S atoms occupying four-fold hollow sites in a c(2 × 2) array, similar to the structure formed for S on Ni (100). For larger H$_2$S exposures at elevated temperature, the growth of Ni_3S_2 islands occurs.

Cluster calculations have also been performed for the bonding of **SO$_2$** and **SO$_3$** to perfect NiO (100) surfaces. SO$_2$ should preferentially adsorb at surface Ni sites, bonding through the S atom in a plane perpendicular to the surface. SO$_3$ exhibits comparable but weaker binding to both surface Ni or O ions. SO$_3$ adsorbed at O-ion sites can be converted to a surface sulfate complex.

There has been virtually no work done on the adsorption of organic molecules on well-characterized d^n transition-metal oxide surfaces. A few experiments have been performed on stoichiometric NiO (100). Both annealed and reduced surfaces were used to study ethylene, **C$_2$H$_4$**, adsorption. The nearly stoichiometric surface only physisorbs C_2H_4 at or below about 200 K. The reduced surface also chemisorbs C_2H_4 at temperatures up to 500 K, the maximum used in the study. At 500 K, the behavior of C_2H_4 is similar to that discussed above for CO in that it reacts with lattice O ions and further reduces the surface. Dissociative chemisorption also occurs, since adsorbed C remains on the surface after interaction with C_2H_4. Pyridine, C_5H_5N, was adsorbed onto NiO (100) surfaces that had been cleaved in air, and surface-enhanced Raman spectra indicated that it bonded to surface Ni ions via its N atom.

6.5.8 Cu_2O

The adsorption of O_2 has been investigated experimentally on both the (100) and (111) surfaces of Cu_2O. Adsorption at room temperature is found to be molecular on nearly stoichiometric (111) surfaces, while it is dissociative on (111) surfaces that had been reduced by ion bombardment. The interaction of O_2 with the (100) surface was determined to be dissociative for several different surface preparation procedures; this may well occur at defect sites, since the (100) surface is polar and does not exhibit a stable surface structure.

The adsorption of H_2O on the polar (100) surface of Cu_2O was measured for substrate temperatures of 110 K and above, and below about 150 K a mixture of molecular and dissociative adsorption was found. Above about 250 K only dissociated H_2O adsorbed. No more than 10 % of a monolayer of dissociated H_2O was seen at either 110 or 300 K, and ion bombarding the surface to produce additional defects did not increase the amount. The latter result is surprising in light of the behavior of virtually all other transition-metal oxides. Although the location of substrate valence-band features in UPS complicated the determination of the adsorbed species, it is believed to be OH^-. In agreement with the behavior of other transition-metal oxides, chemisorbed OH^- does not interact strongly with the Cu 3d features as seen in UPS.

The adsorption of CO has also been measured on the Cu_2O (100) surface. CO adsorbs molecularly, and at 120 K the sticking coefficient is independent of coverage up to 0.4 L exposure, suggesting a precursor state and non-activated adsorption. The low-coverage activation energy was found to be 70 kJ/mol. There appears to be a fairly strong interaction between CO and the Cu 3d orbitals, with some changes apparent in the shape of the Cu 3d features in UPS, and a large decrease in work function (0.6 eV by 1 L exposure; 1.0 eV at saturation) indicative of electron transfer from the adsorbed CO to the surface.

The adsorption of CH_3OH and $CH_2=CHCH_3$ have been studied on both the (111) and (100) surfaces of Cu_2O. Methanol was adsorbed at 90 K, and on both surfaces it heterolytically dissociated by deprotonation, leaving methoxide and hydroxide on the surface. Propene was found to adsorb primarily molecularly at 100 K on both surfaces, although on the (111) surface there was evidence for a small amount of dissociative adsorption, probably to an allyl intermediate. This dissociation was accompanied by desorption of CO and the creation of O-vacancy defects on the Cu_2O (111) surface. Only weak propene adsorption was observed at room temperature.

6.6 Adsorption on high-T_c copper-oxide superconductors

The interaction of molecules with the surfaces of Cu-oxide high-T_c superconductors is extremely important technologically because some of the materials react chemically with ambient gases to form non-superconducting compounds. This effect becomes crucial for thin-film superconductors, where a significant fraction of the film may become non-superconducting as a result of environmental degradation, and for polycrystalline materials, where chemical reactions at grain boundaries and the attendant changes in electrical properties there can interrupt the superconducting pathways.

Problems with the repeatability of measurements of the electrical and magnetic properties of powder samples quickly led researchers to investigate the interaction between the high-T_c oxides and gases present in the atmosphere. The two molecules that were found to be most reactive with the Cu-oxide compounds are H_2O and CO_2. Some experiments looked at the interaction with **H_2O** simply by immersing samples in water. The **$YBa_2Cu_3O_{7-x}$** compounds having $0 \leq x \leq 1$ decompose in water, yielding CuO, $Ba(OH)_2$, non-superconducting Y_2BaCuO_5, and evolving O_2; when finely divided powders are used, 'the water fizzes like seltzer'.[701] The rate of reaction increased dramatically as the temperature of the water was increased. The compound **$YBa_4Cu_2O_{8-x}$**, when immersed in water, 'was destroyed at room temperature in an instant'.[700] Somewhat more gentle conditions of exposure to moist air gave similar results, with the non-superconducting products forming over a time period of from hours to days; in moist air, however, $YBa_4Cu_2O_{8-x}$ reacts much less rapidly than does $YBa_2Cu_3O_{7-x}$. The interaction appears to be primarily with the Ba, which readily forms hydroxide. The **$Bi_2Sr_2CaCu_2O_{8+x}$** compounds do not react readily with H_2O. Photoemission measurements have been made of the interaction of H_2O with UHV-cleaved (100) single-crystal surfaces, and only physisorption at 90 K was seen. However, immersion of powder samples in H_2O for tens of hours did result in decomposition into CuO, $Sr(OH)_2$, $Ca(OH)_2$ and $CuBi_2O_8$.

Carefully controlled surface-science studies have also been performed of the interaction of H_2O with several other superconducting oxides. The results of the exposure of UHV-scraped polycrystalline samples of **$La_{1.8}Sr_{0.2}CuO_4$** are shown

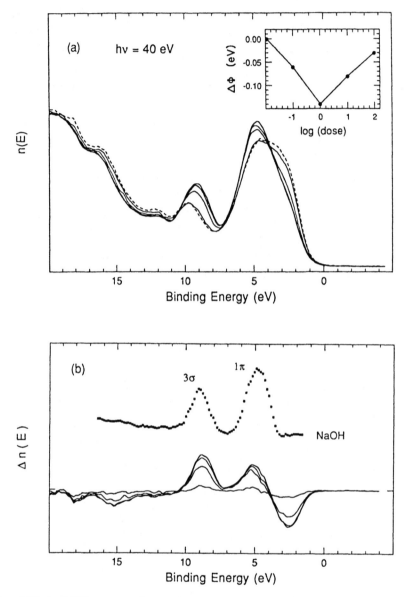

Fig. 6.51 (a) UPS spectra ($\hbar\omega = 40$ eV) for UHV-scraped $La_{1.8}Sr_{0.2}CuO_4$ exposed to increasing amounts of H_2O. The inset shows the change in the work function, $\Delta\Phi$, versus the logarithm of the dose in L; (b) difference spectra for the data in (a) compared to the He II UPS spectrum of solid NaOH. [Ref. 664]

in the UPS spectra of Fig. 6.51 (a). Figure 6.51 (b) compares the UPS difference spectra with the UPS spectrum of solid NaOH, clearly indicating the formation of hydroxide on the surface. Changes in the shape and binding energy of XPS core-level spectra show that the H_2O interacts most directly with the Ba atoms. The

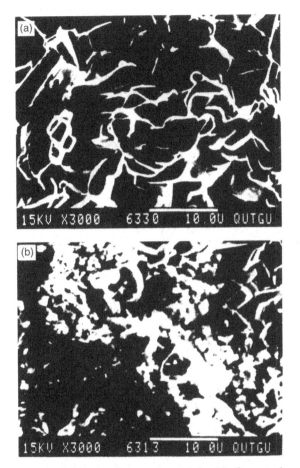

Fig. 6.52 SEM images of an $YBa_2Cu_3O_{7-x}$ surface (a) freshly fractured, and (b) exposed to moisture after fracture. [Ref. 705]

extent of the damage caused to the surfaces of superconducting oxides by H_2O can be seen in a different way in Fig. 6.52, which shows scanning-electron micrographs for polycrystalline $GdBa_2Cu_3O_{7-x}$ (a) immediately after fracture, and (b) after subsequent exposure to moisture in a glove box; both micrographs are taken at the same magnification.

Equally destructive to the superconducting properties of the Cu-oxide compounds is CO_2. The interaction of CO_2 with UHV-scraped $La_{1.8}Sr_{0.2}CuO_4$ as it appears in UPS is shown in Fig. 6.53. In (b) the UPS difference spectra are compared with the UPS spectrum of solid $CaCO_3$, showing that CO_2 reacts with the oxide to form a carbonate. Infrared reflectance, XPS and Raman spectroscopy measurements on $YBa_2Cu_3O_7$ samples exposed to CO_2 were interpreted in terms of the reaction[704]

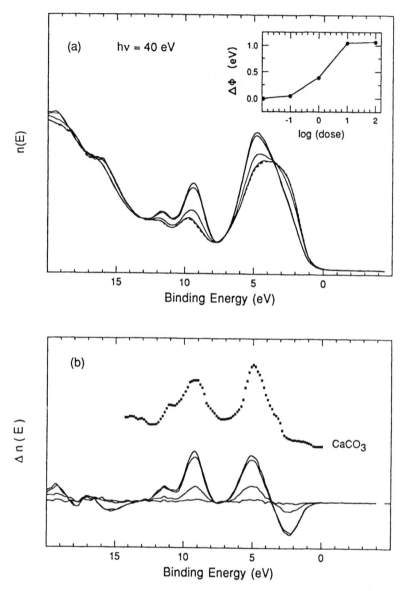

Fig. 6.53 (a) UPS spectra ($\hbar\omega = 40$ eV) for UHV-scraped $La_{1.8}Sr_{0.2}CuO_4$ exposed to increasing amounts of CO_2. The inset shows the change in the work function, $\Delta\Phi$, versus the logarithm of the dose in L; (b) difference spectra for the data in (a) compared to the He II UPS spectrum of solid $CaCO_3$. [Ref. 664]

$$2\,YBa_2Cu_3O_7 + 3\,CO_2 \rightarrow Y_2BaCuO_5 + 5\,CuO + 3\,BaCO_3 + \tfrac{1}{2}\,O_2.$$

This reaction is strongly catalyzed by water vapor, with $Ba(OH)_2$ as an intermediate product, and occurs equally rapidly whether the O content of the oxide is O_7 or

O_6. Thus the most destructive environment for high-T_c superconducting Cu oxides is a mixture of H_2O and CO_2, with temperature also accelerating the reaction.

LEED has been used to monitor the interaction of **CO, O_2, CO_2** and **CH_4** with single-crystal samples of **$Bi_2Sr_2CaCu_2O_{8-x}$** superconductors that were cleaved in UHV.[665] The only molecule that produced any changes in the LEED patterns was CO; the patterns became more diffuse and less intense after CO exposure. However, nothing could be learned about the nature of the CO–surface interaction from LEED. In another study of CO adsorption on UHV-cleaved $Bi_2Sr_2CaCu_2O_{8-x}$ (001) surfaces, molecular adsorption was observed at 26 K, but no adsorption was found at room temperature. The behavior reported for CO_2 in Ref. 665 is very different than that for $YBa_2Cu_3O_{7-x}$, where CO_2 interacts strongly with the surface.

Although the bulk O stoichiometry of the high-T_c oxides is crucial to their superconducting properties and, due to the high mobility of O in the lattices of many of the materials, can change depending upon the O_2 partial pressure in the ambient, exposure of the oxides to **O_2** does not produce any deleterious surface reactions. UPS and XPS spectra exhibit no changes when UHV-cleaved or scraped samples are exposed to O_2. UHV-scraped samples have also been exposed to **CO**, and again only a weak interaction is observed, with UPS spectra indicating that a surface carbonate is formed similar to that for CO_2 exposure, but the sticking coefficient for CO is about 10^{-3} times that for CO_2.

The interaction of metal atoms with the surfaces of high-T_c superconductors, which is extremely important for making electrical contact to them, will be discussed in § 7.4.

6.7 Electron- and photon-stimulated desorption from metal oxides

Although electron-stimulated desorption (ESD) of molecules, atoms and ions from surfaces has been studied for many years, the majority of the work was conducted on metals or semiconductors, and the models that are used to explain the process are specific to those materials.[815] Results that did not agree with those models were observed first in ESD from d^0 metal oxides, and subsequently, using tunable synchrotron radiation, in photon-stimulated desorption (PSD) from the same oxides. The emission of O^+ ions was observed for photon and electron energies above a threshold energy that was too high to be explained by existing models. A theoretical mechanism involving inter-ionic Auger decay, shown schematically for TiO_2 in Fig. 6.54, was postulated by Knotek and Feibelman.[816–818] The energy of the photon or electron couples to the oxide via excitation of an electron from the highest filled level on a surface cation (Ti 3p in the example shown), leaving a hole in that shell. (Emission of O^+ ions as a result of excitation of an electron out of the O 2s core level is not expected to be significant due to the

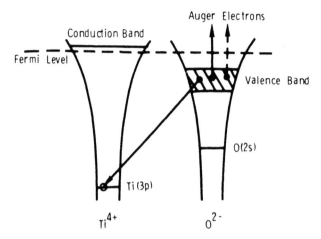

Fig. 6.54 Schematic diagram of the Knotek–Feibelman desorption model. [Ref. 818]

energetics of the steps involved.[816]) Since there are no higher-lying occupied electronic states on the cation, the 3p hole can only be filled by an electron from a neighboring O^{2-} ion. That transition can initiate an Auger process that ejects one or more electrons from the O 2p valence band. If two Auger electrons are emitted, the remaining O^+ ion will desorb by Coulomb repulsion from the surface.

ESD and PSD spectra of lattice O^+ ions that are consistent with the Knotek–Feibelman mechanism have been observed for TiO_2,[127,128,815–821] V_2O_5,[816,822] WO_3,[816] $SrTiO_3$,[815] Na_xWO_3,[823] La_2O_3,[824] CeO_2,[823] $LiNbO_3$,[825] and Er_2O_3.[824] Figure 6.55 shows ESD spectra for O^+ ions from single-crystal TiO_2 and WO_3 and polycrystalline V_2O_5, all of which are d^0 oxides. Very weak O^+ emission begins for incident-electron energies near the O 2p excitation threshold, but significant emission does not begin until the electron energy approaches the threshold for creation of a core hole in the highest-lying cation shell. [The arrows show the location of the peaks in the ELS spectra for each compound; the onset (threshold) for each transition will occur at slightly lower energy than the peak.] Angle-resolved ESD measurements on both atomically flat TiO_2 (110) and faceted TiO_2 (001) surfaces showed that, while the threshold energy for O^+-ion emission is described well by the Knotek–Feibelman model, the magnitude and direction of the emission are highly dependent upon surface geometry. Figure 6.56 shows the angular dependence of the total positive ion yield from TiO_2 (001) as a function of sample annealing temperature. The changes in the patterns correlate with changes in the geometric structure of the surface as different facet planes develop. Angle-resolved PSD of O^+ has also been measured from the V_2O_5 (001) surface; it was found to be strongly peaked along the surface normal, reflecting the local bonding geometry of the topmost O ions in the surface plane.

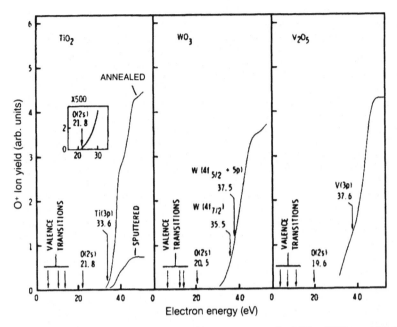

Fig. 6.55 O$^+$- ion yields versus incident-electron energy for TiO$_2$, WO$_3$ and V$_2$O$_5$ surfaces. The arrows denote measured ELS peak energies for each oxide. [Ref. 816]

ESD measurements on single-crystal MgO (100) surfaces that were cleaved in air reported a threshold for O$^+$-ion emission that is too high compared to the Knotek–Feibelman model.[826] The threshold energy for the creation of a Mg 2p core hole in MgO by excitation of an electron to the Mg 3s conduction band is about 54 eV, and yet the O$^+$-ion desorption threshold was observed to occur at 80 eV. While 80 eV is the ionization limit for the free Mg^{2+} ion, electrons excited to energies in the conduction band below the ionization limit should be mobile enough not to immediately recombine with the Mg 2p core hole. Thus inter-ionic Auger decay channels for ion desorption would be expected for photon energies well below 80 eV.

ESD is one of the few techniques that has proven useful for monitoring H on metal oxides. Measurements on annealed TiO$_2$ and SrTiO$_3$ crystals that had not been dosed with either H$_2$ or H$_2$O exhibited emission of H$^+$ ions as well as O$^+$ ions.[636] For both species on both oxides, the primary threshold corresponded to the Ti 3p core-level excitation; however, some H$^+$ ions desorbed above the O 2s threshold as well. The data were interpreted in terms of three different H species on and in the near-surface region of TiO$_2$ and SrTiO$_3$: H bonded to surface Ti ions in a hydride-like bond; H bonded to subsurface O ions in a hydroxyl-like bond; and as part of surface hydroxyls.

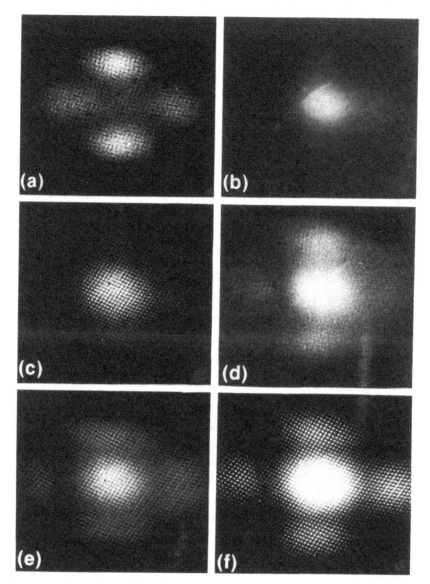

Fig. 6.56 Electron-stimulated-desorption ion-angular-distribution (ESDIAD) patterns of the total positive-ion yield from TiO_2 (001) for (a) the ion-bombarded surface, and for surfaces annealed at (b) 400 K, (c) 500 K, (d) 600 K, (e) 700 K and (g) 1200 K. [Ref. 127]

The Auger-decay channel for positive-ion desorption also occurs for molecules adsorbed on the surfaces of d^0 oxides. This has been verified by measurements on oxides that had been exposed to H_2O[691,815–817,827,828] or F.[816,817,828] Figure 6.57 shows ESD spectra for O^+, H^+ and OH^+ from $SrTiO_3$ surfaces that had been exposed to H_2O While the onset of O^+ desorption corresponds to the Ti 3p

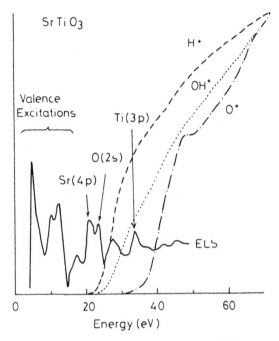

Fig. 6.57 Second-derivative ELS spectrum (solid curve) and ESD spectra for H$^+$, OH$^+$ and
O$^+$ emission from SrTiO$_3$ that had been exposed to H$_2$O. [Ref. 636]

threshold, both H$^+$ and OH$^+$ begin to desorb at much lower incident-electron energies. The H$^+$ threshold is thought to correspond to the Sr 4p excitation, indicated in
the ELS spectra, while OH$^+$ is associated with the O 2s excitation threshold.
Although the energy available for Auger decay into a hole in the O 2s level is not
sufficient to convert a lattice O^{2-} ion into an O$^+$ ion, there is adequate energy available to convert an adsorbed OH$^-$ ion into OH$^+$, which would then desorb by
Coulomb repulsion. Some of the desorbed H$^+$ no doubt also comes from adsorbed
hydroxyls.

Isotope experiments have also been performed using D$_2$O to study the interaction
of water with defective TiO$_2$ surfaces.[815] Whenever such surfaces are exposed to
H$_2$O, both H$^+$ and OH$^+$ desorb in subsequent ESD experiments. To determine to
what extent the H$^+$ signal was from H on or in the surface rather than from the adsorbate, D$_2$O was adsorbed onto a similar surface, and desorbed H$^+$ and D$^+$ were monitored. It was found that the desorbed H$^+$ and D$^+$ signals were equal and were one-
half of the H$^+$ signal obtained when the surface was exposed to H$_2$O. Thus the D$_2$O
molecules must only adsorb at surface sites where an H atom is already adsorbed.

An interesting, although not unexpected, observation has been that PSD spectra
exhibit essentially the same extended fine structure as that seen in x-ray or photon

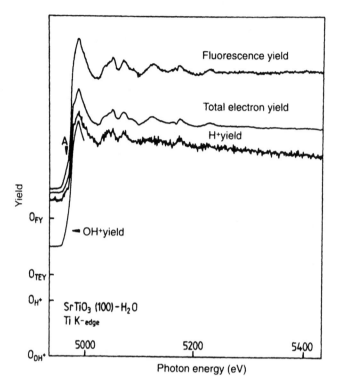

Fig. 6.58 The H⁺ and OH⁺ PSD yields from stepped SrTiO₃ (100) exposed to H₂O. The total electron and fluorescence yields are shown for comparison. [Ref. 691]

absorption spectra.[691,827,828] Since the initial step in the Knotek–Feibelman desorption mechanism is the creation of a core hole, it is reasonable that desorption should mirror the cross-section for creation of that core hole. The oscillations in the core-hole cross-section are the origin of the usual EXAFS structure in x-ray or photon absorption, so one would expect any desorption process that involves the core hole to exhibit similar structure. This is shown for H⁺ desorption from a stepped SrTiO₃ (100) surface for photon energies above the Ti K edge in Fig. 6.58. The H⁺ yield exhibits the same structure as do the fluorescence and total electron yields, the last two of which monitor the EXAFS signal. However, differences between photon absorption and ion desorption spectra have been observed just at threshold.[828–831] Figure 6.59 shows both O⁺- and H⁺-ion-desorption yields as a function of photon energy above the O 1s threshold for MgO (100), along with the fluorescence yield and high-energy transmission ELS spectra. Both the H⁺- and O⁺-ion yields exhibit sharp features just below the main EXAFS absorption edge. These correspond to emission resulting from the creation of excitons on the surface O ions by excitation of an O 1s electron to a localized bound state. The energy

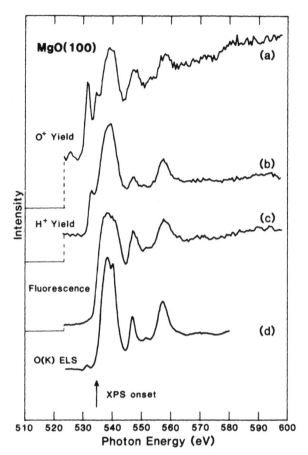

Fig. 6.59 Photon excitation spectra from annealed MgO (100) across the O K threshold: (a) O^+-ion yield, (b) H^+-ion yield, (c) fluorescence signal, and (d) O(K) ELS spectrum. [Ref. 829]

of these excitonic states on the surface is slightly lower than for bulk excitations. Subsequent PSD of H^+ and O^+ ions occurs, although nothing is seen in the EXAFS spectroscopies. This is because the latter sample mainly bulk transitions, and the surface excitons responsible for the ion desorption thus constitute only a tiny fraction of the EXAFS signal.

Although most of the ESD and PSD experiments that have been performed on metal oxides have detected positive-ion emission, one experiment has measured the yield of neutral OH from the (001) surface of TiO_2 in ESD.[832,833] The results are very different than those for desorption of OH^+ ions, as can be seen by comparing Fig. 6.60, which shows the yield of neutral OH as a function of incident-electron energy, with Fig. 6.61 for OH^+ desorbed from TiO_2 (001). For the neutral species, very little structure is seen in the desorption signal at the Ti 3p excitation

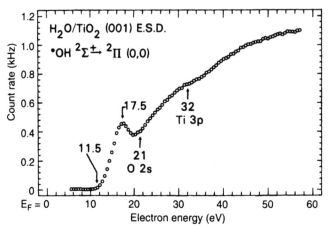

Fig. 6.60 ESD spectrum for neutral OH desorbed after exposure of TiO$_2$ (001) to H$_2$O. The O 2s and Ti 3p excitation thresholds are indicated. [Ref. 832]

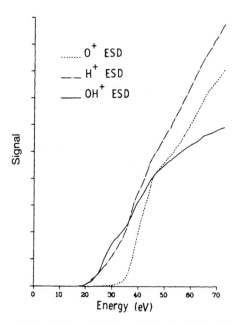

Fig. 6.61 ESD spectra of H$^+$, OH$^+$ and O$^+$ from defective TiO$_2$ (001) that had been exposed to H$_2$O. [Ref. 815]

threshold, and the main emission threshold is at 11.5 eV, about 10 eV lower than the threshold for OH$^+$ desorption. By comparing the data for neutral OH desorption with a cluster calculation of the electronic structure of OH$^-$ adsorbed at O-vacancy defect sites on TiO$_2$ (110), the threshold was identified as corresponding to excitation of an electron from the O 2p non-bonding level to an antibonding

Ti–OH molecular orbital. Thus the desorption of neutral species, which requires less energy than the desorption of positive ions, is associated with valence-band transitions rather than core-level transitions. This study also separated out the contribution of secondary electrons to the total OH yield for electron energies above the desorption threshold, as well as analyzing the contribution of Ti 3d excitation to the neutral OH yield.

An extension of the Knotek–Feibelman model of O^+-ion emission from d^0 oxides has been provided by molecular dynamics simulations of the specific trajectories of O^+ ions desorbed by such a process from TiO_2 (110) surfaces.[834] Using pair-wise sums of spherically symmetric potentials to model ion motion, it was concluded that the most likely ions to desorb are the bridging O ions. Desorption from those sites is fast ($\approx$ 20 fs), whereas desorption of in-plane O ions takes much longer ($\approx$ 100 fs). The in-plane ions have a large probability of being neutralized on the way out of the surface, so they would not be detected. In addition, if they were emitted, they would exhibit distinct angular and energy distributions that are not observed experimentally.

The Knotek–Feibelman Auger-decay model of ESD and PSD is strictly applicable only to d^0 compounds, where no electrons occupy any higher-lying orbitals than the core level from which the initial electron is emitted. However, O^+-ion emission that exhibits threshold effects at cation core-level excitation energies is also observed from d^n oxides. For oxidized Cr metal, where the oxide layer has a composition somewhere between Cr_2O_3 and CrO_2, resonant photoemission and PSD measurements have shown a strong correlation between cation 3p $\rightarrow$ 3d excitations and O^+-ion emission.[546,835] Several competing Auger-decay processes were proposed that are complementary to the Knotek–Feibelman model. The differences between the various Auger-decay channels depend upon whether the initially excited core-level electron acts as a participant in, is a spectator to, or escapes before the subsequent Auger decay. Even more puzzling behavior was found for PSD of O^+ from ion-bombarded, reduced TiO_2 (110) and (001) surfaces, where most of the surface cations have their 3d levels partially occupied.[821] Not only does the O^+ emission exhibit the same threshold as that for stoichiometric TiO_2 surfaces, but the absolute magnitude of the ion yield is three to seven times *larger* than it is for the stoichiometric surfaces. The reasons for this behavior are not understood.

7

Interfaces of metal oxides with metals and other oxides

The adsorption of molecules onto metal-oxide surfaces that was discussed in the preceding chapter is only one aspect of oxide surface chemistry. This chapter will consider the interaction of metal atoms and thin films with oxides in § 7.1–7.6, and the interaction between one metal oxide and an adlayer of another metal oxide in § 7.7.

Metal particles and adlayers on oxide surfaces are relevant to crystal growth, catalysis, gas sensor operation, bonding in composites, etc. Most commercial catalysts consist of small metal particles supported on high-surface-area powders of either SiO_2 or Al_2O_3. In almost all instances those oxides behave as inert surfaces on which the metal can be dispersed in order to obtain maximum exposed surface area. Since there is only a weak interaction between the metal and the support, there has been little impetus to conduct detailed studies of the interaction using single-crystal oxides. A few studies have been performed on single-crystal Al_2O_3, and they are discussed below; SiO_2 is not a metal oxide and hence is not considered here. Metal particles adsorbed onto semiconducting metal oxides such as ZnO and SnO_2 have been found to alter the response of the surface conductivity of those oxides to adsorbed gases, so a number of studies have used model systems in order to understand the processes that are relevant to gas sensor operation. When metals are supported on reducible metal oxides, some very interesting surface chemistry can occur; many of those systems have also been addressed in single-crystal studies.

The interest in oxides on oxides stems primarily from the use of supported oxides in catalysis. This is a very young field from the standpoint of single-crystal studies, so we will have little to say about it. In practice, oxide-supported oxide catalysts are extremely complex. Although the initial preparation of a high-surface-area catalyst may consist of the deposition (usually by wet chemical methods) of one oxide on another (e.g., V_2O_5 on TiO_2), the operation of the catalyst at high temperatures in environments that may be strongly oxidizing or reducing can result in the formation of a mixed oxide containing both types of cation in several oxidation states. Such a complex catalyst could never be modelled be using single crystals. However, valuable insight into the interaction between one oxide and another can be gained from single-crystal studies.

7.1 Metals on oxides

Although most of the interactions between metal atoms and oxide surfaces involve the same physics and chemistry as do those for non-metallic molecules, we discuss metals separately for two reasons. The first is that there has been a great deal of work since 1978 on the interaction between transition- and noble-metal atoms and transition-metal oxides by the catalysis community. A seminal article[836] in 1978, reporting what were referred to as **strong metal/support interactions** (**SMSI**) in supported metallic catalysts, led to extensive studies of metal catalyst particles supported on TiO_2 and other reducible transition-metal oxides.[837,838] The majority of that work was performed on powder-supported catalysts, and the results are thus not particularly easy to interpret in terms that can be modeled theoretically on the atomic scale. But some studies were performed on either single-crystal oxide supports or on thin oxide layers on metal crystals and films. Although in some of those experiments the detailed structure of the surface is still not known, experiments that cannot be performed on powder samples, such as sputter profiling, can be performed on thin-film and single-crystal samples.

The second reason to discuss metals on oxide surfaces separately is that they are models for metal/oxide interfaces. Although most work on metal/oxide interfaces has consisted of the oxidation of metals, there has been increased interest recently in the deposition of metals onto stoichiometric oxides. Usually the metal adsorbate atoms are different than the cations in the oxide, although some work has also been performed on adsorption of the same type of metal atom onto oxide surfaces; the latter is, of course, one of the steps in crystal growth. There is thus interest in the properties of both isolated metal atoms on oxide surfaces, which represent the first stages of interface formation, and thicker metal layers on oxides, which are of more practical interest. In the molecular chemisorption discussed in Chapter 6, rarely was more than a monolayer of adsorbate considered, while for metals the changes that occur as multilayers are deposited onto oxides are of practical importance.

As in the rest of this book, we will concentrate on single-crystal studies of metals on oxides, since our main goal is to understand the basic physics and chemistry of metal/oxide interactions. The discussion of the deposition of metals onto oxides is divided into sections: the remainder of § 7.1 discusses the types of interaction that can occur and some methods of studying them; the following sections then discuss, in order, metals on non-transition-metal oxides, on transition-metal oxides, and finally on high-T_c oxide superconductors. The interaction of metals with the surfaces of high-T_c superconductors is treated separately since the information that has been sought is somewhat different than that for other oxides. Tables 7.2 through 7.4 summarize the literature and contain the relevant references. Chemisorption on oxide-supported metals is discussed briefly in § 7.5 from the standpoint of model catalysts and solid-state gas sensors; no discussion of chemisorption on powder-supported metal catalysts, which comprises the vast majority of the supported catalyst literature, is included. § 7.6 outlines the phenomena that are involved in strong metal/support interactions, using a few examples for illustration. More complete discussions of the SMSI literature can be found in Refs. 837 and 838.

7.1.1 The interaction of metal atoms with oxide surfaces

The ways in which a metal atom can interact with the surface of a metal oxide are potentially extremely varied and complex, and the questions here are only just beginning to be addressed. Experimentally, there is very little information available about the detailed geometric structures adopted by adsorbed metal atoms; information about the initial stages of metal deposition is confined largely to certain aspects of the electronic structure and the degree of charge transfer between the adsorbed metal and the surface. Structural information of a more macroscopic kind, concerning the growth modes of metal films on oxides, is discussed in § 7.1.2.

Aside from catalysis, one of the most important applications of metals on oxide surfaces is to lower the work function. For this purpose, the most effective metals are the alkalis, since the ionization potential for removing the single s electron is relatively small, and it is thus easy to transfer that electron to the oxide. The resulting positive alkali cations on the surface, coupled with their image charges in the oxide, produce a surface dipole layer that can lower the work function (i.e., reduce the potential barrier seen by an electron escaping from the solid) by several electron volts. Examples shown in the tables include the interaction of K with ZnO and NiO, and of Na with TiO_2. Such measurements are only possible, of course, on oxides with sufficient conductivity that the Fermi level can be defined properly.

It is sometimes possible to obtain information about the valence state of cations from the binding energies of their core levels in XPS. In general, oxidation of a metal (i.e., removal of valence electrons to generate a more positive ion) is accompanied by an increase in the binding energy of core levels due to reduced screening of the core hole by the smaller electron density on the ion. For the 3d-transition metals, the binding energy of the 2p core level is usually monitored. Unfortunately, this does not always work, since the binding-energy shift with change in oxidation state may be too small to measure. This is the case, for example, with Na, where the measured binding energies for the 1s level of Na^0 and Na^+ overlap within measurement uncertainties.[839] One way to determine the valence state in cases like this would be to measure the **Auger parameter**, α, which is defined by[839,840]

$$\alpha = BE_{\text{core level}} + KE_{\text{Auger}} - \hbar\omega, \qquad (7.1)$$

where $BE_{\text{core level}}$ is the binding energy of the Na 1s core level and KE_{Auger} is the kinetic energy of a KLL Auger electron (whose initial state is the Na 1s level), both referenced to the same energy. (Sometimes a modified Auger parameter is used, $\alpha' = BE_{\text{core level}} + KE_{\text{Auger}}$, which does not depend on photon energy.[840]) The Auger parameter for Na^+ is 3–6 eV smaller than that for Na^0 metal, so it should be easy to determine the valence state. Not only is this parameter more sensitive in many cases to a change in valence state, but also it should be unaffected by charging problems, since the kinetic energies of Auger and XPS lines should be shifted to the same extent in photon-excited spectra. Unfortunately, Auger-parameter determinations have rarely been made in surface-science studies. (Although not related to surface properties, recent analysis of data for a large number of oxygen-containing compounds has shown that the Auger parameter can give valuable information on the O hole–hole repulsion energies, and hence the nature of the chemical bonds involving oxygen, in oxides.[841])

The alkaline-earth metals are similar to the alkalis in that their two outer s electrons are weakly bound and can readily be transferred to empty states in the substrate. For Al, two s and one p electrons are loosely bound. However, little work has been done on adsorption of those metals on metal oxides, presumably because there are few practical applications of those systems.

Whereas the alkali and alkaline-earth metals readily donate electrons to substrates upon adsorption, resulting in adsorbed 1+ or 2+ ions, respectively, the transition metals may assume any of several valence states when adsorbed on surfaces, depending upon the magnitude of the surface potential at the adsorption site and the availability of electrons or holes in the substrate. Transition-metal atoms having partially filled d orbitals may even be able to *accept* electronic charge from the substrate. The different symmetry of the d orbitals also results in different bonding properties than those of s or sp metals. All of these properties result in rich and often complex adsorption behavior for transition-metal atoms on oxide surfaces.

Given a substantial degree of charge transfer between the adsorbate atom and the substrate, the bonding with the surface is expected to have a strong ionic component. Electron donors, such as alkalis, form cations which presumably interact

Table 7.1. *Reactivity of elements towards oxygen*

Heat of formation of oxide ($-\Delta H_f$ in kJ per mole O)	Elements
0–50	Au, Ag, Pt
50–100	Pd, Rh
100–150	–
150–200	Ru, Cu
200–250	Re, Co, Ni
250–300	Na, Fe, Mo, Sn, Ge, W
300–350	Rb, Cs, Zn
350–400	K, Cr, Nb, Mn
400–450	V
450–500	Si
500–550	Ti, U, Ba, Zr
550–600	Al, Sr, La, Ce
600–650	Mg, Th, Ca, Sc

Data from Ref. 842

ionically with O^{2-} ions on the surface, although there is generally no direct experimental evidence for this. Transition metals can, in principle, adopt more complex modes of bonding: for example, there may be some direct overlap between d orbitals of the adsorbate atom and those of a transition-metal-oxide substrate, giving some degree of metal–metal bonding.

The idea of a metal atom interacting only by electron transfer and bonding to the surface is, however, much too simplistic in general. The adsorption of metals may frequently disrupt the surface structure of the oxide, with a considerable rearrangement of oxygen and metal ions. LEED studies sometimes show reconstruction occurring, as in the case of alkalis on the polar O face of ZnO, discussed below in § 7.2.3. Diffusion of metals into the oxide surface is sometimes claimed, as is the formation of complex mixed-oxide surface phases, perhaps with structures and/or compositions not encountered in the bulk. Another possibility is that metal atoms may extract oxide ions from the substrate. This often happens when metals such as Al and Ti, which have a very high affinity for oxygen, are deposited on oxides of less-reactive elements such as Cu. Surface layers of Al_2O_3 or TiO_2 may then form, with the oxide substrate being reduced. From the point of view of bulk thermodynamics, the possibility of this kind of reaction may be assessed by looking at the relative heats of formation for the oxides of the elements involved. It is essential to normalize these values according to the number of oxygen atoms required, and Table 7.1 shows some relevant data. The elements are ordered according to the heat of formation of the most stable oxide, evaluated as $-\Delta H_f$ in kJ/mol of oxygen atoms. More-reactive metals are ones with a more negative heat of oxide formation, lower down in this table; they should be able to reduce the oxides of metals above them. There are, however, various reasons why the predictions of bulk thermodynamics may not be followed. There is the possibility that surface phases may be formed, with thermodynamic stabilities different from those of bulk oxides. But also it is important to remember that surface reactions of this kind require extensive migration of atoms, a process that may have a large activation energy. Such reactions are therefore more likely at higher temperatures.

7.1.2 Growth modes of metal films on oxides

The result of continued metal deposition is the growth of a film of metal. In XPS and other spectroscopies, peaks originating from the oxide substrate are progressively attenuated, and those from the deposited metal grow in intensity and approach the energies expected for the metallic element. The way in which these changes occur as a function of deposition can, in principle, give information about the mode of film growth, although the data are often not easy to interpret unambiguously. Ordered growth, for example of an epitaxial layer, can be studied by diffraction methods, especially RHEED. When substantial amounts of metal have been deposited, electron microscopy may be the best method of investigation.

There are basically three ways in which thin films of one solid can grow on the surfaces of another solid. (For a more complete discussion of the various types of thin-film growth that have been observed, see Ref. 843) If the interface energy between the film and the substrate is low, and if the temperature is high enough for surface diffusion to occur, then the first monolayer of the adsorbate can grow as a continuous film one atom thick. Subsequent growth of the adsorbate onto the first monolayer can continue in a two-dimensional **layer-by-layer** fashion (which we refer to in the tables as 'layer' growth), or it may occur as three-dimensional islands or clusters, which is referred to as **Stranski–Krastanov** growth (abbreviated 'S–K'). Intermediate cases also occur in which several monolayers grow layer-by-layer before three-dimensional island growth begins. In cases where the interfacial energy between the adsorbate and the substrate is high, three-dimensional clusters may form immediately without the growth of a continuous monolayer; this is referred to as the **Volmer–Weber** (or 'V–W') growth mode.

All three of the above growth modes have been observed for metals deposited onto oxide surfaces, although it is not always easy to differentiate between them, and the tables show examples of apparent discrepancies between the results of different groups. Sometimes these may arise from differences in surface preparation. For example, the growth of many metal films on MgO (100) proceeds differently on air- and UHV-cleaved surfaces. Results such as this show that the interfacial energy between an oxide and a metal may be very sensitive to defects and contamination on the oxide surface.

As a result of the interaction between the adsorbed atoms and the substrate, films will often grow with a preferred orientation that is determined by the geometric structure of the substrate. The most usual growth mode is for a preferred crystal plane of the adsorbate to grow parallel to the surface of the substrate, but with no particular azimuthal alignment with respect to the substrate lattice directions; for fcc metals, this is often the close-packed (111) plane. In some cases, however, the crystal structures of the adsorbate and the substrate are such that the preferred plane will also grow in azimuthal alignment with the substrate, so that all of the crystal directions of the film are determined by the substrate. This growth mode is referred to as **epitaxial**, or sometimes 'heteroepitaxial' since the overlayer is a different material then the substrate. Both of these modes are also observed for metals on oxides.

Very few calculations have been performed for metals on oxides. The ones referenced in the tables, and discussed under the heading of specific systems below, deal with the interaction of isolated atoms and have rather little to say about the growth of films. However, classical image-potential theory has been used to treat the adhesion between metals and oxides in a generalized manner; it has suggested some guidelines as to what might be expected at metal/oxide interfaces.[844,845] It has been suggested that wetting – that is, a low interfacial energy favoring layer growth – will be promoted by a high dielectric constant in the oxide, and by oxide surfaces where ions with high charges are exposed. The defect properties of the oxide were also considered in these calculations, and the somewhat surprising conclusion was reached that, when a range of stoichiometry is possible, the region close to the interface is likely to be oxygen-rich rather than metal-rich.

7.2 Metals on non-transition-metal oxides

The motivation for studying metals on non-transition-metal oxides includes low-work-function thermionic emitters, catalysts and catalyst supports, thermionic energy converters and gas sensors. Table 7.2 summarizes the published reports of experimental and theoretical studies of the interaction of metals with well-characterized single-crystal surfaces of non-transition-metal oxides.

7.2.1 MgO and other alkaline-earth oxides

The simplest metal/oxide systems consist of alkali metals on wide-bandgap oxides such as MgO and CaO. Two such systems have been investigated on single-crystal surfaces. Na on MgO (100) and faceted (111), and on CaO (100). The behavior is

Table 7.2. *Metals on non-transition-metal oxides*

Oxide	Metal	Oxide face	Surface preparation	Results	References
MgO	Na	100	P&A	Weak interaction, at defects, V–W	397
		111	Faceted to (100)	Stronger interaction, at defects, V–W	397
	Mg	100	UHV cleaved	S–K, epitaxial; initially fcc on 90 K substrate	848
			Theory	Strong bond, Mg^{2+}, surface relaxation	254
	Al	100	UHV cleaved	Epitaxial	849
			Air cleaved	S–K epitaxial	849,850
	Y	100	P&A	Epitaxial, Y^{2+} in first monolayer	851,852
	Fe	100	Air cleaved and annealed	Epitaxial Fe (100)	850,853–857
			C contamination	V–W	854
			Ion bombarded	Fe bonds to O^{2-} ions, V–W ?	396
	Ni	100	Air cleaved and annealed	S–K, epitaxial	850
			P&A	Epitaxial Ni (100), S–K ?	858
	Pd	100	UHV and air cleaved	V–W, some epitaxial Pd (100)	859
			Air cleaved and annealed	S–K, epitaxial	850
	Pt	100	Air cleaved and annealed	S–K, epitaxial	850
	Cu	100	UHV cleaved	Epitaxial Cu (100)	858
		100	P&A	Epitaxial Cu (100), S–K, initially Cu^+	91,847,850 860,861
		100	Defect	Preferential adsorption at cation vacancies	441,861
		100	Theory	Strong bond to cation vacancies, Cu^{2+}	862
		111	P&A	Epitaxial, Cu (111), S–K, initially Cu^+	91
	Ag	100	UHV cleaved	Epitaxial Ag (100)	849,858,863
			Air cleaved	V–W, random orientation	849,864
			Air cleaved and annealed	S–K, epitaxial	850
	Au	100	UHV cleaved	Epitaxial Au (100) for T = 373 or 473 K; polycrystalline for T = 78 K; can get Au (111); see § 7.2.1	849,863
			Air cleaved	Random crystallite orientation	849
			Air cleaved and annealed	S–K, epitaxial	850
CaO	Na	100	P&A	Weak interaction, at defects, V–W (?)	730
	Fe	100	P&A	Initially Fe^{2+}, Fe^0 by 2 monolayers	730
Al_2O_3	Al	0001	P&A	Strong interaction, surface reconstruction, S–K	166,168
	Ti	0001	P&A	Strong Ti–O bond, Ti^{2+} or Ti^{3+} for first monolayer, then Ti^0	865
		0001 and $10\bar{1}2$	P&A	Ti oxidized, Al reduced; Ti_3Al formed at high T	421,866
	Nb	0001	P&A	Forms monolayer of oxidized Nb; then Nb^0	421
	Fe	$10\bar{1}2$	P&A	S–K	867
		Cluster	Theory	Covalent Fe 3d – non-bonding O 2p bond	868

Table 7.2. (*cont.*)

Oxide	Metal	Oxide face	Surface preparation	Results	References
Al_2O_3	Rh	0001	P&A	Weak, V–W	869
	Ni	0001	P&A	Weak interaction at RT, Ni^0; surface reduction at 1000 K; spinel formation at 1000 K in O_2	421,870,871
		Cluster	Theory	Ionic bond, $Ni^{1.5+}$	872
				Covalent Ni 3d – non-bonding O 2p bond	868
	Pd	$10\bar{1}2$	P&A, high temperature $(\sqrt{2} \times \sqrt{2})R\,45°$	Diffusion into substrate, then epitaxial (110) or (111)	161
	Pt	0001	P&A	Weak, V–W	422,869, 873–875
		0001 Amorphous–	Theory	Strong interaction with Pt (111), Pt^+–O^{2-} bond Layer	876 873
	Cu	0001	P&A	S–K, initially Cu^+ on (1×1) surface	877–880
				V–W, initially Cu^+, larger clusters on O-deficient surface	165
		Cluster	Theory	Covalent Cu 3d – non-bonding O 2p bond	868
	Ag	Cluster	Theory	Covalent Ag 4d – non-bonding O 2p bond	868
	Ge	$10\bar{1}2$	P&A	V–W	881
$LaAlO_3$	Y	100	P&A	Either S–K or V–W	851,852
ZnO	Na	$000\bar{1}$	P&A	Strong bond, Na is e^- donor, surface reconstruction, perhaps Na diffusion into lattice upon heating	882
	K	$000\bar{1}$	P&A	Strong bond, K is e^- donor, surface reconstruction	882
	Cs	$000\bar{1}$	P&A	Strong bond, Cs is e^- donor, surface reconstruction	882–884
		$10\bar{1}0$ Powder	P&A –	Weak interaction	882–884
				1 monolayer decreases Φ from 4.4 to 1.3 eV	885
	Ca	$000\bar{1}$	P&A	Strong bond, Ca is e^- donor, surface reconstruction	882
	Al	$10\bar{1}0$	P&A	Strong interaction, forms Al oxide	259
	Ni	0001 and $000\bar{1}$	P&A	Layer, stronger on $(000\bar{1})$	8
	Pd	0001 and $000\bar{1}$	P&A	Layer, preferential Pd (111)	886
		$000\bar{1}$	Air cleaved and annealed	V–W	887
	Pt	0001 and $000\bar{1}$	P&A	Layer, epitaxial Pt (111)	875
	Cu	$000\bar{1}$	P&A	Epitaxial Cu (111), S–K	9,888
		0001 and 1010	P&A	Less stable than $(000\bar{1})$	9,889
		$10\bar{1}0$	P&A	Initially Cu^+	861
		$000\bar{1}$	Theory	Cu^{2+} for neutral cluster	722
		0001	Theory	Weak bond	722
	Ag	0001 and $000\bar{1}$	P&A	Layer, preferential Ag (111) orientation	886

Table 7.2. (*cont.*)

Oxide	Metal	Oxide face	Surface preparation	Results	References
ZnO	Au	0001 and 000$\bar{1}$	UHV cleaved	Epitaxial Au (111)	890
			Air cleaved	Polycrystalline Au	890
		10$\bar{1}$0	P&A, stoichiometric, and reduced	Weak interaction	259
SnO$_2$	Pd	110	O-terminated or reduced	V–W, epitaxial Pd (111)	15,443, 891–894
	Sn	110	P&A	Strong interaction, large increase in surface conductivity	892,894

similar in both cases. On stoichiometric (100) surfaces, Na adsorbs at room temperature with very little interaction with the substrate. It is believed that Na adsorbs initially at surface defects, and that three-dimensional islands of Na then grow around the nuclei. The binding energy of the Na 1s core level as measured by XPS was found to decrease by about 0.4 eV from isolated atoms up to coverages equivalent to multilayers on MgO (100). As pointed out above, the measured binding energies for the 1s level of Na^0 and Na$^+$ overlap within measurement uncertainties.[839] Thus there is no unique interpretation of the binding-energy shifts observed when Na is deposited onto MgO or CaO. A decrease in Na 1s binding energy with increasing film thickness could result either from a change in valence state of the Na or from increased screening of the core hole by electrons on adjacent atoms as the Na particles grow in size.

A stronger interaction was found for Na deposited onto faceted MgO (111). The binding energy of the 1s core level for the first Na atoms adsorbed was 0.7 eV larger than that for initial deposition on the (100) surface; it then decreased by 1.0 eV as more Na was deposited, eventually approaching the value for comparable thickness films on (100). Changes in the angle-resolved He II valence-band UPS spectra were observed for the (111) surface after Na deposition, but they were not correlated with any specific changes in electronic structure; more detailed experiments would be required in order to determine the nature of any charge transfer taking place between Na and MgO. The stronger interaction of Na with the faceted surface presumably occurs at edge or corner sites, where the Mg and O ions have coordinations different from that on perfect (100) terraces.

The growth of thin **Mg** films on UHV-cleaved MgO (100) has been studied by using both LEED and RHEED. Substrate temperatures were varied from 90 to 320 K, and the Mg grew as epitaxial islands at all temperatures. For growth temperatures of 170 K and above, the normal hcp Mg phase formed at all coverages;

the epitaxial relationship did, however, depend upon temperature. In contrast, deposition at 90 K produced Mg particles having an fcc structure which was in nearly perfect registry with the MgO substrate. That phase was stable up to coverages of about five monolayers equivalent (i.e., the amount of metal that would form five monolayers if it grew in layer-by-layer fashion); for higher coverages the hcp structure was observed.

In addition to the single-crystal experiments just described, ESR and diffuse reflectance spectroscopy have been used to study the changes that occur when MgO powder samples are exposed to Mg vapor.[846] Deposition of Mg onto the surface changes the sample to a deep blue-violet color, with the appearance of an intense optical absorption band centered at 2.3 eV; the change is also accompanied by the appearance of two overlapping ESR signals. By analogy with the 2 eV ELS peak that is seen on defective MgO (100) surfaces, it was argued that the absorption corresponded to surface F centers. As discussed in § 4.3.2, however, the identification of that ELS transition is open to discussion, so the assumption of a surface F center on the powder is uncertain.

Theoretical defect lattice calculations have been performed of the interaction of Mg atoms with the (100) surface of MgO. The lattice was allowed to relax fully around the Mg adatom, both electronically and geometrically, which resulted in a Madelung potential at the adatom site that was 70 % of that for bulk Mg ions in MgO. That Madelung potential was sufficiently large to stabilize the Mg^{2+} configuration, and the resultant bonding to the surface was very strong, yielding a binding energy of 15.1 eV.

The growth of **Y** films on polished and annealed MgO (100) surfaces has been studied by using a variety of surface spectroscopies. At room temperature the first monolayer of Y is found to oxidize, with the oxide growing epitaxially. The data were interpreted in terms of the formation of Y^{2+} ions (although Y^{2+} is not normally a stable oxidation state of Y). The (1 × 1) LEED patterns observed were interpreted in terms of the formation of an ordered Y–O superstructure, with the O coming form the substrate. That structure can be stabilized in thicker Y layers by further oxidation with O_2.

XPS has been used to study the deposition of **Fe** onto single-crystal CaO surfaces that were cleaned by Ar^+-ion bombardment and annealing, but whose surface structure was not monitored by LEED. The Fe 2p core level binding energy indicated that the initial species adsorbed was Fe^{2+}. The binding energy decreased smoothly with Fe coverage, reaching the value for metallic Fe^0 by two monolayers.

The effect of impurities on the surface of MgO on the growth mode of Fe films has been observed by LEED and Auger. When monolayer amounts of Fe are deposited on air-cleaved MgO (100) surfaces that have been cleaned by annealing in UHV, epitaxial growth is found to occur; there is a good lattice match between

Fe (100) and MgO (100) for a 45° rotation of the [001] axis. Comparison of measured LEED I-V curves with model calculations indicates that the Fe atoms sit directly above surface O ions for the first monolayer. However, when the surface is contaminated with C, three-dimensional island growth is found to occur for all film thicknesses. In experiments on Fe deposition onto air-cleaved MgO (100) surfaces that had been subsequently ion bombarded, it was concluded that the Fe atoms bond to surface O^{2-} ions for low coverages. For temperatures of about 570 K and above, the interdiffusion of Fe and MgO was observed, with the formation of a structurally ordered solid solution in the surface region.

The growth of **Pd** films on both UHV- and air-cleaved MgO (100) substrates occurred via clusters whose predominant orientation was with the Pd (100) plane parallel to the substrate. The size and morphology of the clusters varied depending upon the substrate temperature and the deposition rate. Nucleation appeared to occur at surface defect sites (point defects or steps), and it was found that the highest nucleation rate was obtained for samples cleaved in air, where surface point defects are formed by attack from water vapor in the atmosphere. Sub-monolayer Pd films grown at high temperature on air-cleaved MgO (100) consisted of a large number of 1–3 nm diameter particles having a surprisingly narrow size distribution, as shown in Fig. 7.1.

The interaction of **Cu** atoms with single-crystal MgO surfaces has been addressed both experimentally and theoretically. When Cu was deposited at room temperature in UHV onto well-characterized (100) and (111) surfaces, it was found that the films grew at least partially epitaxially; experiments on the deposition of Ni on MgO (100) led to a similar conclusion. The initially sharp LEED patterns of the clean, stoichiometric surfaces disappeared after the first few Å of Cu deposition, but for coverages $\geq$ 15 Å faint, diffuse Cu (100) or (111) patterns were observed. By monitoring Auger and XPS intensities as a function of deposition time, it was concluded that the growth mechanism was Stranski–Krastanov; i.e., the first monolayer formed continuously, but successive growth was via three-dimensional islands. HREELS measurements of the change in the MgO surface phonon spectrum as a function of Cu deposition, coupled with modelling of the spectrum for such an interface using dielectric theory, lead to the conclusion that the initial layers of Cu interact with the MgO surface to produce a compound similar to Cu_2O rather than growing as metallic Cu.[847]

Cu was also deposited onto defective MgO (100) surfaces that exhibited the 2 eV loss feature discussed in § 4.3.2. As little as 0.5 Å of Cu was sufficient to eliminate the loss peak, leading to the conclusion that the peak corresponds to a surface cation vacancy, and that Cu atoms preferentially adsorb at those vacancy sites. This is in agreement with the results of *ab initio*, unrestricted Hartree–Fock cluster calculations, which showed that either a Cu^0 atom or a Cu^+ ion would bond strongly to Mg

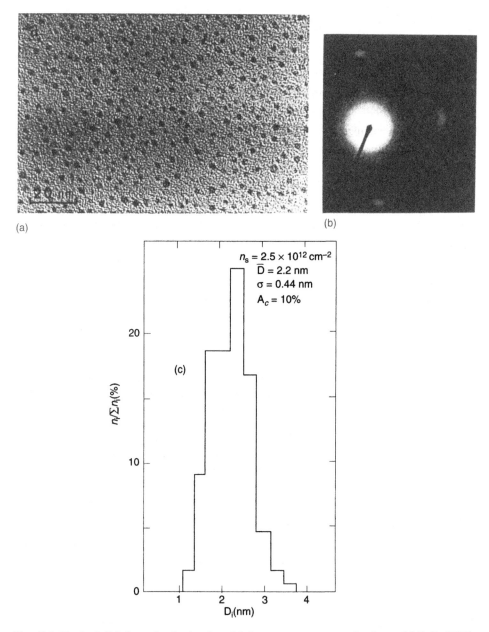

Fig. 7.1 Typical Pd deposit obtained at high temperature on air-cleaved MgO (100): (a) micrograph, (b) diffraction pattern, and (c) size histogram. [Ref. 859]

vacancies on MgO (100), transferring electrons to the substrate to become Cu^{2+} ions. (Relaxation of the surface in the presence of defects or adsorbates was not included in the calculations, however.) The binding energy for a Cu atom at a Mg vacancy site was computed to be 9 eV, and the Cu ion was located 0.5 Å above the normal location of the missing Mg ion. Similar calculations were also performed for Cu adsorption above a surface O ion, which is the site that would be occupied by a Mg ion in bulk MgO. That adsorption site was found to be barely stable, with a binding energy of only 0.2 eV. In judging such calculations that do not include any relaxation of the substrate, one must remember that a moderate amount of relaxation can have a large effect on the Madelung constant at an adsorption site; this in turn affects the relative stability of various valence states of the transition-metal adsorbate.

The initial stages of **Ag** deposition onto MgO (100) have been studied for both air- and UHV-cleaved surfaces. Epitaxial layer-by-layer growth of Ag (100) is found for deposition on UHV-cleaved surfaces at temperatures ranging from 78 to 473 K, with the most highly ordered films occurring for deposition at room temperature and above. Ag films deposited on air-cleaved MgO (100), however, appeared to consist of three-dimensional islands rather than oriented epitaxial growth. This is essentially the same result as for Fe.

The structure of **Au** films grown on UHV-cleaved MgO (100) is more complex than that of Ag or Fe. Good epitaxial (100) films were produced for substrate temperatures of 373 and 473 K, although stacking fault and twinning defects were found much more frequently in Au than in Ag films. Deposition of Au at 78 K resulted in polycrystalline films that did not exhibit any LEED patterns. It was found to be possible to produce oriented (111) Au films on MgO (100) by a two-stage deposition process consisting of initial deposition of one to three monolayers at 78 K, followed by deposition at 473 K. The azimuthal orientation of the resulting films was found to depend on the thickness of the initial low-temperature layer. Au (111) growth was also favored on MgO (100) surfaces that had been heavily electron bombarded prior to deposition. The occurrence of (111) oriented films was thus attributed to the effect of steps or defects on the MgO surface.

7.2.2 Al_2O_3 and $LaAlO_3$

Al_2O_3 is an important substrate for the adsorption of transition metals since Al_2O_3 powder is one of the most commercially important catalyst supports. The alumina used for supports is generally γ-Al_2O_3, but single crystals of that form are not available; so all experiments on well-characterized single-crystal surfaces have been performed on α-Al_2O_3.

The deposition of **Al** onto polished and annealed Al_2O_3 (0001) surfaces has been studied by a variety of surface techniques. The first few monolayers of Al interact strongly with the substrate. On surfaces that exhibited a (1 × 1) LEED pattern when clean, the deposition of Al formed an Al-rich region that exhibited the same $(\sqrt{31} \times \sqrt{31})$ R ± $\tan^{-1}(\sqrt{3}/11)$ LEED patterns that are seen on the clean surface as a function of annealing temperature (see § 2.3.4.2). For average Al coverages greater than five monolayers, metallic Al clusters form on top of the modified interface layer.

Ti also interacts strongly with Al_2O_3 (0001). Ti–O bonds are formed at the interface, with the Ti ions exhibiting a valence between 2+ and 3+. In one study, the ionic layer was determined to be only one monolayer thick, with the second and successive layers consisting of metallic Ti; no determination of the growth morphology of Ti on Al_2O_3 (0001) was made. In another study, annealing of the Ti/Al_2O_3 interface at 1273 K resulted in the formation of a layer of Ti_3Al.

A much weaker interaction is found for **Nb** with Al_2O_3 (0001) surfaces. The initial monolayer of Nb becomes slightly oxidized, donating electrons to the surface. But that reaction stops after the first monolayer, with the deposition of metallic Nb for higher coverages.

The interaction of **Ni** with Al_2O_3 (0001)-(1 × 1) surfaces has been studied as a function of temperature and O_2 ambient. Deposition at room temperature in UHV gave no indication of any reaction at the interface, and the Ni went down as Ni^0. However, partial reduction of the Al_2O_3 surface was found when Ni was deposited at about 1000 K in vacuum. In the presence of a small amount of O_2 (5×10^{-9} Torr), a $NiAl_2O_4$ spinel phase was formed, indicating that the Ni/Al_2O_3 interface is quite complex.

Ab initio, unrestricted Hartree–Fock cluster calculations have been performed of the interaction of Ni with Al_2O_3. No specific crystal surface was modelled, but two different clusters, one having a surface perpendicular to the *c*-axis and the other with a surface that contained the *c*-axis, gave very similar results. An ionic configuration close to $Ni^{1.5+}$ was predicted, with the cation bonding ionically to the surface with an energy of 5.3 eV. No experiments on Ni/Al_2O_3 have been performed to confirm these predictions. SCF-Xα-SW cluster calculations have also been performed for the interaction of **Fe, Ni, Cu** and **Ag** with Al_2O_3. Several different cluster geometries were used, including one simulating the (0001) surface; similar results were found for all of the clusters. This work found that a primarily covalent chemical bond was established between the metal d electrons and the non-bonding O 2p orbitals, with the strength of the bond decreasing from Fe to Ag.

The deposition of **Pd** has been studied on Al_2O_3 ($10\bar{1}2$) surfaces that had been annealed above 1773 K and showed a ($\sqrt{2} \times \sqrt{2}$) R 45° reconstruction attributed to an ordered oxygen-vacancy structure. Three states of growth were suggested: initial deposition of isolated Pd atoms, followed by some Pd diffusion into the substrate, and then at higher coverages the formation of small crystallites oriented epitaxially on either (110) or (111) faces.

Other experiments have been performed of the interaction of **Rh** and **Pt** with the (0001) surface of Al_2O_3. The measurements were performed in UHV, but the surfaces were not characterized by LEED, so their state of perfection is unknown. Rh

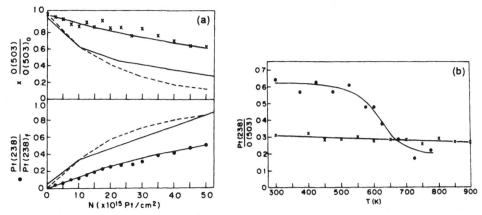

Fig. 7.2 (a) O and Pt Auger peak amplitudes (normalized to the values for bulk Pt and Al_2O_3) versus Pt exposure at 300 K. Data points are shown for Al_2O_3 (0001), while results for amorphous alumina are represented by solid lines. Dashed lines are calculated intensities assuming layer-by-layer growth. (b) Pt-to-O Auger peak ratios as a function of temperature for an initial Pt coverage of 2×10^{15} atoms/cm^2. Circles are for amorphous alumina, crosses are for Al_2O_3 (0001). [Ref. 873]

was found to interact fairly weakly with Al_2O_3 (0001), growing in three-dimensional islands at even the lowest coverages; a similar result was found for Pt. However, when Pt was deposited onto amorphous Al_2O_3 films, layer-by-layer growth was observed. The two growth modes are illustrated in Fig. 7.2 (a), which plots the decrease in the O Auger peak amplitude and the increase in the Pt peak amplitude as a function of the coverage, N, of Pt atoms.[873] The changes in slope in the solid lines for growth on the amorphous alumina substrate occur at the completion of each monolayer of Pt. The absence of such breaks for Pt on Al_2O_3 (0001), along with the smaller dependence of the Auger amplitudes on Pt atom coverage, are indicative of three-dimensional island growth. Figure 7.2 (b) plots the Pt-to-O Auger peak amplitudes as a function of annealing time for the first anneal of each surface. The large decrease in the ratio for Pt on the amorphous substrate results from coalescence of the two-dimensional Pt film into three-dimensional particles; this change is, of course, irreversible. The Pt clusters on stoichiometric Al_2O_3 (0001) were found not to have any particular orientation with respect to the substrate lattice.

Cluster calculations of bonding at the Al_2O_3 (0001)/Pt (111) interface predict a stronger interaction than that observed experimentally for Pt deposition on stoichiometric Al_2O_3 (0001). The calculations show that the surface Al^{3+} ions have empty dangling-bond orbitals that can receive electrons from the Pt atoms at the interface. The resulting Pt^+ ions bond strongly to the substrate O^{2-} ions, whereas neutral Pt atoms do not. The strength of the calculated bonding should result in the

Pt particles growing in registry with the substrate, if not in two-dimensional layers; neither of these is observed experimentally.

Several studies of the interaction of **Cu** with polished and annealed Al_2O_3 (0001) surfaces that were characterized by LEED, XPS and Auger have been performed, and they do not all reach the same conclusions. Most studies reported that the first monolayer of Cu goes down uniformly on the substrate, while further deposition resulted in the growth of three-dimensional islands (i.e., Stranski–Krastanov growth) of metallic Cu. One study, however, which looked at both (1×1) and $(\sqrt{31} \times \sqrt{31})$ R $\pm \tan^{-1}(\sqrt{3}/11)$ surface structures, concluded that Cu grew as clusters at all coverages (i.e., Volmer–Weber growth); the clusters were larger on the O-deficient reconstructed surface. All studies agree that the initial fraction of a monolayer of Cu had a valence close to Cu^+, but that above about one monolayer the Cu was metallic. The amount of charge transfer from the Cu atoms to the substrate is smaller on the slightly reduced $(\sqrt{31} \times \sqrt{31})$ R $\pm \tan^{-1}(\sqrt{3}/11)$ surface than on the oxidized (1×1) surface.

The growth of **Ge** films on well-characterized Al_2O_3 (10$\bar{1}$2) surfaces has also been studied. For the entire substrate temperature range used, 300–900 K, Ge clusters formed even at the lowest coverages that could be measured.

In one study that was concerned primarily with the use of Auger spectroscopy in determining overlayer growth modes, **Y** was deposited onto **LaAlO$_3$** (100) surfaces. All that could be determined was that the growth mode was primarily via islands, but a uniform coverage of the first monolayer (i.e., Stranski–Krastanov growth) could not be ruled out. Attempts were also made to deposit Cu onto $LaAlO_3$, but surface charging complicated any interpretation of the results.

7.2.3 ZnO

In oxides that have enough bulk conductivity that charging does not occur in UPS spectra, the decrease in work function that accompanies alkali metal adsorption and electron transfer to the substrate can be measured directly. For ZnO powder samples, deposition of roughly one monolayer of Cs decreased the work function to 1.3 eV;[885] the effects of Cs on polycrystalline ZnO are important in the operation of thermionic energy converters. In single-crystal studies of the interaction of **Cs, K, Na** and **Ca** with the polar (000$\bar{1}$)-O face of ZnO, and of Cs with the non-polar (10$\bar{1}$0) face, LEED and Auger were used to monitor the surface geometric structure and the amount of adsorbed metal. On the non-polar (10$\bar{1}$0) face, only a weak interaction was found, with no change in the symmetry of the surface; the Cs was completely removed by annealing at 650 K. A much stronger interaction was found for Cs, K and Ca with the (000$\bar{1}$) surface. Since this surface has predominantly incompletely coordinated O ions exposed, the metals readily donate their s electron to the surface and bond ionically with surface O ions. Both (2×2) and $(\sqrt{3} \times \sqrt{3})$ reconstructions were formed during annealing of the metal-covered surface, with the $(\sqrt{3} \times \sqrt{3})$ structure, which corresponded to about one-third of a monolayer coverage of metal, being stable at higher temperatures. One-third monolayer coverage is close to the one-fourth of a monolayer of 1+ cations that would theoretically stabilize the $(000\bar{1})$ surface and eliminate the net surface dipole.[883] All three metals are tightly bound to the surface, and even annealing at 1100 K was not sufficient to remove the last one-third of a monolayer; it was necessary to ion bombard and anneal the surface in order to completely remove the metal and restore the (1×1) clean surface structure.

The behavior of Na on the ZnO $(000\bar{1})$ surface was slightly different. The same stable $(\sqrt{3} \times \sqrt{3})$ LEED pattern was observed after annealing a Na-covered surface at 850 K, but Auger spectra showed no trace of Na remaining on the surface. It was postulated that either the Na had evaporated as an Na oxide, leaving O vacancies on the surface which stabilized the reconstruction, or that the Na ions had diffused substitutionally into the ZnO, with a sufficient concentration remaining in the near-surface region to stabilize the reconstruction, but too small to be seen by Auger. The fact that it was necessary to Ar$^+$-ion bombard the sample for 18 hr at 900 K in order to restore the clean surface structure argues for the latter explanation.

Al also interacts strongly with ZnO surfaces; this is not unexpected due to the strong affinity of Al for oxygen, as shown in Table 7.1. Figure 7.3 plots both the change in sample work function (measured by using a Kelvin probe) and the change in band-bending (measured from the surface photovoltage) as a function of the thickness of the Al layer deposited onto both defective and defect-free ZnO $(10\bar{1}0)$. There is a large decrease in work function for the first 0.5–1 Å of Al, after which the work function rises, eventually reaching the value for bulk Al. The rise is much faster on the reduced surface than on the stoichiometric one; it was suggested that this might be the result of the availability of more O ions on the stoichiometric surface to form Al oxides at the interface, resulting in a thicker interfacial Al oxide layer. No determination of the growth mode of Al on ZnO was made.

When **Ni** is deposited at room temperature onto polished and annealed surfaces of either ZnO (0001) or $(000\bar{1})$, the film grows in layer-by-layer form, although no information is available on the orientation, if any, of the film. UPS was used to monitor changes in the surface electronic structure during Ni deposition and to determine changes in band-bending and chemical composition at the surface. Ni was found to interact more strongly with the $(000\bar{1})$-O face than with the (0001)-Zn face. Figure 7.4 presents He I and He II UPS spectra for ZnO $(000\bar{1})$ before and after the deposition of 3.2 Å of Ni. The motion of the filled Zn 3d band by about 0.6 eV toward E_F upon Ni deposition is indicated by the vertical lines; emission

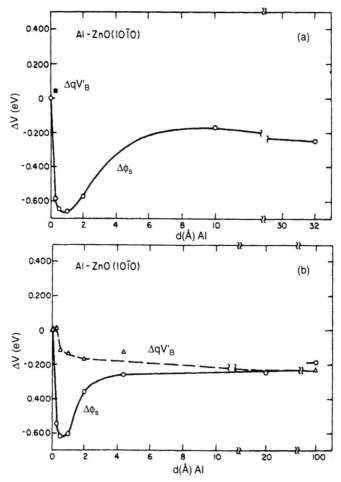

Fig. 7.3 Variations of surface work function, $\Delta\Phi_s$, and band-bending, $\Delta qV_B'$, versus Al overlayer thickness, d, for ZnO $(10\bar{1}0)$ (a) nominally defect-free and (b) containing 0.02 monolayer of O-vacancy point defects. [Ref. 259]

from the Ni 3d band near E_F and other changes in the valence-band structure can also be seen. Negligible band-bending is seen for the (0001) surface. Also, when Ni was deposited on the $(000\bar{1})$-O face, O ions were found to diffuse into the Ni film; no O diffusion was observed for Ni films deposited on the (0001)-Zn face.

There is some disagreement in the literature as to the manner in which **Pd** grows on single-crystal ZnO surfaces at room temperature. Auger and UPS measurements on polished and annealed (0001) and $(000\bar{1})$ surfaces that had not been characterized by LEED indicated layer-by-layer growth, with preferential orientation of the polycrystalline grains having their (111) planes parallel to the surface, but with no azimuthal alignment. On the other hand, deposition of Pd onto ZnO $(000\bar{1})$ surfaces

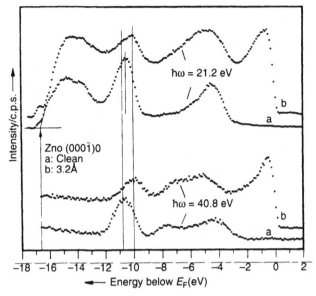

Fig. 7.4 He I and He II UPS spectra for ZnO (000$\bar{1}$) (a) clean and (b) covered with a 3.2 Å film of Ni. The spectra are referenced to the Fermi level. [Ref. 8]

that had been cleaved in air and subsequently annealed in UHV and that exhibited excellent (1 × 1) LEED patterns was interpreted in term of cluster growth at all coverages. Upon annealing, the clusters agglomerated, resulting in a greater area of exposed ZnO substrate and a (111) orientation of the larger Pd crystallites. In any event, the growth of Pd is not epitaxial, which is not surprising considering that there is a 16 % lattice mismatch between the polar ZnO surfaces and Pd (111).

Pt was found to grow layer-by-layer on ZnO (0001) and (000$\bar{1}$) surfaces. Transmission-electron microscopy performed on the (0001) face showed that the Pt grows with its (111) plane parallel to the ZnO surface, but with a 30° rotation between the two lattices. The two-dimensional Pt layers are stable to about 650 K, above which coalescence into three-dimensional particles occurs.

The behavior of **Cu** on ZnO surfaces is particularly interesting since Cu-Zn-O catalysts are extremely important commercially for the production of methanol from CO and H_2. In an attempt to understand the role of Cu in the reaction, several studies have been performed in which Cu atoms were deposited onto the (0001), (000$\bar{1}$) and (10$\bar{1}$0) surfaces of ZnO. For room-temperature deposition, the initial monolayer of Cu forms uniformly on all three surfaces, growing epitaxially as Cu (111) on the (000$\bar{1}$) face. Further deposition results in the formation of three-dimensional islands on top of the initial monolayer (i.e., Stranski–Krastanov growth), with the Cu (111) islands rotationally aligned with respect to the ZnO (000$\bar{1}$) substrate. Upon heating, Cu on the (0001) surface is least stable and most prone to clustering, with Cu on the (000$\bar{1}$) surface being the most stable; the Cu/ZnO (10$\bar{1}$0) system is intermediate in stability.

ELS measurements of Cu adsorption on the ZnO (10$\bar{1}$0) surface suggested that the Cu exists in a Cu$^+$ valence state.[861] Somewhat different results were obtained theoretically by using the semi-empirical quantum-mechanical INDO approach to Cu-Zn-O clusters. When a single Cu atom was substituted for a Zn atom on the (000$\bar{1}$) surface or was placed in either a three-fold hollow or an on-top site on that surface, the charge on the Cu atom for a neutral cluster was essentially the same as that of the Zn ions in the cluster. In other words, either adsorbed or substitutional Cu ions are nearly Cu^{2+}. Only if the cluster was given a net negative charge did the Cu atom become less negative than the Zn ions. The Cu adsorbed on the (000$\bar{1}$) surface was found to be quite stable and tightly bound, while Cu in similar sites on the (0001) surface was found to be only weakly bound. This is in agreement with the experimental observation of the relative stability of Cu layers on those two surfaces discussed above.

Ag deposited on polished and annealed ZnO (0001) and (000$\bar{1}$) surfaces was found to grow in a layer-by-layer mode with preferential orientation of the (111) planes of individual Ag crystallites parallel to the surface, but without any strong azimuthal orientation of the crystallites.

Electron diffraction measurements on thick (200–500 Å) **Au** films grown on both the (0001) and (000$\bar{1}$) surfaces of ZnO showed that the films grew epitaxially on UHV-cleaved samples, with the Au (111) plane parallel to the surface and the Au [$\bar{1}$10] direction parallel to ZnO [11$\bar{2}$0]; however, unoriented polycrystalline Au layers were formed on air-cleaved samples that were not treated in UHV. Au was also deposited on polished and annealed ZnO (10$\bar{1}$0) surfaces both with and without O-vacancy defects created either by high-temperature treatment or ultraviolet irradiation. The growth mode was not determined in these studies, but measurements of the surface band-bending and work function showed that there was little interaction between the Au and either the defective or defect-free surface. Figure 7.5, which is analogous to Fig. 7.3 for Al, plots both the change in sample work function and the change in band-bending as a function of the thickness of the Au layer on both types of surface. In both cases the bands bend down only slightly at the surface, and the work function increases smoothly from the value for the clean surface to that for bulk Au.

7.2.4 SnO$_2$

The interest in metals on SnO$_2$ surfaces arises primarily from its use in gas sensors. Although actual sensors utilize polycrystalline SnO$_2$, some studies have been performed on single-crystal SnO$_2$ (110) surfaces. **Pd** has been deposited onto both stoi-

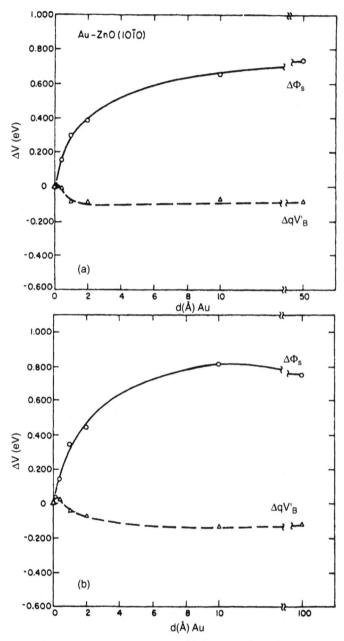

Fig. 7.5 Same as Fig. 7.3, but for Au deposition on ZnO (10$\bar{1}$0). [Ref. 259]

chiometric and reduced SnO_2 (110) surfaces (recall the discussion concerning the ease of O-ion removal from that surface in § 2.3.3.1), with similar behavior seen on both of them; most attention has been paid to the O-terminated stoichiometric surface, however. Pd is found to exhibit cluster growth on SnO_2 (110), with no indication of a uniform monolayer preceding cluster formation. However, by the time the Pd films are about 10 monolayers thick, they exhibit very good (111) LEED patterns that are in azimuthal registry with the substrate; growth at the interface thus appears to be epitaxial. The amount of clustering was found to increase with post-growth annealing temperature as agglomeration occurred. The Pd/SnO_2 interaction is weak, and there is very little change in the surface conductivity of SnO_2 (which is the property that is important in gas sensors) upon Pd deposition. (An exception to this is for deposition onto plasma-oxidized SnO_2 surfaces that have an excess of chemisorbed oxygen species; there Pd deposition results in a larger increase in conductivity.) Only after the Pd layer is thick enough to become continuous and metallic does it form a conducting layer on the surface. There is some question of the stability of Pd layers on SnO_2 (110) under actual sensor operating conditions, however. While the Pd remains in the form of metallic clusters on the surface at temperatures up to 670 K in reducing environments, it forms p-type semiconducting PdO under oxidizing conditions at those temperatures. For temperatures above 670 K and O_2 partial pressures above 600 Pa, Pd^{2+} ions diffuse into the bulk of the SnO_2.

Atomic **Sn** has also been deposited onto stoichiometric SnO_2 (110) surfaces. It behaves very differently than does Pd. No oriented growth is seen, with the substrate LEED patterns merely disappearing as the thickness of the Sn layer increases. However, as little as 0.1 monolayer of Sn produced a fifty-fold increase in the conductivity of the stoichiometric SnO_2 (110) surface. [On plasma-oxidized SnO_2 (110) surfaces, the deposition of Sn results in a thousand-fold increase in surface conductivity.] This effect is clearly due to electronic interactions between the Sn and the substrate and not to conductivity of the Sn overlayer. In fact, there was little additional increase in conductivity as the Sn overlayer became continuous. No such changes in substrate surface conductivity were observed when Sn was deposited onto the ordered but reduced SnO_2 (110) surface, which already contains an excess of Sn. It was proposed that the adsorbed Sn reacts with surface O ions to create vacancy-like defects that probably contain Sn^{2+} ions in an SnO-like site. The charge transfer involved at these defect sites bends the SnO_2 bands down at the surface, resulting in the dramatic increase in surface conductivity.

7.2.5 *Na_xWO_3*

There is one published report of a metal deposited onto sodium tungsten bronze. **Pt** deposited onto a polished and annealed $Na_{0.7}WO_3$ (100) surface that exhibited a

(2×1) LEED pattern went down in a layer-by-layer mode, based upon the observation of discrete changes in slope of Auger amplitudes as a function of deposition time. The Pt did not form an ordered layer on the surface, since the substrate LEED patterns weakened and disappeared with increasing Pt layer thickness.

7.3 Metals on transition-metal oxides

Transition-metal oxides offer an additional dimension for chemisorption compared to non-transition-metal oxides in that the number of electrons on the metal cations can be changed relatively easily. Also, it is often possible to reduce the surface of a transition-metal oxide to a lower oxide upon chemisorption, something that is generally impossible (with the exception of SnO_2) for non-transition-metal oxides.

One of the most important driving forces for the study of metal/transition-metal-oxide surface interactions over the last decade or so has been the strong catalyst metal/support interactions referred to earlier; that work will be briefly summarized in § 7.6 below from the standpoint of its catalytic importance. But the basic single-crystal studies of metal/oxide interactions that grew out of interest in the catalytic systems will be covered in the present section. Table 7.3 summarizes the results obtained for the interaction of metals with well-characterized single-crystal surfaces of transition-metal oxides.

Little work has been done on the interaction of non-transition metals with transition-metal-oxide surfaces; the only oxides that have been studied are TiO_2, $SrTiO_3$ and NiO. The most complex of all metal/oxide systems consists of transition metals on transition-metal oxides. Both the substrate cations and the adsorbate can have several different valence states, and in many cases both types of cation will compete strongly for lattice O ions, resulting in mixed-oxide formation at the interface. This is an area that is just beginning to be addressed, and a great deal of work remains to be done before we can say that we really understand these systems.

7.3.1 TiO_2 and $SrTiO_3$

Both because of interest in its use as a catalyst support and since its electronic structure is simpler than those of most other transition-metal oxides, the majority of metal adsorption studies on transition-metal oxides have been performed on TiO_2. The behavior of $SrTiO_3$ is quite similar in many ways, and so both compounds are discussed in the same section here.

In one study, **Na** was deposited onto polished and annealed surfaces of both planar TiO_2 (110) and stepped TiO_2 (441). In both cases the large decrease in the

Table 7.3. *Metals on transition-metal oxides*

Oxide	Metal	Oxide face	Surface preparation	Results	References
TiO$_2$	Na	110	P&A	Strong interaction, layer, Na$^+$, creates Ti^{3+} ions, reduces Φ by 3.4 eV	19
		441	Stepped	Reduces Φ by 2.3 eV	19
	K	100	P&A	Strong interaction, S–K, K$^+$, creates Ti^{3+} ions, reduces Φ by 3.7 eV	21,22,55
		110	P&A	Strong interaction, creates Ti^{3+}, reduces Φ by > 2 eV	895
	Ti	110 and 100	P&A	Creates reduced surface Ti ions; like O vacancies	23,483,473
	V	110	P&A	Strong interaction, Ti-V-O compound formed at interface, then V^0	896
			Reduced	Weak interaction, V^0	896
	Fe	110	P&A	Layer, strong O–Fe interaction	897
			P&A and reduced	V–W, strong O–Fe interaction, preferential Fe (100)	898
		001	P&A, faceted	Layer, strong interaction	899
	Rh	110	P&A	Weak interaction, V–W, Rh^0	131,270
			Reduced	Rh is e$^-$ acceptor, oxidizes surface, V–W ?	131,270
	Ni	100	Slightly reduced	S–K, preferential Ni (111)	900–902
		110	P&A	S–K, Ni is slight e$^-$ donor	17,903,904
	Pt	110, 100 and 001	P&A, (001) faceted	Not epitaxial, V–W ?	905
		001	P&A, faceted	Layer, some interaction	899
		110	Theory	Pt can be e$^-$ acceptor	906
	Cu	110	P&A	Weak interaction, S–K, preferential Cu (111)	129,907,908
			Reduced	Cu is slight e$^-$ donor	907
			P&A and reduced	V–W on both stoichiometric and reduced surfaces, preferential Cu (111)	940
SrTiO$_3$	Ba	100	P&A or reduced	Monolayer of BaO	909
	Y	100	P&A or reduced	Layer, strong interaction, monolayer of Y$_2$O$_3$	852,909
	Ti	100	P&A	Creates reduced surface Ti ions; similar to O vacancies	909
	Pt	100	Reduced	Pt is e$^-$ acceptor, V–W ?	112,910
	Cu	100	P&A	Weak, V–W	24,909
FeO	Fe	100	Theory	Strong bond, Fe^{2+}, surface relaxation	254
NiO	K	100	P&A	V–W, reduces Φ by 3 eV	34
	Cs	100	P&A	Island growth, sticking coefficient ≈ 0.7	911
	Mo	100	Cleaved	Strong interaction, reduces surface, forms oriented defective MoO$_2$	912
Na$_{0.7}$WO$_3$	Pt	100	P&A, (2 × 1)	Layer	913
ZrO$_2$	Pt	100	P&A	S–K ?, preferential Pt (111), structure not stable in air	914

Table 7.3. (*cont.*)

Oxide	Metal	Oxide face	Surface preparation	Results	References
RuO_2	Ti	110 and 100	P&A	Reduces surface, forms TiO_2	915
			Reduced	Partial oxidation of Ti	915

work function [3.4 eV for the (110) surface; 2.3 eV for (441)] and the appearance of emission in the bulk bandgap in UPS spectra indicated that the Na adsorbed ionically, donating its 3s electron to the Ti^{4+} ions in TiO_2 and creating surface Ti^{3+} ions. The Na/TiO_2 interaction is strong, and growth appeared to occur in nearly a layer-by-layer fashion, at least for low coverages, with a c(4 × 2) LEED pattern appearing for 0.5 monolayer coverage and a (1 × 1) pattern for monolayer coverage. The bonding of Na to the surface was modeled in terms of Na–O interactions and the creation of surface Na_2O units. Since both Na and Ti ions compete for ionic bonding to the O ions, the actual surface structure could well be more complicated than the model proposed.

The interaction of **K** with both TiO_2 (100) and (110) surfaces has been investigated, and the results are qualitatively similar to those for Na. On the (100) surface, it is estimated that for K coverages between 0.25 and 0.75 monolayer each K atom transfers roughly 0.25 electrons into substrate Ti 3d orbitals. This results in a decrease in the work function to 3.7 eV below its clean-surface value. For the (110) surface, the work-function decrease was smaller, but it still dropped by more than 2 eV. For large K doses, K_2O multilayers formed on the (110) surface, with the O supplied by diffusion from the TiO_2 substrate; the K_2O layers were stable to temperatures as high as 900 K. By using SEXAFS from the potassium K-edge to determine the distance and coordination of surrounding O ions, a model was proposed for the detailed position of the K atoms on TiO_2 (100).[55]

A strong adsorbate–substrate interaction is found when **Y** or **Ba** are deposited onto either stoichiometric or reduced $SrTiO_3$ (100) surfaces. The reaction is stronger for Y than for Ba, and it is stronger for reduced than for stoichiometric $SrTiO_3$ surfaces, but the chemistry is similar in all cases. The adsorbate extracts O ions from the substrate, forming a thin layer in which the adsorbate atoms are oxidized. The oxides initially formed for monolayer amounts of adsorbate were identified as Y_2O_3 and BaO based upon XPS core level binding energies; however, it is probable that some sort of mixed oxide is formed in the interface region. As the amount of adsorbate deposited increases, metallic layers of Y and Ba are formed; for Y, the growth mode is thought to be layer-by-layer. Several monolayers are necessary before Y becomes metallic, while the second monolayer of Ba is thought to be metallic.

In order to better understand defects on TiO_2, small amounts of metallic **Ti** have been evaporated onto both the TiO_2 (110) and (100) surfaces; similar behavior is found for Ti deposited onto $SrTiO_3$ (100) surfaces. For sub-monolayer coverages, the changes in electronic structure are essentially the same as those that result from the reduction of the surface by the creation of O-vacancy point defects, which were discussed in § 5.2.2. The Ti adatoms donate some of their electrons to the substrate, presumably into the empty 3d orbitals of surface Ti^{4+} ions, resulting in the creation of reduced surface Ti^{3+} species. Emission from those ions appears in UPS spectra at essentially the same energy in the bulk bandgap as does emission from O-vacancy defects. Other surface-sensitive spectroscopies such as XPS, IPS and ELS also change in a similar manner for both types of reduced surfaces. Detailed experiments have also been carried out of the changes in surface conductivity and work function upon Ti deposition.[483] Those parameters are at least two orders-of-magnitude more sensitive to changes in the surface electronic structure than are electron-spectroscopic determinations; however, the interpretation of the origin of the changes is indirect.

A particularly interesting and reactive metal/oxide system that has been investigated in some detail is **V** on TiO_2 (110). Metallic V has been deposited onto both stoichiometric, nearly perfect TiO_2 (110) surfaces and TiO_2 (110) surfaces containing O-vacancy defects; the strongest interaction is found for V on the stoichiometric surface. Both V and Ti have an extremely strong affinity for O, and adsorbed V attacks the surface Ti–O bonds violently. The excellent (1 × 1) LEED patterns exhibited by the stoichiometric (110) surface completely disappear before even 0.5 monolayer of V has adsorbed. The changes that occur in the valence band density-of-states are shown in Fig. 7.6, which presents UPS spectra for stoichiometric TiO_2 (110) with up to two monolayers of deposited V. Complex changes are seen in the valence band as the amount of deposited V increases. The emission from the O 2p band rapidly loses its two-peaked structure, and the band initially moves away from E_F. In this submonolayer-to-monolayer region of V coverage, it is concluded that some type of Ti–O–V compound is formed. As the V coverage approaches two monolayers, however, the bottom of the valence band moves back toward E_F, and the density-of-states at E_F becomes large, indicating a conducting layer on the surface. The behavior of the sample work function, Φ, as a function of V coverage is shown in Fig. 7.7; it is reminiscent of Fig. 7.3 for Al on ZnO, another system in which there is strong competition between both types of cation for lattice O ions. The sharp initial decrease in Φ corresponds to a surface dipole in which the V ions, if they are assumed to lie above the surface plane, are positive. XPS spectra for small V coverages give a V 2p core-level binding energy charac-

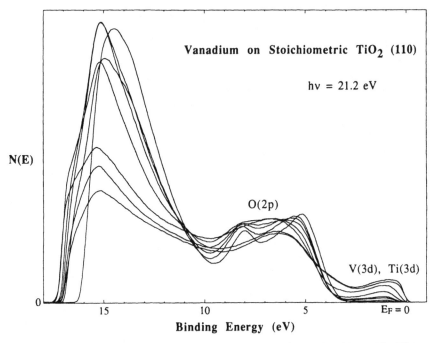

Fig. 7.6 He I UPS spectra for V deposition onto stoichiometric TiO$_2$ (110). The maximum coverage is about 1.5 monolayers equivalent. [Ref. 896]

teristic of one of the lower oxides of V, although the V 2p binding energies in those oxides are not sufficiently different to permit identification of the V valence state. As the amount of V deposited increases, Φ increases toward the value for metallic V; this is consistent with the appearance of a finite density-of-states at E_F in the UPS spectra. A detailed analysis of various spectroscopic results leads to the conclusion that the Ti–O–V layer is only one monolayer thick, with subsequent layers consisting of V metal. [In x-ray diffraction studies of impregnated powder Ti–V–O catalysts, the interface phases identified were TiO$_2$, V$_2$O$_5$ and a TiO$_2$–VO$_2$ solid solution.[916]]

Very different behavior is found for V adsorption onto reduced TiO$_2$ (110) surfaces. Figure 7.8 shows UPS spectra for up to two monolayers of V deposited onto that surface. The changes in the shape and position of the O 2p valence band are much less pronounced, and the work function, which is determined from the low-energy cut-off of the UPS spectra at zero electron kinetic energy (near 17 eV apparent binding energy in Fig. 7.8), does not change. The V 2p core-level binding energy here is close to that for metallic V, even for low V coverages. The model that results for this system is that there is only a weak interaction at the V–TiO$_2$ interface, with even the first monolayer of V being essentially metallic.

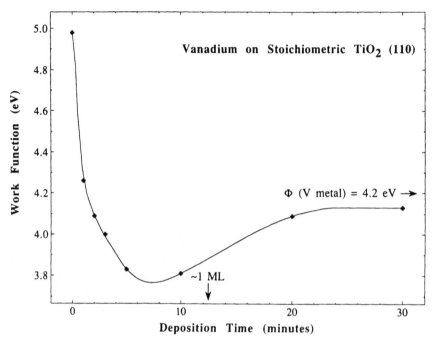

Fig. 7.7 Change of work function during deposition of V onto stoichiometric TiO_2 (110).
(1 ML = 1 monolayer.) [Ref. 896]

Presumably the reduced surface is sufficiently O-deficient that the V atoms interact primarily with surface Ti ions and are unable to bond to O ions.

Results of two experiments on the growth of **Fe** on TiO_2 (110) surfaces have yielded slightly different results. In one study it was concluded that Fe grows in a layer-by-layer mode, regardless of the state of reduction of the surface. However, the valence state of the Fe at the interface does depend upon the surface stoichiometry. On fully oxidized TiO_2 (110) surfaces, the first fraction of a monolayer of Fe is primarily Fe^{3+}, with some Fe^{2+} also present. This results from an interaction between the Fe atoms and the O in the substrate; Ti^{3+} ions are formed simultaneously as the substrate is partially reduced by the Fe. On reduced TiO_2 surfaces, however, Fe goes down as Fe^0 for all coverages. These results are qualitatively similar to those for the interaction of V with TiO_2 (110) surfaces discussed above. Another study, which used low-energy ISS, also found a strong interaction between Fe and TiO_2, but the growth mode at room temperature was determined to be Volmer–Weber. For Fe films thicker than 13 Å, the preferential orientation was (100).

The growth of **Ni** on slightly defective TiO_2 (100) surfaces has been studied by using Auger and secondary-ion mass spectroscopy. The first three monolayers were found to grow in a layer-by-layer fashion, with three-dimensional Ni islands

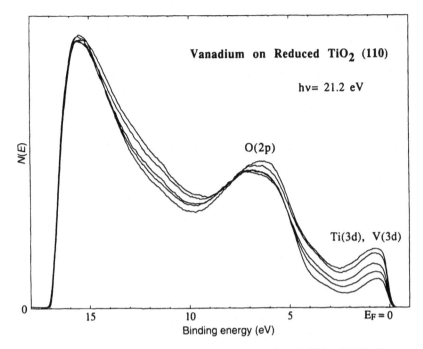

Fig. 7.8 He I UPS spectra for V deposition onto reduced TiO_2 (110). The maximum coverage is about 1.5 monolayers equivalent. [Ref. 896]

appearing for films thicker than three monolayers. Although LEED was used to characterize the substrate before deposition [a (1 × 3) pattern was observed], it was not used to monitor Ni film growth, so it is not known in what orientation, if any, the Ni films grow. Rather unusual results were obtained for deposition of Ni onto more heavily reduced TiO_2 (100) samples.[902] The Ni was found to diffuse into the bulk of the TiO_2, even at room temperature, with the amount diffusing increasing with the density of O vacancies. No island growth was observed to occur on heavily reduced samples. In a separate study of Ni deposition onto polished and annealed TiO_2 (100) surfaces that also exhibited (1 × 3) LEED patterns, a 20 Å Ni film grown on a 413 K substrate exhibited blurred hexagonal LEED patterns, which became sharp after annealing at 463 K.[901] Thus the Ni (111) planes grow parallel to the TiO_2 (100) surface under those conditions.

Somewhat conflicting results have been obtained for the interaction of Ni with TiO_2 (110) surfaces. In one study Ni was deposited onto slightly reduced TiO_2 (110) surfaces.[903] The growth mode was not determined, and efforts were concentrated on determining the amount of charge transfer between the substrate and the Ni adatoms. Making a number of assumptions about electron screening, etc., values near 0.1 electron transferred from surface Ti^{3+} species to each Ni atom at the interface were determined for two TiO_2 (110) surfaces prepared in slightly different ways. When Ni films a few monolayers thick were annealed at 573 K or higher, Ni appeared to diffuse into the bulk of the TiO_2 crystal. In a more recent study, Ni was deposited onto annealed TiO_2 (110) surfaces that had a relatively small density of O-vacancy defects.[17] The Ni was found to exhibit Stranski–Krastanov growth, and the decrease in work function observed during the first 0.5 monolayer was interpreted as due to the transfer of about 0.1 electron from the Ni adatoms to the TiO_2 substrate. Films thicker than about two monolayers exhibited a work function characteristic of bulk Ni. The Stranski–Krastanov growth mode at room temperature was confirmed in a separate LEED and Auger study.[904]

Some of the most extensive studies of the electronic interactions between metals and well-characterized single-crystal TiO_2 surfaces have been performed for **Rh** on TiO_2 (110). The interest in this system stems from its strong metal/support interaction behavior, which will be reviewed in § 7.6 below. Both stoichiometric, nearly perfect (110) surfaces and reduced surfaces containing O-vacancy defects have been used as supports, and very different interactions have been found in the two cases. Rh grows as three-dimensional particles on both surfaces, although transmission-electron microscopy has shown that in the first monolayer the Rh atoms are preferentially aligned parallel to the rows of bridging O ions on the surface.[917]

The stoichiometric TiO_2 (110) surface is relatively inert to Rh. Figure 7.9 (a) shows UPS spectra of that TiO_2 surface before and after deposition of about 0.5 monolayer of Rh at room temperature. The emission from the O 2p valence band is attenuated by the Rh overlayer, but there is no change in the structure of the band and no band-bending occurs. The shape of the emission in the bulk bandgap region upon Rh deposition is very nearly that of Rh metal, so the UPS spectra are essentially a superposition of emission from the substrate and the adsorbate. No charge transfer between Rh and the stoichiometric TiO_2 surface occurs. The Rh core-level XPS peaks exhibit a small shift to lower binding energy as the size of the Rh particles increases, as shown in Fig. 7.10; this is due to the increased screening of the core hole by electrons on neighboring Rh atoms in Rh particles compared to isolated Rh atoms.

A very different interaction of Rh with TiO_2 is seen for the reduced surface. Figure 7.9 (b) presents UPS spectra for heavily reduced TiO_2 (110) before and after the same amount of Rh deposition as in Fig. 7.9 (a). Two important differences between reduced and stoichiometric TiO_2 can be seen. A large amount of band-bending occurs for the reduced surface, as can be seen by the motion of the O 2p valence band toward E_F as the thickness of the Rh layer increases. By one monolayer of Rh, the band has moved up by 0.8 eV at the surface; this is the same band-bending that occurs when the defective TiO_2 (110) surface is completely oxidized by O_2 or SO_2 (see § 6.4.1). The density of states in the TiO_2 bulk bandgap

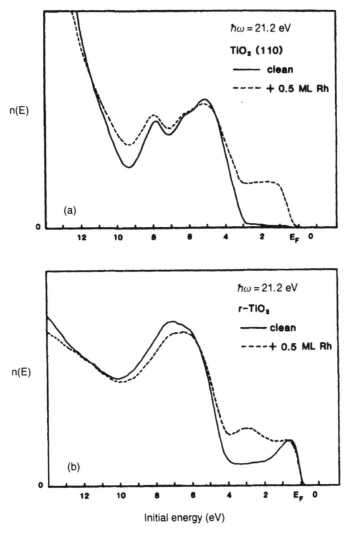

Fig. 7.9 He I UPS spectra for (a) stoichiometric and (b) reduced TiO$_2$ (110) before (solid curves) and after (dashed curves) deposition of 0.5 ML of Rh. [Ref. 270]

region is also very different from that for Rh on the stoichiometric surface. That is expected since emission from the reduced surface Ti species is also present in that region of the spectrum. But the shape of the bandgap density-of-states does not look at all like the superposition of Rh and Ti emission. The increase in the UPS intensity in the higher-binding-energy region of the bandgap is similar to differences that have been observed between the valence bands of pure Ni and Pt and those of bulk Ni$_3$Ti and Pt$_3$Ti alloys, where the intermetallic bond has a partially ionic character, with the Ni or Pt ions possessing a slightly negative charge. Other surface spectroscopic techniques also indicate that charge is transferred from the

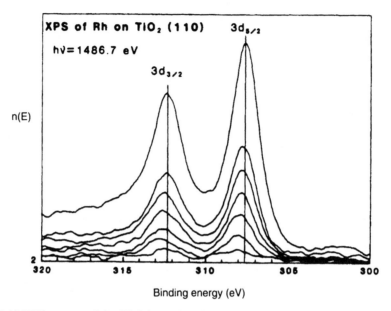

Fig. 7.10 XPS spectra of the Rh 3d core levels for increasing amounts of Rh deposited onto stoichiometric TiO$_2$ (110). The spectra correspond to equivalent Rh coverages of about 0.1, 0.2, 0.3, 0.45, 0.6, 0.75 and 1.5 ML. [Ref. 131]

reduced surface Ti ions to the Rh upon adsorption. An example is shown in Fig. 7.11, which presents XPS spectra of the Ti 2p core levels for a slightly reduced [i.e., not as heavily reduced as in Fig. 7.9 (b)] TiO$_2$ (110) surface before and after deposition of 1.5 monolayers of Rh. The clean surface exhibits a distinct shoulder on the low-binding-energy side of the Ti^{4+} peaks that corresponds to the presence of Ti^{3+} ions on the surface. Rh deposition almost eliminates that shoulder as electrons from the Ti^{3+} ions are transferred to the adsorbed Rh atoms. It is estimated that roughly 0.5 electron is transferred from reduced Ti ions to Rh atoms at surface defect sites for small coverages.

Similar results have been found for the deposition of **Pt** onto slightly reduced, non-stoichiometric SrTiO$_3$ (100) surfaces. As for Rh on TiO$_2$, Pt does not grow epitaxially or layer-by-layer on SrTiO$_3$. From changes in various types of spectra, it was estimated that, for Pt coverages of the order of one monolayer or less, roughly 0.6 electron is transferred from surface Ti^{3+} ions to Pt atoms. A number of assumptions concerning the nature of screening in TiO$_2$, etc. must be made in order to determine the amount of charge transferred, however, so the accuracy of the result is not known.

Structural studies using x-ray photoelectron diffraction have been performed for Pt deposited on the TiO$_2$ (110), (100) and (001) surfaces.[905] By measuring the azimuthal and polar angle dependence of emission from the Pt 4f level, it is possible

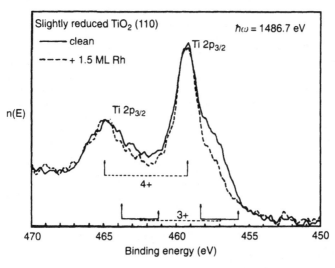

Fig. 7.11 XPS spectra of Ti 2p core levels for slightly reduced TiO₂ (110) before (solid curve) and after (dashed curve) deposition of 1.5 ML of Rh. [Ref. 270]

to determine whether or not the Pt layer is ordered with respect to the single-crystal substrate. Pt deposited at room temperature on all three polished and annealed faces did not exhibit any diffraction features, indicating that the films did not have any epitaxial relation with the substrate. However, annealing the (110) and (100) samples at 823 K after deposition resulted in strong diffraction features characteristic of Pt (111) orientation parallel to the TiO₂ surface. Models were proposed for the geometry of adsorption of the Pt atoms on those surfaces. On the (001) surface, however, annealing did not produce any angular-dependent effects. This is not too surprising, since the (001) surface used in this study was faceted to expose several different crystal planes (see § 2.3.3.3), and Pt might well grow differently on the different facets. (See also § 7.6 below for discussion of TEM results of the reduction of Pt/TiO₂ samples.)

The issue of charge transfer between catalyst metals and reducible transition-metal-oxide supports has been addressed theoretically for the case of Pt on TiO₂. Xα-SCF-SW cluster calculations were performed for two clusters representing different adsorption sites on the TiO₂ surface, and electron transfer from Ti ions to the Pt atom were found in certain cases. The calculations only simulated sites on the perfect TiO₂ (110) surface, however, and there is some question as to the critical size for a cluster in order to accurately represent Pt adsorption on TiO₂ surfaces.[918] But the basic physical idea is most likely correct.

Thin **Cu** films have been deposited onto both stoichiometric and reduced TiO₂ (110) surfaces by two groups, and different results were reported. In one study in which UPS, ELS and other techniques were used, the interaction with stoichiometric

polished and annealed surfaces was found to be weak, with the Cu overlayer atten-
uating the substrate features in the spectra but not otherwise altering them appreciably. The growth mode was thought to be Stranski–Krastanov, with the atoms in
the initial Cu layer in registry with the substrate along the $[1\bar{1}0]$ direction (i.e., the
direction perpendicular to the rows of bridging O ions) but incommensurate in the
other direction, resulting in a slightly contracted hexagonal superlattice and the
growth of Cu (111) films. Some charge transfer to the substrate from the Cu atoms
was inferred, with the effect larger on the reduced surface than on the stoichiometric one. No estimate of the amount of charge transferred from the Cu atoms in the
first monolayer to the substrate was given, but it is certainly no more than a fraction of an electron. In another study that used ISS to characterize the films, the
presence of a substrate signal for Cu thicknesses up to several tens of Å suggested
that the growth mode was Volmer–Weber, even for deposition temperatures as low
as 160 K. Surface defects on TiO_2 were found to have little effect on film growth.
Both studies are in agreement that Cu films grow preferentially with (111) orientation.

A very weak interaction has been found for Cu deposited onto $SrTiO_3$ (100) surfaces. There the growth occurs entirely by island formation; even after the deposition of 36 Å of Cu, only about 80 % of the $SrTiO_3$ surface was covered with Cu.

7.3.2 Cr, Mn and Fe oxides

No experimental work has been reported of metals on single crystals of any oxides
of Cr, Mn, or Fe, although a few studies have been performed on oxidized metal
samples. The information that can be obtained from such systems is not as detailed
as that from well-characterized single crystals, but it at least gives an idea of the
general behavior. **Cu** was deposited onto **oxidized Mn**, with the substrate considered to be MnO (although it no doubt contained other valence states of Mn as
well).[919] The Cu atoms were found to intermix strongly with the substrate for coverages in the range of 1–2 Å. For higher coverages a layer of metallic Cu grew on
top of the intermixed region. Very different behavior was found for the deposition
of Cu onto similarly prepared **oxidized Cr** surfaces, which were assumed to be
Cr_2O_3.[545] The Cu atoms did not interact appreciably with the substrate, and predominantly layer-by-layer growth was deduced.

The interaction of **Fe** atoms with the (100) surface of **FeO** has been treated theoretically by means of defect lattice calculations. The lattice was allowed to relax
both electronically and geometrically around the Fe adatoms, which produced a
Madelung potential at the adatom site that was 70 % of that for bulk Fe^{2+} ions in
FeO. The Fe adatoms were found to bond strongly to the surface and to have a
Fe^{2+} configuration; the binding energy was computed to be 12.7 eV.

7.3.3 NiO

The adsorption of **K** and **Cs** on polished and annealed NiO (100) surfaces has also been studied by using several surface-sensitive techniques. K adsorbs in three-dimensional clusters, while Cs is thought to go down in large islands. The work function decreases by almost 3 eV, but no determination of the charge state of the initially adsorbed K or Cs could be made based upon the experimental data. For Cs the initial sticking coefficient was estimated to be about 0.7.

There has been one reported study of **Mo** adsorption on single-crystal NiO. When cleaved NiO (100) surfaces were exposed to Mo vapor at room temperature, a very strong chemical reaction was found to occur. The Mo oxidized to form a continuous layer of defective MoO_2, whose ($\bar{2}01$) plane was parallel to the NiO (100) surface. When the sample was subsequently annealed at 700 K, Ni crystallites were formed on the surface, showing that the Mo completely reduced the Ni^{2+} ions in the NiO, and the Mo further oxidized. No well defined Ni-Mo oxide phase could be detected in the surface region.

7.3.4 ZrO₂

One important application of zirconia is in oxygen sensors, which are used, for example, in conjunction with catalytic converters for automobile exhaust systems. A surface coating of Pt catalyzes the interconversion of O_2 and oxide ions through the electrode reaction

$$O_2 + 4e^- = 2O^{2-} .$$

The function of the zirconia is then to complete the electrical circuit by conducting the O^{2-} ions between the Pt electrode of each side. An electrical potential, the magnitude of which depends on the difference of oxygen pressure on the two sides, is generated between the electrodes.

The deposition of **Pt** onto polished and annealed (100) surfaces of yttria-stabilized cubic ZrO_2 at room temperature has been investigated. The growth mechanism appears to be on the borderline between two- and three-dimensional for the first few monolayers. Auger spectra taken during deposition in UHV suggest a two-dimensional layered morphology, but when the samples are exposed to air and then measured by transmission-electron microscopy and diffraction, evidence of three-dimensional particles is found. It is postulated that the two-dimensional layers are unstable to either air exposure or heating above about 350 K. The Pt islands are oriented with respect to the ZrO_2 substrate, however, with their (111) planes parallel to the surface.

7.3.5 *RuO*$_2$

The interaction of **Ti** with the RuO_2 (110) and (100) surfaces has been studied by using LEED, XPS and Auger. Ti bonds more strongly to O than does Ru, and the RuO_2 surface is reduced upon Ti adsorption. The TiO_2 layer formed on the surface is found to stabilize the RuO_2 crystal from further reduction at elevated temperatures. When a few monolayers of Ti are deposited on ion-bombarded and reduced RuO_2 surfaces, the oxidation of the Ti is less complete than on stoichiometric surfaces, and evidence for Ti^{3+} ions is seen in XPS.

7.4 Metals on high-T_c superconductors

Since most of the ultimate uses for high-T_c superconductors will require making electrical contact to the oxide, it is very important to understand how different metals interact with the surfaces of those materials. If high-T_c materials are to be incorporated into semiconducting devices, then compatibility with semiconductors is also extremely important. Surface-science techniques have been used to monitor the surface properties of UHV-fractured single-crystal and polycrystalline superconducting oxides as metal and semiconductor atoms are deposited onto their surfaces. The results of the studies that have been reported to date are summarized in Table 7.4. Although particular metal/oxide interfaces may exhibit specific properties, the behavior of the interfaces between metals and high-T_c superconductors obeys broad general principles.

The predominant interfacial interaction that occurs when metals are deposited onto oxide superconductors is competition for the O ions in the lattice and consequent reduction of the surface region of the superconductor. Therefore metals whose oxides have high heats of formation, such as Ti, Al and Mg, will interact more strongly with the surface than will relatively inert metals such as Au or Ag. The removal of O from the lattice creates a non-superconducting layer at the interface that may even be insulating (e.g., TiO_2 or Al_2O_3). In some cases the destruction of superconductivity can extend many lattice constants into the superconductor for only one monolayer of deposited metal. For Ti or Cu on $YBa_2Cu_3O_{7-x}$, for example, one monolayer of metal was found to reduce the substrate sufficiently to destroy its superconductivity to a depth of 30–50 Å below the surface. In superconductors containing Bi, that is the ion that is most easily reduced by the metal overlayer, although Cu ions are also reduced when active metals are deposited on the surface. The reduction of the Bi ions in $Bi_5Sr_3Ca_2Cu_4O_x$ when Mg is deposited onto the surface of a polycrystalline sample can be seen clearly in Fig. 7.12, which presents UPS spectra for the Bi 5d core levels as a function of the amount of Mg deposited.[944] In the case of Fe on $YBa_2Cu_3O_{7-x}$, Ba was found to segregate to the

Table 7.4. *Interaction of metals with high-T$_c$ superconductors*

Oxide	Metal	Results	References
La$_{2-x}$Sr$_x$CuO$_4$	Ti	Reduction, forms insulating layer. no segregation	920
	Fe	Strong reduction of Cu and removal of O from substrate	921,922
	Cu	Strong reduction of Cu in top 40–50 Å of substrate	923
	Au	No reaction	924
YBa$_2$Cu$_3$O$_{7-x}$	Al	Reduction; if evaporated in O or O$^+$ ambient. get Al oxide and little interaction with substrate	925
	Ti	Strong reduction, forms TiO$_2$. ≈ 1 monolayer reduces 30–50 Å of substrate	926,927
		Evaporation at 20 K: slight reduction, but increased reduction of substrate and TiO$_2$ formation upon warming to RT	928
	Cr	Evaporation at 20 K: only thin reacted layer forms, stable upon warming to RT	928
	Fe	Strong reduction, insulating layer, Ba segregation to surface	929
	Pd	Reduction; destroys Cu–O bonds, reaction limited to first few Å of film	930
	Cu	Strong reduction, ≈ 1 monolayer reduces 30–50 Å of substrate	926,930,931
	Ag	Partial reduction of surface	932–934
		No reaction	930
	Au	No reaction	930
	Si	Reduction; if evaporated in O or O$^+$ ambient. get Si oxide and little interaction with substrate	925
	Bi	Reduction; if evaporated in O or O$^+$ ambient. get Bi oxide and little interaction with substrate	925,935
Y$_2$BaCuO$_5$	Ti	Reduction, forms TiO$_2$	927
EuBa$_2$Cu$_3$O$_{7-x}$	Au	At 20 K, forms insulating layer, but no strong chemical reaction	936
	Ga	Reduction, formation of Ga oxide; then Ga islands grow on oxide	937
	As	Arsenic oxide forms, but without breaking Cu–O bonds; stops at 1 monolayer	937
Bi$_2$Sr$_{2-x}$Ca$_{1+x}$ Cu$_2$O$_{8+y}$	Rb	Reduction, but only affects O and Bi	938,939
	Al	Strong reduction of Bi and Cu; if evaporated in O or O$^+$ ambient get Al oxide and little interaction with substrate	925,938
	Ti	Strong reduction, forms TiO$_2$, ≈ 1 monolayer reduces 30–50 Å of substrate	926,927
		Evaporation at 20 K: slight reduction, but increased reduction of substrate and TiO$_2$ formation upon warming to RT	928
	Cr	Complete reduction of Cu^{2+} on surface	941
		Evaporation at 20 K: weak interaction, stable when warmed to RT	928,941
		Deposit 30 Å clusters at 100 K: slight disruption of Bi–O bonds, no reduction of Ca^{2+}; warm to RT, see some reduction of Cu^{2+} (less than for 20 K evaporation and warm)	941
	Cu	Reduction (less than for YBa$_2$Cu$_5$O$_{7-x}$), disrupts Bi–O bonds, some Bi segregation to surface	931
		Evaporate at 20 K: less reduction than for RT evaporation, stable when warmed to RT	928
	Ag	Partial reduction of surface	932,942

Table 7.4. (*cont.*)

Oxide	Metal	Results	References
$Bi_2Sr_{2-x}Ca_{1+x}$ Cu_2O_{8+y}	Au	No interaction at 20 K or 300 K Forms localized non-metallic phase at 100 K	936,943 943
	Si	Reduction; if evaporate in O or O$^+$ ambient, get Si oxide and little interaction with substrate	925
	Bi	Reduction; if evaporate in O or O$^+$ ambient, get Bi oxide and little interaction with substrate	925,935
$Bi_5Sr_3Ca_2Cu_4O_x$	Mg	Strong reduction, form semiconducting layer, reduces Bi and Cu	944
	Ag	Little interaction	944

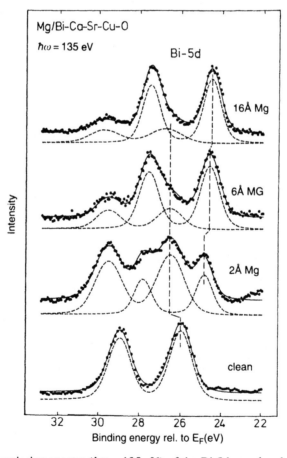

Fig. 7.12 Photoemission spectra ($\hbar\omega = 135$ eV) of the Bi 5d core levels for increasing Mg coverages on $Bi_5Sr_3Ca_2Cu_4O_x$, together with the results of a least-squares analysis (solid lines and dashed subspectra). [Ref 944]

outer surface of the Fe film, even for films as thick as 250 Å. Some Bi segregation to the surface of the metal film was also seen when Cu was deposited onto $Bi_2Sr_{2-x}Ca_{1+x}Cu_2O_{8+y}$.

The disruption of the surface of the superconductor at the metal/oxide interface can be minimized for some active metals by oxidizing the metal atoms before they land on the surface. When Al, Bi or Si were deposited in either UHV or a molecular O_2 ambient onto $YBa_2Cu_3O_{7-x}$ or $Bi_2Ca_{1+x}Sr_{2-x}Cu_2O_{8+y}$, reduction of the substrate took place. However, when the same atoms were evaporated in an ambient that contained atomic O and O^+ ions, they deposited as oxides on the surface and exhibited minimal reaction and disruption of the superconductor surface. (Of course, this technique is of no use if one is trying to make electrical contact to the superconductor!)

It has also been found that the interaction between metal overlayers and oxide superconductors is much less pronounced when the metal is deposited at very low temperatures. When Cu, Ti or Cr were deposited onto samples of $YBa_2Cu_3O_{7-x}$ or $Bi_2Sr_{2-x}Ca_{1+x}Cu_2O_{8+y}$ at 20 K, there was much less disruption and reduction of the surface than there was when the metals were deposited at room temperature. This is perhaps not too surprising, since at that temperature most thermally activated processes are frozen out. What is surprising, however, is that for Cu and Cr the non-reactive interface was stable when the samples were subsequently warmed up to room temperature. (In the case of Ti, the substrate became reduced and a layer of TiO_2 formed upon warming, similar to the situation for deposition at room temperature.)

An interesting phenomenon has been found in the interaction between Au and $Bi_2Ca_{1+x}Sr_{2-x}Cu_2O_{8+y}$. When Au is deposited onto single-crystal superconductor surfaces at either 20 K or 300 K, no reaction is found to occur. But when Au is deposited at 100 K, XPS core-level shifts indicate that the Au interacts with both the Sr and Ba in the superconductor, forming a thin non-metallic layer at the interface. At all temperatures the Au goes down in a layer-by-layer mode. The reason for this very temperature-specific reaction is not understood.

Another approach that has been used to deposit active metals onto oxide superconductors without seriously degrading the electrical properties of the interface consists of depositing the metals as clusters of atoms about 30 Å in diameter. This was done by first condensing a thin Xe buffer layer onto the superconductor surface at 20 K and then depositing Cr on top of the Xe. The Cr goes down in clusters, with 2 Å equivalent of Cr resulting in 30 Å diameter clusters that have metallic character. The sample was then warmed to 100 K, where the Xe evaporates and the Cr clusters settle onto the surface of the superconductor. The disruption of the superconductor surface at low temperature, and upon subsequent warming to room temperature, was found to be even less than that for evaporation of Cr onto the bare surface at 20 K.

For electronic device applications it is important to determine whether oxide superconductors can be deposited directly onto semiconductors without destroying their superconductivity. Some experiments have addressed the interface between Si and $YBa_2Cu_3O_{7-x}$ and $Bi_2Sr_{2-x}Ca_{1+x}Cu_2O_{8+y}$, and others have deposited Ga and As separately onto $EuBa_2Cu_3O_{7-x}$. Si and Ga are observed to form oxides and reduce the superconductors, creating non-superconducting layers on the surface. Arsenic, on the other hand, forms an oxide whose thickness self-limits at about one monolayer and which does not measurably affect the properties of the superconductor.

7.5 Chemisorption on model metal / oxide systems

Any attempt to discuss the chemisorption behavior of metal particles supported on metal oxides comes dangerously close to opening the Pandora's box of catalysis on oxide-supported metals. That literature is incredibly vast, and it is beyond the scope of this book to address it. However, a few experiments have been conducted of chemisorption on metal layers deposited onto well-characterized single-crystal oxide surfaces in an attempt to model actual catalysts and gas sensors, and a brief discussion of some examples of that work is instructive here. For a discussion of the chemisorption behavior of metals supported on oxide powders, the reader can refer to the many books available on catalysis.

One of the reasons that chemisorption on metals supported on single-crystal oxides has received some attention is that the presence of metal particles has been found to alter the gas sensing properties of semiconducting oxides. For example, H_2 does not interact with or dissociate on most metal-oxide surfaces. However, exposure of ZnO or TiO_2 surfaces to atomic H results in a very strong interaction (see Table 6.1). If an oxide surface containing metal particles that dissociate H_2 is exposed to H_2, the H that is formed on the metal can 'spill over' onto the oxide, thus strongly affecting its surface electrical properties.

The interaction of H_2 and O_2 with SnO_2 (110) surfaces onto which about three monolayers equivalent of Pd were deposited have been studied by measuring the surface conductivity in addition to using the usual surface-sensitive spectroscopies.[443,891,945] The addition of the Pd greatly enhances the sensing behavior, as shown in Fig. 7.13, where the clean and Pd-dosed surfaces are compared as a pulse of gas is admitted. For the SnO_2 surface preparation used here, the clean surface is very insensitive to either O_2 or H_2. However, the presence of Pd particles produces large changes in conductivity – as expected, with O_2 being an electron acceptor and H_2 a donor – presumably due to the considerably higher sticking coefficients of both molecules on Pd, their dissociation and subsequent spillover of H and O onto the SnO_2 surface. More complicated processes, including Schottky-barrier

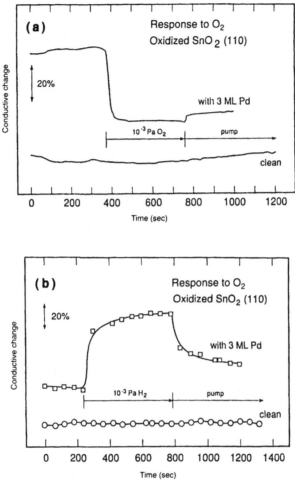

Fig. 7.13 Conductance changes at 400 K for clean, oxidized SnO_2 (110) during (a) O_2 dosing and (b) H_2 dosing. [Ref. 443]

modulation, may also take place, since the conductivity changes, particularly for O_2, are not completely reversible. The sensitivity of Pd/SnO_2 model sensors to CH_4 has also been studied.[630]

The other reason that chemisorption studies have been conducted on single-crystal oxides containing metal adlayers is to better understand catalysis. An example of that type of work involves Cu supported on single-crystal ZnO to model Cu–Zn–O methanation catalysts. Measurements have been made of the chemisorption of CO onto single-crystal ZnO (0001), (000$\bar{1}$) and (10$\bar{1}$0) surfaces onto which sub-monolayer amounts of Cu had been deposited.[9] Cu bonds most strongly to the O-terminated (000$\bar{1}$) surface, and that is also the surface on which the CO bonding is the weakest; the measured heat of adsorption was less than 50 kJ/mol. The strongest CO bonding, exhibiting a $\Delta H_{ads} = 88$ kJ/mol, was with Cu^+ ions on the Zn-terminated (0001) surface that had been annealed in O_2. UPS also indicated that adsorption on the (0001) surface perturbs the CO electronic structure more than adsorption on the other Cu/ZnO faces. CO adsorbed to Cu on the (10$\bar{1}$0) surface and on the (0001) surface that had not been annealed in O_2 with essentially the same bond strength as for CO on Cu metal ($\Delta H_{ads} = 65$ kJ/mol). The high-binding-energy (0001) site is believed to most closely represent the active site for CO adsorption in methanol synthesis on Cu–Zn–O catalysts.

The adsorption of ethanol, C_2H_5OH, on ZnO (000$\bar{1}$) surfaces onto which Pd had been deposited has been studied by using thermal desorption.[779,946,947] The same desorption products are observed as for the clean ZnO (000$\bar{1}$) surface, but desorption begins at lower temperatures and extends to higher temperatures on the Pd-covered surface. The low-temperature desorption indicates that ethanol adsorbs primarily on the Pd particles.

A number of studies have been performed of the catalytic activity of TiO_2 surfaces onto which metal particles had been deposited. Representative systems include: photo- and thermal-assisted water/gas shift reactions on TiO_2 (100) surfaces onto which about one monolayer of Pt had been deposited;[948] the hydrogenation activity for CO/H_2 mixtures over TiO_2 (100) surfaces covered with 1–2 monolayers of Ni;[901] the adsorption of CO_2 and NO on TiO_2 (110) surfaces covered with 1–2 monolayers of Na;[949] and measurements of the adsorption/desorption kinetics of CO on Pd particles epitaxially grown on MgO (100).[950] For additional studies of this type, the reader is referred to the catalysis literature.

7.6 'Strong metal / support interactions'

As mentioned earlier, much of the interest in metals on oxides stems form the observation that strong interactions may exist between small metal catalyst particles and their supports when they are supported on reducible transition-metal oxides. There is an extensive literature devoted to understanding the origin of this so-called **SMSI** effect, and it is not within the scope of this book to review it. The interested reader is referred to review articles for details of that work.[837,838] But it is appropriate to briefly summarize here the aspects of SMSI that are relevant to our understanding of the fundamental interactions between metals and oxides. Our discussion will be confined to results obtained on single-crystal metal-oxide supports.

The original, and perhaps most striking, observation of the SMSI effect was the report that when Pt, Rh, Ru, Pd, Ir or Os were supported on TiO_2 powders, and when the catalysts were reduced at a high temperature in H_2, the chemisorption

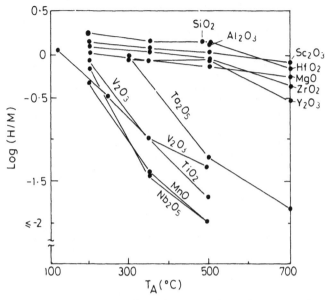

Fig. 7.14 Room-temperature H_2 chemisorption on Ir (1 wt %) supported on various oxides as a function of prior activation in H_2 for 1 hr at each of various temperatures T_A. H/M is the atomic ratio of H adsorbed to Ir in the catalyst. [Redrawn from Ref. 951]

behavior of the metals changed drastically compared to that for the same catalysts not reduced at high temperature.[836] Whereas those metals normally chemisorb both H_2 and CO at room temperature, catalysts that had been 'high-temperature reduced' would no longer adsorb either molecule, although the original amount of metal was still present on the catalyst; this is referred to as the 'SMSI state' of the catalyst. The difference in the behavior of catalyst metals supported on reducible transition-metal oxides as opposed to non-reducible oxides is graphically apparent in Fig. 7.14, which plots the room-temperature H_2 uptake of Ir supported on several different oxides as a function of the reduction temperature of the catalyst.[951] Subsequent heating of the catalyst in O_2 at least partially restores the normal chemisorption behavior of the catalysts. A number of other dramatic changes in catalytic behavior also accompany the existence of the SMSI state, but the H_2 and CO chemisorption properties are the key ones, and they are the ones that have been used by the majority of researchers to verify that the SMSI state has actually been achieved.

One of the first speculations as to the origin of SMSI was that the charge state of the catalyst metal was changed by interaction with the electrons on the transition-metal ions in the support; this different charge state then altered the chemisorption and catalytic properties of the metal. It was this idea that prompted much of the work on charge transfer between metals and metal oxides that was referred to in

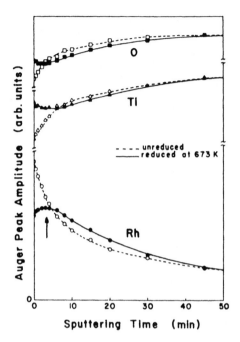

Fig. 7.15 Auger peak amplitude versus Ar+-ion-bombardment time for unreduced (open points, dashed curves) and reduced (solid points and curves) Rh/TiO$_2$ model catalysts. Curves for Ti and O have been shifted up for clarity. [Ref. 130]

§ 7.1.1 above. However, charge transfer alone was found to be insufficient to explain the phenomena, and one of the other ideas that was considered was the physical blocking of many of the catalytically active sites on the metal particles by lower oxides of the support that had migrated onto the surfaces of the particles. This process is referred to as **encapsulation** or **decoration**. It ultimately proved to be the primary cause of SMSI, although there are still some unanswered questions as to the nuances of the effect. One of the most direct experiments that showed the presence and effect of encapsulation consisted of preparing model catalysts composed of monolayer amounts of Rh deposited at room temperature onto well-characterized single-crystal rutile TiO$_2$ (110) supports.[130] The catalysts chemisorbed CO as expected for Rh metal. After the model catalysts were heated in a reducing environment that mimicked the high-temperature reduction treatment used on real catalysts, they no longer chemisorbed CO. The composition as a function of depth below the surface was measured for both reduced and unreduced catalysts by sputter profiling and Auger spectroscopy, and the results shown in Fig. 7.15 were obtained. The unreduced catalysts yielded a profile characteristic of a layer of Rh on top of the TiO$_2$ support, but the reduced catalysts exhibited an initial *increase* in the Rh Auger signal, with an attendant decrease in the Ti and O signals, as the Ti

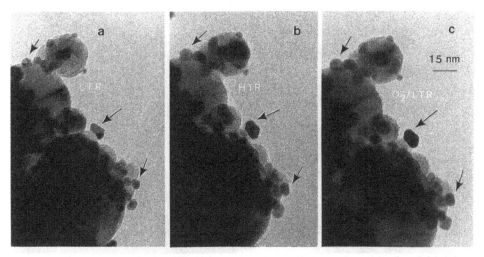

Fig. 7.16 High-resolution TEM micrographs of Rh/TiO$_2$ catalyst (a) after initial low-temperature reduction, (b) after high-temperature reduction, and (c) after subsequent oxidation and low-temperature reduction. Arrows indicate Rh particles where reversible encapsulation by TiO$_x$ can be seen. [Ref. 954]

oxide that had been covering the Rh particles was removed. When the reduced catalyst was sputtered only to the maximum in the Rh Auger signal (indicated by the arrow in Fig. 7.15), so that most of the encapsulating oxide had been physically removed, CO again chemisorbed on the Rh particles.

The encapsulation of Rh particles on TiO$_2$ powder-supported catalysts has also been observed in real space by using high-resolution TEM in the profile-imaging mode.[952–954] Figure 7.16 shows transmission electron micrographs for a Rh/TiO$_2$ (anatase) catalyst before and after high-temperature reduction and after subsequent oxidation and low-temperature reduction. The presence of an amorphous layer covering the Rh particles on the high-temperature-reduced sample can be seen clearly. The process is also seen to be largely reversible upon oxidation.

Although encapsulation is the accepted explanation for most of the SMSI phenomena, it must necessarily be accompanied by electronic and bonding effects, which presumably constitute the driving force for the migration of lower oxides onto the metal particles. The studies of room-temperature deposition of Rh onto both stoichiometric and reduced TiO$_2$ (110) surfaces described in § 7.3.1 above were aimed at determining the nature of the charge-transfer process without the complications induced by the presence of encapsulation.[131,270] Transfer of electrons from the reduced cations of the support to the Rh particles was observed, but it was found that the presence of charge transfer, and the existence of slightly negative Rh atoms on the surface, did *not* have any major influence on CO chemisorption. Thus encapsulation remains the dominant mechanism of SMSI.

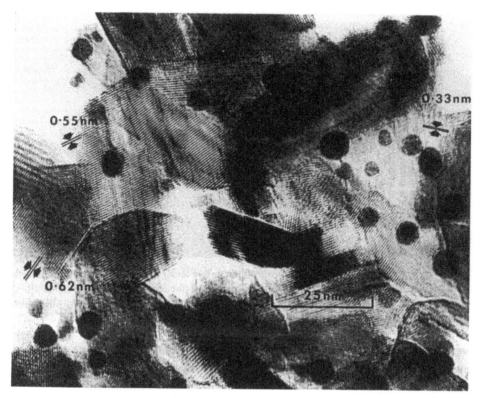

Fig. 7.17 High-resolution TEM micrographs of Pt on TiO_2 reduced at 825 K, showing lattice fringes of Ti_4O_7 and raft-like Pt particles on the substrate. [Ref. 955]

An interesting metal/oxide interaction was found to occur for Pt particles on TiO_2.[955] When supported catalysts were prepared without high-temperature reduction, the Pt occurred as three-dimensional particles on the TiO_2 surface. But after high-temperature reduction, the Pt was found to wet the substrate, spreading out into raft-like structures only a few atomic layers thick. These can be seen in Fig. 7.17, which shows a transmission-electron micrograph of Pt particles supported on TiO_2 after reduction. In addition to the presence of raft-like Pt particles, the substrate lattice fringes indicate that the TiO_2 has been reduced to Ti_4O_7, which is really the support for the particles. This effect has not been observed for other metal/support systems.

The above is necessarily an oversimplified picture of SMSI, and a great many sophisticated experiments have been performed that have provided a very detailed description of the effect for many metal/support systems. The reader is referred to the literature for a full appreciation of the SMSI work.

7.7 Oxides on oxides

As discussed in § 7.1 above, the structures formed when metals are deposited onto the surfaces of metal oxides can be quite complex; the initial monolayer or so may involve the formation of a different oxide, leading effectively to an oxide/oxide interface. Oxide/oxide interfaces are of great importance in a variety of technological areas, including heterogeneous catalysis, adhesion, friction, etc. The role of oxide/oxide interfaces in catalysis has been considered indirectly for a long time, and the literature on oxide-supported oxide powder catalysts is fairly extensive.[358] However, such interfaces are difficult to study by surface-science techniques, and to date there has been very little work in that direction. There is also a large literature on the growth of films of one metal oxide on single-crystal substrates of another oxide where the substrate surface was not well-characterized on the atomic scale before deposition. The primary concern in such studies is the quality of the resulting film, and no attention is paid to what is actually occurring at the interface between the two oxides; the greatest interest currently in this area is the growth of high-T_c superconductor films on insulating substrates. We will only consider here research that has directly addressed the oxide/oxide interface on an atomic scale.

One of the main roles played by the support in an oxide-supported oxide catalyst is believed to be the promotion of the growth of specific crystal faces on the oxide catalyst particles.[956] However, the situation can be far more complicated than that, as has been shown for **MoO$_3$** supported on UHV-cleaved **CoO** (100) surfaces.[957,958] This system is of particular interest to catalytic chemists because of the wide use of Co/Mo/Al$_2$O$_3$ heterogeneous catalysts in the industrial hydroprocessing of petroleum. By using Auger spectroscopy and RHEED, a variety of different surface structures were observed depending upon substrate temperature, annealing time, etc. After MoO$_3$ was deposited onto the CoO surface by evaporation, the diffraction patterns observed included: none (amorphous layer), (2 × 1), ($\sqrt{5} \times \sqrt{5}$) R 26°, (2 × 2), (3 × 1), (3 × 1) R 45° and (4 × 2) R 45°. Partial reduction of the surface was also seen, with the formation of oriented islands of fcc Co under certain conditions, as well as the formation of epitaxial islands of CoMoO$_4$ under different conditions.

Another oxide/oxide interface that has been studied is **BaO** on **MgO** (100).[959] The BaO layers were formed by successive cycles of Ba deposition, O$_2$ exposure and high-temperature annealing, and the resulting surfaces were characterized by using XPS, LEED and ELS. Attempts to produce a theoretically predicted ($\sqrt{2} \times \sqrt{2}$) superstructure[286] for one monolayer of BaO on MgO (100) did not succeed; the LEED patterns observed always had the symmetry of the MgO substrate and very nearly the same spot spacing. The BaO overlayer was determined to be oriented with its (100) plane parallel to the MgO surface and with its [010] and [001] directions rotated 45° with respect to the MgO <100> axes. This orientation provided the best lattice matching between MgO and BaO, since the latter has a lattice constant roughly $\sqrt{2}$ larger than the former. For BaO layers thicker than three monolayers, the lattice spacing determined by LEED is that of the bulk BaO structure. Several theoretical models were considered to explain the observed structure, and, while they shed some light on the nature of the MgO/BaO interface, none were able to completely reproduce the observations because of the non-commensurate relationship between the MgO and BaO lattices.

MgO has also proven to be an excellent substrate for the epitaxial growth of **NiO** and **Fe₃O₄** layers.[960] The lattice mismatch for both overlayers is less than 1 %, and, by using oxygen-plasma-assisted molecular-beam epitaxy, it has been possible not only to grow high quality films of NiO and Fe_3O_4 separately, but also Fe_3O_4/NiO superlattices exhibiting strong crystalline ordering and sharp interfaces. Superlattices having repeat distances as small as 20 Å, which corresponds to one Fe_3O_4 unit cell plus two NiO unit cells, have been produced. The quality of these superlattices can be seen from the low-angle x-ray diffraction spectra for films having six different repeat distances, which are shown in Fig. 7.18.

Theoretical calculations on a different type of system have been performed in an effort to model the silicon lever tip of an atomic force microscope.[961] The interaction of **SiO** molecules with the unrelaxed **MgO** (100) surface was studied by an *ab initio* LCAO–SCF approach. The SiO molecules were assumed to be oriented perpendicular to the surface. For an isolated SiO molecule, the lowest-energy adsorption site corresponded to bonding via the O atom to surface Mg^{2+} ions; this configuration gave an adsorption energy of 20 kJ/mol, compared to only 10 kJ/mol for bonding via the Si atom. There was essentially no electron transfer between the substrate and the molecule. An interesting result was found for higher coverages of SiO on MgO. Since the SiO molecule is quite polar, it was found that layers of SiO molecules tended to orient with adjacent molecules antiparallel, so that one-half of the molecules would be bound to the surface through their O atoms and the other half through their Si atoms. No experiments have been performed on this system.

One model oxide-supported oxide system that has been studied by using single-crystal supports is **vanadia** on **TiO₂**.[896] Stoichiometric rutile TiO_2 (110) surfaces were used, and vanadium oxide was obtained by evaporating metallic V in a UHV surface-analysis system in an O_2 ambient of up to 1×10^{-5} Torr. The measured XPS binding energies of the V 2p core levels indicated that the adsorbed species was a lower oxide of V and not the $3d^0$ oxide V_2O_5, but no distinction could be made between the various possible valence states of the reduced V ion. No reduction of the TiO_2 substrate was seen as determined by XPS of the Ti 2p core levels,

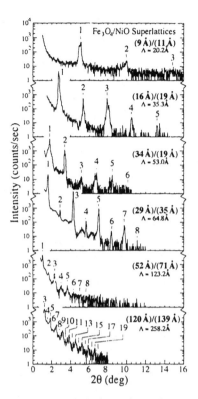

Fig. 7.18 Near-grazing-incidence low-angle x-ray diffraction spectra from Fe_3O_4/NiO superlattices grown on MgO (100) having repeat distances of 20.1 to 258.2 Å. [Ref. 960]

and the changes that occurred in the UPS spectra of the valence-band region were consistent with very little interaction between the vanadia overlayer and the substrate. It was concluded that the vanadia/TiO_2 interaction is weak and that the junction is essentially atomically sharp.

The epitaxial growth of **TiO_2** onto **Al_2O_3** $(11\bar{2}0)$ substrates has also been studied in detail.[962–966] Two types of TiO_2 nuclei form at the interface: 70 % of them are oriented with the TiO_2 (101) plane parallel to the Al_2O_3 surface, with the remainder of the nuclei having their (100) plane parallel to the surface. However, the growth rate of the (100) crystallites is so much faster than that of the (101) grains that they quickly dominate the film morphology. Epitaxial growth of TiO_2, although with a large lattice mismatch, has also been achieved on Al_2O_3 (0001) and $(10\bar{1}2)$.[963,964] **VO_2** films could also be grown epitaxially on all three of the above Al_2O_3 surfaces.[963,964]

Although not an oxide, **CuCl** has been deposited onto polished and annealed **TiO_2** (110) surfaces in an effort to model the role played by CuCl as a promoter in heterogeneous catalysis.[967,968] A large increase in work function (≈ 0.7 eV) indicates that the molecule adsorbs to the surface via its Cu end, and changes in the valence-band UPS spectra provide evidence for hybridization between the Cu 3d and O 2p orbitals. A (4 × 1) reconstruction of the TiO_2 substrate accompanies the adsorption.

There has been a tremendous amount of interest recently in the growth of single-crystal films of high-T_c superconductors on metal-oxide substrates. The goal is to grow high quality films on substrates that will not react with the superconductor, degrading its properties. For example, when **$YBa_2Cu_3O_{7-x}$** films are grown on **sapphire**, which has many of the electrical characteristics of a desirable substrate, the films are generally of poor quality; this is attributed to an interaction between Al ions in the Al_2O_3 and the superconducting film.[969] One approach to remedying this problem is to deposit a thin film of another oxide, which is to act as a barrier layer, onto the sapphire before growing the superconductor. One promising candidate is yttria-stabilized zirconia, which has the formula $(Y_2O_3)_x(ZrO_2)_{1-x}$; this compound maintains a cubic structure over the composition range $0.08 \leq x \leq 0.4$. Ion-channeling and TEM measurements have been used to study the nature of the interface between laser-deposited **yttria-stabilized zirconia** films and **Al_2O_3** $(10\bar{1}2)$ surfaces.[970] The interface was imaged in TEM by using thin cross-section samples, and it was found to be no more than 5 Å in thickness. Both (100) and (110) zirconia planes grew epitaxially on the Al_2O_3 $(10\bar{1}2)$ surface. Lattice imaging of cross-section samples by TEM has also been used to study the interface between yttria-stabilized zirconia and **$YBa_2Cu_3O_{7-x}$** deposited onto it.[970,971] A semiconducting, Ba-rich interface layer about 50 Å thick was found to grow epitaxially on the zirconia; it is speculated that it is a heavily doped and faulted compound derived from $BaZrO_3$.[971] In some cases an amorphous layer 10–20 Å thick forms between the interfacial layer and the crystalline $YBa_2Cu_3O_{7-x}$. In contrast, **$YBa_2Cu_3O_{7-x}$** films grown epitaxially on **$SrTiO_3$** (100) surfaces, where the lattice mismatch is only 1 %, exhibit interfaces that are sharp within 10 Å;[972,973] STM studies showed that the growth mode for that system is Stranski–Krastanov.[974] In experiments on the epitaxial growth of c-oriented **$Bi_2Sr_2CuO_x$** and **$Bi_2Sr_2(Ca_{0.85}Sr_{0.15})Cu_2O_x$** on $SrTiO_3$ (100), films up to about 60 Å thick exhibited a and b lattice constants that were the same as that of the substrate and about 3 % larger than the bulk lattice constants for those oxides.[979]

One of the most promising metal-oxide substrates for the epitaxial growth of high-T_c oxide superconductors is **$LaAlO_3$**, and the interface between (100) single-crystal faces of that oxide and **$YBa_2Cu_3O_{7-x}$** have been studied by TEM.[973,975] That work is an excellent example of the detail that can be achieved in characterizing the interface between one oxide and another. Figure 7.19 shows a high-resolution TEM micrograph of an $YBa_2Cu_3O_{7-x}$/$LaAlO_3$ (100) interface, along with a diffraction pattern taken from an area that contains both the substrate and the film;

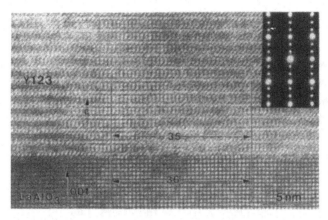

Fig. 7.19 High-resolution TEM micrograph of the interface between a $YBa_2Cu_3O_{7-x}$ film and a $LaAlO_3$ substrate. The inset diffraction pattern taken from an area including both film and substrate confirms epitaxial growth. Counting atomic planes within the marked region shows the presence of an interfacial dislocation in the form of an extra half-plane in the substrate. [Ref. 975]

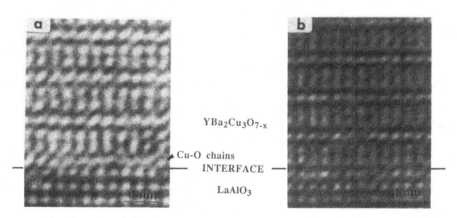

Fig. 7.20 Higher-magnification TEM images of the interface in Fig. 7.19, showing two different interfacial configurations: (a) the first atomic plane in the film next to the interface consists of Cu–O chains, and (b) the first layer is BaO. [Ref. 975]

the latter confirms the epitaxial nature of the film growth. There is a small lattice mismatch between the two materials, however, and that gives rise to an interfacial dislocation, consisting of an extra half-plane of atoms in the substrate, positioned roughly in the middle of the interface in the figure. Even finer structural detail can be seen in Fig. 7.20, which shows two sections of the interface at higher magnification. The micrograph in (a) shows a region where the plane of Cu–O chains nucleated at the interface, and (b) shows a region where a Ba–O layer nucleated.

MgO is also a candidate substrate for the growth of high-T_c superconductors, and films of **$Bi_2(Sr,Ca)_3Cu_2O_x$** films have been grown on cleaved MgO (100).[976] RHEED and TEM were used to characterize the film growth. During the initial stages of growth, the superconductor nucleated at steps on the MgO (100) surface that were present after cleaving, and the most rapid growth direction was along the step direction. The lattice constants do not match particularly well for these two materials, and the resulting films exhibited two equivalent types of misoriented domains. STM has been used to study the growth of **$YBa_2Cu_3O_{7-x}$** on **MgO** (100), where the lattice mismatch is about 9 %, and the superconductor was found to grow in the Volmer–Weber mode with the formation of islands at all coverages.[974]

Both semiconductor/superconductor and superconductor/superconductor interfaces are also of practical interest for use in devices. One interface that has been studied is the growth of **$Bi_2Sr_2CuO_6$** films on single-crystal **$Bi_2Sr_2CaCu_2O_8$**.[977] The $Bi_2Sr_2CaCu_2O_8$ crystal was cleaved in air along its *a–b* plane. The growth was monitored by RHEED, and it was found that the semiconducting $Bi_2Sr_2CuO_6$ grew epitaxially and layer-by-layer, but that the quality of the film was not as good as that of the $Bi_2Sr_2CaCu_2O_8$ substrate. Heterostructures consisting of alternating layers of **$YBa_2Cu_3O_{7-x}$** and **$PrBa_2Cu_3O_{7-x}$** have also been fabricated, and the resulting interfaces have been characterized by TEM.[978] This system is interesting because $PrBa_2Cu_3O_{7-x}$ is non-superconducting and displays a wide range of conductivity as a function of Pr concentration, permitting the fabrication of superconducting–normal–superconducting weak-link heterostructures. Atomically sharp interfaces were observed, and the CuO_2 planes were continuous from one compound to the other.

References

1   A. Fujishima and K. Honda, *Nature* **238**, 37–8 (1972)
2   P. W. Murray, F. M. Leibsle, H. J. Fischer, C. F. J. Flipse, C. A. Muryn and G. Thornton, *Phys. Rev. B* **46**, 12877–9 (1992)
3   R. J. Lad and M. D. Antonik, *Ceramic Trans.* **24**, 359–66 (1991)
4   D. P. Woodruff and T. A. Delchar, *Modern Techniques of Surface Science* (Cambridge University Press, Cambridge, 1986)
5   H. Moormann, D. Kohl and G. Heiland, *Surf. Sci.* **80**, 261–4 (1979)
6   D. Kohl, H. Moorman and G. Heiland, *Surf. Sci.* **73**, 160–2 (1978)
7   J. Marien, *phys. stat. sol. a* **38**, 513–22 (1976)
8   D. Schmeisser and K. Jacobi, *Surf. Sci.* **88**, 138–52 (1979)
9   S. V. Didziulis, K. D. Butcher, S. L. Cohen and E. I. Solomon, *J. Amer. Chem. Soc.* **111**, 7110–23 (1989)
10 K. Wandelt, in *Auger Spectroscopy and Electronic Structure*, ed. G. Cubiotti, G. Mondio and K. Wandelt (Springer-Verlag, Berlin, 1989), pp. 224–36
11 A. Gutmann, G. Zwicker, D. Schmeisser and K. Jacobi, *Surf. Sci.* **137**, 211–41 (1984)
12 K. Jacobi, G. Zwicker and A. Gutman, *Surf. Sci.* **141**, 109–25 (1984)
13 H. Moormann, D. Kohl and G. Heiland, *Surf. Sci.* **100**, 302–14 (1980)
14 K. Sakamaki, S. Matsunaga, K. Itoh, A. Fujishima and Y. Gohshi, *Surf. Sci.* **219**, L531–6 (1989)
15 J. F. Geiger, K. D. Schierbaum and W. Göpel. *Vacuum* **41**, 1629–32 (1990)
16 Y. W. Chung, W. J. Lo and G. A. Somorjai, *Surf. Sci.* **64**, 588–602 (1977)
17 H. Onishi, T. Aruga, C. Egawa and Y. Iwasawa, *Surf. Sci.* **233**, 261–8 (1990)
18 H. Onishi, T. Aruga, C. Egawa and Y. Iwasawa, *Surf. Sci.* **193**, 33–46 (1988)
19 H. Onishi, T. Aruga, C. Egawa and Y. Iwasawa, *Surf. Sci.* **199**, 54–66 (1988)
20 V. E. Henrich, G. Dresselhaus and H. J. Zeiger, *Solid State Commun.* **24**, 623–6 (1977)
21 R. Casanova, K. Prabhakaran and G. Thornton, *J. Phys.: Condens. Matter* **3**, S91–5 (1991)
22 P. J. Hardman, R. Casanova, K. Prabhakaran, C. A. Muryn, P. L. Wincott and G. Thornton, *Surf. Sci.* **269/270**, 677–81 (1992)
23 W. J. Lo, Y. W. Chung and G. A. Somorjai, *Surf. Sci.* **71**, 199–219 (1978)
24 J. E. T. Andersen and P. J. Møller, *Thin Solid Films* **186**, 137–46 (1990)
25 W. J. Lo and G. A. Somorjai, *Phys. Rev. B* **17**, 4942–50 (1978)
26 R. L. Kurtz and V. E. Henrich, *Phys. Rev. B* **25**, 3563–71 (1982)

27   K. E. Smith and V. E. Henrich, *Phys. Rev. B* **32**, 5384–90 (1985)
28   J. M. McKay and V. E. Henrich, *Surf. Sci.* **137**, 463–72 (1984)
29   R. L. Kurtz and V. E. Henrich, *Phys. Rev. B* **28**, 6699–706 (1983)
30   K. E. Smith and V. E. Henrich, *Surf. Sci.* **225**, 47–57 (1990)
31   R. L. Kurtz and V. E. Henrich, *Phys. Rev. B* **36**, 3413–21 (1987)
32   J. M. McKay and V. E. Henrich, *Phys. Rev. B* **32**, 6764–72 (1985)
33   J. Szuber, *phys. stat. sol. a* **74**, K83–7 (1982)
34   S. Kennou, M. Kamaratos and C. A. Papageorgopoulos, *Vacuum* **41**, 22–4 (1990)
35   J. Szuber, *J. Electron Spectros.* **34**, 337–41 (1984)
36   D. F. Shriver, P. W. Atkins and C. H. Langford, *Inorganic Chemistry* (Oxford University Press, Oxford, 1990)
37   A. R. West, *Solid State Chemistry and its Applications* (Wiley, Chichester, 1984)
38   A. F. Wells, *Structural Inorganic Chemistry, 5th edn* (Oxford University Press, Oxford, 1984)
39   N. N. Greenwood and A. Earnshaw, *Chemistry of the Elements* (Pergamon Press, Oxford, 1984)
40   P. A. Cox, *Transition Metal Oxides: An Introduction to Their Electronic Structure and Properties* (Clarendon Press, Oxford, 1992)
41   W. Hayes and A. M. Stoneham, *Defects and Defect Processes in Nonmetallic Solids* (Wiley, New York, 1985)
42   F. Agullo-Lopez, C. R. A. Catlow and P. D. Townsend, *Point Defects in Materials* (Academic Press, London, 1988)
43   O. T. Sorensen (ed.) *Nonstoichiometric Oxides* (Academic Press, New York, 1981)
44   R. J. D. Tilley, *Defect Crystal Chemistry* (Blackie, Glasgow, 1987)
45   C. Pisani, M. Causà, R. Dovesi and C. Roetti, *Prog. Surf. Sci.* **25**, 119–37 (1987)
46   M. A. Van Hove and S. Y. Tong, *Surface Crystallography by LEED* (Springer-Verlag, Berlin, 1979)
47   K. Heinz, *Prog. Surf. Sci.* **27**, 239–326 (1988)
48   B. D. Cullity, *Elements of X-Ray Diffraction* (Addison-Wesley, Reading, 1956)
49   H. Ohtani, C. T. Kao, M. A. Van Hove and G. A. Somorjai, *Prog. Surf. Sci.* **23**, 155–316 (1986)
50   A. J. Dekker, *Solid State Phys.* **6**, 251–311 (1958)
51   G. Binnig, H. Rohrer, C. Gerber and E. Weibel, *Phys. Rev. Lett.* **49**, 57–61 (1982)
52   J. A. Golovchenko, *Science* **232**, 48–53 (1986)
53   G. Binnig, C. F. Quate and C. Gerber, *Phys. Rev. Lett.* **56**, 930–3 (1986)
54   H. Heinzelmann, E. Meyer, P. Grütter, H. R. Hidber, L. Rosenthaler and H. J. Güntherodt, *J. Vac. Sci. Technol. A* **6**, 275–8 (1988)
55   K. Prabhakaran, D. Purdie, R. Casanova, C. A. Muryn, P. J. Hardman, R. Lindsay, P. L. Wincott and G. Thornton, *Phys. Rev. B* **45**, 6969–72 (1992)
56   A. Santoni, D. B. Tran Thoai and J. Urban, *Solid State Commun.* **68**, 1039–41 (1988)
57   A. Zangwill, *Physics at Surfaces* (Cambridge University Press, Cambridge, 1988)
58   C. Klein and C. S. Hurlbut, Jr, *Manual of Mineralogy, 20th edn* (Wiley, New York, 1985)
59   R. G. W. Wyckoff, *Crystal Structures, 2nd edn* (Wiley Interscience, New York, vol. 1, 1963; vol.2 1964).
60   K. O. Legg, M. Prutton and C. Kinniburgh, *J. Phys. C: Solid State Phys.* **7**, 4236–46 (1974)
61   C. G. Kinniburgh, *J. Phys. C: Solid State Phys.* **8**, 2382–94 (1975)
62   M. Prutton, J. A. Walker, M. R. Welton-Cook, R. C. Felton and J. A. Ramsey, *Surf. Sci.* **89**, 95–101 (1979)

63 M. R. Welton-Cook and W. Berndt, *J. Phys. C: Solid State Phys.* **15**, 5691–710 (1982)
64 T. Urano, T. Kanaji and M. Kaburagi, *Surf. Sci.* **134**, 109–21 (1983)
65 D. L. Blanchard, D. L. Lessor, J. P. LaFemina, D. R. Baer, W. K. Ford and T. Guo, *J. Vac. Sci. Technol. A* **9**, 1814–19 (1991)
66 A. Ichimiya and Y. Takeuchi, *Surf. Sci.* **128**, 343–9 (1983)
67 P. Maksym, *Surf. Sci.* **149**, 157–74 (1985)
68 K. H. Rieder, *Surf. Sci.* **118**, 57–65 (1982)
69 G. Brusdeylins, R. B. Doak, J. G. Skofronick and J. P. Toennies, *Surf. Sci.* **128**, 191–206 (1983)
70 P. Cantini and E. Cevasco, *Surf. Sci.* **148**, 37–41 (1984)
71 E. Kolodney and A. Amirav, *Surf. Sci.* **155**, 715–31 (1985)
72 D. R. Jung, M. Mahgerefteh and D. R. Frankl, *Phys. Rev. B* **39**, 11164–7 (1989)
73 J. Cui, D. R. Jung and D. R. Frankl, *Phys. Rev. B* **42**, 9701–4 (1990)
74 D. R. Jung, J. Cui and D. R. Frankl, *J. Vac. Sci. Technol. A* **9**, 1589–94 (1991)
75 Y. Murata, S. Murakami, H. Namba, T. Gotoh and K. Kinosita, *Proc. 7th Intern. Vac. Congr. & 3rd Intern. Conf. Solid Surf.* (Vienna, 1977), pp. 2439–42
76 Y. Murata and S. Murakami, *Inst. Phys. Congr. Ser.* **41**, 218–22 (1978)
77 T. Gotoh, S. Murakami, K. Kinosita and Y. Murata, *J. Phys. Soc. Japan* **50**, 2063–8 (1981)
78 Y. Murata, *J. Phys. Soc. Japan* **51**, 1932–5 (1982)
79 C. Duriez, C. Chapon, C. R. Henry and J. M. Rickard, *Surf. Sci.* **230**, 123–36 (1990)
80 M. Mahgerefteh, D. R. Jung and D. R. Frankl, *Phys. Rev. B* **39**, 3900–4 (1989)
81 H. Nakamatsu, A. Sudo and S. Kawai, *Surf. Sci.* **194**, 265–74 (1988)
82 R. N. Barnett and R. Bass, *J. Chem. Phys.* **67**, 4620–5 (1977)
83 M. R. Welton-Cook and M. Prutton, *Surf. Sci.* **74**, 276–84 (1978)
84 A. J. Martin and H. Bilz, *Phys. Rev. B* **19**, 6593–600 (1979)
85 F. W. deWette, W. Kress and U. Schröder, *Phys. Rev. B* **32**, 4143–57 (1985)
86 G. V. Lewis and C. R. A. Catlow, *J. Phys. C: Solid State Phys.* **18**, 1149–61 (1985)
87 M. Causà, R. Dovesi, C. Pisani and C. Roetti, *Surf. Sci.* **175**, 551–60 (1986)
88 A. Fujimori and N. Tsuda, *Surf. Sci.* **100**, L445–8 (1980)
89 V. E. Henrich, *Surf. Sci.* **57**, 385–92 (1976)
90 M. Causà, R. Dovesi, E. Kotomin and C. Pisani, *J. Phys. C: Solid State Phys.* **20**, 4983–90 (1987)
91 J. W. He and P. J. Møller, *Surf. Sci.* **178**, 934–42 (1986)
92 M. Prutton, J. A. Ramsey, J. A. Walker and M. R. Welton-Cook, *J. Phys. C: Solid State Phys.* **12**, 5271–80 (1979)
93 F. P. Netzer and M. Prutton, *J. Phys. C: Solid State Phys.* **8**, 2401–12 (1975)
94 C. G. Kinniburgh and J. A. Walker, *Surf. Sci.* **63**, 274–82 (1977)
95 M. R. Welton-Cook and M. Prutton, *J. Phys. C: Solid State Phys.* **13**, 3993–4000 (1980)
96 N. Floquet and L. C. Dufour, *Surf. Sci.* **126**, 543–9 (1983)
97 P. A. Cox and A. A. Williams, *Surf. Sci.* **152/153**, 791–6 (1985)
98 R. C. Felton, M. Prutton, S. P. Tear and M. R. Welton-Cook, *Surf. Sci.* **88**, 474–8 (1979)
99 R. J. Lad and V. E. Henrich, *Phys. Rev. B* **38**, 10860–9 (1988)
100 M. A. Langell and N. R. Cameron, *Surf. Sci.* **185**, 105–19 (1987)
101 E. B. Bas, U. Bäninger and H. Mülethaler, *Japan. J. Appl. Phys., Suppl. 2, Pt. 2, Proc. 2nd Intern. Conf. on Solid Surfaces*, pp. 671–3 (1974)
102 R. C. Felton, M. Prutton and J. A. D. Matthew, *Surf. Sci.* **79**, 117–24 (1979)
103 P. W. Tasker, *Adv. in Ceramics* **10**, 176–89 (1984)

104 W. C. Mackrodt and P. W. Tasker, *Chem. Britain* **21**, 13–16 (1985)

105 A. Gibson, R. Haydock and J. P. LaFemina, *J. Vac. Sci. Technol. A* **10**, 2361–6 (1992)

106 P. A. Cox, F. W. H. Dean and A. A. Williams, *Vacuum* **33**, 839–41 (1983)

107 P. W. Tasker, *J. Phys. C: Solid State Phys.* **12**, 4977–84 (1979)

108 A. M. Stonehan and P. W. Tasker, in *Surface and Near-Surface Chemistry of Oxide Materials*, ed. J. Nowotny and L. C. Dufour (Elsevier, Amsterdam, 1988), Chap. 1

109 G. Heiland and P. Kunstmann, *Surf. Sci.* **13**, 72–84 (1969)

110 C. Kittel, *Introduction to Solid State Physics* (Wiley, New York, 1986), Chap. 13

111 V. E. Henrich, G. Dresselhaus and H. J. Zeiger, *Phys. Rev. B* **17**, 4908–21 (1978)

112 Y. W. Chung and W. B. Weissbard, *Phys. Rev. B* **20**, 3456–61 (1979)

113 R. Courths, *phys. stat. sol. b* **100**, 135–48 (1980)

114 N. Bickel, G. Schmidt, K. Heinz and K. Müller, *Phys. Rev. Lett.* **62**, 2009–11 (1989)

115 N. Bickel, G. Schmidt, K. Heinz and K. Müller, *Vacuum* **41**, 46–8 (1990)

116 D. Aberdam, G. Bouchet and P. Ducros, *Surf. Sci.* **27**, 559–70 (1971)

117 D. Aberdam and C. Gaubert, *Surf. Sci.* **27**, 571–85 (1971)

118 M. A. Langell and S. L Bernasek, *J. Vac. Sci. Technol.* **17**, 1296–302 (1980)

119 M. A. Langell and S. L. Bernasek, *Surf. Sci.* **69**, 727–30 (1977)

120 M. A. Langell and S. L. Bernasek, *Prog. Surf. Sci.* **9**, 165–89 (1979)

121 M. A. Langell and S. L. Bernasek, *J. Vac. Sci. Technol.* **17**, 1287–95 (1980)

122 C. J. Schramm, M. A. Langell and S. L. Bernasek, *Surf. Sci.* **110**, 217–26 (1981)

123 H. Nakamatsu, Y. Yamamoto, S. Kawai, K. Oura and T. Hanawa, *Japan. J. Appl. Phys.* **22**, L461–3 (1983)

124 R. G. Egdell, H. Innes and M. D. Hill, *Surf. Sci.* **149**, 33–47 (1985)

125 V. E. Henrich and R. L. Kurtz, *Phys. Rev. B* **23**, 6280–7 (1981)

126 R. H. Tait and R. V. Kasowski, *Phys. Rev. B* **20**, 5178–91 (1979)

127 R. L. Kurtz, *Surf. Sci.* **177**, 526–52 (1986)

128 R. L. Kurtz, R. Stockbauer and T. E. Madey, *Nucl. Instr. Methods B* **13**, 518–24 (1986)

129 P. J. Møller and M. C. Wu, *Surf. Sci.* **224**, 265–76 (1989)

130 H. R. Sadeghi and V. E. Henrich, *J. Catal.* **87**, 279–82 (1984)

131 H. R. Sadeghi and V. E. Henrich, *Appl. Surf. Sci.* **19**, 330–40 (1984)

132 B. L. Maschoff, J. Pan and T. E. Madey, *Surf. Sci.* **259**, 190–206 (1991)

133 P. R. Zschack, J. B. Cohen and Y. W. Chung, *Surf. Sci.* **262**, 395–408 (1992)

134 M. D. Antonik, L. C. Edwards and R. J. Lad, *Proc. Mater. Res. Soc.* **237**, 459–64 (1992)

135 M. D. Antonik and R. J. Lad, *J. Vac. Sci. Technol. A* **10**, 669–73 (1992)

136 C. G. Mason, S. P. Tear, T. N. Doust and G. Thornton, *J. Phys. : Condens. Matter* **3**, S97–102 (1991)

137 L. E. Firment, *Surf. Sci.* **116**, 205–16 (1982)

138 G. E. Poirier, B. K. Hance and J. M. White, *J. Vac. Sci. Technol. B* **10**, 6–15 (1992)

139 E. deFrésart, J. Darville and J. M. Gilles, *Solid State Commun.* **37**, 13–17 (1980)

140 D. F. Cox, T. B. Fryberger and S. Semancik, *Phys. Rev. B* **38**, 2072–83 (1988)

141 E. deFrésart, J. Darville and J. M. Gilles, *Appl. Surf. Sci.* **11/12**, 637–51 (1982)

142 J. W. Erickson and S. Semancik, *Surf. Sci.* **187**, L658–88 (1987)

143 D. F. Cox, T. B. Fryberger and S. Semancik, *J. Vac. Sci. Technol. A* **6**, 828–9 (1988)

144 D. F. Cox, T. B. Fryberger and S. Semancik, *Surf. Sci.* **224**, 121–42 (1989)

145 R. Cavicchi, M. Tarlov and S. Semancik, *J. Vac. Sci. Technol. A* **8,** 2347–52 (1990)

146 D. F. Cox, S. Semancik and P. D. Szuromi, *J. Vac. Sci. Technol. A* **4**, 627–8 (1986)

147 G. L. Shen, R. Casanova, G. Thornton and I. Colera, *J. Phys. : Condens. Matter* **3**, S291–6 (1991)

148 T. J. Godin and J. P. LaFemina, in *Structure and Properties of Interfaces in Materials*, ed. W. T. A. Clark, U. Dahman and C. L. Briant (Material Research Society, Pittsburgh, 1992)
149 E. deFrésart, J. Darville and J. M. Gilles, *Surf. Sci.* **126**, 518–22 (1983)
150 W. E. O'Grady, Lj. Atanasoska, F. H. Pollak and H. L. Park, *J. Electroanal. Chem.* **178**, 61–8 (1984)
151 Lj. Atanasoska, W. E. O'Grady, R. T. Atanasoski and F. H. Pollak, *Surf. Sci.* **202**, 142–66 (1988)
152 F. R. F. Fan and A. J. Bard, *J. Phys. Chem.* **94**, 3761–6 (1990)
153 G. S. Rohrer, V. E. Henrich and D. A. Bonnell, *Science* **250**, 1239–41 (1990)
154 G. S. Rohrer and D. A. Bonnell, in *Chemistry of Electronic Ceramic Materials*, *National Institute of Standards and Technology Special Publication 804* (NIST, Washington, DC, 1990), pp. 447–54
155 G. S. Rohrer, V. E. Henrich and D. A. Bonnell, *Mater. Res. Soc. Symp. Proc.* **209**, 611–16 (1991)
156 P. A. Cox, J. B. Goodenough, P. Tavener, D. Telles and R. G. Egdell, *J. Solid State Chem.* **62**, 360–70 (1986)
157 J. M. Charig, *Appl. Phys. Lett.* **10**, 139–40 (1967)
158 C. C. Chang, in *The Structure and Chemistry of Solid Surfaces*, ed. G. A. Somorjai (John Wiley & Sons, New York, 1969), pp. 77-1 to 77-14
159 T. Hsu and Y. Kim, *Ultramicroscopy* **32**, 103–12 (1990)
160 T. Hsu and Y. Kim, *Surf. Sci.* **243**, L63–6 (1991)
161 E. Gillet, C. Legressus and M. Gillet, *J. chimie physique* **84**, 167–74 (1987)
162 C. C. Chang, *J. Appl. Phys.* **39**, 5570–3 (1968)
163 T. M. French and G. A. Somorjai, *J. Phys. Chem.* **74**, 2489–95 (1970)
164 J. M. Charig and D. K. Skinner, in *The Structure and Chemistry of Solid Surfaces*, ed. G. A. Somorjai (John Wiley & Sons, New York, 1969), pp. 34-1 to 34-20
165 M. Gautier, J. P. Duraud and L. Pham Van, *Surf. Sci.* **249**, L327–32 (1991)
166 M. Vermeersch, R. Sporken, Ph. Lambin and R. Caudano, *Surf. Sci.* **235**, 5–14 (1990)
167 M. Gautier, J. P. Duraud, L. Pham Van and M. J. Guittet, *Surf. Sci.* **250**, 71–80 (1991)
168 M. Arbab, G. S. Chottiner and R. W. Hoffman, *Mater. Res. Soc. Symp. Proc.* **153**, 63–9 (1989)
169 M. Causà, R. Dovesi, C. Pisani and C. Roetti, *Surf. Sci.* **215**, 259–71 (1989)
170 R. J. Lad and V. E. Henrich, *Surf. Sci.* **193**, 81–93 (1988)
171 C. Sanchez, M. Hendewerk, K. D. Sieber and G. A. Somorjai, *J. Solid State Chem.* **61**, 47–55 (1986)
172 R. L. Kurtz and V. E. Henrich, *Surf. Sci.* **129**, 345–54 (1983)
173 R. L Kurtz and V. E. Henrich, *Phys. Rev. B* **26**, 6682–9 (1982)
174 P. J. Lawrence, S. C. Parker and P. W. Tasker, *J. Amer. Ceram. Soc.* **71**, C389–91 (1988)
175 A. R. Lubinsky, C. B. Duke, S. C. Chang, B. W. Lee and P. Mark, *J. Vac. Sci. Technol.* **13**, 189–92 (1976)
176 C. B. Duke, A. R. Lubinsky, B. W. Lee and P. Mark, *J. Vac. Sci. Technol.* **13**, 761–8 (1976)
177 C. B. Duke, *J. Vac. Sci. Technol.* **14**, 870–7 (1977)
178 C. B. Duke, A. R. Lubinsky, S. C. Chang, B. W. Lee and P. Mark, *Phys. Rev. B* **15**, 4865–73 (1977)
179 C. B. Duke and A. R. Lubinsky, *Surf. Sci.* **50**, 605–14 (1975)
180 C. B. Duke, R. J. Meyer, A. Paton and P. Mark, *Phys. Rev. B* **18**, 4225–40 (1978)

181 W. Göpel, J. Pollmann, I. Ivanov and B. Reihl, *Phys. Rev. B* **26**, 3144–50 (1982)

182 Y. R. Wang and C. B. Duke, *Surf. Sci.* **192**, 309–22 (1987)

183 C. B. Duke and Y. R. Wang, *J. Vac. Sci. Technol. A* **6**, 692–5 (1988)

184 C. B. Duke and Y. R. Wang, *J. Vac. Sci. Technol. B* **6**, 1440–3 (1988)

185 C. B. Duke, in *The Chemical Physics of Solid Surfaces and Heterogeneous Catalysis*, ed. D. A. King and D. P. Woodruff (Elsevier, Amsterdam, 1988), pp. 69–118

186 M. F. Chung and H. E. Farnsworth, *Surf. Sci.* **22**, 93–110 (1970)

187 J. D. Levine, A. Willis, W. R. Bottoms and P. Mark, *Surf. Sci.* **29**, 144–64 (1972)

188 S. C. Chang and P. Mark, *Surf. Sci.* **46**, 293–300 (1974)

189 H. van Hove and R. Leysen, *phys. stat. sol. a* **9**, 361–7 (1972)

190 A. J. Skinner and J. P. LaFemina, *Phys. Rev. B* **45**, 3557–64 (1992)

191 J. M. McKay and V. E. Henrich, *Phys. Rev. Lett.* **53**, 2343–6 (1984)

192 C. Palache, H. Berman and C. Frondel, *Dana's System of Mineralogy*, (Wiley, New York, 1944)

193 R. R. Gay, M. H. Nodine, V. E. Henrich, H. J. Zeiger and E. I. Solomon, *J. Amer. Chem. Soc.* **102**, 6752–61 (1980)

194 L. E. Firment and A. Ferretti, *Surf. Sci.* **129**, 155–76 (1983)

195 L. C. Dufour, O. Bertrand and N. Floquet, *Surf. Sci.* **147**, 396–412 (1984)

196 O. Bertrand, N. Floquet and D. Jacquot, *Surf. Sci.* **164**, 305–19 (1985)

197 N. Floquet and O. Bertrand, *Surf. Sci.* **198**, 449–60 (1988)

198 W. P. Ellis and R. L. Schwoebel, *Surf. Sci.* **11**, 82–98 (1968)

199 W. P. Ellis, *J. Chem. Phys.* **48**, 5695–701 (1968)

200 W. P. Ellis and T. N. Taylor, *Surf. Sci.* **75**, 279–86 (1978)

201 W. P. Ellis and T. N. Taylor, *Surf. Sci.* **91**, 409–22 (1980)

202 K. A. Thompson, W. P. Ellis, T. N. Taylor, S. M. Valone and C. J. Maggiore, *Nucl. Instrum. Meth. Phys. Res.* **218**, 475–9 (1983)

203 P. W. Tasker, *Surf. Sci.* **78**, 315–24 (1979)

204 T. N. Taylor and W. P. Ellis, *Surf. Sci.* **107**, 249–62 (1981)

205 L. E. Cox and W. P. Ellis, *Solid State Commun.* **78**, 1033–7 (1991)

206 W. P. Ellis, *Surf. Sci.* **45**, 569–84 (1974)

207 T. N. Taylor and W. P. Ellis, *Surf. Sci.* **77**, 321–36 (1978)

208 K. H. Schulz and D. F. Cox, *Phys. Rev. B* **43**, 1610–21 (1991)

209 J. F. Halet and R. Hoffmann, *J. Amer. Chem. Soc.* **111**, 3548–59 (1989)

210 A. T. Fiory, A. F. Hebard, R. H. Eick, L. F. Schneemeyer, J. V. Waszczak and H. J. Gossmann, *Appl. Phys. Lett.* **58**, 777–9 (1991)

211 R. Claessen, R. Manzke, H. Carstensen, B. Burandt, T. Buslaps, M. Skibowski and J. Fink, *Phys. Rev. B* **39**, 7316–19 (1989)

212 P. A. P. Lindberg, Z. X. Shen, B. O. Wells, D. B. Mitzi, I. Lindau, W. E. Spicer and A. Kapitulnik, *Appl. Phys. Lett.* **53**, 2563–5 (1988)

213 R. Claessen, T. Buslaps, G. Mante, R. Manzke and M. Skibowski, *Vacuum* **41**, 986–8 (1990)

214 S. Nakanishi, N. Fukuoka, K. Nakahigashi, M. Kogachi, H. Sasakura, S. Minamigawa and A. Yanase, *Japan. J. Appl. Phys.* **28**, L71–4 (1989)

215 A. A. Abramov, N. V. Abrosimov, V. A. Grazhulis, A. M. Ionov and A. M. Panchenko, *Surf. Sci.* **242**, 45–9 (1991)

216 V. A. Grazhulis, G. A. Emel'chenko and A. M. Ivanov, *Physica C* **159**, 404–6 (1989)

217 N. L. Allan, P. Kenway, W. C. Mackrodt and S. C. Parker, *J. Phys. : Condens. Matter* **1**, SB119–22 (1989)

218 H. Poelman, J. Vennik and G. Dalmai, *J. Electron Spectros.* **44**, 251–62 (1987)

219 L. Fiermans, P. Clauws, W. Lambrecht, L. Vandenbroucke and J. Vennik, *phys. stat.*

sol. a **59**, 485–504 (1980)

220 H. E. Bishop and J. C. Rivière, in *The Structure and Chemistry of Solid Surfaces*, ed. G. A. Somorjai (John Wiley & Sons, New York, 1969), pp. 37-1 to 37-7

221 P. W. Tasker, *UKAEA Report AERE-M3239* (1982)

222 H. Knözinger and P. Ratnasamy, *Catal. Rev. – Sci. Engin.* **17**, 31–70 (1978)

223 M. N. Colpaert, P. Clauws, L. Fiermans and J. Vennik, *Surf. Sci.* **36**, 513–25 (1973)

224 L. Fiermans and J. Vennik, *Surf. Sci.* **9**, 187–97 (1968)

225 D. A. Bonnell, *Ceramics Trans.* **5**, 315–26 (1989).

226 E. Garfunkel, G. Rudd, D. Novak, S. Wang, G. Ebert, M. Greenblatt, T. Gustafsson and S. H. Garofalini, *Science* **246**, 99–100 (1989)

227 G. Rudd, D. Novak, D. Saulys, R. A. Bartynski, S. Garofalini, K. V. Ramanujachary, M. Greenblatt and E. Garfunkel, *J. Vac. Sci. Technol. B* **9**, 909–13 (1991)

228 D. Anselmetti, R. Wiesendanger, H. J. Güntherodt and G. Grüner, *Europhys. Lett.* **12**, 241–5 (1990)

229 D. Saulys, G. Rudd and E. Garfunkel, *J. Appl. Phys.* **69**, 1707–11 (1991)

230 M. L. Norton, J. G. Mantovani and R. J. Warmack, *J. Vac. Sci. Technol. A* **7**, 2898–900 (1989)

231 M. D. Kirk, J. Nogami, A. A. Baski, D. B. Mitzi, A. Kapitulnik, T. H. Geballe and C. F. Quate, *Science* **242**, 1673–5 (1988)

232 M. D. Kirk, C. B. Eom, B. Oh, S. R. Spielman, M. R. Beasley, A. Kapitulnik, T. H. Geballe and C. F. Quate, *Appl. Phys. Lett.* **52**, 2071–3 (1988)

233 X. L. Wu, Z. Zhang, Y. L. Wang and C. M. Lieber, *Science* **248**, 1211–14 (1990)

234 W. Göpel and G. Neuenfeldt, *Surf. Sci.* **55**, 362–6 (1976)

235 M. Henzler, *Surf. Sci.* **36**, 109–22 (1973)

236 W. H. Cheng and H. H. Kung, *Surf. Sci.* **102**, L21–8 (1981)

237 D. Kohl, M. Henzler and G. Heiland, *Surf. Sci.* **41**, 403–11 (1974)

238 M. Henzler, *Surf. Sci.* **19**, 159–71 (1970)

239 M. Henzler, *Appl. Phys.* **9**, 11–17 (1976)

240 S. E. Gilbert and J. H. Kennedy, *J. Electrochem. Soc.* **135**, 2385–6 (1988)

241 G. S. Rohrer and D. A. Bonnell, *Surf. Sci.* **247**, L195–200 (1991)

242 K. Itaya and E. Tomita, *Surf. Sci.* **219**, L515–20 (1989)

243 R. C. Jaklevic, L. Elie, W. Shen and J. T. Chen, *J. Vac. Sci. Technol. A* **6**, 448–53 (1988)

244 S. C. Langford, M. Zhenyi, L. C. Jensen and J. T. Dickinson, *J. Vac. Sci. Technol. A* **8**, 3470–8 (1990)

245 J. M. Cowley, *Prog. Surf. Sci.* **21**, 209–50 (1986)

246 A. F. Moodie and C. Warble, *Phil. Mag.* **16**, 891–904 (1967)

247 G. Lehmpfuhl and C. Warble, *Ultramicroscopy* **19**, 135–46 (1986)

248 A. F. Moodie and C. E. Warble, *J. Crystal Growth* **10**, 26–38 (1971)

249 D. J. Smith, L. A. Bursill and D. A. Jefferson, *Surf. Sci.* **175**, 673–83 (1986)

250 Z. C. Kang, D. F. Smith and L. Eyring, *Surf. Sci.* **175**, 684–92 (1986)

251 W. Zhou, D. A. Jefferson and W. Y. Liang, *Surf. Sci.* **209**, 444–54 (1989)

252 T. Hsu and Y. Kim, *Surf. Sci.* **258**, 119–30 (1991)

253 Y. Kim and T. Hsu, *Surf. Sci.* **258**, 131–46 (1991)

254 J. Kendrick, E. A. Colbourn and W. C. Mackrodt, *Radiation Effects* **73**, 259–66 (1983)

255 E. A. Colbourn, J. Kendrick and W. C. Mackrodt, *Surf. Sci.* **126**, 550–7 (1983)

256 E. A. Colbourn and W. C. Mackrodt, *Solid State Ionics* **8**, 221–31 (1983)

257 P. W. Tasker and D. M. Duffy, *Surf. Sci.* **137**, 91–102 (1984)

258 W. Göpel, R. S. Bauer and G. Hansson, *Surf. Sci.* **99**, 138–58 (1980)

259 W. Göpel, L. J. Brillson and C. F. Brucker, *J. Vac. Sci. Technol.* **17**, 894–8 (1980)

260 W. Göpel and U. Lampe, *Phys. Rev. B* **32**, 6447–62 (1980)

261 W. C. Mackrodt and R. F. Stewart, *J. Phys. C: Solid State Phys.* **10**, 1431–46 (1977)

262 D. M. Duffy, J. P. Hoare and P. W. Tasker, *J. Phys. C: Solid State Phys.* **17**, L195–9 (1984)

263 E. A. Colbourn and W. C. Mackrodt, *Adv. Ceramics* **10**, 190–204 (1985)

264 H. Nakamatsu, A. Sudo and S. Kawai, *Surf. Sci.* **223**, 193–200 (1989).

265 W. Hirshwald, in *Non-Stoichiometric Compounds: Surfaces, Grain Boundaries and Structural Defects*, ed. J. Nowotny and W. Weppner (Kluwer, Dordrecht, 1989), pp. 203–19

266 G. K. Wehner, *Adv. Electr. Electr. Phys.* **VII**, 239–98 (1955)

267 H. M. Naguib and R. Kelly, *Radiation Effects* **25**, 1–12 (1975)

268 J. B. Malherbe, S. Hofmann and J. M. Sanz, *Appl. Surf. Sci.* **27**, 355–65 (1986)

269 V. E. Henrich, G. Dresselhaus and H. J. Zeiger, *Phys. Rev. Lett.* **36**, 1335–9 (1976)

270 H. R. Sadeghi and V. E. Henrich, *J. Catal.* **109**, 1–11 (1988)

271 M. H. Mohamed, H. R. Sadeghi and V. E. Henrich, *Phys. Rev. B* **37**, 8417–23 (1988)

272 J. L. Sullivan, S. O. Saied and I. Bertoti, *Vacuum*, **42**, 1203–8 (1991)

273 M. A. Langell, *Surf. Sci.* **186**, 323–38 (1987)

274 J. L. Mackay and V. E. Henrich, *Phys. Rev. B* **39**, 6156–68 (1989)

275 P. A. P. Lindberg, Z. X. Shen, W. E. Spicer and I. Lindau, *Surf. Sci. Rep.* **11**, 1–137 (1989)

276 M. Cotter and R. G. Egdell, *J. Solid State Chem.* **66**, 364–8 (1987)

277 J. Herion, G. Scharl and M. Tapiero, *Appl. Surf. Sci.* **14**, 233–48 (1982/83)

278 P. Wynblatt and R. C. McCune, in *Surfaces and Interfaces in Ceramic and Ceramic–Metal Systems*, ed. J. A. Pask and A. G. Evans (Plenum, New York, 1981), pp. 83–95

279 R. C. McCune and P. Wynblatt, *J. Amer. Ceramic Soc.* **66**, 111–17 (1983)

280 R. Souda, T. Aizawa and Y. Ishizawa, *J. Vac. Sci. Technol. A* **8**, 3218–23 (1990)

281 J. E. T. Andersen and P. J. Møller, *Appl. Phys. Lett.* **56**, 1847–9 (1990)

282 M. Cotter, S. Campbell, L. L. Cao, R. G. Egdell and W. C. Mackrodt, *Surf. Sci.* **208**, 267–84 (1989)

283 R. G. Egdell and W. C. Mackrodt, in *Surfaces and Interfaces of Ceramic Materials*, ed. L. C. Dufour, C. Monty and G. Petot-Ervas (Kluwer, Dordrecht, 1989), pp. 185–203

284 E. A. Colbourn and W. C. Mackrodt, *J. Mater. Sci.* **17**, 3021–38 (1982)

285 E. A. Colbourn, W. C. Mackrodt and P. W. Tasker, *J. Mater. Sci.* **18**, 1917–24 (1983)

286 P. W. Tasker, E. A. Colbourn and W. C. Mackrodt, *J. Amer. Ceramic Soc.* **68**, 74–80 (1985)

287 P. Masri and P. W. Tasker, *Surf. Sci.* **149**, 209–25 (1985)

288 P. Masri, P. W. Tasker, J. P. Hoare and J. H. Harding, *Surf. Sci.* **173**, 439–54 (1986)

289 W. C. Mackrodt, P. W. Tasker and E. A. Colbourn, *Surf. Sci.* **152/153**, 940–6 (1985)

290 J. Nunan, J. Cunningham, A. M. Deane, E. A. Colbourn and W. C. Mackrodt, in *Adsorption and Catalysis on Oxide Surfaces*, ed. M. Che and G. C. Bond (Elsevier, Amsterdam, 1985), pp. 83–96

291 P. Masri and P. W. Tasker, *J. Electron Spectros.* **39**, 333–41 (1986)

292 E. A. Colbourn, W. C. Mackrodt and P. W. Tasker, *Physica* **131B**, 41–5 (1985)

293 R. G. Egdell and S. C. Parker, in *Science of Ceramic Interfaces*, ed. J. Nowotny, *Materials Science Monographs* **75** (Elsevier, Amsterdam, 1991), pp. 41–76

294 W. C. Mackrodt and P. W. Tasker, *J. Amer. Ceram. Soc.* **72**, 1576–83 (1989)

295 P. W. Tasker, *Solid State Ionics* **8**, 233–41 (1983)

296 S. Campbell, L. L. Cao, M. Cotter, R. G. Egdell, W. R. Flavell, K. F. Mok and W.

C. Mackrodt, *J. Phys. : Condens. Matter* **1**, SB237–8 (1989)

297 L. L. Cao, R. G. Egdell, W. R. Flavell, K. F. Mok and W. C. Mackrodt, *J. Mater. Chem.* **1**, 785–8 (1991).

298 W. Hirschwald, I. Sikora, F. Stolze and J. Oblakowski, *Surf. Interface Anal.* **14**, 477–81 (1989)

299 S. Y. Liu and H. H. Kung, *Surf. Sci.* **110**, 504–22 (1981)

300 J. M. Blakely and S. M. Mukhopadhyay, in *Surfaces and Interfaces of Ceramic Materials*, ed. L. C. Dufour, C. Monty and G. Petot-Ervas (Kluwer, Dordrecht, 1989), pp. 285–99

301 S. Baik, D. E. Fowler, J. M. Blakely and R. Raj, *J. Amer. Ceramic Soc.* **68**, 281–6 (1985)

302 S. M. Mukhopadhyay, A. P. Jardine and J. M. Blakely, *J. Amer. Ceram. Soc.* **71**, 358–62 (1988)

303 S. Baik, *J. Amer. Ceram. Soc.* **69**, C101–3 (1986)

304 S. Baik and C. L. White, *J. Amer. Ceram. Soc.* **70**, 682–8 (1987)

305 R. C. McCune, W. T. Donlon and R. C. Ku, *J. Amer. Ceram. Soc.* **69**, C196–9 (1986)

306 R. De Gryse, J. P. Landuyt, L. Vandenbroucke and J. Vennik, *Surf. Interface Anal.* **4**, 168–73 (1982)

307 W. C. Mackrodt and P. W. Tasker, *Mater. Res. Soc. Symp. Proc.* **60**, 291–8 (1986)

308 M. J. Davies, P. R. Kenway, P. J. Lawrence, S. C. Parker, W. C. Mackrodt and P. W. Tasker, *J. Chem. Soc. Faraday Trans. 2* **85**, 555–63 (1989)

309 J. Nowotny, in *Surfaces and Interfaces of Ceramic Materials*, ed. L. C. Dufour, C. Monty and G. Petot-Ervas (Kluwer, Dordrecht, 1989)

310 D. A. Bonnell and D. R. Clarke, *J. Amer. Ceramic Soc.* **71**, 629–37 (1988)

311 Y. J. Chabal, *Surf. Sci.* Rep. **8**, 211–358 (1988)

312 H. Froitzheim, in *Electron Spectroscopy for Surface Analysis*, ed. H. Ibach (Springer-Verlag, Berlin, 1977), Chap. 6

313 D. Decroupet, M. Liehr, P. A. Thiry, J. J. Pireaux and R. Caudano, *J. Vac. Sci. Technol. A* **4**, 1304–5 (1986)

314 A. A. Lucas, J. P. Vigneron, Ph. Lambin, P. A. Thiry, M. Liehr, J. J. Pireaux and R. Caudano, *Intern. J. Quantum Chem. : Quantum Chem. Symp.* **19**, 687–705 (1986)

315 N. W. Ashcroft and N. D. Mermin, *Solid State Physics* (Holt, Rinehart and Winston, New York, 1976), Chaps. 22 and 23

316 P. M. A. Sherwood, *Vibrational Spectroscopy of Solids* (Cambridge University Press, Cambridge, 1972)

317 N. R. Summitt, *J. Appl. Phys.* **39**, 3762–7 (1968)

318 H. Ibach and D. L. Mills, *Electron Energy Loss Spectroscopy and Surface Vibrations* (Academic Press, New York, 1982), Chap. 5

319 F. Garcia-Moliner and F. Flores, *Introduction to the Theory of Solid Surfaces* (Cambridge University Press, Cambridge, 1979)

320 P. A. Thiry, M. Liehr, J. J. Pireaux and R. Caudano, *Phys. Rev. B* **29**, 4824–6 (1984)

321 P. A. Cox and A. A. Williams, *J. Electron Spectros.* **39**, 45–58 (1986)

322 C. Oshima, T. Aizawa, R. Souda and Y. Ishizawa, *Solid State Commun.* **73**, 731–4 (1990)

323 M-C. Wu, C. A. Estrada, J. S. Corneille, and D. W. Goodman, *J. Chem. Phys.* **96**, 3892–900 (1992)

324 M. Liehr, P. A. Thiry, J. J. Pireaux and R. Caudano, *Phys. Rev. B* **33**, 5682–97 (1986)

325 M. Liehr, P. A. Thiry, J. J. Pireaux and R. Caudano, *J. Vac. Sci. Technol. A* **2**, 1079–82 (1984)

326 H. Ibach, *Phys. Rev. Lett.* **24**, 1416–18 (1970)

327 Y. Goldstein, A. Many, I. Wagner and J. Gersten, *Surf. Sci.* **98**, 599–12 (1980)

328 A. Many, I. Wagner, A. Rosenthal, J. I. Gersten and Y. Goldstein, *Phys. Rev. Lett.* **46**, 1648–51 (1981)

329 J. I. Gersten, I. Wagner, A. Rosenthal, Y. Goldstein, A. Many and R. E. Kirby, *Phys. Rev. B* **29**, 2458–68 (1984)

330 P. A. Cox, R. G. Egdell, W. R. Flavell and R. Helbig, *Vacuum* **33**, 835–8 (1983)

331 W. R. Flavell, private communication

332 P. A. Cox, R. G. Egdell, C. Harding, W. R. Patterson and P. J. Tavener, *Surf. Sci.* **123**, 179–203 (1982)

333 P. A. Cox, R. G. Egdell, S. Eriksen and W. R. Flavell, *J. Electron Spectros.* **39**, 117–26 (1986)

334 G. Rocker, J. A. Schaefer and W. Göpel, *Phys. Rev. B* **30**, 3704–8 (1984)

335 S. Eriksen and R. G. Egdell, *Surf. Sci.* **180**, 263–78 (1987)

336 L. L. Kesmodel, J. A. Gates and Y. W. Chung, *Phys. Rev. B* **23**, 489–92 (1981)

337 G. Durinck, H. Poelman, P. Clauws, L. Fiermans, J. Vinnik and G. Dalmai, *Solid State Commun.* **80**, 579–81 (1991)

338 A. D. Baden, P. A. Cox, R. G. Egdell, A. F. Orchard and R. J. D. Willmer, *J. Phys. C: Solid State Phys.* **14**, L1081–4 (1981)

339 R. G. Egdell and P. D. Naylor, *Chem. Phys. Lett.* **91**, 200–5 (1982)

340 P. A. Cox, R. G. Egdell and P. D. Naylor, *J. Electron Spectros.* **29**, 247–52 (1983)

341 P. A. Cox, W. R. Flavell, A. A. Williams and R. G. Egdell, *Surf. Sci.* **152/153**, 784–90 (1985)

342 H. Poelman, Ph. D. Thesis, University of Ghent, 1991 (unpublished)

343 D. G. Aitken, P. A. Cox, R. G. Egdell, M. D. Hill and I. Sach, *Vacuum* **33**, 753–6 (1983)

344 P. A. Cox, M. D. Hill, F. Peplinskii and R. G. Egdell, *Surf. Sci.* **141**, 13–30 (1984)

345 P. A. Cox and J. P. Kemp, *Surf. Sci.* **210**, 225–37 (1989)

346 R. G. Egdell, in *Adsorption and Catalysis on Oxide Surfaces,* ed. M. Che and G. C. Bond (Elsevier, Amsterdam, 1985), pp. 173–82

347 P. A. Thiry, M. Liehr, J. J. Pireaux and R. Caudano, *J. Electron Spectros.* **39**, 69–78 (1986)

348 P. A. Thiry, A. Degbomont, J. J. Pireaux, R. Caudano, J. R. Naegele, J. Rebizant and J. C. Spirlet, *J. Less Common Metals* **122**, 31–3 (1986)

349 P. A. Thiry, J. J. Pireaux, R. Caudano, J. R. Naegele, J. Rebizant and J. C. Spirlet, *J. Chem. Soc. , Faraday Trans.* 2 **83**, 1229–33 (1987)

350 R. Fuchs and K. L. Kliewer, *Phys. Rev.* **140**, A2076–88 (1965)

351 T. Inaoka, D. M. Newns and R. G. Egdell, *Surf. Sci.* **186**, 290–308 (1987)

352 J. P. Kemp, P. A. Cox, R. G. Egdell and K. Kang, *J. Phys. : Condens. Matter* **1**, SB123–6 (1989)

353 C. S. Rastomjee, P. A. Cox, R. G. Egdell, J. P. Kemp and W. R. Flavell, *J. Mater. Chem.* **1**, 451–5 (1991)

354 P. A. Cox, R. G. Egdell, W. R. Flavell, J. P. Kemp, F. H. Potter and C. S. Rastomjee, *J. Electron. Spectros.* **54/55**, 1173–82 (1990)

355 G. Lakshmi and F. W. de Wette, *Phys. Rev. B* **22**, 5009–13 (1980)

356 G. Lakshmi and F. W. de Wette, *Phys. Rev. B* **23**, 2035–8 (1981)

357 W. Kress, F. W. de Wette, A. D. Kulkarni and U. Schröder, *Phys. Rev. B* **35**, 5783–94 (1987)

358 H. H. Kung, *Transition Metal Oxides: Surface Chemistry and Catalysis* (Elsevier, Amsterdam, 1989)

359 H. H. Richardson, H-C. Chang, C. Noda and G. E. Ewing, *Surf. Sci.* **216**, 93–104

(1989)

360 W. T. Petrie and J. M. Vohs, *Surf. Sci.* **245**, 315–23 (1991)

361 M. C. Wu, C. A. Estrada and D. W. Goodman, *Phys. Rev. Lett.* **67**, 2910–13 (1991)

362 P. A. Cox, *The Electronic Structure and Chemistry of Solids* (Oxford University Press, Oxford, 1987)

363 C. Satoko, M. Tsukada and H. Adachi, *J. Phys. Soc. Japan* **45**, 1333–40 (1978)

364 M. Tsukada, H. Adachi and C. Satoko, *Prog. Surf. Sci.* **14**, 113–74 (1983)

365 I. Ivanov and J. Pollmann, *Phys. Rev. B* **24**, 7275–96 (1981)

366 P. M. A. Sherwood, *Phys. Rev. B* **41**, 10151–4 (1990)

367 Z. M. Jarzebski and J. P. Marton, *J. Electrochem. Soc.* **123**, 299C–310C (1976)

368 Z. M. Jarzebski and J. P. Marton, *J. Electrochem. Soc.* **123**, 333C–346C (1976)

369 V. C. Lee and H. S. Wong, *J. Phys. Soc. Japan* **45**, 895–8 (1978)

370 J. Robertson, *J. Phys. C: Solid State Phys.* **12**, 4767–76 (1979)

371 W. C. Mackrodt and R. F. Stewart, *J. Phys. C: Solid State Phys.* **12**, 5015–36 (1979)

372 C. Kunz, in *Photoemission in Solids, Topics in Applied Physics, Vol. 27*, ed. L. Ley and M. Cardona (Springer-Verlag, Berlin, 1979), pp. 299–348

373 J. L. Freeouf, *Phys. Rev. B* **7**, 3810–30 (1970)

374 J. M. Ziman, *Principles of the Theory of Solids, 2nd edn* (Cambridge University Press, Cambridge, 1972)

375 O. Madelung, ed.,*Semiconductors: Physics of II-VI and I-VII Compounds, Semimagnetic Semiconductors*, Landoldt Bornstein New Series III, 17b (Springer-Verlag, Berlin, 1982)

376 R. H. Ritchie, *Surf. Sci.* **34**, 1–19 (1973)

377 V. E. Henrich, G. Dresselhaus and H. J. Zeiger, *Phys. Rev. B* **22**, 4764–75 (1980)

378 R. Dorn and H. Lüth, *J. Appl. Phys.* **47**, 5097–8 (1976)

379 Y. Margoninski and D. Eger, *Phys. Lett.* **59A**, 305–6 (1976)

380 Y. Margoninski and D. Eger, *J. Electron Spectros.* **13**, 337–44 (1978)

381 J. Onsgaard, S. M. Barlow and T. E. Gallon, *J. Phys. C: Solid State Phys.* **12**, 925–42 (1979)

382 A. Ebina and T. Takahashi, *Surf. Sci.* **74**, 667–75 (1978)

383 J. D. Levine and P. Mark, *Phys. Rev.* **144**, 751–63 (1966).

384 E. Garrone, A. Zecchina and F. S. Stone, *Phil. Mag. B* **42**, 683–703 (1980)

385 L. Ley, M. Cardona and R. A. Pollack, in *Photoemission in Solids, Topics in Applied Physics,* Vol. 27, ed. L. Ley and M. Cardona (Springer-Verlag, Berlin, 1979)

386 J. Magill, J. Bloem and R. W. Ohse, *J. Chem. Phys.* **76**, 6227–42 (1982)

387 M. J. Madon and S. R. Morrison, *Chemical Sensing with Solid State Devices* (Academic Press, San Diego, 1989).

388 J. Pollmann, in *Festkorperprobleme*, Vol. 20, ed. J. Treusch (Viewey, Brunschweig, 1980)

389 D. H Lee and J. D. Joannopoulos, *Phys. Rev. B* **24**, 6899–907 (1981)

390 A. M. Stoneham, *Theory of Defects in Solids* (Clarendon Press, Oxford, 1975)

391 G. J. M. Janssen and W. C. Nieuwpoort, *Phys. Rev. B* **38**, 3449–58 (1988)

392 M. L. Cohen and V. I. Heine, *Solid State Phys.* **24**, 37–248 (1970)

393 K. H. Johnson, *Adv. Quant. Chem.* **7**, 143–85 (1973)

394 J. C. Slater, *The Self Consistent Field for Molecules and Solids: Quantum Theory of Molecules and Solids,* Vol. 4 (McGraw-Hill, New York, 1974)

395 H. Ibach, ed., *Electron Spectroscopy for Surface Analysis,* (Springer-Verlag, Berlin, 1977)

396 Y. C. Lee, P. Tong and P. A. Montano, *Surf. Sci.* **181**, 559–72 (1987)

397 H. Onishi, C. Egawa, T. Aruga and Y. Iwasawa, *Surf. Sci.* **191**, 479–91 (1987)

398 L. H. Tjeng, A. R. Vos and G. A. Sawatzky, *Surf. Sci.* **235**, 269–79 (1990)

399 V. E. Henrich, G. Dresselhaus and H. J. Zeiger, *Phys. Rev. Lett.* **36**, 158–61 (1976)

400 P. R. Underhill and T. E. Gallon, *Solid State Commun.* **43**, 9–11 (1982)

401 P. A. Cox and A. A. Williams, *Surf. Sci.* **175**, L782–96 (1986)

402 J. W. He and P. J. Møller, *phys. stat. sol. b* **133**, 687–91 (1986)

403 A. M. Stoneham and M. J. L. Sangster, *Phil. Mag. B* **43**, 609–19 (1981)

404 A. M. Stoneham, M. J. L. Sangster and P. W. Tasker, *Phil. Mag. B* **44**, 603–13 (1981)

405 A. B. Laponsky and N. R. Whetten, *Phys. Rev.* **120**, 801–6 (1960)

406 S. Russo and C. Noguera, *Surf. Sci.* **262**, 245–8 (1992)

407 J. P. LaFemina and C. B. Duke, *J. Vac. Sci. Technol. A* **9**, 1847–55 (1991)

408 A. L. Shluger, R. W. Grimes, C. R. A. Catlow and N. Itoh, *J. Phys. : Condens. Matter* **3**, 8027–36 (1991)

409 S. Y. Yousif and H. A. Kassim, *Surf. Sci.* **197**, 509–14 (1988)

410 V. E. Henrich and R. L. Kurtz, *J. Vac. Sci. Technol.* **18**, 416–19 (1981)

411 J. W. He and P. J. Møller, *Chem. Phys. Lett.* **129**, 13–16 (1986)

412 R. R. Sharma and A. M. Stoneham, *J. Chem. Soc. , Faraday Trans. II* **72**, 913–19 (1976)

413 H. A. Kassim, J. A. D. Matthew and B. Green, *Surf. Sci.* **74**, 109–24 (1978)

414 S. Munnix and M. Schmeits, *Phys. Rev. B* **30**, 2202–11 (1984)

415 J. H. Harding (personal communication)

416 P. W. Fowler and P. Tole, *Surf. Sci.* **197**, 457–73 (1988)

417 I. V. Abarenkov and T. Yu. Frenkel, *J. Phys. : Condens. Matter* **3**, 3471–8 (1991)

418 A. R. Protheroe, A. Steinbrunn and T. E. Gallon, *J. Phys. C: Solid State Phys.* **15**, 4951–9 (1982)

419 A. R. Protheroe, A. Steinbrunn and T. E. Gallon, *Surf. Sci.* **126**, 534–42 (1983)

420 R. E. Thomas, A. Shih and G. A. Haas, *Surf. Sci.* **75**, 239–55 (1978)

421 F. S. Ohuchi and M. Kohyama, *J. Amer. Ceram. Soc.* **74**, 1163–87 (1991)

422 E. I. Altman and R. J. Gorte, *Surf. Sci.* **216**, 386–94 (1989)

423 W. J. Gignac, R. S. Williams and S. P. Kowalczyk, *Phys. Rev. B* **32**, 1237–47 (1985)

424 S. Ciraci and I. P. Batra, *Phys. Rev. B* **28**, 982–92 (1983)

425 W. Ranke, *Solid State Commun.* **19**, 685–8 (1976)

426 G. Zwicker and K. Jacobi, *Solid State Commun.* **54**, 701–4 (1985)

427 H. Froitzheim and H. Ibach, *Z. Phys.* **269**, 17–22 (1974)

428 R. Dorn, H. Lüth and M. Büchel, *Phys. Rev. B* **16**, 4675–83 (1977)

429 Y. Margoninski, *Surf. Sci.* **94**, L167–70 (1980)

430 I. Ivanov and J. Pollmann, *Solid State Commun.* **36**, 361–4 (1980)

431 I. Ivanov and J. Pollmann, *J. Vac. Sci. Technol.* **19**, 344–6 (1981)

432 D. H Lee and J. D. Joannopoulos, *J. Vac. Sci. Technol.* **17**, 987–8 (1980)

433 M. Tsukada, E. Miyazaki and H. Adachi, *J. Phys. Soc. Japan* **50**, 3032–9 (1981)

434 R. Kuwabara, H. Adachi and T. Morimoto, *Surf. Sci.* **193**, 271–86 (1988)

435 R. Sekine, H. Adachi and T. Morimoto, *Surf. Sci.* **208**, 177–88 (1989)

436 R. E. Watson, M. L. Perlman and J. W. Davenport, *Surf. Sci.* **115**, 117–40 (1982)

437 P. J. Møller and J. W. He, *Surf. Sci.* **162**, 209–16 (1985)

438 R. P. Holmstrom, J. Lagowski and H. C. Gatos, *Surf. Sci.* **75**, L781–5 (1978)

439 Y. Margoninski and D. Eger, *Surf. Sci.* **80**, 579–85 (1979)

440 R. G. Egdell, S. Eriksen and W. R. Flavell, *Solid State Commun.* **60**, 835–8 (1986)

441 D. F. Cox, T. B. Fryberger, J. W. Erickson and S. Semancik, *J. Vac. Sci. Technol. A* **5**, 1170–1 (1987)

442 R. G. Egdell, S. Eriksen and W. R. Flavell, *Surf. Sci.* **192**, 265–74 (1987)

443 S. Semancik and T. B. Fryberger, *Sensors and Actuators B* **1**, 97–102 (1990)

444 J. M. Themlin, R. Sporken, J. Darville, R. Caudano, J. M. Gilles and R. L. Johnson,

Phys. Rev. B **42**, 11914–25 (1990)

445 S. Munnix and M. Schmeits, *Phys. Rev. B* **27**, 7624–35 (1983)

446 S. Munnix and M. Schmeits, *Solid State Commun.* **43**, 867–71 (1982)

447 S. Munnix and M. Schmeits, *Phys. Rev. B* **33**, 4136–44 (1986)

448 S. Munnix and M. Schmeits, *Surf. Sci.* **126**, 20–4 (1983)

449 S. Munnix and M. Schmeits, *J. Vac. Sci. Technol. A* **5**, 910–13 (1987)

450 D. F. Cox and T. B. Fryberger, *Surf. Sci.* **227**, L105–8 (1990)

451 C. S. Rastomjee, R. G. Egdell, M. J. Lee and T. J. Tate, *Surf. Sci. Lett.* **259**, L769–73 (1991)

452 P. A. Cox, R. G. Egdell, C. Harding, A. F. Orchard, W. R. Patterson and P. J. Tavener, *Solid State Commun.* **44**, 837–9 (1982)

453 P. A. Cox, W. R. Flavell and R. G. Egdell, *J. Solid State Chem.* **68**, 340–50 (1987)

454 A. V. Chadwick, R. M. Geatches and J. D. Wright, *Phil. Mag. A* **64**, 999–1010 (1991)

455 J. M. Imer, F. Patthey, B. Dardel, W. D. Schneider, Y. Baer, Y. Petroff and A . Zettl, *Phys. Rev. Lett.* **63**, 102 (1989)

456 Y. Hwu, L. Lozzi, M. Marsi, S. La Rosa, M. Winokur, P. Davis, M. Onellion, H. Berger, F. Gozzo, F. Levy and G. Margaritondo, *Phys. Rev. Lett.* **67**, 2573–6 (1991).

457 L. C. Davis, *J. Appl. Phys.* **59**, R25–63 (1986)

458 P. A. Cox, *Structure and Bonding* **24**, 59–81 (1975)

459 A. Fujimori and F. Minami, *Phys. Rev. B* **30**, 957–71 (1984)

460 G. A. Sawatzky, in *Core Level Spectroscopy in Condensed Systems*, ed. J. Kanamori and A. Kotani, *Springer Verlag Series in Solid State Sciences, Vol. 81* (Springer-Verlag, Berlin, 1988)

461 G. A. Sawatzky and J. W. Allen, *Phys. Rev. Lett.* **53**, 2339–42 (1984)

462 Z. X. Shen, C. K. Shih, O. Jepsen, W. E. Spicer, I. Lindau and J. W. Allen, *Phys. Rev. Lett.* **64**, 2442–5 (1990)

463 G. K. Wertheim, M. Campagna, J. N. Chazalviel and H. R. Shanks, Chem. *Phys. Lett.* **44**, 50–2 (1976)

464 B. A. de Angelis and M. Schiavello, *Chem. Phys. Lett.* **58** 249–51 (1976)

465 R. Kötz, H. J. Lewerenz and S. Stucki, *J. Electrochem. Soc.* **130**, 825–9 (1983)

466 M. Campagna, G. K. Wertheim, H. R. Shanks, F. Zumsteg and E. Banks, *Phys. Rev. Lett.* **34**, 738–41 (1975)

467 A. Kotani and Y. Toyazawa, *J. Phys. Soc. Japan* **37**, 912–19 (1974)

468 W. Göpel, J. A. Anderson, D. Frankel, M. Jaehnig, K. Phillips, J. A. Schäfer and G. Rocker, *Surf. Sci.* **139**, 333–46 (1984)

469 R. L. Kurtz, R. Stockbauer, T. E. Madey, E. Román and J. L. de Segovia, *Surf. Sci.* **218**, 178–200 (1989)

470 W. Göpel, U. Kirner, H. D. Wiemhöfer and G. Rocker, *Solid State Ionics* **28–30**, 1423–30 (1988)

471 C. A. Muryn, G. Tirvengadum, J. J. Crouch, D. R. Warburton, G. N. Raiker, G. Thornton and D. S. L. Law, *J. Phys. : Condens. Matter* **1**, SB127–32 (1989)

472 M. L. Knotek and J. E. Houston, *Phys. Rev. B* **15**, 4580–6 (1977)

473 A. K. See amd R. A. Bartynski, *J. Vac. Sci. Technol. A* **10**, 2591–6 (1992)

474 F. J. Morin and T. Wolfram, *Phys. Rev. Lett.* **30**, 1214–17 (1973)

475 T. Wolfram and S. Ellialtioglu, *Appl. Phys.* **13**, 21–4 (1977)

476 R. V. Kasowski and R. H. Tait, *Phys. Rev. B* **20**, 5168–77 (1979)

477 M. Tsukada, C. Satoko and H. Adachi, *J. Phys. Soc. Japan* **44**, 1043–4 (1978)

478 M. Tsukada, C. Satoko and H. Adachi, *J. Phys. Soc. Japan* **47**, 1610–19 (1979)

479 S. Munnix and M. Schmeits, *Phys. Rev. B* **28**, 7342–5 (1983)

480 C. R. Wang and Y. S. Xu, *Surf. Sci.* **219**, L537–42 (1989)

481 V. M. Bermudez, *J. Vac. Sci. Technol.* **20**, 741–2 (1982)

482 V. M. Bermudez, *J. Vac. Sci. Technol.* **20**, 51–7 (1982)

483 G. Rocker and W. Göpel, *Surf. Sci.* **181**, 530–58 (1987)

484 K. E. Smith, J. L. Mackay and V. E. Henrich, *Phys. Rev. B* **35**, 5822–9 (1987)

485 K. E. Smith and V. E. Henrich, *J. Vac. Sci. Technol. A* **7**, 1967–71 (1989)

486 K. E. Smith and V. E. Henrich, *Surf. Sci.* **217**, 445–58 (1989)

487 S. Eriksen, P. D. Naylor and R. G. Egdell, *Spectrochimica Acta* **43a**, 1535–8 (1987)

488 Z. Zhang, S. P. Jeng and V. E. Henrich, *Phys. Rev. B* **43**, 12004–11 (1991)

489 P. B. Smith and S. L. Bernasek, *Surf. Sci.* **188**, 241–54 (1987)

490 U. Bardi, K. Tamura, M. Owari and Y. Nihei, *Appl. Surf. Sci.* **32**, 352–62 (1988)

491 C. N. Sayers and N. R. Armstrong, *Surf. Sci.* **77**, 301–20 (1978)

492 G. B. Hoflund, H. K. Lin, A. L. Grogan, Jr. , D. A. Asbury, H. Yoneyama, O. Ikeda and H. Tamura, *Langmuir* **4**, 346–50 (1988)

493 K. Sakamaki, K. Itoh, A. Fujishima and Y. Gohshi, *J. Vac. Sci. Technol. A* **8**, 614–17 (1990)

494 S. E. Gilbert and J. H. Kennedy, *Langmuir* **5**, 1412–15 (1989)

495 S. E. Gilbert and J. H. Kennedy, *Surf. Sci.* **225**, L1–7 (1990)

496 H. Kobayashi and M. Yamaguchi, *Surf. Sci.* **214**, 466–76 (1989)

497 S. Munnix and M. Schmeits, *Phys. Rev. B* **31**, 3369–71 (1985)

498 J. B. Goodenough, *Prog. Solid State Chem.* **5**, 145–399 (1972)

499 K. E. Smith, J. L. Mackay and V. E. Henrich, *J. Vac. Sci. Technol. A* **5**, 689–90 (1987)

500 K. E. Smith and V. E. Henrich, *Phys. Rev. B* **38**, 5965–75 (1988)

501 K. E. Smith and V. E. Henrich, *Solid State Commun.* **68**, 29–32 (1988)

502 K. E. Smith and V. E. Henrich, *Phys. Rev. B* **38**, 9571–80 (1988)

503 J. M. McKay, M. H. Mohamed and V. E. Henrich, *Phys. Rev. B* **35**, 4304–9 (1987)

504 A. Bianconi, S. Stizza and R. Bernardini, *Phys. Rev. B* **24**, 4406–11 (1981).

505 V. E. Henrich, H. J. Zeiger and T. B. Reed, *Phys. Rev. B* **17**, 4121–3 (1978)

506 B. Laks and C. E. T. Gonçalves da Silva, *Surf. Sci.* **71**, 563–74 (1978)

507 R. A. Powell and W. E. Spicer, *Phys. Rev. B* **13**, 2601–4 (1976)

508 B. Cord and R. Courths, *Surf. Sci.* **162**, 34–8 (1985)

509 N. B. Brookes, D. S. L. Law, T. S. Padmore, D. R. Warburton and G. Thornton, *Solid State Commun.* **57**, 473–7 (1986)

510 R. Courths, B. Cord and H. Saalfeld, *Solid State Commun.* **70**, 1047–51 (1989)

511 G. N. Raiker, C. A. Muryn, P. J. Hardman, P. L. Wincott, G. Thornton, D. W. Bullett and P. A. D. M. A. Dale, *J. Phys. : Condens. Matter* **3**, S357–62 (1991)

512 R. Courths, *Ferroelectrics* **26**, 749–52 (1980)

513 N. B. Brookes, F. M. Quinn and G. Thornton, *Physica Scripta* **36**, 711–14 (1987)

514 N. B. Brookes, F. M. Quinn and G. Thornton, *Vacuum* **38**, 405–8 (1988)

515 N. B. Brookes, G. Thornton and F. M. Quinn, *Solid State Commun.* **64**, 383–6 (1987)

516 B. Reihl, J. G. Bednorz, K. A. Müller, Y. Jugnet, G. Landgren and J. F. Morar, *Phys. Rev. B* **30**, 803–6 (1984)

517 V. E. Henrich, Rep. *Prog. Phys.* **48**, 1481–541 (1985)

518 S. Ferrer and G. A. Somorjai, *Surf. Sci.* **97**, L304–8 (1980)

519 S. Ferrer and G. A. Somorjai, *Surf. Sci.* **94**, 41–56 (1980)

520 F. T. Wagner, S. Ferrer and G. A. Somorjai, *Surf. Sci.* **101**, 462–74 (1980)

521 T. Wolfram, E. A. Kraut and F. J. Morin, *Phys. Rev. B* **7**, 1677–94 (1973)

522 S. Ellialtioglu and T. Wolfram, *Phys. Rev. B* **18**, 4509–25 (1978)

523 S. Ellialtioglu , T. Wolfram and V. E. Henrich, *Solid State Commun.* **27**, 321–4 (1978)

524 T. Wolfram and S. Ellialtioglu, in *Theory of Chemisorption*, ed. J. R. Smith

(Springer-Verlag, Berlin, (1980), pp. 149–81

525 T. Wolfram, *J. Vac. Sci. Technol.* **18**, 428–32 (1981)
526 M. Tsukada, C. Satoko and H. Adachi, *J. Phys. Soc. Japan*, **48**, 200–10 (1980)
527 G. Toussaint, M. O. Selme and P. Pecheur, *Phys. Rev. B* **36**, 6135–41 (1987)
528 M. O. Selme, G. Toussaint and P. Pecheur, in *Non-Stoichiometric Compounds: Surfaces, Grain Boundaries and Structural Defects*, ed. J. Nowotny and W. Weppner (Kluwer, Dordrecht, 1989), pp. 173–86
529 B. Cord and R. Courths, *Surf. Sci.* **152/153**, 1141–6 (1985)
530 R. Courths, H. Höchst, P. Steiner and S. Hüfner, *Ferroelectrics* **26**, 745–8 (1980)
531 R. Courths, P. Steiner, H. Höchst and S. Hüfner, *Appl. Phys.* **21**, 345–52 (1980)
532 V. H. Ritz and V. M. Bermudez, *Phys. Rev. B* **24**, 5559–75 (1981)
533 V. Cháb and J. Kubátová, *Appl. Phys.* A **39**, 67–71 (1986)
534 L. Kasper and S. Hüfner, *Phys. Lett.* **81A**, 165–8 (1981)
535 S. Shin, S. Suga, M. Taniguchi, M. Fujisawa, H. Kanzaki, A. Fujimori, H. Daimon, Y. Ueda, K. Kosuge and S. Kachi, *Phys. Rev. B* **41**, 4993–5009 (1990)
536 G. A. Sawatzky and D. Post, *Phys. Rev. B* **20**, 1546–55 (1979)
537 R. De Gryse, J. P. Landuyt, A. Vermeire and J. Vennik, *Appl. Surf. Sci.* **6**, 430–3 (1980)
538 Z. Zhang and V. E. Henrich, *Surf. Sci.* **321**, 133–44 (1994)
539 V. M. Bermudez, R. T. Williams, J. P. Long, R. K. Reed and P. H. Klein, *Phys. Rev. B* **45**, 9266–71 (1992)
540 N. Beatham, I. L. Fragala, A. F. Orchard and G. Thornton, *J. Chem. Soc. Faraday II* **76**, 929–35 (1980)
541 K. E. Smith and V. E. Henrich, *J. Vac. Sci. Technol. A* **6**, 831–2 (1988)
542 K. E. Smith, Ph. D. Thesis, Yale University (1988), unpublished.
543 B. Laks and C. E. T. Gonçalves da Silva, *J. Phys. C: Solid State Phys.* **10**, L99–101 (1977)
544 N. Beatham, A. F. Orchard and G. Thornton, *J. Phys. Chem. Solids* **42**, 1051–5 (1981)
545 V. di Castro, C. Frulani, G. Polzonetti and C. Cozza, *J. Electron Spectros.* **46**, 297–302 (1988)
546 E. Bertel, R. Stockbauer, R. L. Kurtz, D. E. Ramaker and T. E. Madey, *Phys. Rev. B* **31**, 5580–3 (1985)
547 D. E. Eastman and J. L. Freeouf, *Phys. Rev. Lett.* **34**, 395–8 (1975)
548 R. J. Lad and V. E. Henrich, *J. Vac. Sci. Technol. A* **6**, 781–2 (1988)
549 A. Fujimori, N. Kimizuka, T. Akahane, T. Chiba, S. Kimura, F. Minami, K. Siratori, M. Taniguchi, S. Ogawa and S. Suga, *Phys. Rev. B* **42**, 7580–6 (1990)
550 B. Hermsmeier, J. Osterwalder, D. J. Friedman and C. S. Fadley, *Phys. Rev. Lett.* **62**, 478–81 (1989)
551 B. Hermsmeier, J. Osterwalder, D. J. Friedman, B. Sinkovic, T. Tran and C. S. Fadley, *Phys. Rev. B* **42**, 11895–913 (1990)
552 S. P. Jeng, R. J. Lad and V. E. Henrich, *Phys. Rev. B* **43**, 11971 (1991)
553 K. Akimoto, Y. Sakisaka, M. Nishijima and M. Onchi, *J. Phys. C: Solid State Phys.* **11**, 2535–48 (1978)
554 J. P. Kemp, S. T. P. Davies and P. A. Cox, *J. Phys. : Condens. Matter* **1**, 5313–18 (1989)
555 P. S. Bagus, C. R. Brundle, T. J. Chuang and K. Wandelt, *Phys. Rev. Lett.* **19**, 1229–32 (1977)
556 C. R. Brundle, T. J. Chuang and K. Wandelt, *Surf. Sci.* **68**, 459–68 (1977)
557 S. Vasudevan, M. S. Hegde and C. N. R. Rao, *J. Solid State Chem.* **29**, 253–7 (1979)
558 G. Grenet, Y. Jugnet, T. M. Duc and M. Kibler, *J. Chem. Phys.* **72**, 218–20 (1980)

559 S. R. Kelemen, A. Kaldor and D. J. Dwyer, *Surf. Sci.* **121**, 45–60 (1982)

560 A. Fujimori, M. Saeki, N. Kimizuka, M. Taniguchi and S. Suga, *Phys. Rev. B* **34**, 7318–28 (1986)

561 K. Siratori, S. Suga, M. Taniguchi, K. Soda, S. Kimura and A. Yanase, *J. Phys. Soc. Japan* **55**, 690–8 (1986)

562 M. Hendewerk, M. Salmeron and G. Somorjai, *Surf. Sci.* **172**, 544–56 (1986)

563 A. Fujimori, N. Kimizuka, M. Taniguchi and S. Suga, *Phys. Rev. B* **36**, 6691–4 (1987)

564 R. J. Lad and V. E. Henrich, *J. Vac. Sci. Technol. A* **7**, 1893–7 (1989)

565 R. J. Lad and V. E. Henrich, *Phys. Rev. B* **39**, 13478–85 (1989)

566 S. F. Alvarado, M. Erbudak and P. Munz, *Phys. Rev. B* **14**, 2740–5 (1976)

567 N. B. Brookes, D. S. L. Law, D. R. Warburton, P. L. Wincott and G. Thornton, *J. Phys. : Condens. Matter* **1**, 4267–72 (1989)

568 Z. X. Shen, J. W. Allen, P. A. P. Lingberg, D. S. Dessau, B. O. Wells, A. Borg, W. Ellis, J. S. Kang, S. J. Oh, I. Lindau and W. E. Spicer, *Phys. Rev. B* **42**, 1817–28 (1990)

569 S. P. Jeng and V. E. Henrich, *Solid State Commun.* **75**, 1013–17 (1990)

570 Y. Sakisaka, K. Akimoto, M. Nishijima and M. Ouchi, *Solid State Commun.* **24**, 105–7 (1977)

571 S. J. Oh, J. W. Allen, I. Lindau and J. C. Mikkelsen, Jr. , *Phys. Rev. B* **26**, 4845–56 (1982)

572 M. R. Thuler, R. L. Benbow and Z. Hurych, *Phys. Rev. B* **27**, 2082–8 (1983)

573 V. C. Lee and H. S. Wong, *J. Phys. Soc. Japan* **50**, 2351–4 (1981)

574 P. W. Palmberg, R. E. De Wames and L. A. Vredevoe, *Phys. Rev. Lett.* **21**, 682– 5 (1968)

575 M. Prutton, W. D. Doyle and K. Legg, in *Magnetism and Magnetic Materials: 1971,* ed. C. D. Graham, Jr and J. J. Rhyne (American Institute of Physics, New York, 1972), pp. 1430–3

576 T. Suzuki, N. Hirota, H. Tanaka and H. Watanabe, *J. Phys. Soc. Japan* **30**, 888 (1971)

577 K. Hayakawa, K. Namikawa and S. Miyake, *J. Phys. Soc. Japan* **31**, 1408–17 (1971)

578 R. E. De Wames and T. Wolfram, *Phys. Rev. Lett.* **22**, 137–9 (1969)

579 J. A. Walker, C. G. Kinniburgh and J. A. D. Matthew, *Surf. Sci.* **68**, 221–8 (1977)

580 B. Koiller and L. M. Falicov, *Phys. Rev. B* **27**, 346–53 (1983)

581 R. P. Furstenau, G. McDougall and M. A. Langell, *Surf. Sci.* **150**, 55–79 (1985)

582 J. J. Scholz and M. A. Langell, *Surf. Sci.* **164**, 543–57 (1985)

583 M. W. Roberts and R. St C. Smart, *J. Chem. Soc. Faraday Trans. 1* **80**, 2957–68 (1984)

584 J. Ghijsen, L. H. Tjeng, J. van Elp, H. Eskes, J. Westerink, G. A. Sawatzky and N. T. Czyzyk, *Phys. Rev. B* **38**, 11322–30 (1988)

585 M. R. Thuler, R. L. Benbow and Z. Hurych, *Phys. Rev. B* **26**, 669–77 (1982)

586 J. Ghijsen, L. H. Tjeng, H. Eskes, G. A. Sawatzky and R. L. Johnson, *Phys. Rev. B* **42**, 2268–74 (1990)

587 Z. X. Shen, R. S. List, D. S. Dessau, F. Parmigiani, A. J. Arko, R. Bartlett, B. O. Wells, I. Lindau and W. E. Spicer, *Phys. Rev. B* **42**, 8081–5 (1990)

588 H. Eskes, L. H. Tjeng and G. A. Sawatzky, *Phys. Rev. B* **41**, 288–99 (1990)

589 R. S. List, A. J. Arko, Z. Fisk, S. W. Cheong, S. D. Conradson, J. D. Thompson, C. B. Pierce, D. E. Peterson, R. J. Bartlett, N. D. Shinn, J. E. Schirber, B. W. Veal, A. P. Paulikas and J. C. Campuzano, *Phys. Rev. B* **38**, 11966–9 (1988)

590 R. G. Egdell and W. R. Flavell, *Z. Phys. B – Condensed Matter* **74**, 279–82 (1989)

591 A. Balzarotti, M. De Crescenzi, N. Motta, F. Patella, A. Sgarlata, P. Paroli, G.

Balestrino and M. Marinelli, *Phys. Rev. B* **43**, 11500–3 (1991)

592 J. G. Tobin, C. G. Olson, C. Gu, J. Z. Liu, F. R. Sohal, M. J. Fluss, R. H. Howell, J. C. O'Brien, H. B. Radousky and P. A. Sterne, *Phys. Rev. B* **45**, 5563–76 (1992)

593 N. G. Stoffel, Y. Chang, M. K. Kelly, L. Dottl, M. Onellion, P. A. Morris, W. A. Bonner and G. Margaritondo, *Phys. Rev. B* **37**, 7952–5 (1988)

594 S. Kishida, H. Tokutaka, F. Toda, H. Fujimoto, W. Futo, K. Nishimori and N. Ishihara, *Japan. J. Appl. Phys.* **29**, L438–40 (1990)

595 C. Calandra, F. Manghi and T. Minerva, *Vacuum* **41**, 982–5 (1990)

596 L. E. Firment, A. Ferretti, M. R. Cohen and R. P. Merrill, *Langmuir* **1**, 166–9 (1985)

597 T. H. Fleisch, G. W. Zajac, J. O. Schreiner and G. J. Mains, *Appl. Surf. Sci.* **26**, 488–97 (1986)

598 H. Vincent and M. Marezio, in *Low-Dimensional Electronic Properties of Molybdenum Bronzes and Oxides*, ed. C. Schlenker (Kluwer, Dordrecht, 1989), p. 49

599 H. Höchst, R. D. Bringans and H. R. Shanks, *Phys. Rev. B* **26**, 1702–12 (1982)

600 R. G. Egdell and M. D. Hill, *Chem. Phys. Lett.* **88**, 503–6 (1982)

601 M. D. Hill and R. G. Egdell, *J. Phys. C: Solid State Phys.* **16**, 6205–20 (1983)

602 R. L. Benbow and Z. Hurych, *Phys. Rev. B* **17**, 4527–36 (1978)

603 H. Höchst, R. D. Bringans, H. R. Shanks and P. Steiner, *Solid State Commun.* **37**, 41–4 (1980)

604 G. Hollinger, P. Petrosa, J. P. Doumerc, F. J. Himpsel and B. Reihl, *Phys. Rev. B* **32**, 1987–91 (1985)

605 R. D. Bringans, H. Höchst and H. R. Shanks, *Phys. Rev. B* **24**, 3481–9 (1981)

606 R. G. Egdell (personal communication)

607 P. M. G. Allen, P. J. Dobson and R. G. Egdell, *Solid State Commun.* **55**, 701–4 (1985)

608 D. W. Bullett, *J. Phys. C: Solid State Phys.* **16**, 2197–207 (1983)

609 R. D. Bringans, H. Höchst and H. R. Shanks, *Vacuum* **31**, 473–5 (1981)

610 R. D. Bringans, H. Höchst and H. R. Shanks, *Surf. Sci.* **111**, 80–6 (1981)

611 M. A. Langell and S. L. Bernasek, *Phys. Rev. B* **23**, 1584–93 (1981)

612 A. Fujimori, F. Minami, T. Akahane and N. Tsuda, *J. Phys. Soc. Japan*, **49**, 1820–3 (1980)

613 G. Hollinger, F. J. Himpsel, N. Martensson, B. Reihl, J. P. Doumerc and T. Akahane, *Phys. Rev. B* **27**, 6370–5 (1983)

614 M. Tsukada, N. Tsuda and F. Minami, *J. Phys. Soc. Japan* **49**, 1115–22 (1980)

615 R. R. Daniels, G. Margaritondo, C. A. Georg and F. Lévy, *Phys. Rev. B* **29**, 1813–18 (1984)

616 W. P. Ellis, A. M. Boring, J. W. Allen, L. E. Cox, R. D. Cowan, B. B. Pate, A. J. Arko and I. Lindau, *Solid State Commun.* **72**, 725–9 (1989)

617 D. F. Cox and K. H. Schulz (personal communication)

618 D. L. Meixner, D. A. Arthur and S. M. George, *Surf. Sci.* **261**, 141–54 (1992)

619 D. A. Arthur, D. L. Meixner, M. Boudart and S. M. George, *J. Chem. Phys.* **95**, 8521–31 (1991)

620 S. A. Pope, M. F. Guest, I. H. Hillier, E. A. Colbourn, W. C. Mackrodt and J. Kendrick, *Phys. Rev. B* **28**, 2191–8 (1983)

621 E. A. Colbourn and W. C. Mackrodt, *Surf. Sci.* **117**, 571–80 (1982)

622 M. P. Guse, R. J. Blint and A. B. Kunz, *Int. J. Quant. Chem.* **11**, 725–32 (1977)

623 H. Kobayashi, M. Yamaguchi and T. Ito, *J. Phys. Chem.* **94**, 7206–13 (1990)

624 P. Esser and W. Göpel, *Surf. Sci.* **97**, 309–18 (1980)

625 K. L. D'Amico, M. R. McClellan, M. J. Sayers, R. R. Gay, F. R. McFeely and E. I. Solomon, *J. Vac. Sci. Technol.* **17**, 1080–4 (1980)

626 W. Göpel, *Prog. Surf. Sci.* **20**, 9–103 (1985)

627 A. B. Anderson and J. A. Nichols, *J. Amer. Chem. Soc.* **108**, 4742–6 (1986)

628 W. Göpel and G. Rocker, *J. Vac. Sci. Technol.* **21**, 389–97 (1982)

629 D. F. Cox, in *Low-Temperature CO-Oxidation Catalysts for Long-Life CO₂ lasers*, ed. D. R. Schryer and G. B. Hoflund, *NASA Conference Publication 3076* (NASA, Washington, DC, 1990), pp. 263–75

630 R. Huck, U. Böttger, D. Kohl and G. Heiland, *Sensors and Actuators* **17**, 355–9 (1989)

631 G. Heiland and D. Kohl, in *Chemical Sensor Technology*, ed. T. Seiyama (Kodansha, Tokyo, 1988), pp. 15–38

632 G. Rocker and W. Göpel, *Surf. Sci.* **175**, L675–80 (1986)

633 W. Göpel, G. Rocker and R. Feierabend, *Phys. Rev. B* **28**, 3427–8 (1983)

634 T. Wolfram, F. J. Morin and R. Hurst, in *Electrocatalysis on Non-Metallic Surfaces*, ed. A. D. Franklin, *NBS Special Publ. 455* (U. S. G. P. O., Washington, DC, 1976), pp. 21–52

635 M. L. Knotek, *Surf. Sci.* **91**, L17 – 22 (1980)

636 M. L. Knotek, *Surf. Sci.* **101**, 334–40 (1980)

637 S. P. Mehandru and A. B. Anderson, *J. Amer. Chem. Soc.* **110**, 2061–5 (1988)

638 J. M. Rickard, M. Perdereau and L. C. Dufour, *Proc. 7th Intern. Vac. Congr. & 3rd Intern. Conf. Solid Surfaces* (Vienna, 1977), pp. 847–9

639 L. C. Dufour, N. Floquet and B. de Rosa, in *Reactivity of Solids*, ed. P. Barret and L. C. Dufour (Elsevier, Amsterdam, 1985), pp. 47–52

640 A. Boudriss and L. C. Dufour, in *Non-Stoichiometric Compounds: Surfaces, Grain Boundaries and Structural Defects*, ed. J. Nowotny and W. Weppner (Kluwer, Dordrecht, 1989), pp. 311–20

641 N. Floquet, P. Dufour and L. C. Dufour, *J. Microsc. Spectrosc. Electron.* **6**, 473–81 (1981)

642 G. T. Surratt and A. B. Kunz, *Phys. Rev. B* **19**, 2352–8 (1979)

643 G. G. Wepfer, G. T. Surratt, R. S. Weidman and A. B. Kunz, *Phys. Rev. B* **21**, 2596–601 (1980)

642 R. Dorn and H. Lüth, *Surf. Sci.* **68**, 385–91 (1977)

645 W. Göpel, *Surf. Sci.* **62**, 165–82 (1977)

646 W. Göpel, *J. Vac. Sci. Technol.* **15**, 1298–310 (1978)

647 F. Runge and W. Göpel, *Z. Physik. Chemie* **123**, 173–92 (1980)

648 V. M. Allen, W. E. Jones and P. D. Pacey, *Surf. Sci.* **220**, 193–205 (1989)

649 V. E. Henrich, G. Dresselhaus and H. J. Zeiger, *J. Vac. Sci. Technol.* **15**, 534–7 (1978)

650 M. L. Knotek, in *Electrode Materials and Processes for Energy Conversion and Storage* (Electrochemical Society, Princeton, NJ, 1977), pp. 234–46

651 J. M. Pan, B. L. Maschhoff, U, Diebold and T. E. Madey, *J. Vac. Sci. Technol. A* **10**, 2470–6 (1992)

652 V. M. Bermudez and V. H. Ritz, Chem. *Phys. Lett.* **73**, 160–6 (1980)

653 S. Ferrer and G. S. Somorjai, *J. Appl. Phys.* **52**, 4792–44 (1981)

654 N. Floquet and O. Bertrand, *Solid State Ionics* **32/33**, 234–43 (1989)

655 R. L. Kurtz and V. E. Henrich, *J. Vac. Sci. Technol. A* **2**, 842–3 (1984)

656 J. S. Foord and R. M. Lambert, *Surf. Sci.* **169**, 327–36 (1986)

657 J. M. McKay and V. E. Henrich, *J. Vac. Sci. Technol. A* **5**, 722–3 (1987)

658 J. M. Blaisdell and A. B. Kunz, *Solid State Commun.* **40**, 745–7 (1981)

659 J. M. Blaisdell and A. B. Kunz, *Phys. Rev. B* **29**, 988–95 (1984)

660 S. P. Mehandru and A. B. Anderson, *J. Electrochem. Soc.* **133**, 828–32 (1986)

661 V. M. Allen, W. E. Jones and P. D. Pacey, *Surf. Sci.* **199**, 309–19 (1988)

662 R. L. Kurtz, S. W. Robey, R. L. Stockbauer, D. Mueller, A. Shih, L. Toth, A. K.

Singh and M. Osofsky, *Vacuum* **39**, 611–15 (1989)
663 V. I. Nefedov, A. N. Sokolov, M. A. Tyzykhov, N. N. Oleinikov, Y. A. Yeremina and M. A. Kolotyrkina, *J. Electron Spectros.* **49**, 47–60 (1989)
664 R. L. Kurtz, R. Stockbauer, T. E. Madey, D. Mueller, A. Shih and L. Toth, *Phys. Rev. B* **37**, 7936–9 (1988)
665 H. Tokutaka, S. Kishida, H. Fujimoto, K. Nishimori and N. Ishihara, *Surf. Sci.* **242**, 50–3 (1991)
666 R. L. Stockbauer, S. W. Robey, R. L. Kurtz, D. Mueller, A. Shih, A. K. Singh, L. Toth and M. Osofsky, in *Proceedings of the American Vacuum Society Special Conference on High T_c Superconducting Thin Films, Devices, and Applications* (American Institute of Physics, New York, 1989), pp. 276–82
667 W. H. Cheng and H. H. Kung, *Surf. Sci.* **122**, 21–39 (1982)
668 E. Longo, J. A. Varela, A. N. Senapeschi and O. J. Whittemore, *Langmuir* **1**, 456–61 (1985)
669 X. D. Peng and M. A. Barteau, *Surf. Sci.* **233**, 283–92 (1990)
670 X. D. Peng and M. A. Barteau, *Langmuir* **7**, 1426–31 (1991)
671 C. F. Jones, R. A. Reeve, R. Rigg, R. L. Segall, R. St. C. Smart and P. S. Turner, *J. Chem. Soc. , Faraday Trans. 1* **80**, 2609–17 (1984)
672 D. R. Mueller, A. Shih, E. Roman, T. E. Madey, R. L. Kurtz and R. L. Stockbauer, *J. Vac. Sci. Technol. A* **6**, 1067–71 (1988)
673 D. R. Mueller, R. L. Kurtz, R. L. Stockbauer, T. E. Madey and A. Shih, *Surf. Sci.* **237**, 72–86 (1990)
674 A. L. Pushkarchuk, P. P. Mardilovich, A. I. Trokhimets, G. M. Zhidomirov and S. C. Gagarin, *phys. stat. sol. b* **124**, 699–705 (1984)
675 A. L. Pushkarchuk, P. P. Mardilovich, A. I. Trokhimets, S. C. Gagarin and G. M. Zhidomirov, *phys. stat. sol. b* **129**, K181–5 (1985)
676 G. Zwicker and K. Jacobi, *Surf. Sci.* **131**, 179–94 (1983)
677 G. Zwicker, K. Jacobi and J. Cunningham, *Intern. J. Mass Spectros. Ion Processes* **60**, 213–23 (1984)
678 J. A. Rodriguez, *Langmuir* **4**, 1006–20 (1988)
679 J. A. Rodriguez and C. T. Campbell, *Surf. Sci.* **197**, 567–93 (1988)
680 S. Bourgeois and M. Perdereau, in *Secondary Ion Mass Spectrometry: SIMS VI*, ed. A. Benninghoven, A. M. Huber and H. W. Werner (Wiley, Chichester, 1988), pp. 1041–3
681 S. Bourgeois, C. Gimenez and M. Perdereau, in *Reactivity of Solids*, ed. P. Barret and L. C. Dufour (Elsevier, Amsterdam, 1985), pp. 931–4
682 T. K. Sham and M. S. Lazarus, Chem. *Phys. Lett.* **68**, 426–32 (1979)
683 F. J. Bustillo, E. Román and J. L. de Segovia, *Vacuum* **41**, 19–21 (1990)
684 M. S. Lazarus and T. K. Sham, Chem. *Phys. Lett.* **92**, 670–4 (1982)
685 P. B. Smith, S. L. Bernasek, J. Schwartz and G. S. McNulty, *J. Amer. Chem. Soc.* **108**, 5654–5 (1986)
686 C. A. Muryn, P. J. Hardman, J. J. Crouch, G. N. Raiker, G. Thornton, D. S. L. Law, *Surf. Sci.* **251/252**, 747–52 (1991)
687 F. J. Bustillo, E. Román and J. L. de Segovia, *Vacuum* **39**, 659–61 (1989)
688 T. Kawai, M. Tsukada, H. Adachi, C. Satoko and T. Sakata, *Surf. Sci.* **81**, L640–4 (1979)
689 J. M. Kowalski, K. H. Johnson and H. L. Tuller, *J. Electrochem. Soc.* **127**, 1969–73 (1980)
690 C. Webb and M. Lichtensteiger, *Surf. Sci.* **107**, L345–9 (1981)
691 I. W. Owen, N. B. Brookes, C. H. Richardson, D. R. Warburton, F. M. Quinn, D. Norman and G. Thornton, *Surf. Sci.* **178**, 897–906 (1986)

692 G. Thornton, *J. Phys. : Condens. Matter* **1**, SB111–17 (1989)

693 P. A. Cox, K. Joyce, N. A. Lucas and M. Surman, *Vacuum* **38**, 428–9 (1988)

694 N. C. Debnath and A. B. Anderson, *Surf. Sci.* **128**, 61–9 (1983)

695 M. A. Langell and R. P Furstenau, *Appl. Surf. Sci.* **26**, 445–60 (1986)

696 D. F. Cox and K. H. Schulz, *Surf. Sci.* **256**, 67–76 (1991)

697 N. P. Bansal and A. L. Sandkuhl, *Appl. Phys. Lett.* **52**, 323–5 (1988)

698 J. Dominec, L. Smrcka, P. Vasek, S. Geurten, O. Smrckova, D. Sykorova and
     B. Hájek, *Solid State Commun.* **65**, 373–4 (1988)

699 D. Zhuang, M. Xiao, Z. Zhang, S. Yue, H. Zhao and S. Shang, *Solid State Commun.*
     65, 339–41 (1988)

700 I. Nakada, S. Sato, Y. Oda and T. Kohara, *Japan. J. Appl. Phys.* **26**, L697–8 (1987)

701 R. L. Barns and R. A. Laudise, *Appl. Phys. Lett.* **51**, 1373–5 (1987)

702 M. F. Yan, R. L. Barnes, H. M. O'Bryan, Jr, P. K. Gallagher, R. C. Sherwood and S.
     Jin, *Appl. Phys. Lett.* **51**, 532–4 (1987)

703 S. L. Qiu, M. W. Ruckman, N. B. Brookes, P. D. Johnston, J. Chen, C. L. Lin, M.
     Strongin, B. Sinkovic, J. E. Crow and C. S. Jee, *Phys. Rev. B* **37**, 3747–50 (1988)

704 R. G. Egdell, W. R. Flavell and P. C. Hollamby, *J. Solid State Chem.* **79**, 238–49
     (1989)

705 S. Myhra, P. R. Chalker, P. T. Moseley and J. C. Riviere, *Physica C* **165**, 270–8
     (1990)

706 W. R. Flavell, J. H. Laverty. D. S. L. Law, R. Lindsay, C. A. Muryn, C. F. J. Flipse,
     G. N. Raiker, P. L. Wincott and G. Thornton, *Phys. Rev. B* **41**, 11623–6 (1990)

707 S. G. Jin, Z. Z. Zhu, L. M. Liu and Y. L. Huang, *Solid State Commun.* **74**, 1087–90
     (1990)

708 P. Vasek, P. Svoboda, O. Smrcková and D. Sykorová, *Solid State Commun.* **71**,
     403–6 (1989)

709 G. Pacchioni, G. Cogliandro and P. S. Bagus, *Surf. Sci.* **255**, 344–54 (1991)

710 R. Dovesi, R. Orlando, F. Ricca and C. Roetti, *Surf. Sci.* **186**, 267–78 (1987)

711 C. Pisani, R. Dovesi, R. Nada and S. Tamiro, *Surf. Sci.* **216**, 489–504 (1989)

712 E. A. Colbourn and W. C. Mackrodt, *Surf. Sci.* **143**, 391–410 (1984)

713 S. A. Pope, I. H. Hillier, M. F. Guest, E. A. Colbourn and J. Kendrick, *Surf. Sci.*
     139, 299–315 (1984)

714 A. Lakhlifi and C. Girardet, *Surf. Sci.* **241**, 400–15 (1991)

715 M. Causà, E. Kotomin, C. Pisani and C. Roetti, *J. Phys. C: Solid State Phys.* **20**,
     4991–7 (1987)

716 R. Dovesi, C. Roetti, M. Causà and C. Pisani, in *Strucutre and Reactivity of
     Surfaces,* ed. C. Morterra, A. Zecchina and G. Costa (Elsevier, Amsterdam, 1989),
     pp. 385–93

717 C. T. Au, W. Hirsch and W. Hirschwald, *Surf. Sci.* **197**, 391–401 (1988)

718 W. Hirsch, D. Hofmann and W. Hirschwald, *Proc. 8th Intern. Cong. Catal.* (Berlin,
     1984), pp. IV-251 – IV-262

719 W. Hotan, W. Göpel and R. Haul, *Surf. Sci.* **83**, 162–80 (1979)

720 K. L. D'Amico, F. R. McFeely and E. I. Solomon, *J. Amer. Chem. Soc.* **105**, 6380–3
     (1983)

721 H. Lüth, G. W. Rubloff and W. D. Grobman, *Solid State Commun.* **18**, 1427–30
     (1976)

722 J. A. Rodriguez and C. T. Campbell, *J. Phys. Chem.* **91**, 6648–58 (1987)

723 R. C. Baetzold, *J. Chem. Phys.* **89**, 4150–5 (1985)

724 A. B. Anderson and J. A. Nichols, *J. Amer. Chem Soc.* **108**, 1385–8 (1986)

725 S. F. Jen and A. B. Anderson, *Surf. Sci.* **223**, 119–30 (1989)

726 M. Watanabe and T. Kadowaki, *Appl. Surf. Sci.* **28**, 147–66 (1987)

727 D. F. Cox and K. H. Schulz, *Surf. Sci.* **249**, 138–48 (1991)
728 C. T. Au, W. Hirsch and W. Hirschwald, *Surf. Sci.* **199**, 507–17 (1988)
729 C. D. Steinspring and J. M. Cook, *J. Electron Spectros.* **32**, 113–22 (1983)
730 Y. C. Lee, P. A. Montano and J. M. Cook, *Surf. Sci.* **143**, 423–41 (1984)
731 B. J. Hopkins and P. A. Taylor, *J. Phys. C: Solid State Phys.* **9**, 571–8 (1976)
732 P. E. Chandler, P. A. Taylor and B. J. Hopkins, *Surf. Sci.* **82**, 500–10 (1979)
733 J. A. Rodriguez, *Surf. Sci.* **222**, 383–403 (1989)
734 J. A. Rodriguez and C. T. Campbell, *Surf. Sci.* **194**, 475–504 (1988)
735 G. Zwicker, W. Ranke, K. Jacobi and D. Poss, *Ber. Bunsenges. Phys. Chem.* **88**, 364–7 (1984)
736 C. A. Muryn, D. Purdie, P. Hardman, A. L. Johnson, N. S. Prakash, G. N. Raiker, G. Thornton and D. S. L. Law, *Faraday Disc. Chem. Soc.* **89**, 77–89 (1990)
737 D. R. Warburton, D. Purdie, C. A. Muryn, K. Prabhakaran, P. L. Wincott and G. Thornton, *Surf. Sci.* **269/270**, 305–9 (1992)
738 E. Román and J. L. de Segovia, *Surf. Sci.* **251/252**, 742–6 (1991)
739 A. Steinbrunn, C. Lattaud, H. Reteno and J. C. Colson, in *Physical Chemistry of the Solid State: Applications to Metals and their Compounds*, ed. P. Lacombe (Elsevier, Amsterdam, 1984), pp. 551–9
740 A. Steinbrunn and C. Lattaud, *Surf. Sci.* **155**, 279–95 (1985)
741 J. Otamiri, A. Andersson and S. A. Jansen, *Langmuir* **6**, 365–70 (1990)
742 H. Kuhlenbeck, G. Odörfer, R. Jaeger, G. Illing, M. Menges, T. Mull, H. J. Freund, M. Pöhlchen, V. Staemmler, S. Witzel, C. Scharfschwerdt, K. Wennemann, T. Liedke and M. Neumann, *Phys. Rev. B* **43**, 1969–86 (1991)
743 A. Steinbrunn, P. Dumas and J. C. Colson, *Surf. Sci.* **74**, 201–15 (1978)
744 A. Steinbrunn, P. Dumas and J. C. Colson, in *Reactivity of Solids,* ed. J. Wood, O. Lindqvist, C. Helgesson and N. G. Vannerberg (Plenum, New York, 1977), pp. 625–9
745 P. Dumas, A. Steinbrunn and J. C. Colson, *Thin Solid Films* **79**, 267–76 (1981)
746 S. P. Mehandru, A. B. Anderson and J. F. Brazdil, *J. Amer. Chem. Soc.* **110**, 1715–19 (1988)
747 K. J. Børve and L. G. M. Pettersson, *J. Phys. Chem.* **95**, 7401–5 (1991)
748 K. J. Børve, *J. Chem. Phys.* **95**, 4626–31 (1991)
749 X. D. Peng and M. A. Barteau, *Langmuir* **5**, 1051–6 (1989)
750 X. D. Peng and M. A. Barteau, *Catal. Lett.* **7**, 395–402 (1990)
751 W. T. Petrie and J. M. Vohs, *Surf. Sci.* **259**, L750–6 (1991)
752 G. W. Rubloff, H. Lüth and W. D. Grobman, *J. Vac. Sci. Technol.* **13**, 333 (1976)
753 J. M. Vohs and M. A. Barteau, *J. Phys. Chem.* **94**, 882–5 (1990)
754 J. M. Vohs and M. A. Barteau, *J. Phys. Chem.* **91**, 4766–76 (1987)
755 W. H. Cheng, S. Akhter and H. H. Kung, *J. Catal.* **82**, 341–50 (1983)
756 S. Akhter, W. H. Cheng, K. Lui and H. H. Kung, *J. Catal.* **85**, 437–56 (1984)
757 K. Lui, M. Vest, P. Berlowitz, S. Akhter and H. H. Kung, *J. Phys. Chem.* **90**, 3183–7 (1986)
758 J. M. Vohs and M. A. Barteau, *Surf. Sci.* **176**, 91–114 (1986)
759 W. Hirschwald and D. Hofmann, *Surf. Sci.* **140**, 415–24 (1984)
760 J. Tobin, W. Hirschwald and J. Cunningham, *Spectrochimica Acta B* **40**, 725–37 (1985)
761 C. T. Au, W. Hirsch and W. Hirschwald, *Surf. Sci.* **221**, 113–30 (1989)
762 S. Akhter, K. Lui and H. H. Kung, *J. Phys. Chem.* **89**, 1958–64 (1985)
763 G. W. Rubloff, H. Lüth and W. D. Grobman, *Chem. Phys. Lett.* **39**, 493–6 (1976)
764 J. M. Vohs and M. A. Barteau, *Surf. Sci.* **221**, 590–608 (1989)
765 W. Mokwa, D. Kohl and G. Heiland, *Surf. Sci.* **117**, 659–67 (1982)

766 J. M. Vohs and M. A. Barteau, *J. Phys. Chem.* **95**, 297–302 (1991)

767 P. Berlowitz and H. H. Kung, *J. Amer. Chem. Soc.* **108**, 3532–34 (1986)

768 J. M. Vohs and M. A. Barteau, *J. Phys. Chem.* **93**, 8343–54 (1989)

769 G. W. Rubloff, W. D. Grobman and H. Lüth, *Phys. Rev. B* **14**, 1450–7 (1976)

770 J. M. Vohs and M. A. Barteau, *J. Catal.* **113**, 497–508 (1988)

771 J. M. Vohs and M. A. Barteau, *Langmuir* **5**, 965–72 (1989)

772 H. Lüth, G. W. Rubloff and W. D. Grobman, *Surf. Sci.* **74**, 365–72 (1978)

773 J. M. Vohs and M. A. Barteau, *Surf. Sci.* **197**, 109–22 (1988)

774 J. M. Vohs and M. A. Barteau, *Surf. Sci.* **201**, 481–502 (1988)

775 J. M. Vohs and M. A. Barteau, *J. Electron Spectros.* **49**, 87–96 (1989)

776 D. Pöss, W. Ranke and K. Jacobi, *Surf. Sci.* **105**, 77–94 (1981)

777 W. Thoren, D. Kohl and G. Heiland, *Surf. Sci.* **162**, 402–10 (1985)

778 A. Jablonski and K. Wandelt, *Surf. Interface Anal.* **17**, 611–27 (1991)

779 H. Jacobs, W. Mokwa, D. Kohl and G. Heiland, *Vacuum* **33**, 869–70 (1983)

780 K. S. Kim and M. A. Barteau, *Surf. Sci.* **223**, 13–32 (1989)

781 E. Román, F. J. Bustillo and J. L. de Segovia, *Vacuum* **41**, 40–2 (1990)

782 K. S. Kim and M. A. Barteau, *J. Mol. Catal.* **63**, 103–17 (1990)

783 H. Idress, K. Pierce and M. A. Barteau, *J. Amer. Chem. Soc.* **113**, 715–16 (1991)

784 K. S. Kim and M. A. Barteau, *Langmuir*, **6**, 1485–8 (1990)

785 H. Idriss, K. S. Kim and M. A. Barteau, *Structure-Activity and Selectivity Relationships in Heterogeneous Catalysis*, ed. R. K. Grasselli and A. W. Sleight (Elsevier, Amsterdam, 1991), pp. 327–35

786 K. S. Kim and M. A. Barteau, *J. Catal.* **125**, 353–75 (1990)

787 H. Yamada and Y. Yamamoto, *Surf. Sci.* **134**, 71–90 (1983)

788 P. B. Smith, S. L. Bernasek and J. Schwartz, *Surf. Sci.* **204**, 374–86 (1988)

789 T. Chang, S. L. Bernasek and J. Schwartz, *Langmuir* **7**, 1413–18 (1991)

790 T. Chang, S. L. Bernasek and J. Schwartz, *J. Amer. Chem. Soc.* **111**, 758–60 (1989)

791 R. P. Furstenau and M. A. Langell, *Surf. Sci.* **159**, 108–32 (1985)

792 F. Ohuchi, L. E. Firment, U. Chowdhry and A. Ferretti, *J. Vac. Sci. Technol. A* **2**, 1022–3 (1984)

793 U. Chowdhry, A. Ferretti, L. E. Firment, C. J. Machiels, F. Ohuchi, A. W. Sleight and R. H. Staley, *Appl. Surf. Sci.* **19**, 360–72 (1984)

794 A. B. Anderson and N. K. Ray, *J. Amer. Chem. Soc.* **107**, 253–4 (1985)

795 S. P. Mehandru, A. B. Anderson, J. F. Brazdil and R. K. Grasselli, *J. Phys. Chem.* **91**, 2930–4 (1987)

796 K. H. Schulz and D. F. Cox, *Surf. Sci.* **262**, 318–34 (1992)

797 D. F. Cox and K. H. Schulz, *J. Vac. Sci. Technol. A* **8**, 2599–604 (1990)

798 S. P. Mehandru, A. B. Anderson and J. F. Brazdil, *J. Chem. Soc. : Faraday Trans. I* **83,** 463–75 (1987)

799 A. B. Anderson, D. W. Ewing, Y. Kim, R. K. Grasselli, J. D. Burrington and J. F. Brazdil, *J. Catal.* **96**, 222–33 (1985)

800 R. Memming, in *Electrochemistry II*, ed. E. Steckhan (Springer-Verlag, Berlin, 1988), pp. 79–112

801 X. D. Peng and M. A. Barteau, *Surf. Sci.* **224**, 327–47 (1989)

802 V. C. Lee and H. S. Wong, *J. Phys. Soc. Japan* **45**, 1657–63 (1978)

803 R. Huzimura, Y. Yanagisawa, K. Matsumura and S. Yamabe, *Phys. Rev. B* **41**, 3786–93 (1990)

804 N. J. Tro, D. A. Arthur and S. M. George, *J. Chem. Phys.* **90**, 3389–95 (1989)

805 N. J. Tro, D. R. Haynes, A. M. Nishimura and S. M. George, *J. Chem. Phys.* **91**, 5778–85 (1989)

806 D. R. Haynes, K. R. Helwig, N. J. Tro and S. M. George, *J. Chem. Phys.* **93**,

2836–47 (1990)

807 G. Heiland, E. Mollwo and F. Stöckmann, in *Solid State Physics,* ed. F. Seitz and
D. Turnbull (Academic Press, New York, 1959) *Vol. 8,* pp. 191–323

808 W. Hirschwald, *Current Topics in Materials Science* **7**, 143–482 (1981)

809 G. L. Griffin and J. T. Yates, *J. Chem. Phys.* **77**, 3744–50 (1982)

810 G. Yaron, A. Many and Y. Goldstein, *J. Appl. Phys.* **58**, 3508–14 (1985)

811 G. Yaron, J. Levy, Y. Goldstein and A. Many, *J. Appl. Phys.* **59**, 1232–7 (1986)

812 M. Bowker, H. Houghton, K. C. Waugh, T. Giddings and M. Green, *J. Catal.* **84**,
252–5 (1983)

813 V. E. Henrich (unpublished)

814 G. D. Parfitt, *Prog. Surf. Membrane Sci.* **11**, 181–226 (1978)

815 D. Menzel, in *Interactions on Metal Surfaces*, ed. R. Gomer (Springer-Verlag, New
York, 1975), pp. 101–42

816 M. L. Knotek and P. J. Feibelman, *Phys. Rev. Lett.* **40**, 964–7 (1978)

817 M. L. Knotek, V. O. Jones and V. Rehn, *Phys. Rev. Lett.* **43**, 300–3 (1979)

818 M. L. Knotek and P. J. Feibelman, *Surf. Sci.* **90**, 78–90 (1979)

819 R. L. Kurtz, *J. Vac. Sci. Technol. A* **4**, 1248–9 (1986)

820 R. L. Kurtz, R. Stockbauer and T. E. Madey, in *Desorption Induced by Electronic
Transitions: DIET III*, ed. R. H. Stulen and M. L. Knotek (Springer-Verlag, Berlin,
1988), pp. 109–14

822 J. F. van der Veen, F. J. Himpsel, D. E. Eastman and P. Heimann, *Solid State
Commun.* **36**, 99–104 (1980)

821 R. L. Kurtz, R. Stockbauer and T. E. Madey, in *Desorption Induced by Electronic
Transitions: DIET II*, ed. W. Brenig and D. Menzel (Springer-Verlag, Berlin, 1985),
pp. 89–93

823 R. L. Benbow, M. N. Thuler and Z. Hurych, *Phys. Rev. Lett.* **49**, 1264–7 (1982)

824 G. Loubriel, M. L. Knotek, R. H. Stulen, B. E. Koel and C. C. Parks, *J. Vac. Sci.
Technol. A* **1**, 1145–8 (1983)

825 A. Klekamp, H. Donnerberg, W. Heiland and K. J. Snowdon, *Surf. Sci.* **200**, L465–9
(1988)

826 T. Gotoh, S. Takagi and G. Tominaga, *Vacuum* **41**, 213–14 (1990)

827 M. L. Knotek, V. O. Jones and V. Rehn, *Surf. Sci.* **102**, 566–77 (1981)

828 M. L. Knotek, R. H. Stulen, G. M. Loubriel, V. Rehn, R. A. Rosenberg and
C. C. Parks, *Surf. Sci.* **133**, 291–304 (1983)

829 R. L. Kurtz, R. Stockbauer, R. Nyholm, S. A. Flodström and F. Senf, *Phys. Rev. B*
35, 7794–7 (1987)

830 R. L. Kurtz, R. Stockbauer, R. Nyholm, S. A. Flodström and F. Senf, *J. Vac. Sci.
Technol. A* **5**, 1111–12 (1987)

831 R. L. Kurtz, R. Stockbauer, F. Senf, R. Nyholm and S. A. Flodström, in *Desorption
Induced by Electronic Transitions: DIET III*, ed. R. H. Stulen and M. L. Knotek
(Springer-Verlag, Berlin, 1988), pp. 258–61

832 V. M. Bermudez and M. A. Hoffbauer, *Phys. Rev. B* **30**, 1125–8 (1984)

833 F. L. Hutson, D. E. Ramaker, V. M. Bermudez and M. A. Hoffbauer, *J. Vac. Sci.
Technol. A* **3**, 1657–61 (1985)

834 R. E. Walkup and R. L. Kurtz, in *Desorption Induced by Electronic Transitions:
DIET III*, ed. R. H. Stulen and M. L. Knotek (Springer-Verlag, Berlin, 1988),
pp. 160–6

835 E. Bertel, R. Stockbauer, R. L. Kurtz, T. E. Madey and D. E. Ramaker, in
Desorption Induced by Electronic Transitions: DIET II, ed. W. Brenig and
D. Menzel (Springer-Verlag, Berlin, 1985), pp. 84–8

836 S. J. Tauster, S. C. Fung and R. L. Garten, *J. Amer. Chem. Soc.* **100**, 170–5 (1978)

837 G. L. Haller and D. E. Resasco, *Adv. Catal.* **36**, 173–235 (1989)
838 S. J. Tauster, *Acc. Chem. Res.* **20**, 389–94 (1987)
839 C. D. Wagner, W. M. Riggs, L. E. Davis, J. F. Moulder and G. E. Muilenberg, *Handbook of X-Ray Photoelectron Spectroscopy* (Perkin Elmer, Eden Prairie, MN, 1979)
840 C. D. Wagner, L. H. Gale and R. H. Raymond, *Anal. Chem.* **51**, 466–77 (1979)
841 G. Moretti, *J. Electron Spectros.* **58**, 105–18 (1992)
842 J. A. Dean (ed.) *Lange's Handbook of Chemistry, 13th edn*, (McGraw-Hill, New York, 1972).
843 C. Argile and G. E. Rhead, *Surf. Sci. Reports* **10**, 277–356 (1989)
844 A. M. Stoneham and P. W. Tasker, *J. de Physique* **49**, *Suppl. 10*, C5-99–C5-113 (1988)
845 A. M. Stoneham and P. W. Tasker, in *Ceramic Microstructures 86*, ed. J. A. Pask and A. G. Evans (Plenum, New York, 1987), pp. 155–65
846 A. Zecchina, D. Scarano, L. Marchese, S. Coluccia and E. Giamello, *Surf. Sci.* **194**, 531–4 (1988)
847 T. Conrad, J. M. Vohs, P. A. Thiry and R. Caudano, *Surf. Interface Anal.* **16**, 446–51 (1990)
848 A. P. Janssen, R. C. Schoonmaker and A. Chambers, *Surf. Sci.* **49**, 143–60 (1975)
849 A. K. Green, J. Dancy and E. Bauer, *J. Vac. Sci. Technol.* **7**, 159–63 (1979)
850 K. Takayanagi, K. Yaki and G. Honjo, *Thin Solid Films* **48**, 137–52 (1978)
851 J. E. T. Andersen and P. J. Møller, *Surf. Sci.* **258**, 247–58 (1991)
852 J. E. T. Andersen and P. J. Møller, *Phys. Rev. B* **44**, 13645–54 (1991)
853 T. Kanaji, K. Asano and S. Nagata, *Vacuum* **23**, 55–9 (1973)
854 T. Kanaji, T. Kagotani and S. Nagata, *Thin Solid Films* **32**, 217–19 (1976)
855 R. A. Hubert and J. M. Gilles, *Appl. Surf. Sci.* **22/23**, 631–7 (1985)
856 R. Hubert, J. Darville and J. M. Gilles, *Phys. Scripta* **T4**, 179–82 (1983)
857 T. Urano and T. Kanaji, *J. Phys. Soc. Japan* **57**, 3403–10 (1988)
858 D. G. Lord and M. Prutton, *Thin Solid Films* **21**, 341–56 (1974)
859 C. R. Henry, C. Chapon, C. Duriez and S. Giorgio, *Surf. Sci.* **253**, 177–89 (1991)
860 I. Alstrup and P. J. Møller, *Appl. Surf. Sci.* **33/34**, 143–51 (1988)
861 J. W. He and P. J. Møller, *Surf. Sci.* **180**, 411–20 (1987)
862 N. C. Bacalis and A. B. Kunz, *Phys. Rev. B* **32**, 4857–65 (1985)
863 P. W. Palmberg, T. N. Rhodin and C. J. Todd, *Appl. Phys. Lett.* **11**, 33–5 (1967)
864 Y. Shigeta and K. Maki, *Japan. J. Appl. Phys.* **16**, 845–6 (1977)
865 Y. S. Chaug, N. J. Chou and Y. H. Kim, *J. Vac. Sci. Technol. A* **5**, 1288–91 (1987)
866 J. H. Selverian, F. S. Ohuchi and M. R. Notis, *Mater. Res. Soc. Symp. Proc.* **167**, 335–40 (1990)
867 H. Poppa, C. A. Papageorgopoulos, F. Marks and E. Bauer, *Z. Physik D* **3**, 279–89 (1986)
868 K. H. Johnson and S. V. Pepper, *J. Appl. Phys.* **53**, 6634–7 (1982)
869 E. I. Altman and R. J. Gorte, *Surf. Sci.* **195**, 392–402 (1988)
870 Q. Zhong and F. S. Ohuchi, *J. Vac. Sci. Technol. A* **8**, 2107–12 (1990)
871 Q. Zhong and F. S. Ohuchi, *Mater. Res. Soc. Symp. Proc.* **153**, 71–6 (1989)
872 A. D. Zdetsis and A. B. Kunz, *Phys. Rev. B* **32**, 6358–62 (1985)
873 E. I. Altman and R. J. Gorte, *J. Catal.* **110**, 191–6 (1988)
874 A. Fritsch and P. Légaré, *Surf. Sci.* **184**, L355–60 (1987)
875 S. Roberts and R. J. Gorte, *J. Chem. Phys.* **93**, 5337–44 (1990)
876 A. B. Anderson, Ch. Ravimohan and S. P. Mehandru, *Surf. Sci.* **183**, 438–48 (1987)
877 Q. L. Guo and P. J. Møller, *Vacuum* **41**, 1114–17 (1990)
878 Q. L. Guo and P. J. Møller, *Surf. Sci.* **244**, 228–36 (1991)

879 P. J. Møller and Q. L. Guo, *Thin Solid Films* **201**, 267–79 (1991)

880 S. Varma, G. S. Chotiner and M. Arbab, *J. Vac. Sci. Technol. A* **10**, 2857–62 (1992)

881 G. P. Malafsky, *Surf. Sci.* **249**, 159–70 (1991)

882 P. A. Taylor and B. J. Hopkins, *J. Phys. C: Solid State Phys.* **11**, L643–6 (1978)

883 R. Leysen, B. J. Hopkins and P. A. Taylor, *J. Phys. C: Solid State Phys.* **8**, 907–16 (1975)

884 P. A. Taylor, R. Leysen and B. J. Hopkins, *Solid State Commun.* **17**, 983–6 (1975)

885 R. A. Powell and W. E. Spicer, *J. Appl. Phys.* **48**, 4311–14 (1977)

886 W. Gaebler, K. Jacobi and W. Ranke, *Surf. Sci.* **75**, 355–67 (1978)

887 H. Jacobs, W. Mokwa, D. Kohl and G. Heiland, *Surf. Sci.* **160**, 217–34 (1985)

888 C. T. Campbell, K. A. Daube and J. M. White, *Surf. Sci.* **182**, 458–76 (1987)

889 C. T. Campbell, in *Catalysis 1987*, ed. J. W. Ward (Elsevier, Amsterdam, 1987), pp. 783–90

890 E. F. Wassermann and K. Polacek, *Appl. Phys. Lett.* **7**, 259–60 (1970)

891 T. B. Fryberger, J. W. Erickson and S. Semancik, *Surf. Interface Anal.* **14**, 83–9 (1989)

892 J. W. Erickson, T. B. Fryberger and S. Semancik, *J. Vac. Sci. Technol. A* **6**, 1593–8 (1988)

893 H. Jacobs, W. Mokwa, D. Kohl and G. Heiland, *Fresenius Z. Anal. Chem.* **319**, 634 (1984)

894 R. Cavicchi and S. Semancik, *Surf. Sci.* **257**, 70–8 (1991)

895 R. J. Lad and L. S. Dake, *Proc. Mater. Res. Soc.* **238**, 823–8 (1992)

896 Z. Zhang and V. E. Henrich, *Surf. Sci.* **277**, 263–72 (1992)

897 J. Deng, D. Wang, X. Wei, R. Zhai and H. Wang, *Surf. Sci.* **249**, 213–22 (1991)

898 J. M. Pan and T. E. Madey, *J. Vac. Sci. Technol. A* **11**, 1667–74 (1993)

899 D. Brugniau, S. D. Parker and G. E. Rhead, *Thin Solid Films* **121**, 247–57 (1984)

900 S. Bourgeois, D. Diakité, F. Jomard, M. Perdereau and R. Poirault, *Surf. Sci.* **217**, 78–84 (1989)

901 C. C. Kao, S. C. Tsai and Y. W. Chung, *J. Catal.* **73**, 136–46 (1982)

902 S. Bourgeois, F. Jomard and M. Perdereau, *Surf. Sci.* **249**, 194–98 (1991)

903 C. C. Kao, S. C. Tsai, M. K. Bahl, Y. W. Chung and W. J. Lo, *Surf. Sci.* **95**, 1–14 (1980)

904 M. C. Wu and P. J. Møller, in *The Structure of Surfaces III*, ed. S. Y. Tong, M. A. Van Hove, K. Takayanagi and X. D. Xie (Springer-Verlag, Berlin, 1991), pp. 652–9

905 K. Tamura, U. Bardi and Y. Nihei, *Surf. Sci.* **216**, 209–21 (1989)

906 J. A. Horsely, *J. Amer. Chem. Soc.* **101**, 2870–4 (1979)

907 M. C. Wu and P. J. Møller, *Surf. Sci.* **224**, 250–64 (1989)

908 M. C. Wu and P. J. Møller, *Surf. Sci.* **235**, 228–34 (1990)

909 D. M. Hill, H. M. Meyer III and J. H. Weaver, *J. Appl. Phys.* **65**, 4943–50 (1989)

910 M. K. Bahl, S. C. Tsai and Y. W. Chung, *Phys. Rev. B* **21**, 1344–8 (1980)

911 S. Kennou, M. Kamaratos and C. A. Papageorgopoulos, *Surf. Sci.* **256**, 312–16 (1991)

912 S. Bourgeois, D. Devillard, T. Guindet, F. Jomard and A. Steinbrunn, *Vacuum* **41**, 1097–8 (1990)

913 U. Bardi, A. Santucci and G. Rovida, *Surf. Sci.* **162**, 337–41 (1985)

914 S. Roberts and R. J. Gorte, *J. Phys. Chem.* **95**, 5600–4 (1991)

915 Lj. Atanasoska, R. T. Atanasoski, F. H. Pollak and W. E. O'Grady, *Surf. Sci.* **230**, 95–112 (1990)

916 J. P. Nogier, J. Thoret, N. Jammul and J. Fraissard, *Appl. Surf. Sci.* **47**, 287–92 (1991)

917 H. R. Sadeghi, D. E. Resasco, V. E. Henrich and G. L. Haller, *J. Catal.* **104**, 252–5

(1987)

918 V. E. Henrich, *J. Catal.* **88**, 519–22 (1984)

919 V. di Castro and G. Polzonette, *Chem. Phys. Lett.* **139**, 215–18 (1987)

920 H. M. Meyer III, D. M. Hill, S. G. Anderson, J. H. Weaver and D. W. Capone II, *Appl. Phys. Lett.* **51**, 1750–2 (1987)

921 J. H. Weaver, Y. Gao, T. J. Wagener, B. Flandermeyer and D. W. Capone II, *Phys. Rev. B* **36**, 3975–8 (1987)

922 D. M. Hill, H. M. Meyer III, J. H. Weaver, B. Flandermeyer and D. W. Capone II, *Phys. Rev. B* **36**, 3979–82 (1987)

923 D. M. Hill, Y. Gao, H. M. Meyer III, T. J. Wagener, J. H. Weaver and D. W. Capone II, *Phys. Rev. B* **37**, 511–14 (1988)

924 H. M. Meyer III, T. J. Wagener, D. M. Hill, Y. Gao, S. G. Anderson, S. D. Krahn, J. H. Weaver, B. Flandermeyer and D. W. Capone II, *Appl. Phys. Lett.* **51**, 1118–20 (1987)

925 D. M. Hill, H. M. Meyer,III, J. H. Weaver and D. L. Nelson, *Appl. Phys. Lett.* **53**, 1657–9 (1988)

926 H. M. Meyer III, D. M. Hill, T. J. Wagener, Y. Gao, J. H. Weaver, D. W. Capone II and K. C. Goretta, *Phys. Rev. B* **38**, 6500–12 (1988)

927 H. M. Meyer III, J. H. Weaver and K. C. Goretta, *J. Appl. Phys.* **67**, 1995–2002 (1990)

928 Y. Kimachi, Y. Hidaka, T. R. Ohno, G. H. Kroll and J. H. Weaver, *J. Appl. Phys.* **69**, 3176–81 (1991)

929 Y. Gao, T. J. Wagener, J. H. Weaver, B. Flandermeyer and D. W. Capone II, *Appl. Phys. Lett.* **51**, 1032–4 (1987)

930 T. J. Wagener, Y. Gao, I. M. Vitomirov, C. M. Aldao, J. J. Joyce, C. Capasso, J. H. Weaver and D. W. Capone II, *Phys. Rev. B* **38**, 232–9 (1988)

931 D. M. Hill, H. M. Meyer III, J. H. Weaver, C. F. Gallo and K. C. Goretta, *Phys. Rev. B* **38**, 11331–6 (1988)

932 H. M. Meyer III, D. M. Hill, T. J. Wagener, J. H. Weaver, C. F. Gallo and K. C. Goretta, *J. Appl. Phys.* **65**, 3130–5 (1989)

933 Ph. Niedermann, A. P. Grande, J. K. Grepstad, J. M. Triscone, M. G. Karkut, O. Brunner, L. Antagnazza, W. Sadowski, H. J. Scheel and Ø. Fischer, *J. Appl. Phys.* **68**, 1777–81 (1990)

934 Z. H. Gong, R. Fagerberg, F. Vassenden, J. K. Grepstad and R. Høier, *Appl. Phys. Lett.* **60**, 498–500 (1992)

935 H. M. Meyer III, D. M. Hill, J. H. Weaver, D. L. Nelson and K. C. Goretta, *Appl. Phys. Lett.* **53**, 1004–6 (1988)

936 D. S. Dessau, Z. X. Shen, B. O. Wells, W. E. Spicer, R. S. List, A. J. Arko, R. J. Bartlett, Z. Fisk, S. W. Cheong, D. B. Mitzi, A. Kapitulnik and J. E. Schirber, *Appl. Phys. Lett.* **57**, 307–9 (1990)

937 F. Maeda, H. Sugahara, M. Oshima and O. Michikami, *Appl. Phys. Lett.* **59**, 363–5 (1991)

938 P. A. P. Lindberg, B. O. Wells, Z. X. Shen, D. S. Dessau, I. Lindau, W. E. Spicer, D. B. Mitzi and A. Kapitulnik, *J. Appl. Phys.* **67**, 2667–70 (1990)

939 P. A. P. Lindberg, Z. X. Shen, B. O. Wells, D. S. Dessau, D. B. Mitzi, I. Lindau, W. E. Spicer and A. Kapitulnik, *Phys. Rev. B* **39**, 2890–3 (1989)

940 U. Diebold, J. M. Pan and T. E. Madey, *Phys. Rev. B* **47**, 3868–76 (1993)

941 T. R. Ohno, J. C. Patrin, H. M. Meyer III, J. H. Weaver, Y. Kimachi and Y. Hidaka, *Phys. Rev. B* **41**, 11677–80 (1990)

942 P. A. P. Lindberg, Z. X. Shen, I. Lindau, W. E. Spicer, C. B. Eom and T. H. Geballe, *Appl. Phys. Lett.* **53**, 529–31 (1988)

943 Y. Hwu, L. Lozzi, S. La Rosa, M. Onellion, H. Berger, F. Gozzi, F. Lévy and

G. Margaritondo, *Appl. Phys. Lett.* **59**, 979–81 (1991)

944 E. Weschke, C. Laubschat, M. Domke, M. Bodenbach, G. Kaindl, J. E. Ortega and R. Miranda, *Z. Phys. B* **74**, 191–5 (1989)

945 T. B. Fryberger and S. Semancik, *Sensors and Actuators B* **2**, 305–9 (1990)

946 H. Jacobs, W. Mokwa, D. Kohl and G. Heiland, *Surf. Sci.* **126**, 368–73 (1983)

947 W. Mokwa, D. Kohl and G. Heiland, *Fresenius Z. Anal. Chem.* **314**, 315–18 (1983)

948 S. C. Tsai, C. C. Kao and Y. W. Chung, *J. Catal.* **79**, 451–61 (1983)

949 H. Onishi, T. Aruga, C. Egawa and Y. Iwasawa, *J. Chem. Soc. , Faraday Trans I* **85**, 2597–604 (1989)

950 C. Duriez, C. R. Henry and C. Chapon, *Surf. Sci.* **253**, 190–204 (1991)

951 S. J. Tauster, S. C. Fung, R. T. K. Baker and J. A. Horsley, *Science* **211**, 1121–5 (1981)

952 A. D. Logan, E. J. Braunschweig, A. K. Datye and D. J. Smith, *Langmuir* **4**, 827–30 (1988)

953 E. J. Braunschweig, A. D. Logan, A. K. Datye and D. J. Smith, *J. Catal.* **118**, 227–37 (1989)

954 E. J. Braunschweig, A. D. Logan and A. K. Datye, *Mater. Res. Soc. Symp. Proc.* **111**, 35–40 (1988)

955 R. T. K. Baker, E. B. Prestridge and R. L. Garten, *J. Catal.* **56**, 390–406 (1979)

956 J. C. Volta, in *Adsorption and Catalysis on Oxide Surfaces,* ed. M. Che and G. C. Bond (Elsevier, Amsterdam, 1985), pp. 331–42

957 A. Steinbrunn and M. Bordignon, *Bull. Soc. Chim. Belg.* **96**, 941–9 (1987)

958 A. Steinbrunn and M. Bordignon, *Solid State Ionics* **32/33**, 911–17 (1989)

959 M. Cotter, S. Campbell, R. G. Egdell and W. C. Mackrodt, *Surf. Sci.* **197**, 208–24 (1988)

960 D. M. Lind, S. D. Berry, G. Chern, H. Mathias and L. R. Testardi, *Phys. Rev. B* **45**, 1838–50 (1992)

961 E. Kotomin, A. Shluger, M. Causà, R. Dovesi and F. Ricca, *Surf. Sci.* **232**, 399–406 (1990)

962 Y. Gao, K. L. Merkle, H. L. M. Chang, T. J. Zhang and D. J. Lam, *J. Mater. Res.* **6**, 2417–26 (1991)

963 H. L. M. Chang, H. You, J. Guo and D. J. Lam, *Appl. Surf. Sci.* **48/49**, 12–18 (1991)

964 H. L. M. Chang, Y. Gao, J. Guo, C. M. Foster, H. You, T. J. Zhang and D. J. Lam, *J. de Physique IV, Coll.* **C2**, 953–60 (1991)

965 Y. Gao, K. L. Merkle, H. L. M. Chang, T. J. Zhang and D. J. Lam, *Mater. Res. Soc. Symp Proc.* **221**, 59–64 (1991)

966 Y. Gao, K. L. Merkle, H. L. M. Chang, T. J. Zhang and D. J. Lam, *Mater. Res. Soc. Symp Proc.* **209**, 685–90 (1991)

967 M. C. Wu, Q. L. Guo and P. J. Møller, *Vacuum* **41**, 1418–21 (1990)

968 M. C. Wu and P. J. Møller, *Chem. Phys. Lett.* **171**, 136–40 (1990)

969 A. Stamper, D. W. Greve, D. Wong and T. E. Schlesinger, *Appl. Phys. Lett.* **52**, 1746–8 (1991)

970 X. D. Wu, R. E. Muenchausen, N. S. Nogar, A. Pique, R. Edwards, B. Wilkens, T. S. Ravi, D. M. Hwang and C. Y. Chen, *Appl. Phys. Lett.* **58**, 304–6 (1991)

971 D. M. Hwang, Q. Y. Ying and H. S. Kwok, *Appl. Phys. Lett.* **58**, 2429–31 (1991)

972 R. K. Singh, J. Narayan, A. K. Singh and J. Krishnaswamy, *Appl. Phys. Lett.* **54**, 2271–3 (1989)

973 R. Ramesh, A. Inam, D. M. Hwang, T. S. Ravi, T. Sands, X. X. Xi, X. D. Wu, Q. Li, T. Venkatesan and R. Kilaas, *J. Mater. Res.* **6** , 2264–71 (1991)

974 X. Y. Zheng, D. H. Lowndes, S. Zhu, J. D. Budai and R. J. Warmack, *Phys. Rev. B* **45**, 7584–7 (1992)

975 S. N. Basu, A. H. Carim and T. E. Mitchell, *J. Mater. Res.* **6**, 1823–8 (1991)

976 J. Fujita, T. Yoshitake, T. Satoh, T. Ichihashi and H. Igarashi, *IEEE Trans. Magnetics* **27**, 1205–10 (1991)

977 T. Matsumoto, T. Kawai, K. Kitahama, S. Kawai, I. Shigaki and Y. Kawate, *Appl. Phys. Lett.* **58**, 2039–41 (1991)

978 A. Inam, R. Ramesh, C. T. Rogers, B. Wilkens, K. Remschnig, D. Hart and J. Barner, *IEEE Trans. Magnetics* **27**, 1603–6 (1991)

979 M. Kawai, S. Watanabe and T. Hanada, *J. Crystal Growth* **112**, 745–52 (1991)

980 M. F. Hochella, Jr, C. M. Eggleston, V. B. Elings, G. A. Parks, G. E. Brown, Jr, C. M. Wu and K. Kjoller, *Amer. Mineralogist* **74**, 1233–46 (1989)

981 M. F. Hochella, Jr, *Rev. Minerol.* **23**, 87–132 (1990)

982 P. A. Johnsson, C. M. Eggleston and M. F. Hochella, Jr, *Amer. Minerologist* **76**, 1442–5 (1991)

983 C. M. Eggleston and M. F. Hochella, Jr, *Amer. Mineralogist* **77**, 911–22 (1992)

Compound index

Subject index

Printed in the United States
By Bookmasters